Novel Developments in Condensed Matter Physics

Volume I

Novel Developments in Condensed Matter Physics
Volume I

Edited by **Jaron Finley**

NY RESEARCH
PRESS
New York

Published by NY Research Press,
23 West, 55th Street, Suite 816,
New York, NY 10019, USA
www.nyresearchpress.com

Novel Developments in Condensed Matter Physics: Volume I
Edited by Jaron Finley

International Standard Book Number: 978-1-63238-352-5 (Hardback)

Printed in China.

Contents

Preface

One of the earliest researches of condensed states of matter occurred accidently by an English chemist Humphry Davy, when he detected that out of the 40 chemical elements, 26 had metallic properties such as luster, ductility and high electrical and thermal conductivity. This indicated that the atoms in Dalton's atomic theory were not indivisible as Dalton claimed, instead they had inner structure. Condensed Matter Physics owes its origin to the notion of emergence; wherein complex assemblies of particles behave in diverse ways with respect to their individual constituents. The laws of quantum mechanics, electromagnetism and statistical mechanics have aided the discipline by multiple researches worldwide.

This book is a compilation of advanced researches in the field. Some of the most innovative concepts and sub-themes in condensed matter physics are tunneling control of transmission coefficient and group index in a quantum dot nanostructure, phase- and polarization-controlled two-photon rabi oscillation of the biexciton state in a semiconductor quantum dot, investigation on growth, structural, spectral, optical, and mechanical properties of an organic nonlinear optical material: morpholinium hydrogen tartrate, modeling of dual-junction solar cells and the influence of surface morphology of buffer layer on the critical current density in ybco coated conductors.

I hope that this book will raise condensed matter physics a notch higher and aid its development. At last, I would like to thank all the researchers and contributors of this book, my publisher and my family for their endless support.

Editor

Understanding Multiferroic Hexagonal Manganites by Static and Ultrafast Optical Spectroscopy

Yu Ting Wang,[1] Chih Wei Luo,[1] and Takayoshi Kobayashi[1,2]

[1] *Department of Electrophysics, National Chiao Tung University, Hsinchu 30010, Taiwan*
[2] *Advanced Ultrafast Laser Research Center, University of Electro-Communications, Chofugaoka 1-5-1, Chofu, Tokyo 182-8585, Japan*

Correspondence should be addressed to Chih Wei Luo; cwluo@mail.nctu.edu.tw

Academic Editor: R. N. P. Choudhary

Multiferroic hexagonal manganites $ReMnO_3$ studied by optics are reviewed. Their electronic structures were revealed by static linear and nonlinear spectra. Two transitions located at ~1.7 eV and ~2.3 eV have been observed and attributed to the interband transitions from the lower-lying Mn^{3+} $d_{xy}/d_{x^2-y^2}$ and d_{xz}/d_{yz} states to the Mn^{3+} $d_{3z^2-r^2}$ state, respectively. These so-called d-d transitions exhibit a blueshift as decreasing temperatures and an extra blueshift near T_N. This dramatic change indicates that the magnetic ordering seriously influences the electronic structure. On the other hand, the ultrafast optical pump-probe spectroscopy has provided the important information on spin-charge coupling and spin-lattice coupling. Because of the strongly correlation between electronic structure and magnetic ordering, the amplitude of the initial rising component in $\Delta R/R$ shows striking changes at the vicinity of T_N. Moreover, the coherent optical and acoustic phonons were observed on optical pump-probe spectroscopy. Both the amplitude and dephasing time of coherent phonons also exhibit significant changes at T_N, which provide the evidence for spin-lattice interaction in these intriguing materials.

1. Introduction

Multiferroic materials [1–4] with the coexistence of various ferroic order (ferromagnetic, ferroelectric, or ferroelastic) have attracted great attention in condensed matter research due to their great potential for applications in the fields of oxide electronics, spintronics, and even the green energy devices for reducing the power consumption. With multiferroic oxides (in single phase), the coupling interaction among various order parameters causes the so-called magnetoelectric effect [5–10]. In this, a polarization can be induced by applying a magnetic field or a magnetization can be induced by applying an electric field. Although the properties of electricity and magnetism were combined into one common discipline by the Maxwell equations, the electric and magnetic orderings in solid-state materials are always considered separately. This is because that the electric charges of electrons and ions are responsible for the charge effects, whereas the electron spins govern the magnetic properties.

Prellier et al. [11] classified the characteristics of multiferroic materials into three categories. The first is Bi-based compounds, such as $BiMnO_3$ and $BiFeO_3$. The second is perovskites and related $ReMnO_3$ compounds (Re = Y, Ho, Er, Tm, Yb, and Lu). The third is $ReMn_2O_5$ (Re is rare earth element). Additionally, Cheong and Mostovoy [12] summarized the classification of ferroelectrics into "proper" and "improper" according to their mechanisms of inversion symmetry breaking. In proper ferroelectrics, the main driving force toward the polar state was associated with the electronic pairing. On the other hand, the induced polarization in improper ferroelectrics involves a more complex lattice distortion or other accidental orderings. For example, the hexagonal manganite $ReMnO_3$ compounds show a lattice distortion to enlarge their unit cell, which is the so-called geometric ferroelectricity [13–15].

The aim of this paper is to understand the multiferroic hexagonal manganites studied by some optical techniques. In Section 2, the overall features of manganites are introduced.

The structure of rare earth manganite is strongly depending on the size of rare earth and also temperature. The environment of Mn ions significantly affects its electronic structure which was investigated by static linear and nonlinear spectra in Section 3. Section 4 is the main part of this paper. The ultrafast spectroscopies demonstrate the interaction between electron, phonon, and spin systems on multiferroic hexagonal manganites. Finally, the summary and future outlook will be given in Section 5.

2. The Characteristics of Multiferroic Manganites $ReMnO_3$

Recently, multiferroic rare earth manganites ($ReMnO_3$) have attracted a great deal of attention due to their manifestations of intriguing and significant couplings between the magnetic and electric order parameters. Generally, $ReMnO_3$ has two kinds of multiferroic structures, that is, hexagonal and orthorhombic. The rare earth (Re^{3+}) ionic size shrinks from 1.11 Å to 0.94 Å with decreasing the atomic number and reducing the electrons residing in the 4f orbit, called the lanthanum contraction effect. The rare earth ions (Re^{3+}) have closely chemical properties because the same electronic configurations of $5s^2 5p^6$ for outermost electrons. The coexistence of ferroic orders in $ReMnO_3$ with hexagonal (smaller ionic radius of rare earth Re = Sc, Y, and from Ho to Lu) or orthorhombic (larger ionic radius of rare earth Re = La to Dy) (shown in Figure 1) structure yields complex physics in the intimate interactions among charge, orbital, lattice, and spin degrees of freedom. It also has fascinating physical properties which might lead to promising applications. The orthorhombic structure with a *Pnma* space group forms the stable crystal structure of $ReMnO_3$ perovskite oxides with Re = La to Dy, which belongs to the magnetic ferroelectrics induced by magnetic ordering [8, 16]. Furthermore, another family of compounds with smaller ionic size (for Re = Y, Ho, Er, Yb, Lu, etc.) forms the stable hexagonal structure with $P6_3cm$ space group. The critical point of the structure transition is located near $YMnO_3$ (Y^{3+} = 1.06 Å) and $HoMnO_3$ (Ho^{3+} = 1.05 Å). Consequently, the crystal structures of these two families have been transformed from hexagonal to orthorhombic (or vice versa) by many experimental techniques, for example, high-temperature and high-pressure processes [17, 18], chemical solution deposition (CSD) [19], the metal-organic chemical vapor deposition (MOCVD) [20], molecular beam epitaxy (MBE) [21], sputtering [22], and pulsed laser deposition (PLD) [16, 23].

In the following, we are going to focus on recent studies on the multiferroic hexagonal manganites ($ReMnO_3$, Re = rare earth) which have properties similar to various rare-earth elements, both electric transportation or magnetic behaviors. For example, the hexagonal $HoMnO_3$ has ferroelectric ordering at Curie temperature T_C = 875 K, the antiferroicmagnetic (AFM) Mn^{3+} ordering at Néel temperature T_N = 76 K, and the magnetic Ho^{3+} ordering at T_{Ho} = 4.6 K, in which the magnetically active ions are the high spin Mn^{3+} ($3d^4$, $S = 2$) and Ho^{3+} ($4f^{10}$) [5, 14, 33, 40–45]. The ferroelectricity in magnetoelectric $ReMnO_3$ is caused by

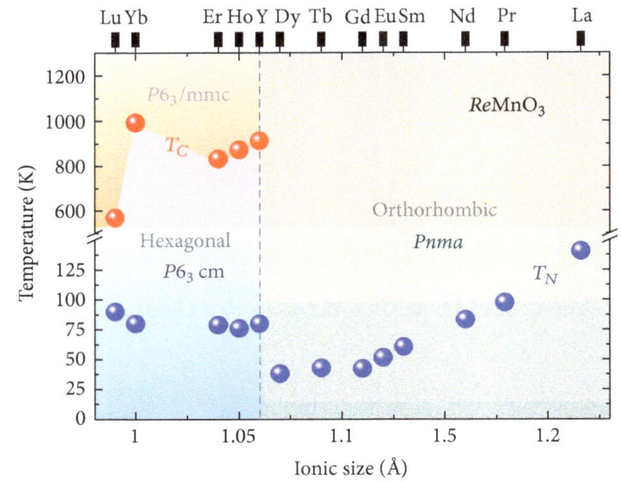

FIGURE 1: Lattice structure in $ReMnO_3$ as a function of the size of the rare-earth elements [11, 24–29].

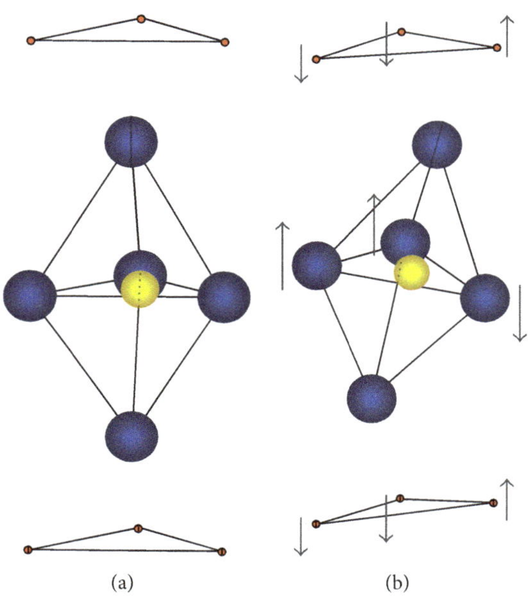

FIGURE 2: Schematic of MnO_5 bipyramid above and below T_C, respectively. (a) Centrosymmetric structure with space group $P6_3/mmc$ and (b) noncentrosymmetric with $P6_3cm$ space group. The arrows indicate atomic displacements with respect to the centrosymmetric structure [14, 30]. (Red: Re, blue: O, yellow: Mn).

a buckling of the layered MnO_5 polyhedral and is accompanied by displacements of the Re ions, which lead to a net electric polarization. In 2004, through X-ray diffraction and first principle density functional calculations, Van Aken et al. [14] demonstrated the previous scenario in the hexagonal manganite $YMnO_3$, as shown in Figure 2 which presents the "geometric" generation of polarization and describes the tilting of a rigid MnO_5 block with a magnetic Mn atom at the center. The magnetic structure has been studied by various experimental methods, such as neutron diffraction [45, 46] or second harmonic generation (SHG) [33, 47]. In

$P6_3cm$ hexagonal manganites, each Mn^{3+} ion is surrounded by five O^{2-} ions, forming triangular planar sublattices in the basal plane (a-b plane). The magnetic order of Mn^{3+} is mainly dominated by AFM in-plane Mn-O-Mn superexchange interaction. Therefore, the triangular lattice of the Mn atoms exhibits strong geometrical frustration effect [48]. Furthermore, when the temperature cools to near T_N, the strong superexchange leads to a 120° arrangement between neighboring Mn^{3+} spins in the basal plane which breaks the triangular frustration. In addition, the spins of Mn rotate with an angle of 90° at T_{SR} (e.g., for h-$HoMnO_3$, $T_{SR} \sim 33$ K) due to the onset of the AFM order of the Re moments, indicating the interaction between Re^{3+} and Mn^{3+} spins. Another magnetic transition at T_{Re} involves the complete magnetic order of the Re^{3+} ions (e.g., for h-$HoMnO_3$, $T_{Ho} \sim 5$ K).

In addition, there are many studies showing the anomaly around T_N, such as the dielectric constant [49], specific heat measurements [50], and the lattice constant measurements [51], which are attributed to the indirect coupling via lattice strain (spin-lattice coupling). Recently, Lee et al. [52] observed the giant magnetoelastic effect in hexagonal $YMnO_3$ and $LuMnO_3$ by using the high-resolution neutron and synchrotron powder diffraction experiments. They further claimed that the magnetoelectric (ME) coupling can be interpreted by the giant magnetoelastic effect very well in $ReMnO_3$, that is, the magnetic long-range ordering induced the very large displacements of all atoms at T_N. In other words, the magnetoelastic coupling allowed the atoms to undergo an isostructural transition, which has been regarded as the primary source of the magnetoelectric phenomena. In hexagonal $ReMnO_3$, this magnetoelastic effect is two orders of magnitude larger than those appearing in other magnetic materials. Therefore, the gigantic magnetoelastic coupling plays a crucial role in dominating many physical properties in hexagonal $ReMnO_3$.

Femtosecond pump-probe spectroscopy has been established as a protocol for studying the interactions between electrons, phonons, and magnons [35, 53–59] and is therefore employed with these multiferroic $ReMnO_3$ materials to gain insight into the magnetoelastic effect. By measuring the transient reflectivity ($\Delta R/R$) or transmittance ($\Delta T/T$), we can observe the intricate correlation processes among the charge, lattice, and spin degrees of freedom, which are explicitly recognized in different characteristic time scales. Thus, the microscopic mechanism of the magnetoelectric coupling or magnetoelastic behavior in this kind of strongly correlated materials can be revealed.

3. Electronic Structure Revealed by Optical Spectroscopy

To understand the extensive physical properties in multiferroic $ReMnO_3$ materials, it is crucial to realize the electronic band structure. With respect to the crystal structure of hexagonal $ReMnO_3$ manganites with MnO_5 bipyramids, in contrast with the MnO_6 octahedron of orthorhombic structure which split to t_{2g} and e_g, the crystal field symmetry leads the quintet d levels of Mn^{3+} ion splits into two low lying doublets of e_{2g} and e_{1g} (d_{xz}/d_{yz} and $d_{xy}/d_{x^2-y^2}$) states and one singlet a_{1g} ($d_{3z^2-r^2}$) state. In general, optical spectroscopy is one of the powerful tools for investigating the electronic band structure in materials. Therefore, this structural difference between the hexagonal and orthorhombic phases could be disclosed by the optical spectra [60, 61].

3.1. Linear Optical Absorption Spectra. The first optical measurements in hexagonal $YMnO_3$ were performed by Kritayakirana et al. in 1969 [62], who observed that a near infrared absorption edge moves to higher energy with lowering temperature. For the case of hexagonal $ScMnO_3$ and $ErMnO_3$, the spectra of the real and imaginary parts of the dielectric function were also measured by Kalashnikova and Pisarev [63]. An abnormally drastic absorption peak in the region of 1.57–1.59 eV is the most pronounced feature in the dielectric spectra. This peak exhibits asymmetry and can be decomposed into two components by peak fitting with the Gaussian functions. Furthermore, the absorption intensity in the region above 2.2–2.4 eV increases with photon energy. In addition, hexagonal $HoMnO_3$ [64] and $LuMnO_3$ [31] also exhibit similar results. It is noteworthy that the near infrared absorption peak exhibits a substantial blueshift of ~0.16 eV from 300 K to 10 K as shown in Figure 3. Recently, Choi et al. [32] reported a systematic study on the hexagonal $ReMnO_3$ (R = Gd, Tb, Dy, and Ho), and the overall features in the spectra are similar to previous results. The most surprising part of those results is that the temperature-dependent absorption peak at ~1.7 eV shows an unexpectedly large blueshift near T_N among the rare-earth ions of Re = Lu, Gd, Tb, Dy, and Ho, as shown in Figure 4(a).

The assignment of the optical absorption peak at ~1.7 eV has also been an issue. One interpretation is that it comes from the charge transfer transition from the O 2p to the Mn 3d states [30, 63, 65, 66]. The photoemission spectroscopy (PES) study proposed that the O 2p is the highest occupied state, while the Mn $d_{3z^2-r^2}$ state is mostly unoccupied. Other occupied Mn 3d states (d_{xz}/d_{yz}, and $d_{xy}/d_{x^2-y^2}$) lie deep below E_F [65]. The result of the first-principle electronic structure calculations also suggested that the top of the valence band is mainly determined by the O 2p states, and the lowest unoccupied band is formed by the $d_{3z^2-r^2}$ of Mn [30, 66]. The other interpretation, which is also an extensively convincing explanation, is that it comes from on-site d-d transition between the Mn 3d levels [31, 34, 64, 67]. The transition was attributed to the d-d transition between occupied e_{2g} and e_{1g} (d_{xz}/d_{yz}, and $d_{xy}/d_{x^2-y^2}$) orbitals and the unoccupied a_{1g} ($d_{3z^2-r^2}$) orbit, as shown in Figure 4(b). Because of the selection rule, transitions between d orbitals are forbidden. However, the selection rule is relaxed through the strong hybridization between the O 2p bands and transition metal d bands, which was observed in the hexagonal phase but not in the orthorhombic phase [60]. Furthermore, the strong peak at ~5 eV is attributed to the charge transfer transition from the hybridized O 2p level to the Mn $d_{3z^2-r^2}$ level.

The other feature of the near infrared absorption peak is that it shows the blueshift as decreasing temperatures accompanying the dramatic changes near T_N, which indicates

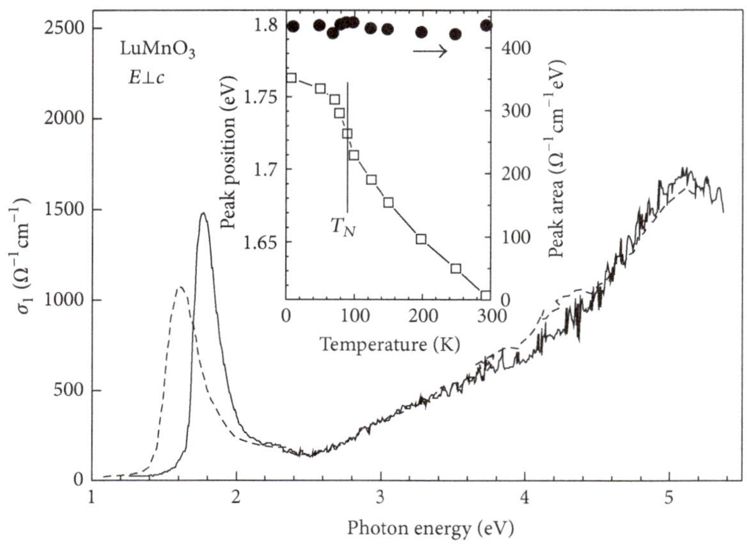

FIGURE 3: Electronic conductivity of $LuMnO_3$ at 300 K (dashed line) and 10 K (solid line). Inset: temperature-dependent peak position (open squares) and spectral weight (solid circles) of the absorption peak around 1.7 eV [31]. Reprinted figure with permission from Souchkov et al. [31]. Copyright (2003) by the American Physical Society.

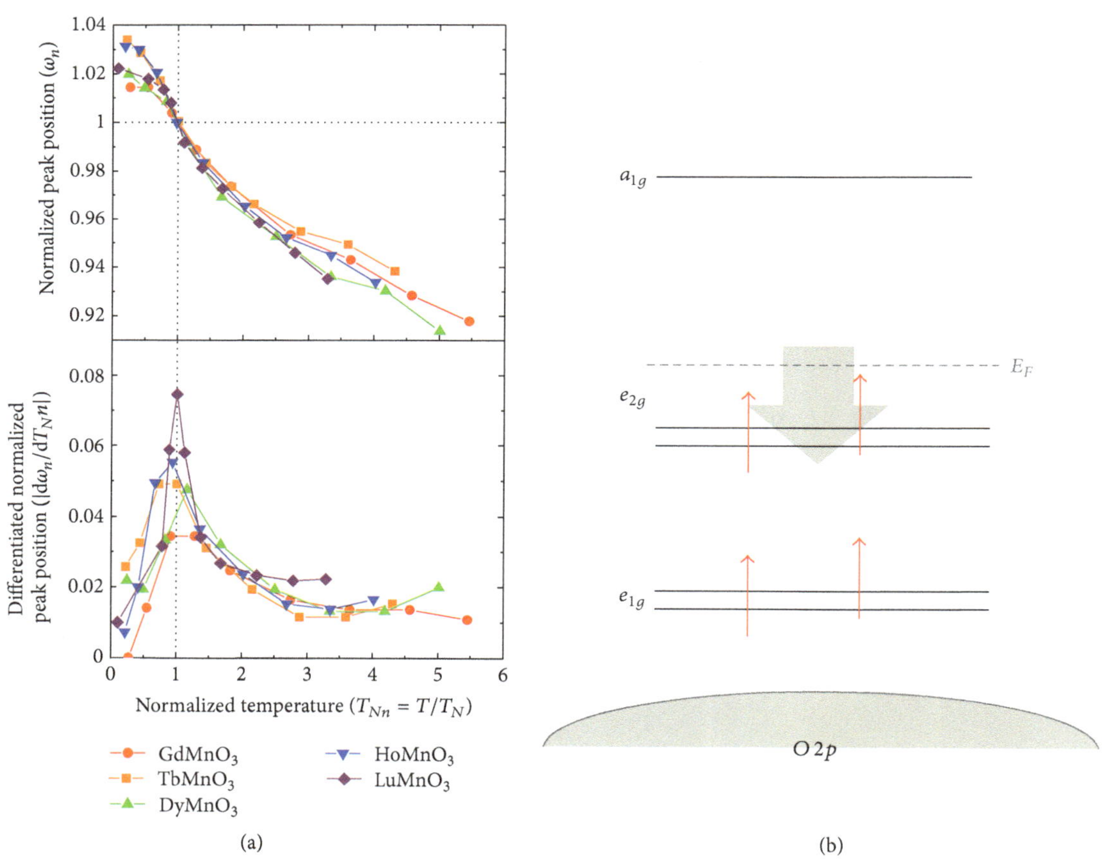

FIGURE 4: (a) Upper panel: temperature dependence of the peak positions normalized to the peak position at T_N as a function of the normalized temperature T_{Nn} ($\equiv T/T_N$). Lower panel: absolute value of normalized peak position differentiated to normalized temperature [32]. Reprinted figure with permission from Choi et al. [32]. Copyright (2008) by the American Physical Society. (b) Schematic diagram of the d-d transition in $ReMnO_3$. The red arrows indicate electrons reside in the d orbitals, and the gray arrow indicates lowering of e_{2g} band caused by the temperature reduction.

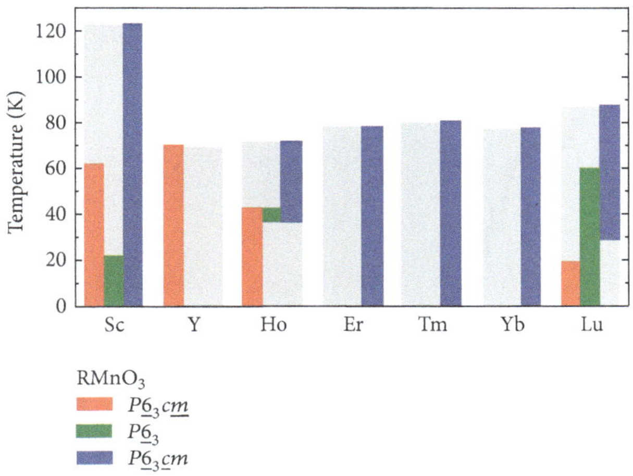

FIGURE 5: Magnetic symmetry of the hexagonal manganites. Sc, Ho, and Lu show a coexistence of magnetic phases with temperature intervals being sample specific. (Rare-earth spin ordering below 6 K was not considered) [33]. Reprinted figure with permission from Fiebig et al. [33]. Copyright (2000) by the American Physical Society.

the effective coupling between the electronic structure and the magnetic ordering. Souchkov et al. [31] attributed this temperature-dependent blueshift to the effects of exchange interactions between the Mn^{3+} ions. The superexchange between Mn^{3+} neighbors leads to lower $d_{xy}/d_{x^2-y^2}$ levels, whereas the $d_{3z^2-r^2}$ level is rarely affected (see Figure 4(b)). Moreover, together with the temperature-dependent and magnetic-dependent results [32] further implies that the magnetic ordering seriously influences the electronic structure.

3.2. Second Harmonic Generation Spectra. From the linear absorption spectra, the strong absorption caused by the d-d transition from e_{2g} to a_{1g} was clearly observed in hexagonal manganites. But the transition from e_{1g} to a_{1g} is another question. Recently, the second harmonic generation (SHG) was successfully used to explore the hidden localized d-d transitions, even those embedded in the much stronger absorption peaks [68]. The nonlinear polarization induced by optical waves is given by

$$\mathbf{P}_i(2\omega) = \varepsilon_0\left(\chi_{ijk}^{(i)} + \chi_{ijk}^{(c)}\right)\mathbf{E}_j(\omega)\,\mathbf{E}_k(\omega), \qquad (1)$$

where $\mathbf{P}_i(2\omega)$ is the nonlinear polarization induced in a medium. $\mathbf{E}_j(\omega)$ and $\mathbf{E}_k(\omega)$ are the electric field components of the fundamental light. $\chi_{ijk}^{(i)}$ and $\chi_{ijk}^{(c)}$ are the second-order nonlinear susceptibility tensor which are uniquely defined by the crystalline and magnetic structure of the crystal, respectively. The subscripts i, j, and k correspond to the Cartesian coordinate axes x, y, and z. i is the direction of the nonlinear polarization generated; j and k are the directions of applied optical field.

There are two types of optical SHG in multiferroic hexagonal manganites. One is caused by the noncentrosymmetric ferroelectric ordering of charges and the nonlinear susceptibility is time invariant (i-type). The other is due to the centrosymmetric antiferromagnetic ordering of spins and the susceptibility is time noninvariant (c-type) [34]. Table 1 lists the allowed susceptibility tensor components in i-type and c-type. Thus, the SHG contribution from ferroelectric and antiferromagnetic can be distinguished easily. Figure 5 summarizes the magnetic symmetries of all the hexagonal $ReMnO_3$ compounds as derived from the SHG experiments [33].

TABLE 1: Nonvanished susceptibility tensor components for hexagonal $ReMnO_3$ with antiparallel coupling of corresponding Mn^{3+} spins along the z-axis.

	Space group	Tensor components χ_{ijk}
Nonmagnetic SH (i-type)	$P6_3cm$	$xxz = xzx = yyz = yzy$, $zxx = zyy$, zzz
Magnetic SH (c-type)	$P\underline{6_3}c\underline{m}$	$yxx = xyx = xxy = -yyy$
	$P\underline{6_3}\underline{c}m$	$xyy = yxy = yyx = -xxx$
	$P\underline{6_3}$	sum of $P\underline{6_3}c\underline{m}$ and $P\underline{6_3}\underline{c}m$

The polarization $\mathbf{P}^{\mathrm{FE}}(2\omega)$ with the space group $P6_3cm$ below the Curie temperature T_C is

$$\begin{aligned}\mathbf{P}_x^{\mathrm{FE}}(2\omega) &= 2\varepsilon_0\chi_{xxz}^{(i)}\mathbf{E}_x(\omega)\,\mathbf{E}_z(\omega),\\ \mathbf{P}_y^{\mathrm{FE}}(2\omega) &= 2\varepsilon_0\chi_{xxz}^{(i)}\mathbf{E}_y(\omega)\,\mathbf{E}_z(\omega),\\ \mathbf{P}_z^{\mathrm{FE}}(2\omega) &= \varepsilon_0\chi_{zxx}^{(i)}\left[\mathbf{E}_x^2(\omega) + \mathbf{E}_y^2(\omega)\right] + \varepsilon_0\chi_{zzz}^{(i)}\mathbf{E}_z^2(\omega).\end{aligned} \qquad (2)$$

The contribution from antiferromagnetic $\mathbf{P}^{\mathrm{AFM}}(2\omega)$ below T_N for the space group $P\underline{6_3}c\underline{m}$ (e.g., $YMnO_3$ [68, 69]) is

$$\begin{aligned}\mathbf{P}_x^{\mathrm{AFM}}(2\omega) &= 2\varepsilon_0\chi_{yyy}^{(c)}\mathbf{E}_x(\omega)\,\mathbf{E}_y(\omega),\\ \mathbf{P}_y^{\mathrm{AFM}}(2\omega) &= \varepsilon_0\chi_{yyy}^{(c)}\left[\mathbf{E}_x^2(\omega) - \mathbf{E}_y^2(\omega)\right],\\ \mathbf{P}_z^{\mathrm{AFM}}(2\omega) &= 0.\end{aligned} \qquad (3)$$

Figure 6(a) exhibits the SHG spectra of zxx component in $YMnO_3$ at 6 K [34]. The strong peak is centered at 2.7 eV, which could be assigned to ${}^5\Gamma_2({}^5\Gamma_6) \rightarrow {}^5\Gamma_1({}^5\Gamma_1)$ (corresponds to $e_{1g} \rightarrow a_{1g}$) transition. Figure 6(b) right panel shows the spectrum of χ_{yyy}. Peaks at 2.46 eV represent the ${}^5\Gamma_1({}^5\Gamma_6) \rightarrow {}^5\Gamma_1({}^5\Gamma_1)$ (corresponds to $e_{1g} \rightarrow a_{1g}$) transition. These assignments have been proved by several studies [47, 68, 69]. In the inset of Figure 6(b), the temperature-dependent intensity of 2.46 eV disappears once the temperature is higher than T_N, which strongly indicates the closed correlation between the antiferromagnetic order parameter and SHG.

Consequently, the electronic structure of hexagonal manganites has been clearly depicted by linear and nonlinear optical spectroscopy, which further provides the basis for the excitation and relaxation of electrons in pump-probe spectroscopy.

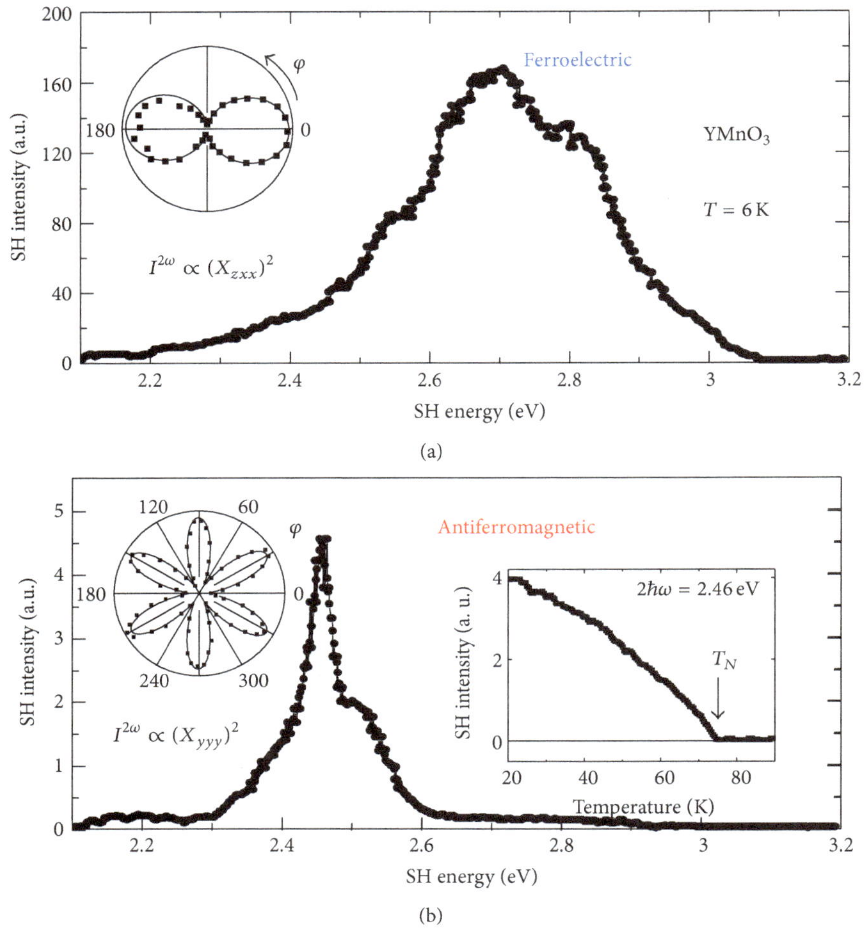

FIGURE 6: SHG spectra of $YMnO_3$ due to (a) the ferroelectric ordering and (b) antiferromagnetic ordering; inset: temperature dependence of the SHG intensity of 2.46 eV [34]. Reprinted figure with permission from Fröhlich et al. [34]. Copyright (1998) by the American Physical Society.

4. Ultrafast Pump-Probe Spectroscopy

Ultrafast optical pump-probe spectroscopy has proven to be a powerful tool to investigate the dynamics of electron-electron interaction, electron-phonon interaction, and electron-spin interaction in various materials (see, e.g., [70–72]). To understand the underlying mechanisms in strongly correlated materials, the interactions between these order parameters should be derived first. Because of the different characteristic time scales and characteristic behavior for various degrees of freedom existing in these complex materials, ultrafast spectroscopy can determine the coupling between each order of parameters.

As shown in Figure 7, the general idea of the pump-probe technique is that changes of a sample induced by a pump pulse are obtained by detecting the change in the reflectivity (ΔR) or transmissivity (ΔT) of probe pulses as a function of probe delay time. The femtosecond time evolutions are derived by delaying the relative arrival time between the pump and probe pulses using a mechanical delay line. The fluence of a probe pulse is usually much weaker than that of a pump pulse in order to avoid second excitation in the samples. The polarizations of pump and probe pulses are set perpendicular to avoid interference between pump and probe beams [73].

Generally, this pump-probe spectroscopy composes two kinds of signals of the decay background and the oscillation, which relates to the relaxation of carriers and the dynamics of phonons or spins, respectively. By the Fourier transform, the frequency of phonon modes can be doubly obtained from the oscillation signals in pump-probe spectroscopy. Moreover, the phonon dynamics, that is, the time-evolution of oscillation signals, can be revealed by short-time Fourier transform (STFT), which cannot be reached by common Raman spectroscopy. Thus, the pump-probe spectroscopy is a powerful tool to provide the information in time- and frequency-domain simultaneously.

4.1. Spin-Charge Coupling. Since the magnetic ordering and the electronic structure of manganites are strongly correlated, as found in temperature-dependent absorption spectra, there is a strong coupling between charge and spin, which can be revealed by the time-resolved femtosecond spectroscopy

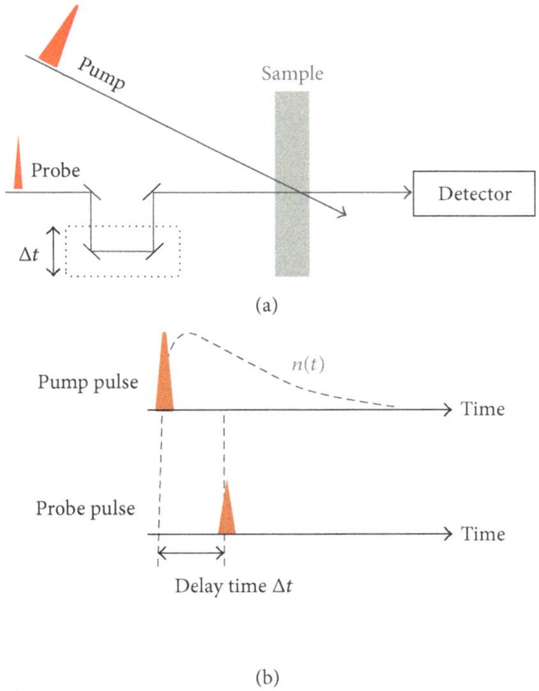

FIGURE 7: (a) Schematic diagram of pump-probe technique. Time delay (Δt) between pump and probe pulses can be controlled by a mechanical delay line. (b) The fundamental principle of pump-probe experiments. Time varying refractive index $n(t)$ changes of the sample induced by pump pulses were observed by detecting the intensity variations of probe pulses.

[72]. Shih et al. observed the spin-charge coupling by various photon energy excitations on hexagonal $HoMnO_3$ single crystals [35]. As shown in Figure 8(a), the $\Delta R/R$ amplitude as a function of temperature shows a significant drop to cross zero amplitude line to become negative at some characteristic temperatures. The excitation photon energy used in the experiment changes from 1.52 to 1.68 eV, which is exactly the range covering the d-d transition energy E_{dd} (transition from e_{2g} to a_{1g}) in the hexagonal $ReMnO_3$. When the photon energy is larger than E_{dd}, the electrons in the e_{2g} orbit can transfer to the unoccupied a_{1g} orbital by absorbing pumping photons. Conversely, this onsite transition would be completely blocked when the transition energy E_{dd} becomes larger than the pumping photon energy. Thus, the $\Delta R/R$ amplitude vanishes at a certain temperature T_0, which drops with increasing photon energy, indicating the blueshift of E_{dd}. Furthermore, Shih et al. found that the optical pump-probe measurement is sensitive to short-range magnetic order. In Figure 8(b), the E_{dd} as a function of T_0 demonstrates a gradual increase in the extent of antiferromagnetic (AFM) ordering. Moreover, this ordering starts to deviate from linearity around T_N (~76 K), suggesting the prevalence of global long-range AFM ordering. The AFM correlation evolves with temperature as it is close to T_N, shown as $d(\Delta R/R)_p/dT$ in Figure 8(c). Later, Wu et al. obtained similar results for hexagonal $YMnO_3$ thin films [74]. Furthermore, they found that the strain-induced changes of T_N, which are difficult to be observed by the temperature-dependent magnetization measurements, can be clearly revealed by the shift of $d(\Delta R/R)_p/dT$.

As a result, the amplitude of the initial rising component of $\Delta R/R$ corresponding to the Mn^{3+} d-d transition is caused by strong coupling between the electronic structure and magnetic ordering. Moreover, the emergence of long-range and short-range magnetic orders can be clearly observed by optical pump-probe spectroscopy.

4.2. Spin-Lattice Coupling. Optically induced lattice vibration can be observed through optical pump-probe spectroscopy through a periodic oscillation. This periodic oscillation in reflectivity (or transmissivity), so-called coherent acoustic phonons, is caused by interference between the reflected probe beams from sample surface and the wave front of the propagating strain pulse generated upon the pump pulse [75, 76]. On the other hand, vibration with frequencies corresponding to optical phonon modes in various materials has been observed by pump-probe spectroscopy. A number of mechanisms have been proposed to explain these oscillations, for example, the displacive excitation of coherent phonons (DECP) [77] and the impulsive stimulated Raman scattering (ISRS) [78, 79]. The simplest idea is that a pump pulse creates a standing wave which was generated by the coherent ionic motion. Then a following probe pulse senses the changes of refractive index which is modulated by the standing wave. In general, the reflectivity (or transmissivity) oscillation observed in pump-probe spectroscopy always corresponds to the optical phonon modes observed in Raman spectroscopy. In conclusion, the measurements of coherent acoustic or optical phonons dynamics at various temperatures are useful to study the spin-lattice coupling in multiferroic materials.

4.2.1. Coherent Optical Phonons. Figure 9 shows the reflectivity oscillation of ~3.6 THz (~120 cm^{-1}) in $LuMnO_3$ [36, 37]. According to Raman studies, this vibration mode can be attributed to optical phonon of A_1 symmetry [80–82]. This fully symmetric mode corresponds to the Lu1 and Lu2 atomic motions in opposite directions along the c-axis, as depicted in Figure 9(d) [36]. In temperature-dependent measurements, the optical phonon almost disappears below T_N, as shown in Figures 9(e) and 9(f) [37]. Additionally, the structural change from highly symmetry space group $P6_3/mmc$ to the tilted and distorted $P6_3cm$ ferroelectric structure [14] in hexagonal manganites (see Figure 1) takes place at the Curie temperature (T_C), which is much higher than T_N. Since there is no structural transition near T_N, the atomic motions of the optical phonon mode would be affected by the appearance of magnetic ordering. In other words, the disappearance of optical phonon mode might be attributed to the magneto-elastic coupling. In temperature-dependent Raman studies, the lowest frequency A_1 mode also shows anomaly at T_N in $YMnO_3$ [82], $ErMnO_3$ [83], and $LuMnO_3$ [80, 82] which support close relation between phonon and spin systems.

4.2.2. Coherent Acoustic Phonons. Coherent acoustic phonons have been observed in many kinds of hexagonal

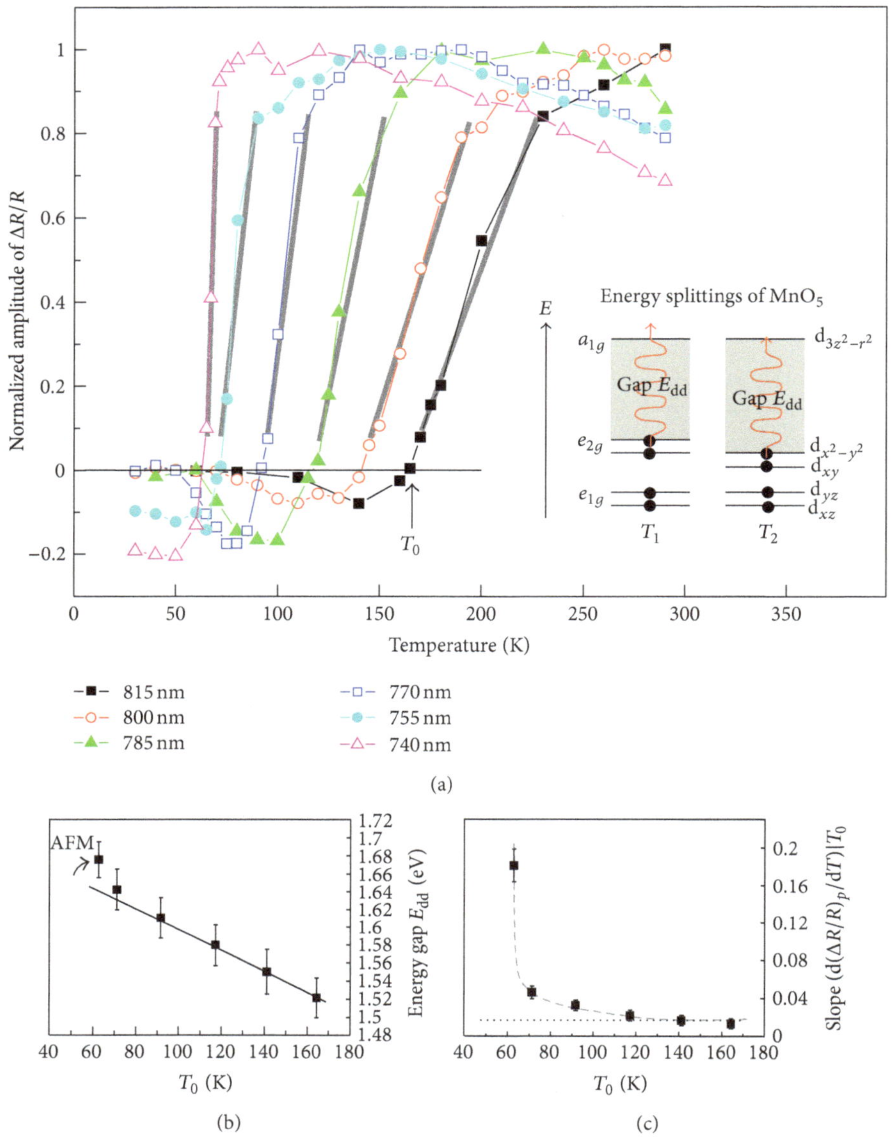

FIGURE 8: (a) Normalized amplitude of $\Delta R/R$ as a function of temperature at various wavelengths. The insets illustrate splitting of the Mn^{3+} energy levels in the local environment MnO_5 for photon energy above and below E_{dd} ($T_1 > T_2$). (b) Energy gap E_{dd} as a function of T_0. E_{dd} is estimated from the wavelengths in (a). The dashed line emphasizes linear behavior in the high-temperature range. (c) The slope of temperature-dependent normalized amplitude of $\Delta R/R$ in (a) (gray thick lines) as a function of T_0 at various wavelengths. The dashed line emphasizes behavior of the slope [35]. Reprinted figure with permission from Shih et al. [35]. Copyright (2009) by the American Physical Society.

manganites such as $LuMnO_3$ [37, 84], $YMnO_3$ [39], and $HoMnO_3$ [38]. The damped oscillation behavior in reflectivity changes with time delay t could be described as follows [76]:

$$\frac{\Delta R}{R} \propto \left(\frac{\Delta R}{R}\right)_0 \cos\left(\frac{4\pi n v_s}{\lambda} t - \phi\right) e^{-(v_s/\xi)t}, \quad (4)$$

where λ is the wavelength of the probe pulses, n is the refraction index of samples without strain, v_s is the strained layer propagating sound speed, and ξ is the penetration depth of probe pulses. The frequency of the oscillation is given as $\omega = 4\pi n v_s/\lambda$ and the dephasing time is ξ/v_s.

According to (4), the dephasing time is proportional to the penetration depth of probe pulses ξ and the inverse sound velocity $1/v_s$. However, the skin depth is much shorter than the propagating length of the strain pulse. Therefore, the dephasing time is dominated by the finite penetration depth ξ of the probe [84]. The dephasing time of acoustic phonons shows strong temperature dependence (Figure 10), and the anomalous increase of the dephasing time near T_N might be because of the photon energy being close to the edge of d-d transition. In other words, the penetration depth increases and the absorption of probe photons for the d-d

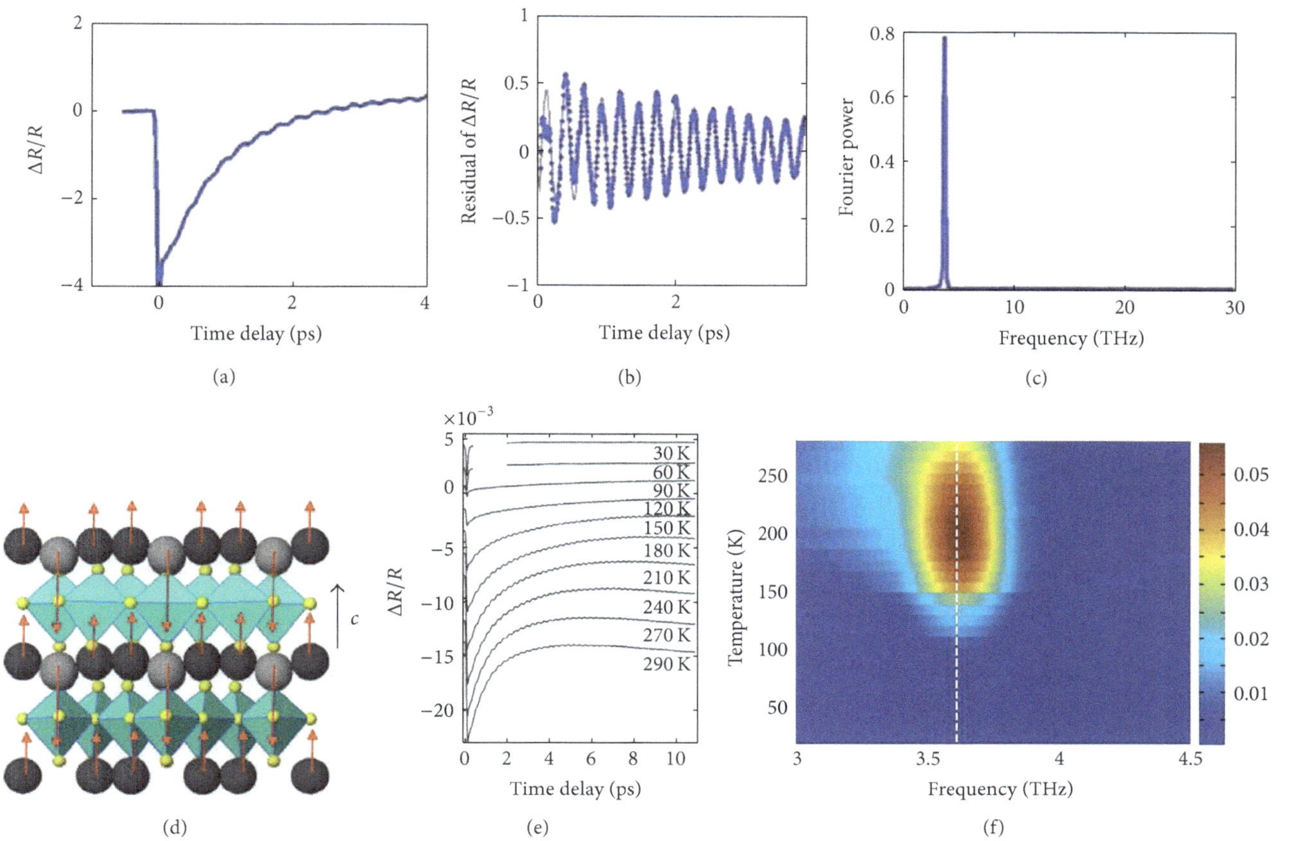

FIGURE 9: $LuMnO_3$ measured at 293 K. (a) $\Delta R/R$ as a function of time delay. (b) Oscillation component after subtraction of the slowly varying response. (c) Fourier power spectrum of the data with frequency 3.61 THz. (d) Schematic of the coherently excited A_1 phonon mode [36]. Reprinted figure with permission from Lou et al. [36]. Copyright (2009) by the American Physical Society. (e) $\Delta R/R$ at various temperatures. (f) Fourier transformed amplitudes of coherent optical phonon [37]. Figures were reproduced with permission from Jang et al. [37].

transition decreases (due to the blueshift of E_{dd} as mentioned in Section 3.1) as the temperature decreases.

The amplitude and period of the oscillation component in $\Delta R/R$ show strong temperature dependence as shown in Figure 11. Both of these parameters are clearly anomalous near the magnetic ordering temperature T_N. The amplitude of the oscillation component in $\Delta R/R$ gradually increases with decreasing temperature and reaches the maximum near a characteristic temperature (~190 K for $YMnO_3$ [39] and ~200 K for $LuMnO_3$ [37]). Then it decreases when the temperature approaches T_N, as shown in Figure 11(a). This may imply the appearance of AFM short-range ordering, as mentioned in [35]. On the other hand, the period of the oscillation component in $\Delta R/R$ also slightly shrinks with decreasing temperature. Surprisingly, it dramatically shrinks around T_N as shown in Figure 11(b). This strongly indicates the close correlation between AFM ordering and the propagation of thermal stress along c-axis, which is consistent with the results in heat capacity measurements [13, 85]. In addition, the ferroelectric polarization along the c-axis caused by polyhedron tilting of MnO_5 and distortion of the rare-earth ions [14] is well known. Thus, the propagation of a strain pulse with the modulation of dielectric constant (or ferroelectric polarization) along c-axis is indeed affected by the appearance of AFM ordering. Neutron diffraction experiments also have demonstrated the extra displacement of the rare-earth and oxygen atoms along c-axis at T_N [52]. Finally, these results in amplitude and period of the oscillation component in $\Delta R/R$ provide evidence for magnetoelastic coupling in these intriguing materials.

5. Summary and Outlook

In this paper, the recent developments on the electronic structures and ultrafast dynamics of hexagonal manganites have been reviewed. The electronic structure can be clearly revealed by the static linear optical absorption spectra and nonlinear second harmonic generation spectra. The temperature-dependent d-d transition energy exhibits an unexpected change near the spin ordering temperature T_N. This strongly indicates the coupling between electronic structure and antiferromagnetism in hexagonal manganites. Fortunately, the photon energy of the pulse laser used in pump-probe spectroscopy accidentally fits well the d-d transition (from e_{2g} to a_{1g}) energy. Therefore, the pump-probe

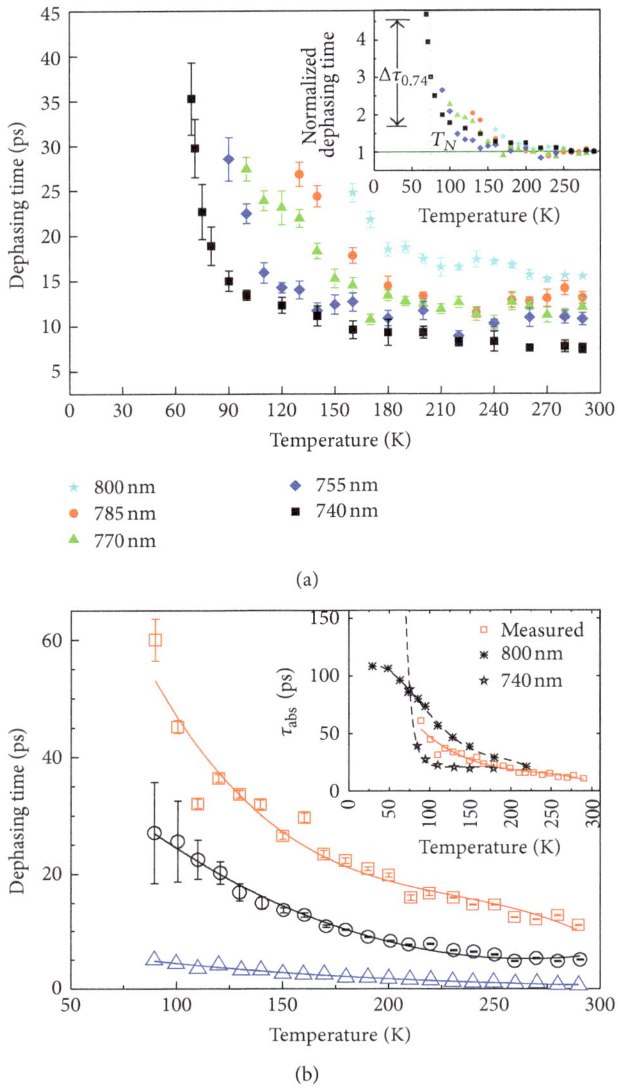

FIGURE 10: (a) Temperature dependence dephasing time of the oscillation component in $\Delta R/R$ at various wavelengths of $HoMnO_3$. Inset: the normalized dephasing time as a function of temperatures at various wavelengths [38]. Figures were reproduced with permission from Shih et al. [38]. (b) Temperature dependence of dephasing times: red squares for acoustic phonons, black circles for optical phonons, and blue triangles for electronic relaxation time [37]. Figures were reproduced with permission from Jang et al. [37].

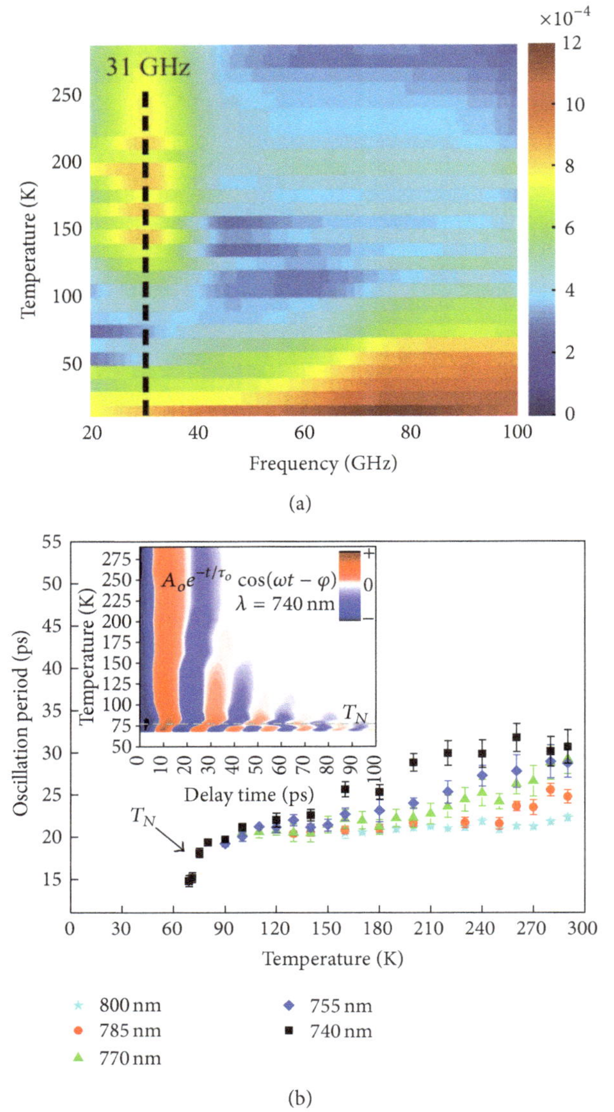

FIGURE 11: (a) Fourier transform amplitudes for coherent oscillation of $YMnO_3$ [39]. Reprinted with permission from Jang et al. [39]. Copyright 2010, American Institute of Physics. (b) Temperature dependence oscillation period in $\Delta R/R$ at various wavelengths of $HoMnO_3$. Inset: the evolution of the oscillation component in $\Delta R/R$ at various temperatures [38]. Figures were reproduced with permission from Shih et al. [38].

measurements can reveal the ultrafast dynamics through the spin-charge coupling and spin-phonon coupling.

However, those ultrafast dynamic experiments only present the charge dynamics between e_{2g} and a_{1g}. If the laser spectrum could cover the whole range of d-d transition, that is, containing the transition from e_{1g} to a_{1g} and e_{2g} to a_{1g}, it could disclose all of the charge dynamics in these strongly magnetic influenced d orbitals. The optical parametric amplifier (OPA) is a suitable light source for this kind of study. Consequently, if all of the charge dynamics in the d orbitals could be completely revealed, it would shed light on the microscopic mechanism of the magnetoelectric coupling or magnetoelastic effect in this kind of strongly correlated material.

Acknowledgments

This work was supported by the National Science Council of Taiwan, under Grants nos. NSC98-2112-M-009-008-MY3, NSC101-2112-M-009-016-MY2 and by the MOEATU Program at NCTU of Taiwan, Taiwan.

References

[1] M. Fiebig, "Revival of the magnetoelectric effect," *Journal of Physics D*, vol. 38, no. 8, pp. R123–R152, 2005.

[2] D. I. Khomskii, "Multiferroics: different ways to combine magnetism and ferroelectricity," *Journal of Magnetism and Magnetic Materials*, vol. 306, no. 1, pp. 1–8, 2006.

[3] W. Eerenstein, N. D. Mathur, and J. F. Scott, "Multiferroic and magnetoelectric materials," *Nature*, vol. 442, no. 7104, pp. 759–765, 2006.

[4] C. W. Nan, M. I. Bichurin, S. Dong, D. Viehland, and G. Srinivasan, "Multiferroic magnetoelectric composites: historical perspective, status, and future directions," *Journal of Applied Physics*, vol. 103, no. 3, Article ID 031101, 2008.

[5] T. Lottermoser, T. Lonkai, U. Amann, D. Hohlwein, J. Ihringer, and M. Fiebig, "Magnetic phase control by an electric field," *Nature*, vol. 430, no. 6999, pp. 541–544, 2004.

[6] B. Lorenz, Y. Q. Wang, and C. W. Chu, "Ferroelectricity in perovskite $HoMnO_3$ and $YMnO_3$," *Physical Review B*, vol. 76, no. 10, Article ID 104405, 2007.

[7] B. Lorenz, Y. Q. Wang, Y. Y. Sun, and C. W. Chu, "Large magnetodielectric effects in orthorhombic $HoMnO_3$ and $YMnO_3$," *Physical Review B*, vol. 70, no. 21, Article ID 212412, 2004.

[8] T. Kimura, T. Goto, H. Shintani, K. Ishizaka, T. Arima, and Y. Tokura, "Magnetic control of ferroelectric polarization," *Nature*, vol. 426, no. 6962, pp. 55–58, 2003.

[9] M. Fiebig, T. Lottermoser, D. Fröhlich, A. V. Goltsev, and R. V. Pisarev, "Observation of coupled magnetic and electric domains," *Nature*, vol. 419, no. 6909, pp. 818–820, 2002.

[10] N. Hur, S. Park, P. A. Sharma, J. S. Ahn, S. Guha, and S. W. Cheong, "Electric polarization reversal and memory in a multiferroic material induced by magnetic fields," *Nature*, vol. 429, no. 6990, pp. 392–395, 2004.

[11] W. Prellier, M. P. Singh, and P. Murugavel, "The single-phase multiferroic oxides: from bulk to thin film," *Journal of Physics*, vol. 17, no. 30, pp. R803–R832, 2005.

[12] S. W. Cheong and M. Mostovoy, "Multiferroics: a magnetic twist for ferroelectricity," *Nature Materials*, vol. 6, no. 1, pp. 13–20, 2007.

[13] T. Katsufuji, S. Mori, M. Masaki, Y. Moritomo, N. Yamamoto, and H. Takagi, "Dielectric and magnetic anomalies and spin frustration in hexagonal $RMnO_3$ (R = Y, Yb, and Lu)," *Physical Review B*, vol. 64, no. 10, Article ID 104419, 2001.

[14] B. B. Van Aken, T. T. M. Palstra, A. Filippetti, and N. A. Spaldin, "The origin of ferroelectricity in magnetoelectric $YMnO_3$," *Nature Materials*, vol. 3, no. 3, pp. 164–170, 2004.

[15] C. J. Fennie and K. M. Rabe, "Ferroelectric transition in $YMnO_3$ from first principles," *Physical Review B*, vol. 72, no. 10, Article ID 100103, 2005.

[16] T. H. Lin, C. C. Hsieh, H. C. Shih et al., "Anomalous magnetic ordering in b-axis-oriented orthorhombic $HoMnO_3$ thin films," *Applied Physics Letters*, vol. 92, no. 13, Article ID 132503, 2008.

[17] M. N. Iliev, M. V. Abrashev, H. G. Lee et al., "Raman spectroscopy of orthorhombic perovskitelike $YMnO_3$ and $LaMnO_3$," *Physical Review B*, vol. 57, no. 5, pp. 2872–2877, 1998.

[18] J. S. Zhou and J. B. Goodenough, "Unusual evolution of the magnetic interactions versus structural distortions in $RMnO_3$ perovskites," *Physical Review Letters*, vol. 96, no. 24, Article ID 247202, 2006.

[19] W. C. Yi, C. S. Seo, S. I. Kwun, and J. G. Yoon, "Temperature dependence of capacitance/current-voltage characteristics of highly (0001)-oriented $YMnO_3$ thin films on Si," *Applied Physics Letters*, vol. 77, no. 7, pp. 1044–1046, 2000.

[20] H. N. Lee, Y. T. Kim, and S. H. Choh, "Comparison of memory effect between $YMnO_3$ and $SrBi_2Ta_2O_9$ ferroelectric thin films deposited on Si substrates," *Applied Physics Letters*, vol. 76, no. 8, pp. 1066–1068, 2000.

[21] S. Imada, T. Kuraoka, E. Tokumitsu, and H. Ishiwara, "Ferroelectricity of $YMnO_3$ thin films on Pt(111)/Al_2O_3(0001) and Pt(111)/Y_2O_3(111)/Si(111) structures grown by molecular beam epitaxy," *Japanese Journal of Applied Physics*, vol. 40, no. 2 A, pp. 666–671, 2001.

[22] Y. T. Kim, I. S. Kim, S. I. Kim, D. C. Yoo, and J. Y. Lee, "Atomic structure of random and c-axis oriented $YMnO_3$ thin films deposited on Si and Y_2O_3/Si substrates," *Journal of Applied Physics*, vol. 94, no. 8, pp. 4859–4862, 2003.

[23] T. Yoshimura, N. Fujimura, and T. Ito, "Ferroelectric properties of c-oriented $YMnO_3$ films deposited on Si substrates," *Applied Physics Letters*, vol. 73, no. 3, pp. 414–416, 1998.

[24] T. Lonkai, D. G. Tomuta, U. Amann et al., "Development of the high-temperature phase of hexagonal manganites," *Physical Review B*, vol. 69, no. 13, Article ID 134108, 2004.

[25] N. Fujimura, T. Ishida, T. Yoshimura, and T. Ito, "Epitaxially grown $YMnO_3$ film: new candidate for nonvolatile memory devices," *Applied Physics Letters*, vol. 69, no. 7, pp. 1011–1013, 1996.

[26] T. Kimura, S. Ishihara, H. Shintani et al., "Distorted perovskite with egl configuration as a frustrated spin system," *Physical Review B*, vol. 68, no. 6, Article ID 060403, 2003.

[27] T. Goto, T. Kimura, G. Lawes, A. P. Ramirez, and Y. Tokura, "Ferroelectricity and giant magnetocapacitance in perovskite rare-earth manganites," *Physical Review Letters*, vol. 92, no. 25, Article ID 257201, 2004.

[28] M. Tachibana, T. Shimoyama, H. Kawaji, T. Atake, and E. Takayama-Muromachi, "Jahn-Teller distortion and magnetic transitions in perovskite $RMnO_3$(R = Ho, Er, Tm, Yb, and Lu)," *Physical Review B*, vol. 75, no. 14, Article ID 14425, 2007.

[29] N. Iwata and K. Kohn, "Dielectric anomalies at magnetic transitions of hexagonal rare earth manganese oxides $RMnO_3$," *Journal of the Physical Society of Japan*, vol. 67, no. 9, pp. 3318–3319, 1998.

[30] J. E. Medvedeva, V. I. Anisimov, M. A. Korotin, O. N. Mryasov, and A. J. Freeman, "Effect of Coulomb correlation and magnetic ordering on the electronic structure of two hexagonal phases of ferroelectromagnetic $YMnO_3$," *Journal of Physics Condensed Matter*, vol. 12, no. 23, pp. 4947–4958, 2000.

[31] A. B. Souchkov, J. R. Simpson, M. Quijada et al., "Exchange interaction effects on the optical properties of $LuMnO_3$," *Physical Review Letters*, vol. 91, no. 2, Article ID 027203, 2003.

[32] W. S. Choi, S. J. Moon, S. S. A. Seo et al., "Optical spectroscopic investigation on the coupling of electronic and magnetic structure in multiferroic hexagonal $RMnO_3$ (R = Gd, Tb, Dy, and Ho) thin films," *Physical Review B*, vol. 78, no. 5, Article ID 054440, 2008.

[33] M. Fiebig, D. Fröhlich, K. Kohn et al., "Determination of the magnetic symmetry of hexagonal manganites by second harmonic generation," *Physical Review Letters*, vol. 84, no. 24, pp. 5620–5623, 2000.

[34] D. Fröhlich, S. Leute, V. V. Pavlov, and R. V. Pisarev, "Nonlinear optical spectroscopy of the two-order-parameter compound $YMnO_3$," *Physical Review Letters*, vol. 81, no. 15, pp. 3239–3242, 1998.

[35] H. C. Shih, T. H. Lin, C. W. Luo et al., "Magnetization dynamics and the Mn^{3+} *d-d* excitation of hexagonal $HoMnO_3$ single crystals using wavelength-tunable time-resolved femtosecond spectroscopy," *Physical Review B*, vol. 80, no. 2, Article ID 024427, 2009.

[36] S. T. Lou, F. M. Zimmermann, R. A. Bartynski, N. Hur, and S. W. Cheong, "Femtosecond laser excitation of coherent optical phonons in ferroelectric $LuMnO_3$," *Physical Review B*, vol. 79, no. 21, Article ID 214301, 2009.

[37] K. J. Jang, J. Lim, J. Ahn et al., "Ultrafast IR spectroscopic study of coherent phonons and dynamic spin-lattice coupling in multiferroic $LuMnO_3$," *New Journal of Physics*, vol. 12, no. 2, Article ID 023017, 2010.

[38] H. C. Shih, L. Y. Chen, C. W. Luo et al., "Ultrafast thermoelastic dynamics of $HoMnO_3$ single crystals derived from femtosecond optical pump-probe spectroscopy," *New Journal of Physics*, vol. 13, no. 5, Article ID 053003, 2011.

[39] K. J. Jang, H. G. Lee, S. Lee et al., "Strong spin-lattice coupling in multiferroic hexagonal manganite $YMnO_3$ probed by ultrafast optical spectroscopy," *Applied Physics Letters*, vol. 97, no. 3, Article ID 031914, 2010.

[40] B. G. Ueland, J. W. Lynn, M. Laver, Y. J. Choi, and S. W. Cheong, "Origin of electric-field-induced magnetization in multiferroic $HoMnO_3$," *Physical Review Letters*, vol. 104, no. 14, Article ID 147204, 2010.

[41] M. Fiebig, C. Degenhardt, and R. V. Pisarev, "Interaction of frustrated magnetic sublattices in $ErMnO_3$," *Physical Review Letters*, vol. 88, no. 2, Article ID 027203, 2002.

[42] O. P. Vajk, M. Kenzelmann, J. W. Lynn, S. B. Kim, and S. W. Cheong, "Magnetic order and spin dynamics in ferroelectric $HoMnO_3$," *Physical Review Letters*, vol. 94, no. 8, Article ID 087601, 2005.

[43] H. Lueken, "A magnetoelectric effect in $YMnO_3$ and $HoMnO_3$," *Angewandte Chemie*, vol. 47, no. 45, pp. 8562–8564, 2008.

[44] D. Khomskii, "Classifying multiferroics: mechanisms and effects," *Physics*, vol. 2 , article 20, 2009.

[45] A. Muñoz, J. A. Alonso, M. J. Martínez-Lope, M. T. Casáis, J. L. Martínez, and M. T. Fernández-Díaz, "Magnetic structure of hexagonal $RMnO_3$ (R = Y, Sc): thermal evolution from neutron powder diffraction data," *Physical Review B*, vol. 62, no. 14, pp. 9498–9510, 2000.

[46] A. Muñoz, J. A. Alonso, M. J. Martínez-Lope, M. T. Casáis, J. L. Martínez, and M. T. Fernández-Díaz, "Evolution of the magnetic structure of hexagonal $HoMnO_3$ from neutron powder diffraction data," *Chemistry of Materials*, vol. 13, no. 5, pp. 1497–1505, 2001.

[47] M. Fiebig, C. Degenhardt, and R. V. Pisarev, "Magnetic phase diagram of $HoMnO_3$," *Journal of Applied Physics*, vol. 91, no. 10, pp. 8867–8869, 2002.

[48] T. Lonkai, D. G. Tomuta, J. U. Hoffmann, R. Schneider, D. Hohlwein, and J. Ihringer, "Magnetic two-dimensional short-range order in hexagonal manganites," *Journal of Applied Physics*, vol. 93, no. 10, pp. 8191–8193, 2003.

[49] Z. J. Huang, Y. Cao, Y. Y. Sun, Y. Y. Xue, and C. W. Chu, "Coupling between the ferroelectric and antiferromagnetic orders in $YMnO_3$," *Physical Review B*, vol. 56, no. 5, pp. 2623–2626, 1997.

[50] D. G. Tomuta, S. Ramakrishnan, G. J. Niewenhuys, and J. A. Mydosh, "The magnetic susceptibility, specific heat and dielectric constant of hexagonal $YMnO_3$, $LuMnO_3$ and $ScMnO_3$," *Journal of Physics Condensed Matter*, vol. 13, no. 20, pp. 4543–4552, 2001.

[51] C. Dela Cruz, F. Yen, B. Lorenz et al., "Strong spin-lattice coupling in multiferroic $HoMnO_3$: thermal expansion anomalies and pressure effect," *Physical Review B*, vol. 71, no. 6, Article ID 060407(R), 2005.

[52] S. Lee, A. Pirogov, M. Kang et al., "Giant magneto-elastic coupling in multiferroic hexagonal manganites," *Nature*, vol. 451, no. 7180, pp. 805–809, 2008.

[53] D. Y. Cho, J. Y. Kim, B. G. Park et al., "Ferroelectricity driven by y d0-ness with rehybridization in $YMnO_3$," *Physical Review Letters*, vol. 98, no. 21, Article ID 217601, 2007.

[54] D. Talbayev, S. A. Trugman, A. V. Balatsky, T. Kimura, A. J. Taylor, and R. D. Averitt, "Detection of coherent magnons via ultrafast pump-probe reflectance spectroscopy in multiferroic $Ba_{0.6}Sr_{1.4}Zn_2Fe_{12}O_{22}$," *Physical Review Letters*, vol. 101, no. 9, Article ID 097603, 2008.

[55] A. V. Kimel, B. A. Ivanov, R. V. Pisarev, P. A. Usachev, A. Kirilyuk, and T. Rasing, "Inertia-driven spin switching in antiferromagnets," *Nature Physics*, vol. 5, no. 10, pp. 727–731, 2009.

[56] C. W. Luo, I. H. Wu, P. C. Cheng et al., "Quasiparticle dynamics and phonon softening in FeSe superconductors," *Physical Review Letters*, vol. 108, no. 25, Article ID 257006, 2012.

[57] C. W. Luo, I. H. Wu, P. C. Cheng et al., "Ultrafast dynamics and phonon softening in $Fe_{1+y}Se_{1-x}Te_x$ single crystals," *New Journal of Physics*, vol. 14, no. 10, Article ID 103053, 2012.

[58] A. V. Kimel, R. V. Pisarev, F. Bentivegna, and T. Rasing, "Time-resolved nonlinear optical spectroscopy of Mn^{3+} ions in rare-earth hexagonal manganites $RMnO_3$ (R = Sc, Y, Er)," *Physical Review B*, vol. 64, no. 20, Article ID 201103(R), 2001.

[59] A. V. Kimel, R. V. Pisarev, F. Bentivegna, and Th. Rasing, "Ultrafast optical spectroscopy of hexagonal manganites $RMnO_3$ (R = Y, Er, Sc)," *Ferroelectrics*, vol. 279, no. 1, pp. 135–176, 2002.

[60] W. C. Yi, S. I. Kwun, and J. G. Yoon, "Study on the electronic structure of hexagonal and orthorhombic $YMnO_3$," *Journal of the Physical Society of Japan*, vol. 69, no. 8, pp. 2706–2707, 2000.

[61] A. Filippetti and N. A. Hill, "Coexistence of magnetism and ferroelectricity in perovskites," *Physical Review B*, vol. 65, no. 19, Article ID 195120, 2002.

[62] K. Kritayakirana, P. Berger, and R. V. Jones, "Optical spectra of ferroelectric-antiferromagnetic rare earth manganates," *Optics Communications*, vol. 1, no. 3, pp. 95–98, 1969.

[63] A. M. Kalashnikova and R. V. Pisarev, "Electronic structure of hexagonal rare-earth manganites $RMnO_3$," *JETP Letters*, vol. 78, no. 3, pp. 143–147, 2003.

[64] R. C. Rai, J. Cao, J. L. Musfeldt, S. B. Kim, S. W. Cheong, and X. Wei, "Spin-charge coupling and the high-energy magnetodielectric effect in hexagonal $HoMnO_3$," *Physical Review B*, vol. 75, no. 18, Article ID 184414, 2007.

[65] J. S. Kang, S. W. Han, J. G. Park et al., "Photoemission and x-ray absorption of the electronic structure of multifferroic $RMnO_3$ (R = Y, Er)," *Physical Review B*, vol. 71, no. 9, Article ID 092405, 2005.

[66] M. Qian, J. Dong, and D. Y. Xing, "Optical properties of the ferroelectromagnet $YMnO_3$ studied from first principles," *Physical Review B*, vol. 63, no. 15, Article ID 155101, 2001.

[67] W. S. Choi, D. G. Kim, S. S. A. Seo et al., "Electronic structures of hexagonal $RMnO_3$ (R = Gd, Tb, Dy, and Ho) thin films: optical spectroscopy and first-principles calculations," *Physical Review B*, vol. 77, no. 4, Article ID 045137, 2008.

[68] M. Fiebig, V. V. Pavlov, and R. V. Pisarev, "Second-harmonic generation as a tool for studying electronic and magnetic

structures of crystals: review," *Journal of the Optical Society of America B*, vol. 22, no. 1, pp. 96–118, 2005.

[69] C. Degenhardt, M. Fiebig, D. Fröhlich, T. Lottermoser, and R. V. Pisarev, "Nonlinear optical spectroscopy of electronic transitions in hexagonal manganites," *Applied Physics B*, vol. 73, no. 2, pp. 139–144, 2001.

[70] V. M. Axt and T. Kuhn, "Femtosecond spectroscopy in semiconductors: a key to coherences, correlations and quantum kinetics," *Reports on Progress in Physics*, vol. 67, no. 4, pp. 433–512, 2004.

[71] S. K. Sundaram and E. Mazur, "Inducing and probing nonthermal transitions in semiconductors using femtosecond laser pulses," *Nature Materials*, vol. 1, no. 4, pp. 217–224, 2002.

[72] R. D. Averitt and A. J. Taylor, "Ultrafast optical and far-infrared quasiparticle dynamics in correlated electron materials," *Journal of Physics Condensed Matter*, vol. 14, no. 50, pp. R1357–R1390, 2002.

[73] C. W. Luo, Y. T. Wang, F. W. Chen et al., "Eliminate coherence spike in reflection-type pump-probe measurements," *Optics Express*, vol. 17, no. 14, pp. 11321–11327, 2009.

[74] K. H. Wu, H. J. Chen, Y. T. Chen et al., "Marked enhancement of Néel temperature in strained $YMnO_3$ thin films probed by femtosecond spectroscopy," *Europhysics Letters*, vol. 94, no. 2, Article ID 27006, 2011.

[75] C. Thomsen, J. Strait, Z. Vardeny, H. J. Maris, J. Tauc, and J. J. Hauser, "Coherent phonon generation and detection by picosecond light pulses," *Physical Review Letters*, vol. 53, no. 10, pp. 989–992, 1984.

[76] C. Thomsen, H. T. Grahn, H. J. Maris, and J. Tauc, "Surface generation and detection of phonons by picosecond light pulses," *Physical Review B*, vol. 34, no. 6, pp. 4129–4138, 1986.

[77] H. J. Zeiger, J. Vidal, T. K. Cheng, E. P. Ippen, G. Dresselhaus, and M. S. Dresselhaus, "Theory for displacive excitation of coherent phonons," *Physical Review B*, vol. 45, no. 2, pp. 768–778, 1992.

[78] R. Merlin, "Generating coherent THz phonons with light pulses," *Solid State Communications*, vol. 102, no. 2-3, pp. 207–220, 1997.

[79] G. A. Garrett, T. F. Albrecht, J. F. Whitaker, and R. Merlin, "Coherent THz phonons driven by light pulses and the Sb problem: what is the mechanism?" *Physical Review Letters*, vol. 77, no. 17, pp. 3661–3664, 1996.

[80] A. Ghosh, J. R. Sahu, S. V. Bhat, and C. N. R. Rao, "A Raman study of multiferroic $LuMnO_3$," *Solid State Sciences*, vol. 11, no. 9, pp. 1639–1642, 2009.

[81] A. P. Litvinchuk, M. N. Iliev, V. N. Popov, and M. M. Gospodinov, "Raman and infrared-active phonons in hexagonal $HoMnO_3$ single crystals: magnetic ordering effects," *Journal of Physics Condensed Matter*, vol. 16, no. 6, pp. 809–819, 2004.

[82] J. Vermette, S. Jandl, A. A. Mukhin et al., "Raman study of the antiferromagnetic phase transitions in hexagonal $YMnO_3$ and $LuMnO_3$," *Journal of Physics Condensed Matter*, vol. 22, no. 35, Article ID 356002, 2010.

[83] J. Vermette, S. Jandl, and M. M. Gospodinov, "Raman study of spin-phonon coupling in $ErMnO_3$," *Journal of Physics*, vol. 20, no. 42, Article ID 425219, 2008.

[84] D. Lim, R. D. Averitt, J. Demsar, A. J. Taylor, N. Hur, and S. W. Cheong, "Coherent acoustic phonons in hexagonal manganite $LuMnO_3$," *Applied Physics Letters*, vol. 83, no. 23, pp. 4800–4802, 2003.

[85] P. A. Sharma, J. S. Ahn, N. Hur et al., "Thermal conductivity of geometrically frustrated, ferroelectric $YMnO_3$: extraordinary spin-phonon interactions," *Physical Review Letters*, vol. 93, no. 17, Article ID 177202, p. 1, 2004.

Fabrication and Optoelectrical Properties of IZO/Cu_2O Heterostructure Solar Cells by Thermal Oxidation

Cheng-Chiang Chen, Lung-Chien Chen, and Yi-Hsuan Lee

Department of Electro-optical Engineering, National Taipei University of Technology, 1, Section 3, Chung-Hsiao E. Road, Taipei 106, Taiwan

Correspondence should be addressed to Lung-Chien Chen, ocean@ntut.edu.tw

Academic Editor: Nigel Wilding

Indium zinc oxide (IZO)/cupper oxide (Cu_2O) is a nontoxic nature and an attractive all-oxide candidate for low-cost photovoltaic (PV) applications. The present paper reports on the fabrication of IZO/Cu_2O heterostructure solar cells which the Cu_2O layers were prepared by oxidation of Cu thin films deposited on glass substrate. The measured parameters of cells were the short-circuit current (I_{sc}), the open-circuit voltage (V_{oc}), the maximum output power (P_m), the fill factor (FF), and the efficiency (η), which had values of 0.11 mA, 0.136 V, 5.05 μW, 0.338, and 0.56%, respectively, under AM 1.5 illumination.

1. Introduction

Cu_2O has long been considered an attractive to silicon, and there are other semiconductors being favored for the fabrication of cheaper solar cells for terrestrial applications [1–4]. The Cu_2O is intrinsically P-type, has a band gap (Eg) of 2.17 eV [5], and is expected to have a maximum theoretical efficiency of 20% [6, 7]. Its advantages include: (1) its non-toxic nature, (2) abundance of the starting material (copper), and (3) cheap production processing. Cuprous oxide is of particular interest in the field of solar energy, because of their low cost, abundance of the starting material (copper), and its nontoxic nature and the light-to-electricity conversion efficiency [8]. IZO/Cu_2O is an attractive all-oxide candidate for low-cost photovoltaic (PV) applications. The indium zinc oxide (IZO) films have been attracting a lot of attention because of good conductivity [9], high optical transparency [10], low deposition temperature, and very smooth surface roughness [11] in comparison with ITO films.

Therefore, Cu_2O thin films can be prepared by various techniques, such as reactive sputtering [12], MOCVD [13], electrochemical deposition [14–17], solution chemical [18–20], and direct oxidation of Cu sheets [21], of which the magnetron sputtering presents many advantages as low-cost approach, large scale, and easy control. Some groups reported ZnO/Cu_2O solar cells prepared [21, 22]. The efficiency was to be 1-2%. In this work, the effect of the conditions of deposition of the P-type Cu_2O thin films by the radiofrequency magnetron sputtering method is studied; in particular, the effect of changing the temperature and time of annealing to obtain the different performances of P-type Cu_2O thin layers on crystal quality was investigated. Finally, the fabrication and optoelectronic performance of an IZO/Cu_2O solar cell is considered.

2. Experimental

In the study, devices that had cupper (Cu) layers of one micron thick were deposited with a purity of 99.995% on a glass substrate by magnetron reactive sputtering from high-purity Cu targets in argon (Ar) gas at a flow rate of 15 sccm and a stable pressure of 3×10^{-3} Torr. The sputtering system consisted of a vacuum chamber with a target and a substrate copper holder, a turbo molecular pump parallel to a rotary pump that provides a high vacuum, a radio frequency (RF) power supply at 13.56 MHz, and mass flow controllers that maintain a steady gas flow rate. The target was cleaned by presputtering with Ar plasma for 5 min prior to each deposition process. In all our Cu deposition

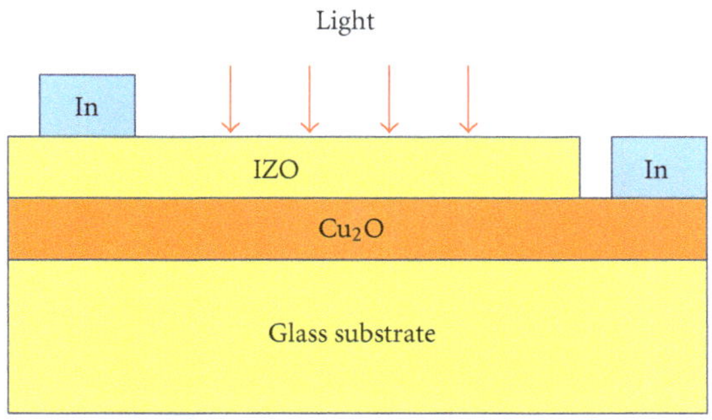

FIGURE 1: Schematic crosssection of the completed structure.

experiments the RF power and the gas pressure were kept constant at 50 W and 3.1×10^{-3} Torr, respectively. The Cu_2O layers were obtained by oxidation of Cu layers annealing at various temperatures for 10 min in air. For easy and convenient characterization of the same material, Cu_2O films were prepared on glass to measure the electronic characteristics by Hall measurements and their absorbance from transmittance spectra. The films' crystallinity was studied by X-ray diffraction (XRD) and exhibits a polycrystalline structure. Indium-oxide-doped zinc oxide (IZO, In : Zn = 1 : 9) was deposited onto the Cu_2O layer by sputtering, and indium electrodes were formed by the evaporation onto both the surface of the IZO layer and the Cu_2O layer, respectively, to complete IZO/Cu_2O heterostructure solar cells. Figure 1 shows the crosssection of the completed structure. Additionally, the current density-voltage (J-V) characteristics were determined using a Keithley 2420 programmable source meter under irradiation by a 100 W xenon lamp. Finally, the irradiation power density on the surface of the sample was calibrated as 100 W/m^2.

3. Results and Discussion

In order to obtain a better understanding of the thermal oxidation mechanism, phase identification was performed. Figure 2 shows the XRD spectrum of the measured Cu_2O films at various temperatures of 300, 400, and 500°C. The XRD diffraction shows that single phase of Cu_2O films is obtained for growth at different thermal oxidation temperature with diffraction peaks at 29.58° and 43.35° corresponding to the (110) and (111) planes of the cubic-structured Cu_2O, respectively. As oxidation temperature increases, the CuO peak decreases and eventually vanished at 500°C. The strong Cu peaks at temperatures below 300°C [23], the sudden stop of mass gain in the using thermogravimetric analysis results, together with the SEM images, imply that a self passivation mechanism may involve in the formation of a compact Cu_2O layer. We employed the X-ray diffraction technique to get main crystalline phases and the possible orientation of crystalline in the films prepared at optimum conditions.

Figure 3 shows the FESEM micrographs of the Cu_2O layers with thermal treatment at various temperatures. The micrographs indicate that the surface of the film

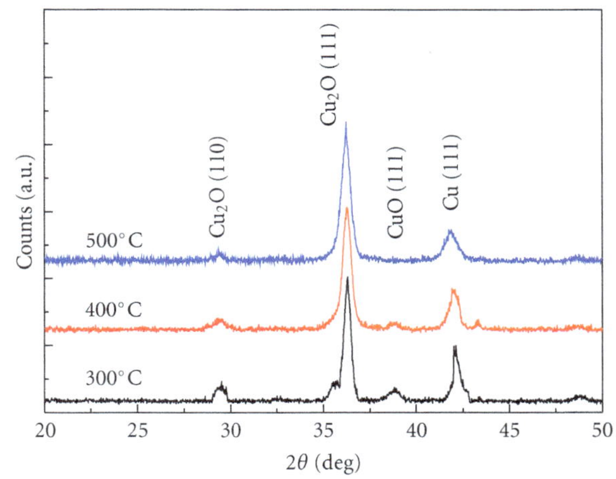

FIGURE 2: X-ray diffraction patterns of the samples with thermal oxidation at various temperatures.

consisted of small particles. The average particle sizes were approximately (a) 70, (b) 100, and (c) 200 nm for 300, 400, and 500°C annealing, respectively. The particle sizes of layer are increases with oxidation temperature increase. That shows thin oxides formed, particularly developed in the grain boundary areas, implying a product of fast-diffusion processes, which might be responsible for the small value of activation energy at oxidation of 500°C [24].

Figure 4 plots both the carrier concentration and the mobility as a function of thermal oxidation annealing temperatures. As the thermal oxidation annealing temperatures are elevated to 500°C, the carrier concentration is likely to decrease, and at the same time, the mobility increases to 5 cm^2/Vs. That is, a film with relatively high mobility is obtained by thermal oxidation annealing of 400°C samples, although the temperature required to obtain the same mobility by thermal oxidation annealing appears to be slightly higher than in the case that the film is directly formed by oxidation at elevated temperatures such as 400 to 500°C. The increase of mobility shown in Figure 4 may be attributed to this increase of grain size.

Figure 5(a) shows the results of the absorption measurements for the Cu_2O layers with treatment at various annealing temperatures for 10 min. The layers have a very high absorption in visible region resulting in good materials for the solar energy devices. According to this figure, the absorption increased continuously as the annealing temperature increased from 300 to 500°C owing to that the Cu content decreased. Figure 5(b) shows the results of the absorption measurements for the IZO layers. The IZO layer yields an absorption edge at ~3.25 eV. Therefore, the transparent semiconductor IZO layer is suitable as the upper layer for the IZO/Cu_2O solar cells.

Figure 6 shows the I-V characteristics of the IZO/Cu_2O heterostructure with and without any illumination, respectively. The cell performance was measured under AM 1.5 illumination with a solar intensity of 10 mW/cm^2 at 25°C. The cell has an active area of 0.3×0.3 cm^2 and no antireflective coating. The Cu_2O layers were prepared by

FIGURE 3: SEM images of the samples with thermal oxidation at various temperatures.

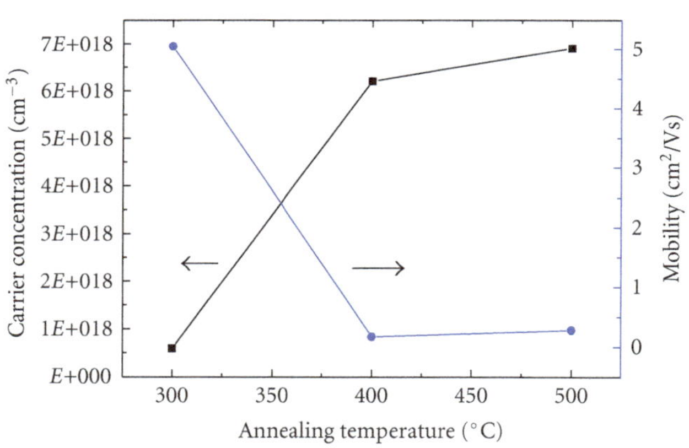

FIGURE 4: The carrier concentration and the mobility as a function of thermal oxidation annealing temperatures.

oxidation of Cu thin films at 500°C for 10 min. IZO/Cu_2O solar cells exhibited the following static parameters: I_{sc} of 0.11 mA (J_{sc} of 1.22 mA/cm^2) and V_{oc} of 0.136 V. As is well known, the fill factor (FF) can be described by [25]

$$FF = \frac{I_m V_m}{I_{sc} V_{oc}}, \quad (1)$$

where I_m is the maximum output current, and V_m is the maximum output voltage. Therefore, using the values of I_m and V_m deduced from Figure 6, the value of FF results is equal to 0.338. Similarly, the conversion efficiency (η) defined by [25]

$$\eta = \frac{I_m V_m}{P_{inc}} \quad (2)$$

with P_{inc} is the incident power, results are to be 0.56%. Low FF value and poor conversion efficiency are owing to the high series resistance (~760 Ω, in this study). The series resistance is mainly caused by the bulk resistance of materials, contacts, and interconnections. However, in this work, the high value of the series resistance may be caused by the Cu_2O layers and the contact resistance between the electrodes and the IZO layer and the Cu_2O layer.

4. Conclusions

The present paper reports on the fabrication of IZO/Cu_2O heterostructure solar cells in which the Cu_2O layers were prepared by oxidation of Cu thin films deposited on glass substrate. The Cu_2O layers have a very high absorption in visible region resulting in good materials for the solar

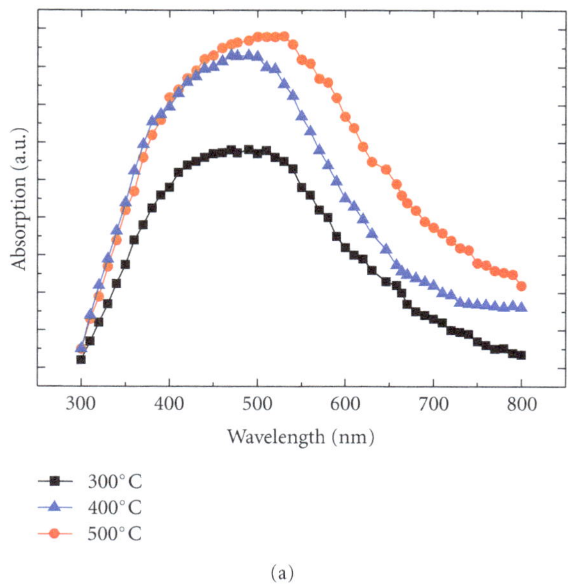

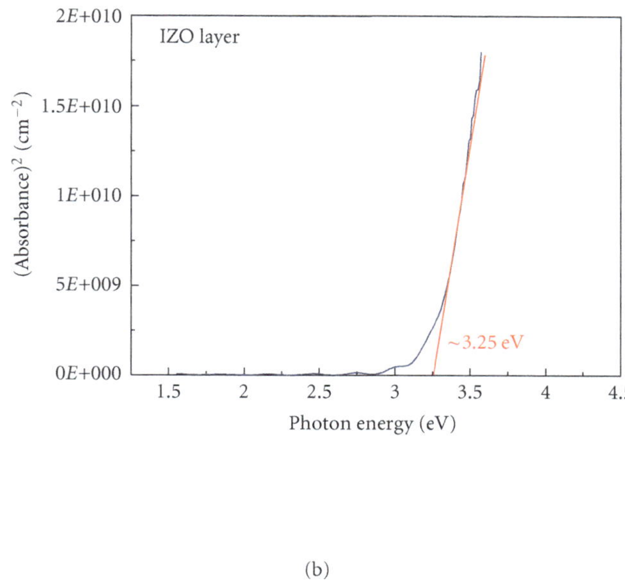

FIGURE 5: Absorbance spectra for (a) Cu_2O films with thermal oxidation at various temperatures and (b) the relationship between absorption squared versus photon energy.

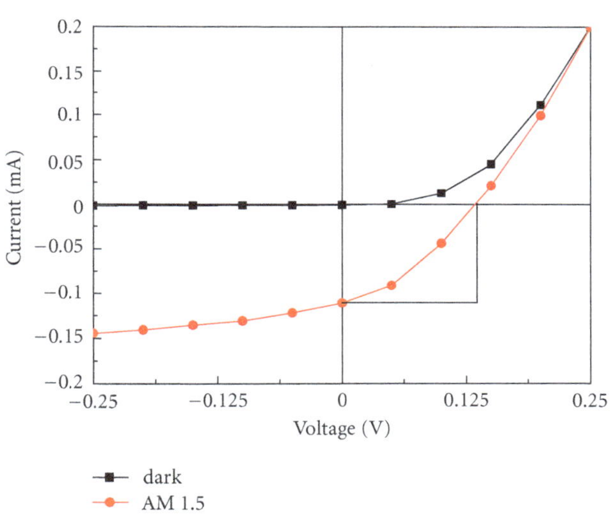

FIGURE 6: I-V curves evaluated for IZO/Cu_2O heterostructure solar cells in the dark and under an illumination.

energy devices. The measured parameters of cells were the short-circuit current (I_{sc}), the open-circuit voltage (V_{oc}), the maximum output power (P_m), the fill factor (FF), and the efficiency (η), which had values of 0.11 mA, 0.136 V, 5.05 μW, 0.338, and 0.56%, respectively, under AM 1.5 illumination. Therefore, IZO/Cu_2O is a nontoxic nature and an attractive all-oxide candidate for low-cost photovoltaic (PV) applications in the future.

Acknowledgment

The authors would like to thank the National Science Council of the Republic of China for financially supporting this research under Contract no. NSC 100-2215-E-027-057.

References

[1] C. Noguet, M. Tapiero, C. Schwab et al., "Cuprous oxide as a photovoltaic converter," in *Proceedings of the Photovoltaic Solar Energy Conference*, p. 1170, Luxemburg, 1977.

[2] A. A. Berezin and F. L. Weichman, "Photovoltaic effect in cuprous oxide-copper junctions in relation to the optical absorption spectrum of cuprous oxide," *Solid State Communications*, vol. 37, no. 2, pp. 157–160, 1981.

[3] M. Tapiero, C. Noguet, J. P. Zielinger, C. Schwab, and D. Pierrat, "Conversion photovoltaïgque dans Cu_2O," *Applied Physics Reviews*, vol. 14, p. 231, 1979.

[4] L. C. Olsen, F. W. Addis, and R. C. Bohara, "Investigation of Cu_2O solar cells," in *Proceedings of the 14th IEEE Photovoltaic Specification Conference*, pp. 462–467, January 1980.

[5] P. Poizot, S. Laruelle, S. Grugeon, L. Dupont, and J. M. Tarascon, "Nano-sized transition-metal oxides as negative-electrode materials for lithium-ion batteries," *Nature*, vol. 407, no. 6803, pp. 496–499, 2000.

[6] S. S. Jeong, A. Mittiga, E. Salza, A. Masci, and S. Passerini, "Electrodeposited ZnO/Cu_2O heterojunction solar cells," *Electrochimica Acta*, vol. 53, no. 5, pp. 2226–2231, 2008.

[7] A. Mittiga, E. Salza, F. Sarto, M. Tucci, and R. Vasanthi, "Heterojunction solar cell with 2% efficiency based on a Cu_2O substrate," *Applied Physics Letters*, vol. 88, no. 16, Article ID 163502, 2006.

[8] E. Y. Wang, D. Trivich, H. Sawaiha, and G. Thomas, "Cuprous oxide Schottky photovoltaic cells as potential solar energy converters," in *Proceedings of the Complex International Conference*, p. 643, Dharan, Saudi Arabia, 1975.

[9] T. Minami, T. Kakumu, and S. Takata, "Preparation of transparent and conductive In_2O_3–ZnO films by radio frequency magnetron sputtering," *Journal of Vacuum Science & Technology A*, vol. 14, no. 3, article 1704, 5 pages, 1996.

[10] N. Naghavi, A. Rougier, C. Marcel, C. Guery, J. B. Leriche, and J. M. Tarascon, "Characterization of indium zinc oxide thin

films prepared by pulsed laser deposition using a $Zn_3In_2O_6$ target," *Thin Solid Films*, vol. 360, no. 1-2, pp. 233–240, 2000.

[11] A. Kaijo, *Display Imaging*, vol. 4, p. 143, 1996.

[12] S. Izhizuka, S. Kato, T. Maruyama, and K. Akimoto, "Nitrogen doping into Cu_2O thin films deposited by reactive radio-frequency magnetron sputtering," *Japanese Journal of Applied Physics*, vol. 40, pp. 2765–2768, 2001.

[13] S. Jeong and E. S. Aydil, "Heteroepitaxial growth of Cu_2O thin film on ZnO by metal organic chemical vapor deposition," *Journal of Crystal Growth*, vol. 311, no. 17, pp. 4188–4192, 2009.

[14] K. Nakaoka and K. Ogura, "Electrochemical preparation of p-type cupric and cuprous oxides on platinum and gold substrates from copper(II) solutions with various amino acids," *Journal of the Electrochemical Society*, vol. 149, no. 11, pp. C579–C585, 2002.

[15] A. K. Mukhopadhyay, A. K. Chakraborty, A. P. Chatterjee, and S. K. Lahiri, "Galvanostatic deposition and electrical characterization of cuprous oxide thin films," *Thin Solid Films*, vol. 209, no. 1, pp. 92–96, 1992.

[16] S. Joseph and P. V. Kamath, "Electrochemical deposition of Cu_2O on stainless steel substrates: promotion and suppression of oriented crystallization," *Solid State Sciences*, vol. 10, no. 9, pp. 1215–1221, 2008.

[17] T. Fujiwara, T. Nakaue, and M. Yoshimura, "Direct fabrication and patterning of Cu_2O film by local electrodeposition method," *Solid State Ionics*, vol. 175, no. 1–4, pp. 541–544, 2004.

[18] L. Zhang, H. Li, Y. H. Ni, J. Li, K. M. Liao, and G. C. Zhao, "Porous cuprous oxide microcubes for non-enzymatic amperometric hydrogen peroxide and glucose sensing," *Electrochemistry Communications*, vol. 11, p. 812, 2009.

[19] C. H. Kuo and M. H. Huang, *Journal of the American Chemical Society*, vol. 130, p. 12815, 2008.

[20] G. Jimenez-Cadena, E. Comini, M. Ferroni, and G. Sberveglieri, "Synthesis of Cu_2O bi-pyramids by reduction of $Cu(OH)_2$ in solution," *Materials Letters*, vol. 64, no. 3, pp. 469–471, 2010.

[21] A. Mittiga, E. Salza, F. Sarto, M. Tucci, and R. Vasanthi, "Heterojunction solar cell with 2% efficiency based on a Cu_2O substrate," *Applied Physics Letters*, vol. 88, no. 16, Article ID 163502, 2006.

[22] T. Minami, T. Miyata, K. Ihara, Y. Minamino, and S. Tsukada, "Effect of ZnO film deposition methods on the photovoltaic properties of ZnO–Cu_2O heterojunction devices," *Thin Solid Films*, vol. 494, no. 1-2, pp. 47–52, 2006.

[23] Y. Zhu, K. Mimura, and M. Isshiki, "Oxidation Mechanism of Cu_2O to CuO at 600–1050°C," *Oxidation of Metals*, vol. 62, no. 3/4, pp. 207–222, 2004.

[24] W. Gao, H. Gong, J. He, A. Thomas, L. Chan, and S. Li, "Oxidation behaviour of Cu thin films on Si wafer at 175–400°C," *Materials Letters*, vol. 51, p. 78, 2001.

[25] Sze S. M., *Physics of Semiconductor Devices*, John Wiley & Sons, New York, NY, USA, 2nd edition, 1981.

High-Temperature Electroluminescence of InGaN/GaN Light-Emitting Devices with Multiple Quantum Barriers

Ya-Fen Wu

Department of Electronic Engineering, Ming Chi University of Technology, Taishan, New Taipei City 243, Taiwan

Correspondence should be addressed to Ya-Fen Wu, yfwu@mail.mcut.edu.tw

Academic Editor: Donghui Li

We investigate the high-temperature characteristics of InGaN/GaN multiple quantum well light-emitting devices with and without multiple quantum barriers (MQBs) in depth. The electroluminescence measurements were carried out over a temperature range from 200 to 380 K and an injection current level from 1 to 100 mA. Enhanced carrier confinement and stronger carrier localization in the active layer are achieved for the sample with MQBs. Furthermore, it is found that the external quantum efficiency of the sample possessing MQBs is higher than that of the sample with GaN barriers. The MQB structure improves the high-temperature operation of light-emitting devices.

1. Introduction

The progress in the epitaxial growth and device processing of GaN-based heterostructures has made this material system a promising light emitter operated in the wavelength regime ranging from infrared to ultraviolet [1–3]. Much interest has been focused on InGaN/GaN multiple quantum wells (MQWs) because they are key structures for the high-brightness blue-green light-emitting diodes (LEDs) and laser diodes (LDs) [3, 4]. The understanding of optical properties of MQW structure is very important for the further improvement in the performance of the devices. In spite of the existence of high-density ($\sim 10^{10}/\mathrm{cm}^2$) misfit dislocations, the InGaN MQW LEDs shows very bright emission characteristics. Thus, origins of the high quantum efficiency have been receiving much attention. The quantum confinement effects on the InGaN alloy well and efficient carrier capturing by the localized radiative recombination centers in the quantum-dot-like states have been claimed to be important for origins of the high emission efficiency. It has been known that potential fluctuation, originating from segregation effect in InGaN, is one of the important reasons for the exciton localization effect in InGaN/GaN MQWs.

Because of large lattice and thermal expansion coefficient mismatches between the sapphire substrate and the epitaxial layers, the InGaN-based heterostructures display misfit dislocation in nanocrystalline structures, which lead to nonradiative recombination centers. The performance and power output of such devices are limited by the thermally activated loss of electrons overflowing from the active layer to the nonradiative recombination centers, thus reducing the emission efficiency. Moreover, for the lighting applications, the InGaN/GaN MQW devices are confronted by the requirements of high current injection. High current operation in LEDs causes self-heating and raises the junction temperature, further degrading the quantum efficiency. To improve the carrier blocking effect, an AlGaN thin film layer has been used in the multiple quantum wells (MQWs) region of high-power UV/blue/green LEDs to achieve high device efficiency [5, 6]. An alternative approach is to use multiple quantum barrier (MQB) structures to inhibit hot carrier overflow [7–9]. This technique was based on the use of a superlattice which consisted of alternating thin layers of narrow and wide bandgap semiconductors materials. The MQB structure has been proposed to enhance carrier confinement of semiconductors lasers by virtually increasing the barrier height, using the interference of the reflected electron wave from the boundaries of stacked superlattices [7–9].

In this study, we investigate the high-temperature characteristics of InGaN/GaN MQWs light-emitting devices with GaN barriers and with MQB structures. The MQW samples were prepared by metal-organic vapor phase epitaxy system. The electroluminescence (EL) measurements were carried out over a high-temperature range from 200 to 380 K and an injection current level from 1 to 100 mA. From the measured results of EL spectra, it is found that the sample with MQBs exhibits a less temperature-sensitive characteristic and a stronger localization effect. The external quantum efficiency of the sample with MQBs is significantly higher than that of the sample with GaN barriers at room temperature. The experimental results indicate that the MQB structures could not only improve the carrier recombination, but also profoundly improve the high-temperature operation of light-emitting devices.

2. Experiment

Figure 1 shows a schematic diagram of the layered structure investigated in this work. The samples were grown by metal-organic vapor phase epitaxy (MOVPE) on *c*-plane sapphire substrates. Ammonia (NH_3), trimethylgallium (TMG), trimethylindium (TMI), silane (SiH_4), and bis(cyclopentadienyl)magnesium (Cp_2Mg) were used either as precursors or as dopants. The layer structure of the sample with GaN barriers consisted of a 20-nm-thick GaN buffer layer, a 3-μm-thick Si-doped n-type GaN layer, an undoped GaN layer possessing five periods of $In_{0.18}Ga_{0.82}N$/GaN MQWs, and a 100-nm-thick Mg-doped p-type GaN layer. The doping concentrations of n- and p-type GaN were nominally about 5×10^{18} cm^{-3} and 1×10^{19} cm^{-3}, respectively. The thicknesses of the InGaN wells and the GaN barriers in the MQW structures were 2 nm and 11 nm, respectively. For the sample with MQBs, the layer structure was similar, but the MQWs were five periods of $In_{0.15}Ga_{0.85}N$/GaN MQWs, and the barrier layer was replaced by a five-period $In_{0.005}Ga_{0.995}N$/GaN (1 nm/1 nm) heterolayers. In the MQW and MQB structures, the Indium composition was determined by photoluminescence and X-ray diffraction measurements [10]. The light-emitting diode chips were fabricated over a broad area ($350 \times 350\ \mu m^2$) using standard photolithography processes. Ti/Al/Ti/Au and Ni-Au were employed for the n- and p-type electrodes, respectively. For temperature-dependent EL measurements, the samples were mounted in a closed-cycle He cryostat, where the temperature (T) was varied from 200 to 380 K and using a current source operated from 1 to 100 mA. The luminescence signal was dispersed through a 0.5 m monochromator and detected using a Si photodiode by standard lock-in amplification technique.

3. Results and Discussion

Figure 2 shows the evolution of EL spectra for the samples over a temperature range from 200 to 380 K at the injection current level 10 mA. We observed InGaN-related emissions with peak energies 2.69 and 2.93 eV at room temperature for samples with GaN barriers and with MQBs, respectively. Referring to the conduction band profile of MQB structure shown in Figure 3, the MQB structure is engineered to align a forbidden energy miniband of the MQB

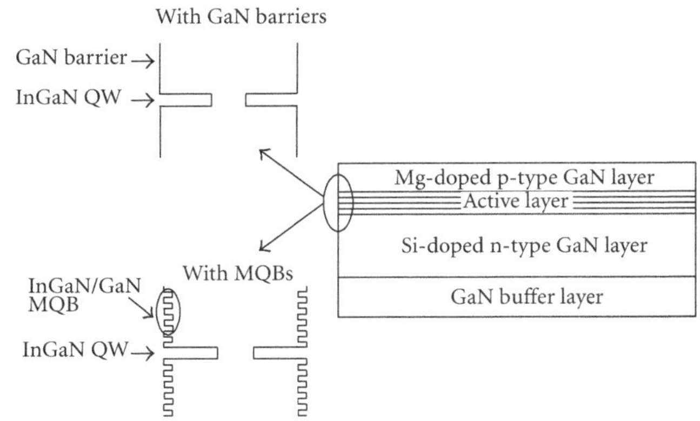

Figure 1: Schematic structure of InGaN/GaN multiple quantum well light-emitting devices (a) with GaN barrier and (b) with multiquantum barriers.

superlattice with the conduction band minimum to enhance the conduction band discontinuity. By using the transfer matrix method to solve the Schrödinger equation, one can obtain the reflectivity probability against incident electron energy. The result illustrates that a band of nonallowed electron states can be created above the classical barrier height U [9, 11]. This effectively enhances the reflectivity of electrons incident on the barriers and hence, increases the band discontinuity by dU. The higher EL peak energy of this sample with MQBs gives the evidence of the increased conduction band discontinuity. Besides, the higher EL spectra intensity observed for the sample with MQBs in Figure 2(b) could be attributed to the improvement of carrier confinement and inhibited carrier leakage in the active layer. With increasing temperature, both samples exhibit redshift of emission peaks due to the bandgap shrinkage which is resulting from the temperature induced dilation of the lattice and electron-lattice interaction. From the figure, we observe the amounts of redshift from 200 to 380 K for samples with GaN barriers and with MQBs are 18.5 meV and 15.2 meV, respectively, implying a less temperature-sensitive characteristic of the MQB sample. An S-shaped temperature dependence of the EL peak energy (redshift-blueshift-redshift) has been reported to be a fingerprint of the existence of the localization effect based on the band-tail-filling model for thermalized carriers with a Gaussian distribution of energy potential fluctuation and dispersion [12, 13]. In our measurements the blueshift is not observed since the temperature-dependent EL spectra were measured in the high-temperature range. The blueshift of peak energy induced resulting from the increasing temperature which enables carriers at the localized energy states to populate the higher levels is compensated and overcome by the redshift of the emission peak wavelength due to the band gap narrowing caused by increasing temperature.

To further investigate the luminescence mechanism, the driving current-dependent EL measurements were carried out. Figure 4 describes the EL spectra of the sample with GaN barriers in the driving current range 1 to 100 mA at different temperatures. With increasing injection current,

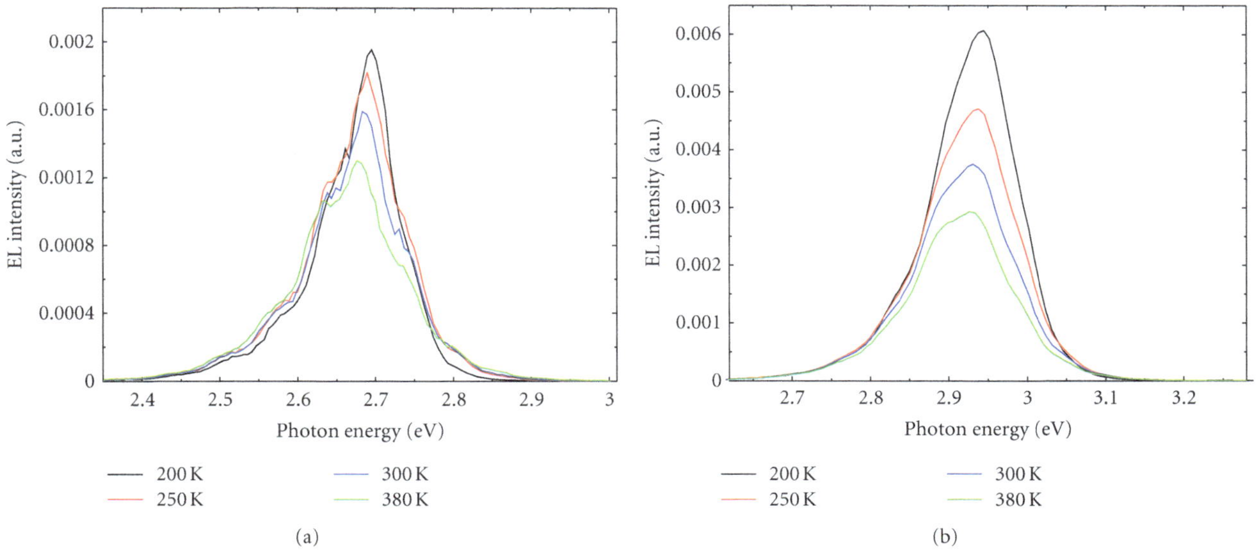

FIGURE 2: EL spectra of InGaN/GaN MQWs in the temperature range from 200 to 380 K for (a) sample with GaN barriers and (b) sample with MQBs. The driving current is 10 mA.

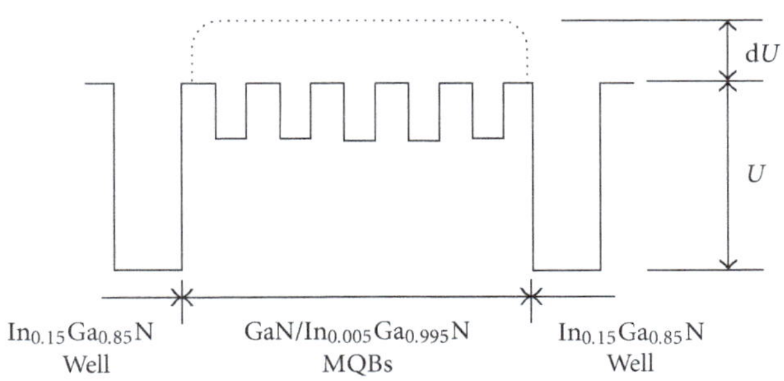

FIGURE 3: Schematic diagram showing potential energy profile of the MQB structure, where U and dU are the classical and enhanced barrier height, respectively.

the band-filling effect of the localized energy states caused by indium composition fluctuation in the MQW structures and the screening of piezoelectric field induced quantum-confined Stark effect (QCSE) have been reported for nitride-based structures [14–16]. The QCSE exists in GaN-based MQWs due to the strong piezoelectric fields arising from the strain caused by lattice mismatch between the barrier and well material. Generally, the strain-induced polarization field tilts the potential profile, which results in a redshift of the emission energy. Besides, this field induces a spatial separation of the electron and hole wave functions in the well. The wave function overlap decreases, and the interband recombination rate decreases. When the samples are driven with a current source, the carrier screening of the QCSE flattens the potential across the MQWs [17]. Both band-filling and QCSE-screening effects should induce an increase in luminescence intensity and a blueshift of emission peak of the EL spectra. The measured results for the sample with GaN barriers at 200 K are shown in Figure 4(a). As the injection current increases from 1 to 100 mA, the intensities of the EL spectra increase. There is no change in the emission peak energy as the current is lower than 60 mA, and then a redshift appears as the current is further increased to 100 mA. It is reasonable to infer that the blueshift of EL peak energy predicted by raising the driving current is compensated and overcome by the thermally activated bandgap shrinkage, thus a shift to lower energy side for the EL emission peaks is observed. The redshifts of EL peak energy shown in Figures 4(b) and 4(c) are more visible because the sample was operated at higher temperatures, and the heating effect is evident. Furthermore, at injection current 100 mA, the EL spectra intensity becomes saturated, which is ascribed to the thermally-activated nonradiative recombination of the injected carriers.

A different injection current-dependent behavior of EL spectra is observed for the sample with MQBs presented in Figure 5. Owing to the higher indium content of this sample the effect of the indium-rich regions on the EL spectra is remarkable. Referring to the curves shown in Figures 5(a), 5(b), and 5(c) which measured at 200, 300 and 380 K, respectively, at low injection currents only a change in shape was observed, and no significant shift of peak energy of the spectra was observed. As the injection current increases to 80 mA, in Figure 5(a), the EL intensity saturates and in Figures 5(b) and 5(c) it starts to degrade. The current where the intensity drops is lower than that of the sample with GaN barriers. Thanks to crystal quality decreases with the increased indium content, the degree of the overall potential fluctuations in the InGaN composition and the roughness of the quantum well interface are enhanced. This sample exhibits more nonradiative recombination centers, which in turn, causing a faster intensity quenching rate at high injection current levels. However, even operated at high injection

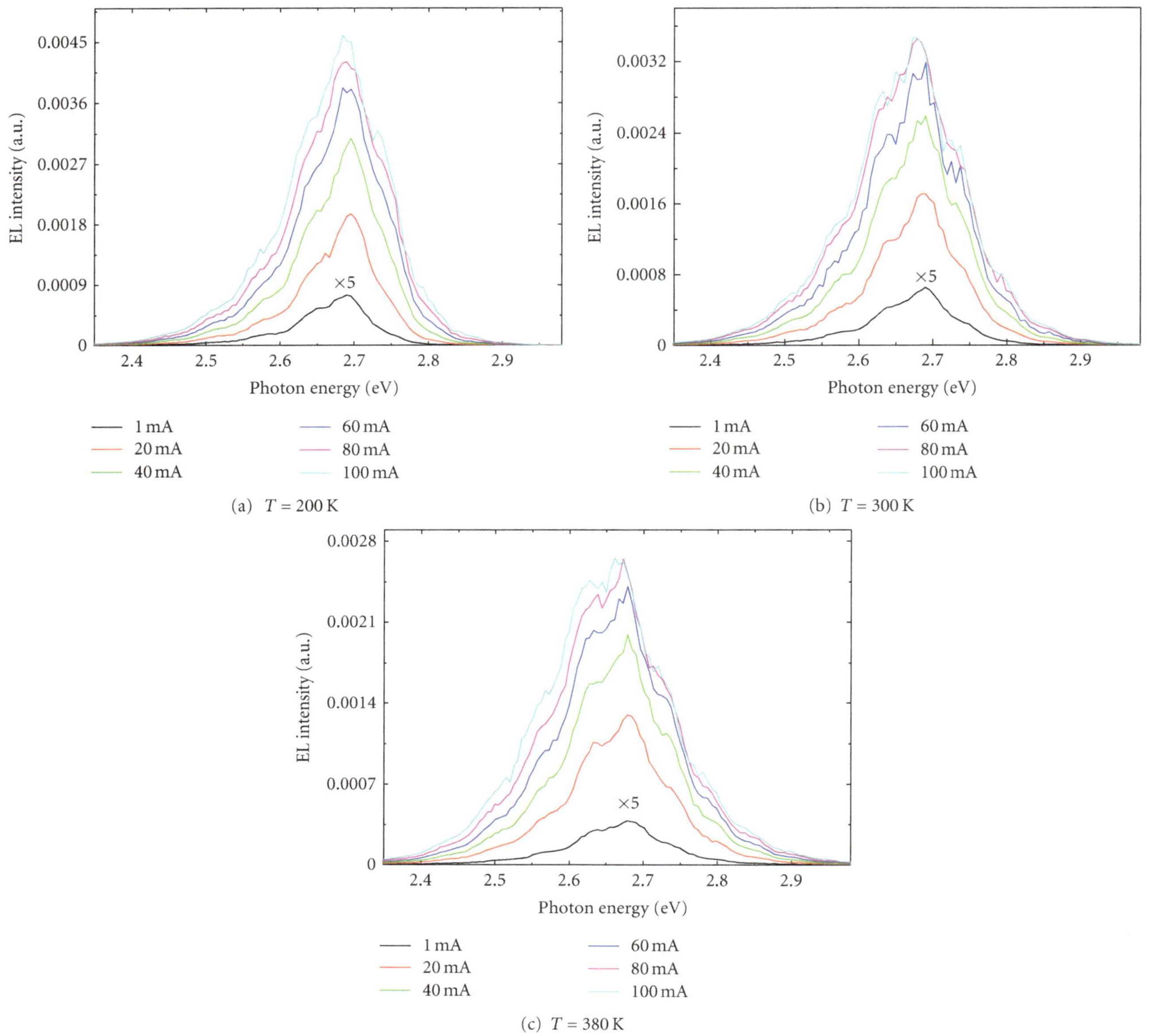

FIGURE 4: The EL spectra of sample with GaN barriers in the driving currents range from 1 to 100 mA at temperature (a) 200 K, (b) 300 K, and (c) 380 K.

current, the EL intensity of MQB sample is higher than that of the sample with GaN barriers at room temperature.

In Figures 5(b) and 5(c), there are anomalous behaviors of driving current dependence of EL spectra. Interestingly, with increasing currents, the dominant emission peak changed from higher energy to lower energy indicates a radiative recombination of excitons localized in the quantum-dot-like indium-rich regions. Since the sample is with higher indium content, more localized states may exist and yield a stronger localization effect of it. The higher energy side emission is thought to be the radiative recombination of excitons confined in the regular MQWs, and the lower energy side emission is from the recombination of localized excitons in the MQWs. We decompose the EL signals into two Gaussian-shaped peaks, and the intensities of the emission peaks from higher energy and lower energy are denoted as I_{high} and I_{low}, respectively. In the high injection current range the heating effect is apparent, leading to the increasing nonradiative recombination of carriers. Injected carriers that are localized in indium-rich regions prevent them from reaching nonradiative recombination sites. The higher energy emission is quenched quickly, and the lower energy emission becomes dominant in the EL emission spectra. More evidences are shown in Figure 6, where the EL intensity of emissions from higher energy (I_{high}) and lower energy (I_{low}) are presented separately as a function of driving current at different temperatures. The influence of I_{low} on the EL intensity becomes conspicuous as the

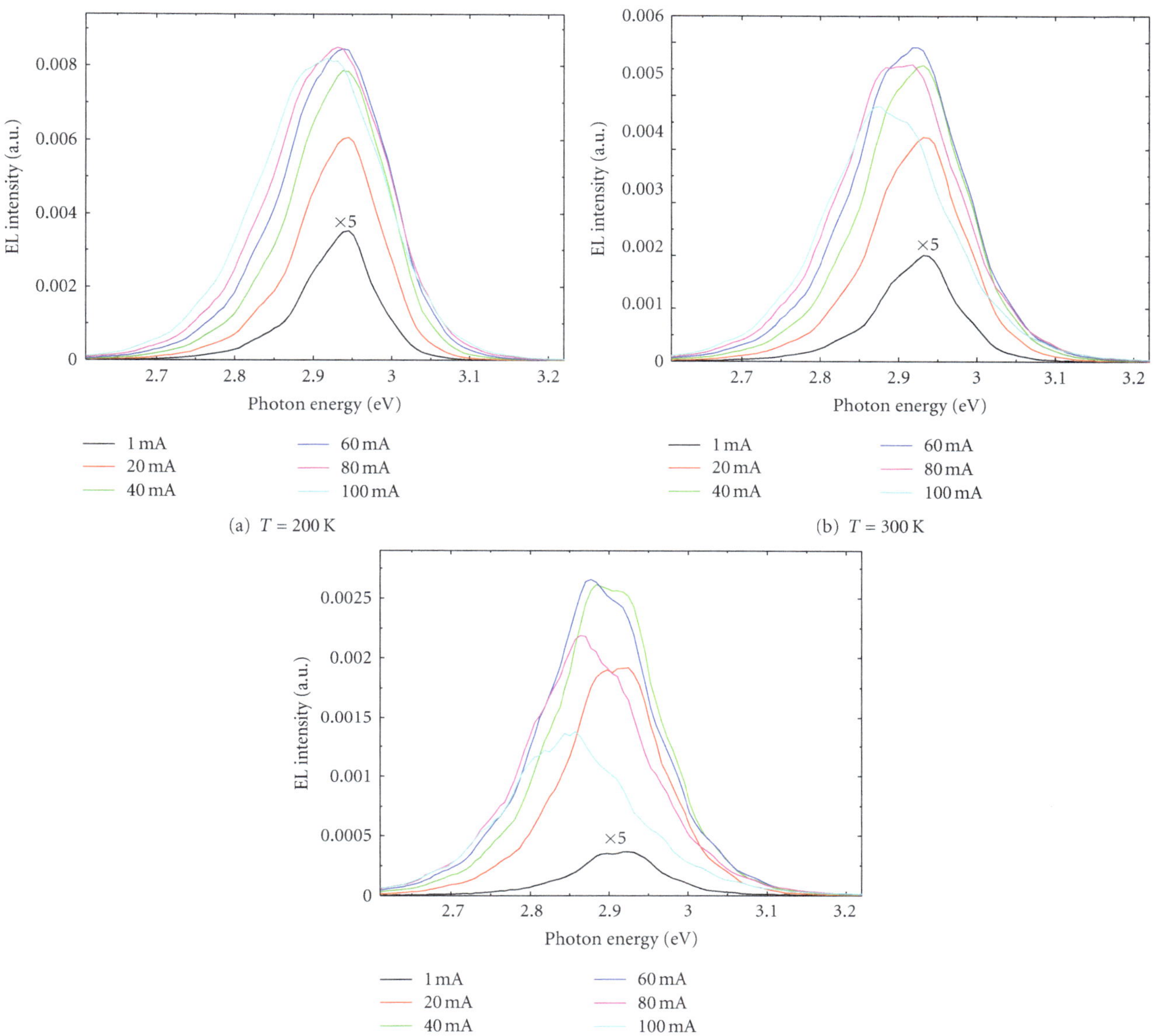

(a) $T = 200$ K

(b) $T = 300$ K

(c) $T = 380$ K

FIGURE 5: The EL spectra of sample with MQBs in the driving currents range 1 to 100 mA at temperature (a) 200 K, (b) 300 K, and (c) 380 K.

current goes up, especially in the high-temperature region. It implies that the localization effect in the sample with MQBs is evident. The stronger localization effect, together with the higher conduction band discontinuity, lead to the higher EL intensity for the sample with MQBs.

The effect of MQB structures on the performance of our samples is further investigated by discussing the EL efficiency of the samples. The EL external quantum efficiency of the InGaN/GaN MQW light-emitting diode is also influenced by the increasing temperatures and increasing driving currents. One of the major problems of InGaN/GaN light-emitting devices is the suppression of carrier overflow from the active layer into the p-layer. Because the MQB structure effectively increases the band discontinuity, it is expected to prevent the electrons from overflowing as operated at high temperatures and high driving currents. In order to show detailed variations of the EL efficiency as a function of injection current at different temperatures for the samples, the integrated EL intensity is divided by current (denoted by $\eta_{\text{eff-ex}}$), which is proportional to the EL external quantum efficiency. For the injection current 10 mA, the calculated $\eta_{\text{eff-ex}}$ ratio of sample with MQBs to sample with GaN barriers as a function of temperature is shown in Figure 7(a). It is clearly seen that the efficiency of the sample with MQBs is greatly improved. The external quantum efficiency of sample with MQBs exhibits up to 3.5 times higher than that of the sample with GaN barriers, implying that an improved light-emitting device performance can be achieved by the adoption of an appropriate MQB structure. As the injection current is raised from 10 to 100 mA, the $\eta_{\text{eff-ex}}$ ratio gradually

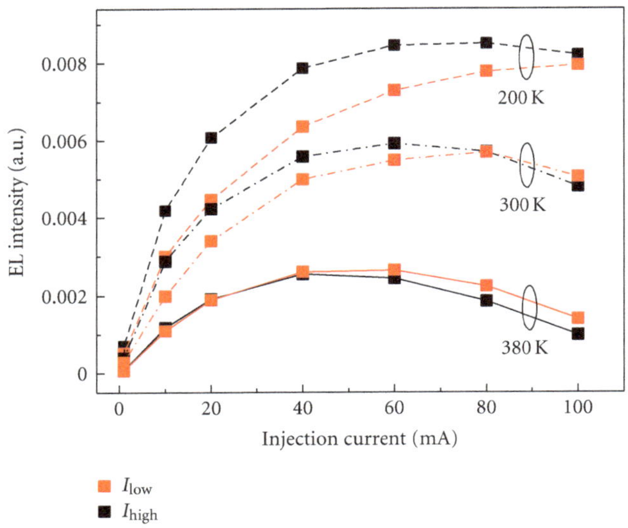

FIGURE 6: Driving current-dependent EL intensity of emissions from higher energy (I_{high}) and lower energy (I_{low}) at temperature 200, 300, and 380 K, respectively.

dropped, as shown in Figure 7(b). Nevertheless, the external quantum efficiency of sample with MQBs is greater than the sample with GaN barriers until the injection current is larger than 80 mA, and the temperature is higher than 340 K. Due to the intensified nonradiative recombination of carriers for the MQB sample, the luminescence efficiency exhibits a rapid degradation in the high temperature and high driving current range. As discussed above, the presence of more inhomogeneous composition of indium in this sample paves the way for localized carriers from the MQBs and quantum well to contribute to rapid quenching of efficiency at high temperature. However, as far as the external quantum efficiency at room temperature and high driving current up to 100 mA, the sample with MQBs essentially exhibits the better device characteristics than the sample with GaN barriers.

4. Conclusion

In this paper, we investigate the performance of InGaN/GaN MQW LEDs with GaN barriers and with MQBs by the measurements of EL spectra over a broad range of driving current 1 to 100 mA and temperature range 200 to 380 K. From the detailed analysis of the experimental results, the sample with MQBs exhibits an enhancement of carrier confinement and a stronger localization effect in the active layer coincides with the increase of effective barrier heights and the higher indium content of the sample. Moreover, from the evaluation of injection current-dependent EL external quantum efficiency at room temperature, the sample with MQBs exhibits a greater efficiency as the driving current increases from 10 to 100 mA. It is concluded that the MQB structure improves the performance of InGaN/GaN MQW light-emitting devices.

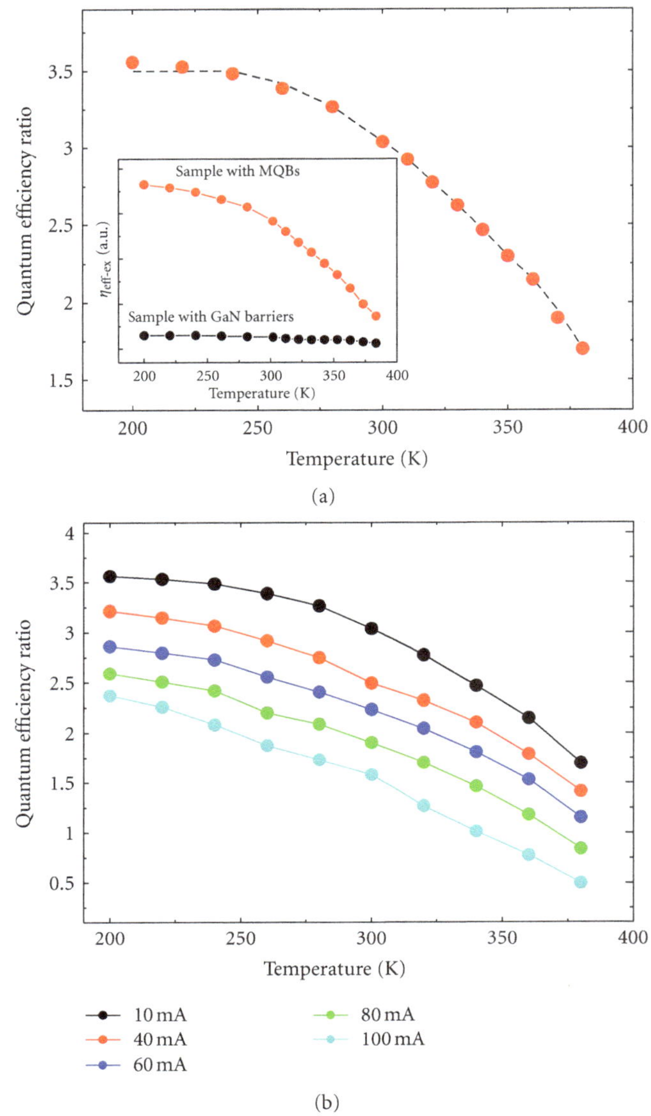

FIGURE 7: (a) Temperature-dependent EL quantum efficiency ratio of the sample with MQBs to the sample with GaN barriers. The inset shows the dependence of the EL efficiency on temperature for the samples. The driving current is 10 mA. (b) Temperature-dependent EL quantum efficiency ratio of the samples in the driving current range 10 to 100 mA.

Acknowledgment

The author would like to thank Mr. Jen-Cheng Wang for his technical assistance.

References

[1] S. Nakamura, S. Pearton, and G. Fasol, *The Blue Laser Diode*, Springer, Berlin, Germany, 2000.

[2] E. F. Schubert and J. K. Kim, "Solid-state light sources getting smart," *Science*, vol. 308, no. 5726, pp. 1274–1278, 2005.

[3] H. Wang, Z. Ji, S. Qu et al., "Influence of excitation power and temperature on photoluminescence in InGaN/GaN multiple

quantum wells," *Optics Express*, vol. 20, no. 4, pp. 3932–3940, 2012.

[4] M. Brendel, A. Kruse, H. Jönen et al., "Auger recombination in GaInN/GaN quantum well laser structures," *Applied Physics Letters*, vol. 99, no. 3, Article ID 031106, 2011.

[5] R. M. Lin, S. F. Yu, S. J. Chang, T. H. Chiang, S. P. Chang, and C. H. Chen, "Inserting a p-InGaN layer before the p-AlGaN electron blocking layer suppresses efficiency droop in InGaN-based light-emitting diodes," *Applied Physics Letters*, vol. 101, no. 8, Article ID 081120, 4 pages, 2012.

[6] C. H. Jang, J. K. Sheu, C. M. Tsai, S. C. Shei, W. C. Lai, and S. J. Chang, "Effect of thickness of the p-AlGaN electron blocking layer on the improvement of ESD characteristics in GaN-Based LEDs," *IEEE Photonics Technology Letters*, vol. 20, no. 13, pp. 1142–1144, 2008.

[7] K. S. Teng, M. Brown, A. Kestle et al., "An investigation of multi-quantum barriers for band offset engineering in AlGaInP/GaInP lasers," *Applied Surface Science*, vol. 190, no. 1–4, pp. 284–287, 2002.

[8] M. P. Ackland, P. R. Dunstan, M. R. Brown, K. S. Teng, S. P. Wilks, and R. Cobley, "Characterisation of the influence of multi-quantum barrier reflectors within GaInP/AlGaInP quantum well lasers using near-field imaging techniques," *Applied Surface Science*, vol. 255, no. 3, pp. 649–652, 2008.

[9] T. E. Nee, J. C. Wang, H. T. Shen, and Y. F. Wu, "Effect of multiquantum barriers on performance of InGaN/GaN multiple-quantum-well light-emitting diodes," *Journal of Applied Physics*, vol. 102, no. 3, Article ID 033101, 7 pages, 2007.

[10] J. C. Lee, "Investigation of nanostructure properties in InGaN/GaN multiple quantum wells by X-ray diffraction analysis," *Solid State Sciences*, vol. 13, no. 5, pp. 1013–1016, 2011.

[11] Y. Y. Lin, R. W. Chuang, S. J. Chang et al., "GaN-based LEDs with a chirped multiquantum barrier structure," *IEEE Photonics Technology Letters*, vol. 24, no. 18, pp. 1600–1602, 2012.

[12] Y. F. Wu, J. C. Lee, T. E. Nee, and J. C. Wang, "Carrier localization effect on luminescence spectra of IIIV heterostructures," *Journal of Luminescence*, vol. 131, no. 7, pp. 1267–1271, 2011.

[13] C. Y. Chen, C. Hsieh, C. H. Liao et al., "Effects of overgrown p-layer on the emission characteristics of the InGaN/GaN quantum wells in a high-indium light-emitting diode," *Optics Express*, vol. 20, no. 10, pp. 11321–11335, 2012.

[14] T. Akasaka, H. Gotoh, T. Saito, and T. Makimoto, "High luminescent efficiency of InGaN multiple quantum wells grown on InGaN underlying layers," *Applied Physics Letters*, vol. 85, no. 15, pp. 3089–3091, 2004.

[15] N. Khan and J. Li, "Effects of compressive strain on optical properties of $In_xGa_{1-x}N$/GaN quantum wells," *Applied Physics Letters*, vol. 89, no. 15, Article ID 151916, 3 pages, 2006.

[16] D. Y. Lee, S. H. Han, D. J. Lee et al., "Effect of an electron blocking layer on the piezoelectric field in InGaN/GaN multiple quantum well light-emitting diodes," *Applied Physics Letters*, vol. 100, no. 4, Article ID 041119, 4 pages, 2012.

[17] S. De, A. Layek, S. Bhattacharya et al., "Quantum-confined stark effect in localized luminescent centers within InGaN/GaN quantum-well based light emitting diodes," *Applied Physics Letters*, vol. 101, no. 12, Article ID 121919, 5 pages, 2012.

4

Mix and Inject: Reaction Initiation by Diffusion for Time-Resolved Macromolecular Crystallography

Marius Schmidt

Physics Department, University of Wisconsin-Milwaukee, WI, USA

Correspondence should be addressed to Marius Schmidt; m-schmidt@uwm.edu

Academic Editor: Abbas Ourmazd

Time-resolved macromolecular crystallography unifies structure determination with chemical kinetics, since the structures of transient states and chemical and kinetic mechanisms can be determined simultaneously from the same data. To start a reaction in an enzyme, typically, an initially inactive substrate present in the crystal is activated. This has particular disadvantages that are circumvented when active substrate is directly provided by diffusion. However, then it is prohibitive to use macroscopic crystals because diffusion times become too long. With small micro- and nanocrystals diffusion times are adequately short for most enzymes and the reaction can be swiftly initiated. We demonstrate here that a time-resolved crystallographic experiment becomes feasible by mixing substrate with enzyme nanocrystals which are subsequently injected into the X-ray beam of a pulsed X-ray source.

1. Enzymology

The "unspent vital force" was the historical and metaphysical expression for a biological catalyst. It was eradicated in 1890 by an article worth almost 100 pages by O'Sullivan and Tompson [1], which probably denotes the beginning of what we know now as enzymology. The enzyme was originally thought to be the leaven in sour dough and then materialized as a chemical, a macromolecule, or something that can be manipulated, tested, and investigated. Similarities with chemical catalysts were discovered. This finally led to the nowadays well-established and important field of enzymology. In 1926 Sumner reported the purification and crystallization of the first enzyme, urease [2]. At that time, X-ray structure determination has been established starting in the first decade of the 20th century. Although the father of X-ray crystallography, Max von Laue, did not believe that it would be possible to determine atomic structures of enzymes with X-ray crystallography [3], the first structures of proteins were determined at the end of the 50s and the beginning of the 60s of the last century [4, 5]. Only then biologists learned to think in terms of 3D atomic structure of biological molecules. Structure-function relationships were discovered. Enzymes could be studied by directly seeing how, where, and in which orientation effectors were binding to the enzyme. Phenomena such as allosteric inhibition of enzyme activity first discovered by enzyme kineticists finally could be understood on the atomic length scale. This provided new venues, for example, to design new drugs to cure diseases. However, the atomic structures were still static. Enzymologists knew that catalysis must be accompanied by structural transitions of the catalyst. At thermal equilibrium, in crystallography, conformational heterogeneity and structural fluctuations are hidden in the Debye-Waller factor [6]. The protein is flexible, even in the crystal. Catalytic function, structural flexibility, and heterogeneity are intimately related. Once the atomic structure is locked into place, the enzyme is dead and nonfunctional [7]. However, in most cases protein crystals are catalytically active and the crystalline lattice does not constrain the dynamics of the enzyme that would otherwise impair its function. At the end of the 20th century, crystallography evolved to the point that the catalytic action of proteins could be observed directly in the crystal. Time-resolved crystallography made this possible [8, 9]. Ultrafast time-resolved crystallography is a pump-probe method that requires short, extremely intense X-ray pulses to achieve the best time resolution. Snapshots of a protein in action in the crystal can be taken as a function of time after the reaction is initiated and assembled into a movie. This movie shows how the protein ensemble evolves in the crystal. The crystallographic experiment becomes a kinetic experiment. The rise and decay of short lived (transient) states can be observed during the reaction. Since it is a crystallographic experiment, the structures of the transient states can be determined from the same data. Chemical, kinetic mechanisms that connect the states can be tested and become falsifiable [10–12]. Each new transient state structure is biologically interesting and may be a new drug target. It can be manipulated to alter the chemical, kinetic mechanism of a specific reaction with the goal, for example, to bring that reaction to a full stop in order to eradicate a pathogen.

2. The Challenge

The long-term goal is to rapidly and routinely investigate all kinds of biologically and pharmaceutically interesting enzymatic reactions with time-resolved crystallography. There are limitations that prevented the maturation of this technology to its full potential so far. The three most prominent ones are (i) the limited pulse intensity at the synchrotron, (ii) limited crystal quality, and (iii) the necessity to mount single crystals for exposure to the X-rays. Whereas the former (i) allows for improvements, for example, by microfocusing to reach single pulse capability even with small crystals, the latter two are by far the most challenging of all, and there is no satisfactory solution for this to date. When a single crystal is mounted, typically in a capillary for room temperature studies, it can be exposed many times to the X-ray pulses. The structure can be solved by rotating the crystal to collect the integrated reflection intensities. However, rotation is almost always prohibitive with fast reactions, simply because the crystal cannot be rotated fast enough through the Ewald sphere. One way out is the usage of the Laue method where a bandwidth of X-rays is used to collect the intensities from stills. With the Laue method one needs good crystals with low mosaicities. Otherwise the reflections become too streaky and their intensities cannot be extracted [13, 14]. Poor crystal quality, which might potentially get worse when the reaction proceeds, already excludes many, otherwise interesting enzymes. If an enzymatic reaction is to be investigated, the reaction can be initiated only once in a particular volume of the crystal, because it is typically noncyclic or irreversible. With fast detectors the reaction might be followed up using subsequent X-ray exposures and readout between the pulses. Readout times of large area detectors are constantly improving [15–17] with the goal of MHz (!) readout rates. However, when the reaction is over, the product must be removed, new substrate loaded and the crystal reset to cover more of reciprocal space to assemble a complete dataset. This can be achieved by either probing another, previously unaffected crystal volume, by using flow cells [18] to wash away product and load new substrate, or by mounting another crystal charged with substrate all together. The latter two methods are tedious and prevent rapid results. For cyclic reactions such as dissociation and geminate rebinding reactions in heme proteins the situation is different. One waits until the reaction comes to an end, resets the crystal, reinitiates the reaction, and repeats the experiment until radiation damage is so severe that either the crystal is destroyed or the kinetics are altered, whichever comes first [19]. These are the reasons that so few reactions that were investigated by time-resolved crystallography are actually enzymatic reactions. Most of them are cyclic reactions in heme proteins and light activated signaling proteins [20–30]. Nevertheless, lessons learned from these experiments are priceless. Structural relaxations become directly accessible and transient state structures explain a vast number of results previously taken by spectroscopic approaches. Most importantly, chemical kinetics are united with structure determination, something that the early crystallographers would not have believed in their wildest dreams. However, approaches to enzymatic reactions remain sparse [31–34]. To change this, it will be shown in the following that any X-ray source can be used that provides enough flux into single ultrashort pulses to probe tiny crystals so that the time-resolved experiments do not depend on the availability of large crystals. In particular, new generation X-ray sources, XFELs, provide sufficient flux and are monochromatic, and pulses are short enough that diffraction patterns become insensitive to radiation damage.

3. Nanocrystallography at XFEL

The 4th generation X-ray sources deliver on the order of 10^{12} photons in a single femtosecond, monochromatic X-ray pulse. Since the pulses are so short, diffraction occurs before the specimen is destroyed. Accordingly, the diffraction patterns are free of radiation damage. In addition, the conventional rotation method can be given up in favor of injecting small crystals directly into the beam. During the femtosecond pulse duration all motions of the crystal are frozen in time. One obtains a snapshot of the crystal akin to a still exposure, but with the crystal in random orientation. It has been shown recently that a complete data set with sufficient resolution can be collected by injecting a large number of these small, randomly oriented crystals [35]. Data reduction differs from the conventional rotation method in so far that a large number (>100,000) of randomly oriented diffraction patterns containing only partial reflections are available from which the integrated reflection intensities must be reconstructed [36]. Due to its enormous spatial coherence, the XFEL beam can be focused on a nanometer sized focal spot. In this case crystals with edge length on the order of nanometers can be injected into the beam and yield analyzable diffraction patterns. This allows structure determination even from crystals naturally grown within cells [37, 38]. This technology has been named "serial femtosecond X-ray crystallography" (SFX) [39].

A time-resolved crystallographic experiment may benefit from these ultrashort pulses [40], provided the reaction itself can be initiated on a similar time scale. In these cases the ultimate time resolution given by the X-ray pulse can potentially be reached. Typical examples include the mentioned photoflash experiments on heme proteins, where a small molecule such as carbon monoxide can be flashed away from the heme moiety by an ultra-short laser pulse. With enzymes, however, it is exceptionally difficult to initiate a reaction on this ultrashort time scale. The reason for this is that substrate would have to be provided to the enzyme in an ultrafast fashion. This turns out to be very difficult if not impossible to do on the sub-μs time scale. In an ideal case, inactive, so-called caged substrate is soaked into the crystals and binds directly to the active center. Then diffusion times are extremely short. The caged substrate is activated by a short laser pulse. Typical activation rates vary widely from nanoseconds to milliseconds [41, 42]. Chemical expertise is necessary to design and produce new caged substances tailored to the reaction to be investigated, which, in many cases, are not routinely available. Also, the quantum yield for activation is typically low and a single laser shot might not be sufficient to activate a significant fraction of the caged substrate. Cryophotolysis combined with temperature-controlled freeze-trap cycles can be used to trap intermediates even in the fastest enzymes [43]. Still, the ultimate goal is to let the reaction in an enzyme advance freely and at ambient temperatures. A number of such experiments were conducted at the synchrotron with the Laue method to date [31, 34, 44–48]. An enzyme that crystallizes excellently as well as a caged substrate whose activation time complies with the time-resolution is required. For routine applications it is highly desirable that the reaction is initiated in a different way than using a sample-specific caged substrate. The simplest way is to mix crystals and active substrate and let the substrate diffuse into the crystal. Once substrate is present at the active site, the reaction starts. Assuming that crystals can be mixed with substrate fast enough, objections to this approach are causally connected: (i) time resolution would depend on diffusion times and (ii) zero time, the point in time where the reaction starts, would be dependent on the position in the crystal. Both objections are void, if diffusion times are much faster than the characteristic lifetimes of the intermediates to be investigated. We will explore this possibility further in the following sections. With methods of linear algebra such as singular value decomposition the time information can be exploited in a sense that it might be possible to correct for the zero time distribution and push time resolution to the limit.

4. Velocities in Enzymes and the ccNIR Reaction

Reaction times in enzymes vary enormously. The fastest enzyme is probably the catalase [49] with catalytic rates k_{cat} on the order of 10^6. The turnover time is $1/k_{cat}$ on the order of a microsecond. Other enzymes such as the hydroxymethylbilane synthetase react on the order of hours [50]. Most other enzymes are lying in between with catalytic rates of 100 1/s or so [51]. This means that at least millisecond time resolution is required to characterize these reactions adequately. In Figure 1 the reaction catalyzed by the enzyme cytochrome-c nitrite reductase (ccNIR) is shown as an example. This enzyme is one of the most important enzymes in the biologically highly relevant nitrogen cycle and reduces nitrite to ammonia in a 6-electron transfer reaction. Three intermediates are formed during the catalytic cycle. These are the ccNIR-nitric oxide, the ccNIR-hydroxylamine, and finally the ccNIR-ammonia complexes, all structures of which are unknown due to the short lifetimes of the corresponding transient states. Recently, we determined the structure of the ccNIR of *Shewanella oneidensis* [52]. Its turnover time at room temperature is about 1.2 ms (k_{cat} = 824 1/s). Several intermediates accumulate during the reaction. A transient state experiment needs a time resolution that is on the submillisecond time scale.

The ccNIR of *S. oneidensis* is difficult to work with. Crystals are labile and sensitive to radiation damage. After about 10 monochromatic X-ray exposures, the diffraction patterns fade away. A time-resolved crystallographic experiment on this ccNIR is very challenging. It requires great chemical expertise to produce suitable electron donors attached to the ccNIR in the crystal that can be activated by short laser flashes for the reaction to proceed in a time-resolved fashion. These experiments are underway. However, the following proposes time-resolved crystallographic experiments where chemical expertise is kept at a minimum, the extent of reaction initiation is large, the crystal quality can be moderate, radiation damage is zero, and time resolution is good enough to capture the reaction intermediates. This experiment has the potential to transform enzymology because it produces entire time series (movies) from reactions in enzymes rapidly. Most importantly it has the potential to become a routine procedure to investigate all kinds of enzymatic reactions. The requirement is that micro and nanocrystals can be obtained from the enzyme and that the substrate is soluble, which is true for most enzymes.

5. "Mix and Inject" at a Pulsed X-Ray Source

The traditional approach to enzymology is to mix enzyme with substrate and follow the reaction with time-resolved methods such as time-resolved absorption spectroscopy. For this, special devices are constructed called either continuous-flow or stopped-flow devices [51]. The enzyme is mixed with substrate and injected into a cuvette. The evolution of the mixture in the cuvette is probed, for example, with visible light. Newer mixing devices allow time resolutions as good as about 150 μs [53]. The design of the mixer must be modified to be used for time-resolved crystallography in order to inject nano- and microcrystals into the X-ray beam (Figure 2(a)). The goal is to control both mixing and injecting and synchronize it with the arrival time of the ultrashort X-ray pulse. The time delay Δt could be selected by either controlling the residence time of crystal and substrate in the mixer or by varying the distance ΔL of the mixer from the pulsed X-ray beam given the velocity of the ejected droplet. Consequently, injector technology is a key to this experiment. Injector technology to deliver specimen into the XFEL beam progressed rapidly in the last few years [54–57]. However, a device that performs both mixing and synchronized injection is needed. A prerequisite to the success of this method is that not only mixing can be performed rapidly, but also diffusion times into the crystals are kept at a minimum. If diffusion times are longer than the turnover time of the enzyme, the time resolution becomes so poor that it would be impossible to conduct a transient state kinetic experiment.

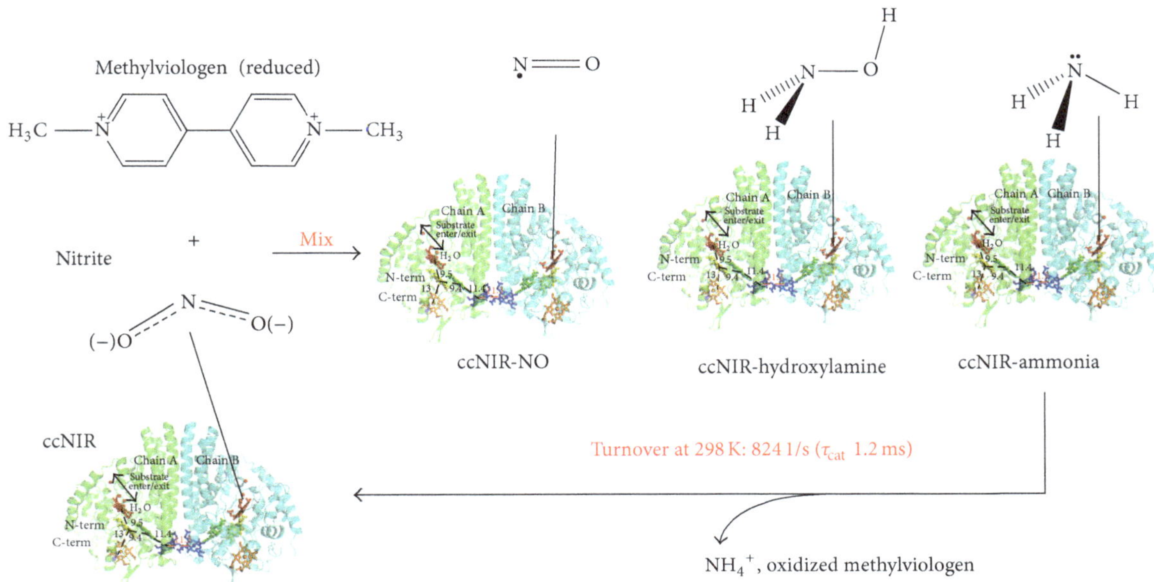

Figure 1: Catalytic cycle of the ccNIR. Initially, nitrite binds to a specific c-heme in the ccNIR. Methylviologen serves as electron donor for the 6-electron transfer reaction to produce ammonia from nitrite. The NO, hydroxylamine, and ammonia complexes are shown. The catalytic rate is 824 1/s at room temperature.

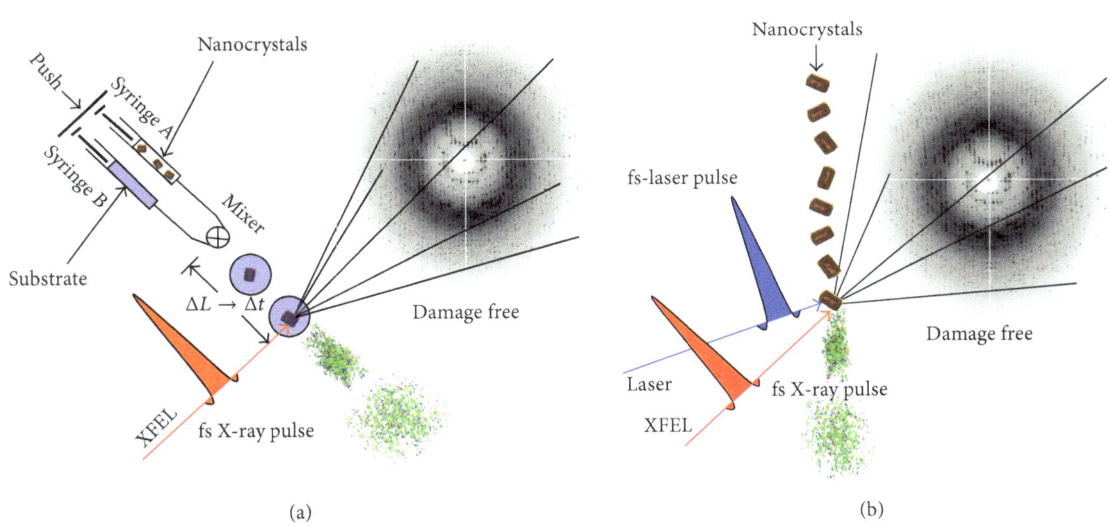

Figure 2: (a) Schematic setup for a mix and inject time-resolved femtosecond serial crystallographic experiment at the XFEL. Nano- and microcrystals are in syringe A and substrate in syringe B. After mixing, the crystal/substrate droplet is injected into the XFEL beam. The crystal might disintegrate after the fs X-ray pulse, but diffraction occurs earlier. The diffraction pattern is damage free. (b) Time-resolved pump-probe experiment with serial femtosecond crystallography. Setup can be used to stimulate photoreceptors for photoflash experiments or to activate caged substrates. Conceptual similarities between (a) and (b) are obvious.

6. Diffusion Times

Diffusion of (a soluble) substrate in solution follows Fick's 2nd law (1),

$$D\nabla^2 c = \frac{\partial c}{\partial t}, \qquad (1)$$

where D is the diffusion coefficient, and ∇^2 is the Laplace operator, c is the time-dependent concentration of the diffusing species. Equation (1) is a second order partial differential equation that can be solved by separation of variables. In special cases an analytic solution can be found subject to certain boundary conditions. In most other cases solutions are only

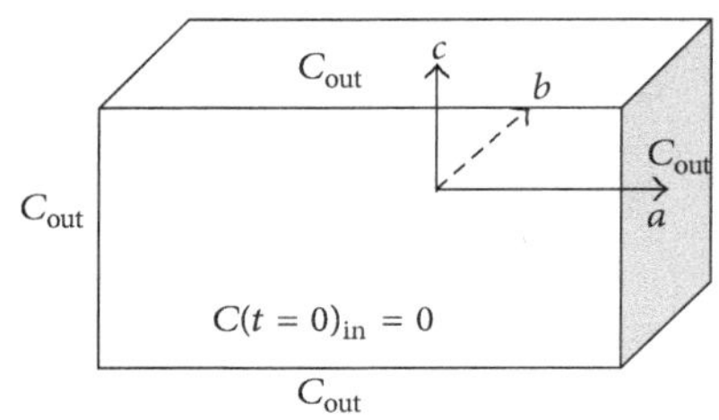

FIGURE 3: Shoe-box like crystal with sides $2a$, $2b$, and $2c$. The substrate concentration is C_{out} outside the crystal at any time. Inside, C_{in} is initially zero and evolves as a function of time.

found by numerical methods. The simple case of substrate diffusing into a shoe-box like crystal with sides $2a$, $2b$, and $2c$ is depicted in Figure 3. The situation is idealized by assuming the following: (i) the change of the outside concentration of the substrate is negligible at all times, (ii) binding of substrate to the enzyme is neglected, so that the substrate diffuses freely into the crystal, and (iii) mixing of crystal with substrate is instantaneous at time $t = 0$, so that the concentrations c of the substrate inside and outside are $C_{out} = C$, so that the concentration of substrate outside the crystal is C_{out} and the concentration C inside is zero initially. With these boundary conditions, the solution to Fick's 2nd law is (2) [58]:

$$C(x, y, z, t) = C_{out}\left[1 - \frac{64}{\pi^3}\sum_{l=0}^{\infty}\sum_{m=0}^{\infty}\sum_{n=0}^{\infty}\frac{(-1)^{l+m+n}}{(2l+1)(2m+1)(2n+1)} \times \cos\frac{(2l+1)\pi x}{2a}\cos\frac{(2m+1)\pi y}{2b} \times \cos\frac{(2n+1)\pi z}{2c} \times \exp^{-(D\pi^2/4)[((2l+1)^2/a^2)+((2m+1)^2/b^2)+((2n+1)^2/c^2)]\cdot t}\right]. \quad (2)$$

Equation (2) is reminiscent of a Fourier sum. The concentration of substrate inside the crystal is dependent on the initial outside concentration C_{out} and a time-dependent exponential term that contains the characteristic time scale τ_D of the diffusion problem (3):

$$\tau_D = \frac{4}{D\pi^2\left[\left((2l+1)^2/a^2\right)+\left((2m+1)^2/b^2\right)+\left((2n+1)^2/c^2\right)\right]}, \quad (3)$$

where τ_D contains the diffusion coefficient D and accounts for the time after which the concentration of substrate reached $1 - 1/e = 0.63$ of the initial amplitude of the corresponding Fourier term. The slowest τ_D is that of the $l, m, n = 0$ term, and everything else is faster. The $l, m, n = 0$ term accounts for $64/3\pi^3 = 69\%$ of the initial concentration. Consequently, this term accounts for almost all the concentration. The corresponding τ_D is the slow limit for the diffusion time. Most

TABLE 1: Diffusion times τ_D for various crystal sizes from calculation, simulation, and experiment.

Crystal size	τ_D
$400 \times 400 \times 1600\,\mu m^3$	16 s (3) 24 s [60] <1 min [59]
$300 \times 400 \times 500\,\mu m^3$	9.5 s
$10 \times 20 \times 30\,\mu m^3$	15 ms
$3 \times 4 \times 5\,\mu m^3$	1 ms
[a]$1 \times 2 \times 3\,\mu m^3$	150 μs
$0.5 \times 0.5 \times 0.5\,\mu m^3$	17 μs
[b]$0.1 \times 0.2 \times 0.3\,\mu m^3$	1.5 μs

[a]With much smaller crystals mixing times might be slower than diffusion times.
[b]*S. oneidensis* ccNIR crystals would have about 4000 unit cells.

importantly, τ_D depends on the square of the crystal half sides a, b, and c. This is apprehensible, since diffusion into a volume is dependent on the surface exposed. Equation (3) is used to calculate the evolution of substrate concentration in the center of the crystal at $x, y, z = 0$ (see Table 1) using the diffusion coefficient $D = 5\times10^{-6}\,cm^2/s$ for a typical substrate, glucose, previously measured in a crystal [59]. Results are compared to those obtained from numerical simulation and experiment. Table 1 shows that diffusion times into macroscopic crystals are on the order of 20 s, as found from our simple equation and by a more complex and accurate numerical simulation [60], which has been corroborated by a crystallographic experiment [59]. This shows that our calculation is in the right ballpark, and we can use it to determine diffusion times with various crystal sizes. If the crystal size shrinks by one order of magnitude, diffusion times accelerate by 2 orders of magnitude according to (3). Diffusion times into a $3\times4\times5\,\mu m^3$ crystal are already in the fast millisecond time regime (1 ms). For a $1 \times 2 \times 3\,\mu m^3$ crystal diffusion times become submillisecond (150 μs). This would be sufficient to initiate a time-resolved crystallographic experiment with the ccNIR. With lysozyme atomic structures were already successfully determined from crystals of similar sizes using fs X-ray pulses from the LCLS [35], and even smaller crystals can be used [61]. With $0.5 \times 0.5 \times 0.5\,\mu m^3$ nanocrystals diffusion times would approximately match the fastest mixing times, which is excellent for almost all reactions in biological macromolecules.

7. Simulations

In a time-resolved experiment, intermediate states can only be observed if they are populated sufficiently by active enzyme molecules. The concentrations of the intermediates are determined by the magnitudes of rate coefficients of the underlying chemical, kinetic mechanism. It is the goal of any time-resolved experiment to determine (i) the physical properties, for example, the atomic structures or the spectra, of the intermediate states, (ii) the time-dependent concentrations of these intermediates, and (iii) the magnitudes of all rate coefficients that specify the chemical, kinetic

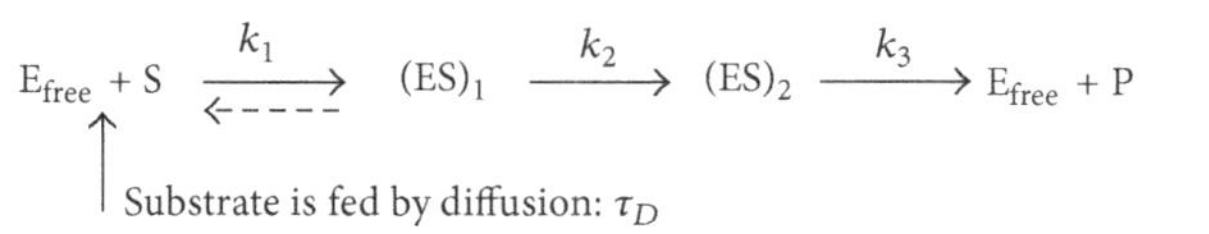

FIGURE 4: Chemical kinetic mechanism of a hypothetical enzymatic reaction where substrate is fed by diffusion. Two transient states, enzyme-substrate complexes $(ES)_1$ and $(ES)_2$ are successively populated until the free enzyme is released. The dashed arrow denotes a potential equilibrium that is ignored here.

mechanism of the reaction. To illustrate this, a chemical, kinetic mechanism is depicted in Figure 4 that features the free enzyme and 2 intermediate states. Given the magnitude of the rate coefficients in this mechanism, one can now assess how fast diffusion of substrate S into the crystal has to be that both intermediates are well separated in time. Only then can the time information be exploited to assess mixtures, separate these mixtures into pure states, and ultimately determine the structures of the pure, admixture-free intermediates from the time-resolved crystallographic data. Rate coefficient k_1 in the mechanism is a second order rate coefficient that accounts for substrate binding to form enzyme-substrate complex 1 $(ES)_1$. k_2 and k_3 are catalytic rate coefficients for conversion from $(ES)_1$ to $(ES)_2$ and finally to product P and free enzyme, respectively. The free enzyme can reenter the catalytic cycle at any time. Since k_1 is a second order rate coefficient the rate itself depends sensibly on both, the concentration of free enzyme and the concentration of substrate. The magnitude of $k_1 = 1000\,\text{L mmol}^{-1}\,\text{s}^{-1}$ accounts for tight substrate binding. The other two rate coefficients are single step, first-order rate coefficients. From the value of 4000 1/s for k_2 one expects that $(ES)_1$ decays on a 1/4000 s = 250 μs time scale. k_3 is much smaller and is the rate limiting step of the enzyme. k_3 is chosen (800 1/s) that it matches approximately the catalytic rate of the ccNIR (see above). It is also assumed that the supply of substrate is constant at $c_0 = 50$ mmol/L. Concentrations of proteins in crystals are up to 100 mmol/L. The concentration of the ccNIR is about 25 mmol/L, which is chosen here for the simulation. The characteristic diffusion time τ_D is varied by changing the crystal size. For simplicity, the concentration of substrate at the center of the crystal is calculated from $c(t) = c_0(1 - e^{-(1/\tau_D)(t)})$ with τ_D given by (3) with $l, m, n = 0$ and the diffusion coefficient of glucose in the crystal mentioned above. $c(t)$ is fed to the free enzyme to start the reaction. A certain time after reaction initiation a steady-state must emerge in the simulations, because the assumption is that C_{out} is constant in time. The rate equations are coupled differential equations, which were solved for the concentrations of the particular intermediates and the free enzyme by numerical methods. Figure 5 shows the result.

Although the simulations are quite simple, the results are illuminating. In Figure 5(a) the reaction initiation is much faster than the turnover time, it is even faster than the binding rate to the free enzyme. This is an ideal case, which requires a caged substrate that can be quickly activated. Free enzyme and the enzyme-substrate complexes $(ES)_1$ and $(ES)_2$ are well separated in time. The steady state forms after one turnover. It consists of 83% $(ES)_2$ which is contaminated by a small amount (17%) of $(ES)_1$. The structures of the pure intermediates in conjunction with a suitable chemical, kinetic mechanism are likely to be extractable from time-resolved crystallographic movie using established methods [11]. In Figure 5(b), diffusion times are comparable to the turnover time. The concentration profiles for free enzyme, $(ES)_1$ and $(ES)_2$, do not differ very much from those in Figure 5(a). This is because only a very little substrate concentration is necessary to bind to the free enzyme. However, the free enzyme is pulled down later (see vertical dashed lines), which is still faster than the turnover time. $(ES)_1$ accumulates properly. As long as the conditions are such that sufficient $(ES)_1$ accumulates before it decays to $(ES)_2$, the intermediates are separable. Again the steady state forms, when changes in the substrate or the free enzyme concentrations are negligible. When diffusion times are much longer than the turnover time (Figure 5(c)), it is not possible to use the time information to separate the intermediates. Formation of $(ES)_1$ is on the same time scale than $(ES)_2$ because the velocity (rate) of the formation of $(ES)_1$ depends on both substrate and free enzyme. The substrate concentration, however, is very small up to 100 ms. At the same time $(ES)_1$ is already decaying, and the ratio $(ES)_1/(ES)_2$ is roughly given by k_3/k_2 at any time. The steady state is delayed (dashed blue arrow) until the free enzyme concentration is negligible or the substrate reaches a constant level. In all instances, probing the steady state is possible even with conventional monochromatic crystallography. No or very little time-resolution is required. Under steady state conditions the Michaelis complex, which is here a mixture of $(ES)_2$ and $(ES)_1$, can be extracted from the crystallographic data. This is already very valuable information for the enzymologist. The simulations show that for transient state experiments where $(ES)_1$ and $(ES)_2$ states are well separated in time, small crystals must be used. A size of $3 \times 4 \times 5\,\mu\text{m}^3$ is sufficient (Figure 5(b)) whereas the experiment likely fails with crystals that are an order of magnitude larger. Of course, all of this depends on, in addition to the magnitude of the rate coefficients in the kinetic mechanism, the diffusion coefficient of the substrate in the crystal and the highest substrate concentration C_{out} that can be prepared and used under the experimental conditions.

8. Discussion

Figure 2 shows that cyclic (reversible) and noncyclic (nonreversible) reactions as well as mix and inject and pump-probe experiments are on the same footing. With somewhat larger crystals and correspondingly lower time resolution, mix and inject experiments are already at the reach of the synchrotron. Time resolutions on the order of 5 ms could be achieved with 10 μm sized crystals which would easily tolerate the dose deposited by the 100 ps long synchrotron X-ray pulse [62], and "diffract before destroy" would not be necessary. However, the single pulse capability of the X-ray source mentioned earlier is required. The best time-resolution can only be achieved with the smallest crystals and an XFEL is required. If these crystals are hit by the XFEL beam, they are destroyed. Doses absorbed [35] are several orders of

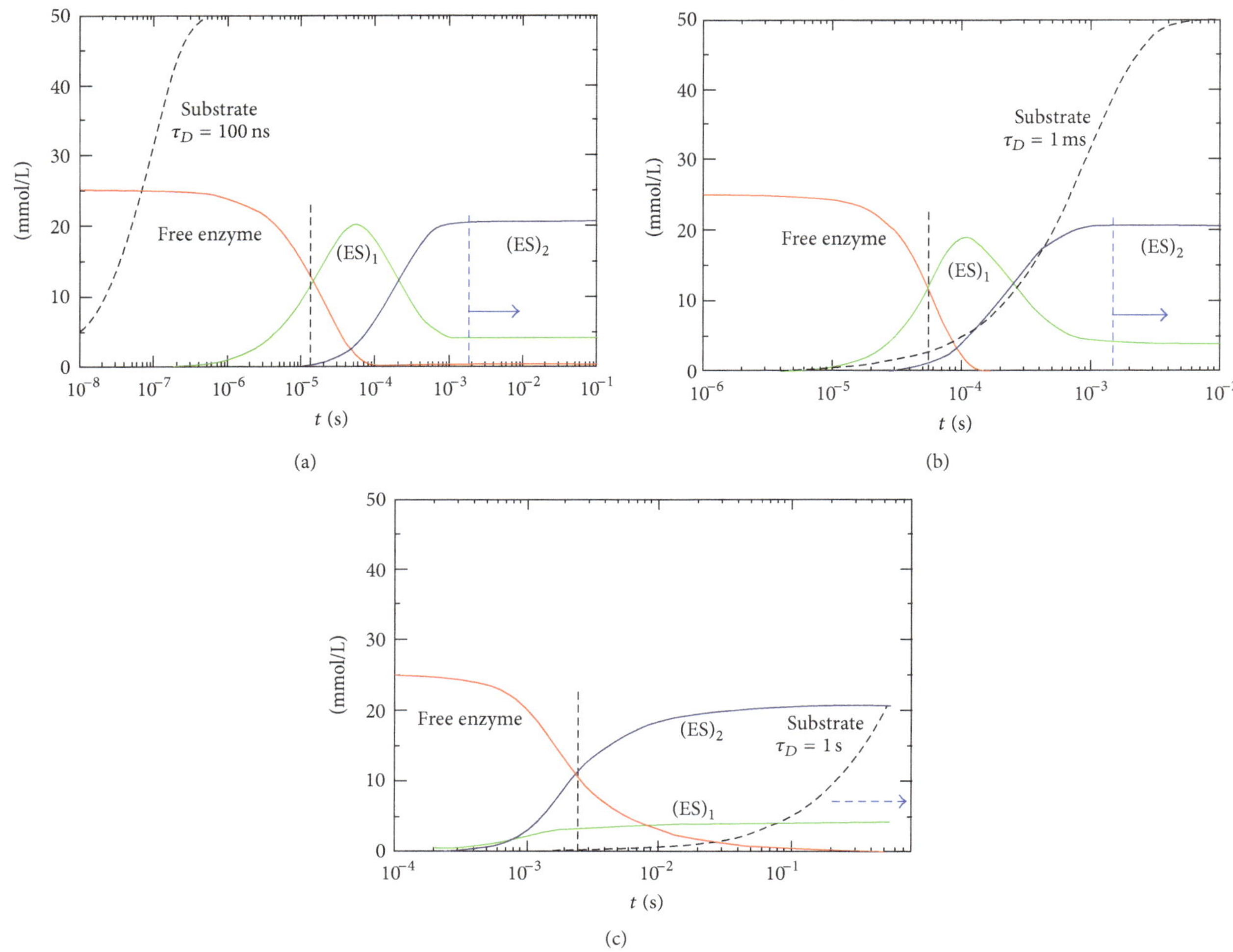

FIGURE 5: Concentration profiles of free enzyme (red), $(ES)_1$ complex (green), and $(ES)_2$ complex (light blue); substrate in black. Vertical black dashed lines: relaxation time of the free enzyme, vertical blue dashed lines: beginning of steady state. (a) Caged substrate, activation, and diffusion times much faster than binding, instantaneous activation, (b) mixing, crystal size $3 \times 4 \times 5\ \mu m^3$, and diffusion times on the order of the of the turnover time, (c) mixing, crystal size $100 \times 120 \times 140\ \mu m^3$, and diffusion time much slower than the turnover time.

magnitude higher than the commonly accepted limit even at cryogenic temperatures, but yet diffraction patterns show no damage [63]. Mix-and-inject transient state kinetic experiments that employ diffusion of substrate into crystals become feasible. Time-resolution is still limited to sub-milliseconds, which is however comparable to using caged substrates. Smaller crystals are also advantageous for pump-probe experiments where reaction is initiated by a femtosecond laser flash (Figure 2(b)). There is only a limited number of photons available in the flash. Since protein crystals are unusually optically thick, only a thin layer of molecules is activated at the surface. If, however, the crystals are very small, the laser flash can initiate a reaction uniformly throughout the entire crystal. For example, photoactive yellow (PYP) protein has a molar absorption coefficient of 45,500 $cm^2\ mmol^{-1}$ at peak absorbance. Given the concentration of 96 mmol/L of PYP in the crystal, one absorption unit is reached with 2.3 μm crystals. Unlike larger crystals, the wavelength of the exciting laser can be tuned directly into the absorption maximum (449 nm) and, still, reaction initiation would be achieved uniformly. Even if the crystal lattice is not commensurable with the macromolecular reaction and the crystal loses integrity at a certain point in time of the reaction, it would still be possible to up to this point in time, because the crystal is discarded in any case and a new one is injected. This holds for both mix-and-inject as well as for pump-probe experiments at the XFEL and potentially, with larger crystals, at the synchrotron. Since the XFEL pulse is quite monochromatic and all the diffraction patterns are stills, all reflections are partials. For each reflection, the intensity is obtained from a sharp intersection of the Ewald sphere with the, in reciprocal space umbrella shaped crystal mosaicity [14] or, for very small crystals, with their shape transforms [36]. This results in an extremely low reflection overlap, even in the presence of higher mosaicity [61]. Consequently, the time-resolved experiment might also work with relatively low quality crystals.

Although some information about the microscopic rate coefficients of the mechanism can be obtained, a comprehensive mechanism remains underdetermined from only one crystallographic time series. There are more rate coefficients than observables that can be extracted from the time series [64]. An additional parameter such as the temperature can be varied to obtain additional observables. This requires rapid and routine production of entire time series from the beginning to the end of a reaction. With an appropriate design of a mixing device [57] longer time scales become also accessible without sacrificing the advantage of serially providing crystals to the X-ray beam. Before injection, the nanocrystal substrate mixture could reside for an extended period of time within the mixing device's flow, whose velocity can be controlled over a large range to achieve appropriate time delays [57]. When the temperature in the mixing device is also controlled, five-dimensional crystallography [64] becomes possible, and the energetics of the reaction can be investigated in addition. That includes entropy and enthalpy values of the barriers of activation in the chemical, kinetic mechanism. Free energy landscapes of enzymatic reactions become assessable by the time-resolved and temperature-dependent X-ray data.

When mixing times approach diffusion times, it is not very useful to reduce the crystal size further. Given the assumptions in this paper and the best mixers, this limit would be reached approximately with 0.5 μm–1 μm sized crystals (see Table 1). The ultimate experiment would be a time-resolved experiment with a "crystal" reduced to the size of a single molecule [65–69]. The substrate would be close-by almost instantaneously. The best time-resolution possible is then determined mainly by the mixing time. This suggests that mixers with the fastest mixing times must be available to accommodate such an experiment. Alternatively, caged substrates, which then require appropriate rapid activation rates, can be used in a pump-probe type experiment (Figure 2(b)). To date, no single enzyme molecule was successfully injected into the XFEL beam. The number of photons scattered would be on the order of 200 for a moderately sized enzyme molecule. An extensive number of sparsely populated diffraction patterns must be analyzed with special embedding techniques on an extremely low signal-to-noise level [66, 67]. A small ensemble of single molecules in random orientation may be well prepared in the mixing chamber and injected. There are, indeed, approaches to analyze the diffraction patterns obtained from this ensemble [70, 71]. Angular intensity correlations can be determined from these diffraction patterns [72]. From a large number of correlations, finally, the diffraction volume of a single molecule can be calculated [73]. The time-resolved experiment has the advantage that the structure of the dark or reference state is known, which might be used to facilitate the analysis of these sparse data. However, the experimental and the theoretical framework first have to be implemented and refined (see [74]).

Acknowledgment

The author thanks Vukica Srajer for critical reading of the paper. The work on time-resolved crystallography is supported by Grant NSF-0952643 (Career), work on the ccNIR is supported by NSF-1121770, and work on time-resolved structures from solution is supported by NSF-1158138.

References

[1] C. O'Sullivan and F. W. Tompson, "LX.—Invertase: a contribution to the history of an enzyme or unorganised ferment," *Journal of the Chemical Society, Transactions*, vol. 57, pp. 834–931, 1890.

[2] J. B. Sumner, "The isolation and crystallization of the enzyme urease: preliminary paper," *The Journal of Biological Chemistry*, vol. 69, pp. 435–441, 1926.

[3] M. von Laue, *Roentgenstrahlinterferenzen*, Physik und Chemie und ihre Anwendungen in Einzeldarstellungen, Akademische Verlagsgesellschaft, Leipzig, Germany, 1942.

[4] M. F. Perutz, "Relation between structure and sequence of hæmoglobin," *Nature*, vol. 194, no. 4832, pp. 914–917, 1962.

[5] J. C. Kendrew, G. Bodo, H. M. Dintzis, R. G. Parrish, H. Wyckoff, and D. C. Phillips, "A three-dimensional model of the myoglobin molecule obtained by X-ray analysis," *Nature*, vol. 181, no. 4610, pp. 662–666, 1958.

[6] M. Schmidt, K. Achterhold, V. Prusakov, and F. G. Parak, "Protein dynamics of a β-sheet protein," *European Biophysics Journal*, vol. 38, no. 5, pp. 687–700, 2009.

[7] F. G. Parak, K. Achterhold, S. Croci, and M. Schmidt, "A physical picture of protein dynamics and conformational changes," *Journal of Biological Physics*, vol. 33, no. 5-6, pp. 371–387, 2007.

[8] K. Moffat, "Time-resolved macromolecular crystallography," *Annual Review of Biophysics and Biophysical Chemistry*, vol. 18, pp. 309–332, 1989.

[9] K. Moffat, "Time-resolved biochemical crystallography: a mechanistic perspective," *Chemical Reviews*, vol. 101, no. 6, pp. 1569–1581, 2001.

[10] M. Schmidt, S. Rajagopal, Z. Ren, and K. Moffat, "Application of singular value decomposition to the analysis of time-resolved macromolecular X-ray data," *Biophysical Journal*, vol. 84, no. 3, pp. 2112–2129, 2003.

[11] M. Schmidt, "Structure based enzyme kinetics by time-resolved X-ray crystallography," in *Ultrashort Laser Pulses in Medicine and Biology*, W. Zinth, M. Braun, and P. Gilch, Eds., Biological and Medical Physics, Biomedical Engineering, Springer, Berlin, Germany, 2008.

[12] M. Schmidt, R. Pahl, V. Vukica Srajer et al., "Protein kinetics: structures of intermediates and reaction mechanism from time-resolved X-ray data," *Proceedings of the National Academy of Sciences of the United States of America*, vol. 101, no. 14, pp. 4799–4804, 2004.

[13] Z. Ren and K. Moffat, "Quantitative-analysis of synchrotron laue diffraction patterns in macromolecular crystallography," *Journal of Applied Crystallography*, vol. 28, pp. 461–481, 1995.

[14] Z. Ren, D. Bourgeois, J. R. Helliwell, K. Moffat, V. Šrajer, and B. L. Stoddard, "Laue crystallography: coming of age," *Journal of Synchrotron Radiation*, vol. 6, no. 4, pp. 891–917, 1999.

[15] D. Greiffenberg and A. Collaboration, "The AGIPD detector for the European XFEL," *Journal of Instrumentation*, vol. 7, 2012.

[16] H. T. Philipp, M. Hromalik, M. Tate, L. Koerner, and S. M. Gruner, "Pixel array detector for X-ray free electron laser experiments," *Nuclear Instruments and Methods in Physics Research A*, vol. 649, no. 1, pp. 67–69, 2011.

[17] H. Graafsma, "Requirements for and development of 2 dimensional X-ray detectors for the European X-ray Free Electron

Laser in Hamburg," *Journal of Instrumentation*, vol. 4, no. 12, Article ID P12011, 2009.

[18] G. Kuriso, A. Sugimoto, Y. Kai, and S. Harada, "A flow cell suitable for time-resolved X-ray crystallography by the Laue method," *Journal of Applied Crystallography*, vol. 30, no. 1, part 5, pp. 555–556, 1997.

[19] M. Schmidt, V. Srajer, N. Purwara, and S. Tripathia, "The kinetic dose limit in room-temperature time-resolved macromolecular crystallography," *Journal of Synchrotron Radiation*, vol. 19, part 2, pp. 264–273, 2012.

[20] J. E. Knapp, R. Pahl, V. Šrajer, and W. E. Royer, "Allosteric action in real time: time-resolved crystallographic studies of a cooperative dimeric hemoglobin," *Proceedings of the National Academy of Sciences of the United States of America*, vol. 103, no. 20, pp. 7649–7654, 2006.

[21] V. Šrajer, T. Y. Teng, T. Ursby et al., "Photolysis of the carbon monoxide complex of myoglobin: nanosecond time-resolved crystallography," *Science*, vol. 274, no. 5293, pp. 1726–1729, 1996.

[22] F. Schotte, M. Lim, T. A. Jackson et al., "Watching a protein as it functions with 150-ps time-resolved X-ray crystallography," *Science*, vol. 300, no. 5627, pp. 1944–1947, 2003.

[23] J. M. Key, V. Srajer, R. Pahl, and K. Moffat, "Time-resolved crystallographic studies of the heme domain of the oxygen sensor FixL: structural dynamics of ligand rebinding and their relation to signal transduction," *Biochemistry*, vol. 46, no. 16, pp. 4706–4715, 2004.

[24] H. Ihee, S. Rajagopal, V. Srajer et al., "Visualizing reaction pathways in photoactive yellow protein from nanoseconds to seconds," *Proceedings of the National Academy of Sciences of the United States of America*, vol. 102, no. 20, pp. 7145–7150, 2005.

[25] D. Bourgeois, B. Vallone, A. Arcovito et al., "Extended subnanosecond structural dynamics of myoglobin revealed by Laue crystallography," *Proceedings of the National Academy of Sciences of the United States of America*, vol. 103, no. 13, pp. 4924–4929, 2006.

[26] M. Schmidt, K. Nienhaus, R. Pahl et al., "Ligand migration pathway and protein dynamics in myoglobin: a time-resolved crystallographic study on L29W MbCO," *Proceedings of the National Academy of Sciences of the United States of America*, vol. 102, no. 33, pp. 11704–11709, 2005.

[27] S. Rajagopal, S. Anderson, V. Srajer, M. Schmidt, R. Pahl, and K. Moffat, "A structural pathway for signaling in the E46Q mutant of photoactive yellow protein," *Structure*, vol. 13, no. 1, pp. 55–63, 2005.

[28] S. Tripathi, V. Srajer, N. Purwar, R. Henning, and M. Schmidt, "pH dependence of the photoactive yellow protein photocycle investigated by time-resolved crystallography," *Biophysical Journal*, vol. 102, no. 2, pp. 325–332, 2012.

[29] B. Perman, V. Šrajer, Z. Ren et al., "Energy transduction on the nanosecond time scale: early structural events in a xanthopsin photocycle," *Science*, vol. 279, no. 5358, pp. 1946–1950, 1998.

[30] F. Schotte, H. S. Cho, V. R. Kaila et al., "Watching a signaling protein function in real time via 100-ps time-resolved Laue crystallography," *Proceedings of the National Academy of Sciences of the United States of America*, vol. 109, no. 47, pp. 19256–19261, 2012.

[31] B. L. Stoddard, B. E. Cohen, M. Brubaker, A. D. Mesecar, and D. E. Koshland, "Millisecond laue structures of an enzyme-product complex using photocaged substrate analogs," *Nature Structural Biology*, vol. 5, no. 10, pp. 891–897, 1998.

[32] I. Schlichting, J. Berendzen, K. Chu et al., "The catalytic pathway of cytochrome P450cam at atomic resolution," *Science*, vol. 287, no. 5458, pp. 1615–1622, 2000.

[33] I. Schlichting, G. Rapp, J. John, A. Wittinghofer, E. F. Pai, and R. S. Goody, "Biochemical and crystallographic characterization of a complex of c-Ha-ras p21 and caged GTP with flash photolysis," *Proceedings of the National Academy of Sciences of the United States of America*, vol. 86, no. 20, pp. 7687–7690, 1989.

[34] E. M. H. Duke, S. Wakatsuki, A. Hadfield, and L. N. Johnson, "Laue and monochromatic diffraction studies on catalysis in phosphorylase b crystals," *Protein Science*, vol. 3, no. 8, pp. 1178–1196, 1994.

[35] S. Boutet, L. Lomb, G. J. Williams et al. et al., "High-resolution protein structure determination by serial femtosecond crystallography," *Science*, vol. 337, no. 6092, pp. 362–364, 2012.

[36] R. A. Kirian, X. Wang, U. Weierstall et al., "Femtosecond protein nanocrystallography—data analysis methods," *Optics Express*, vol. 18, no. 6, pp. 5713–5723, 2010.

[37] L. Redecke, K. Nass, D. P. DePonte et al. et al., "Natively inhibited Trypanosoma brucei cathepsin B structure determined by using an X-ray laser," *Science*, vol. 339, no. 6116, pp. 227–230, 2012.

[38] R. Koopmann, K. Cupelli, L. Redecke et al. et al., "In vivo protein crystallization opens new routes in structural biology," *Nature Methods*, vol. 9, no. 3, pp. 259–262, 2012.

[39] H. N. Chapman, P. Fromme, A. Barty et al., "Femtosecond X-ray protein nanocrystallography," *Nature*, vol. 470, no. 7332, pp. 73–78, 2011.

[40] R. Neutze and K. Moffat, "Time-resolved structural studies at synchrotrons and X-ray free electron lasers: opportunities and challenges," *Current Opinion in Structural Biology*, vol. 22, no. 5, pp. 651–659, 2012.

[41] S. R. Adams and R. Y. Tsien, "Controlling cell chemistry with caged compounds," *Annual Review of Physiology*, vol. 55, pp. 755–784, 1993.

[42] M. Goelder and R. Givens, *Dynamic Studies in Biology: Phototriggers, Photoswitches and Caged Biomolecules*, Wiley VCH, 2005.

[43] T. Ursby, M. Weik, E. Fioravanti, M. Delarue, M. Goeldner, and D. Bourgeois, "Cryophotolysis of caged compounds: a technique for trapping intermediate states in protein crystals," *Acta Crystallographica Section D*, vol. 58, no. 4, pp. 607–614, 2002.

[44] I. Schlichting, S. C. Almo, G. Rapp et al., "Time-resolved X-ray crystallographic study of the conformational change in Ha-Ras p21 protein on GTP hydrolysis," *Nature*, vol. 345, no. 6273, pp. 309–315, 1990.

[45] B. L. Stoddard, P. Koenigs, N. Porter, K. Petratos, G. A. Petsko, and D. Ringe, "Observation of the light-triggered binding of pyrone to chymotrypsin by Laue X-ray crystallography," *Proceedings of the National Academy of Sciences of the United States of America*, vol. 88, no. 13, pp. 5503–5507, 1991.

[46] E. M. H. Duke, A. Hadfield, S. Walters, S. Wakatsuki, R. K. Bryan, and L. N. Johnson, "Time-resolved diffraction studies on glycogen phosphorylase b," *Philosophical Transactions of the Royal Society A*, vol. 340, no. 1657, pp. 245–261, 1992.

[47] A. J. Scheidig, E. F. Pai, I. Schlichting et al., "Time-resolved crystallography on H-ras p21," *Philosophical Transactions of the Royal Society A*, vol. 340, no. 1657, pp. 263–272, 1992.

[48] Y. O. Jung, J. H. Lee, J. Kim et al., "Volume-conserving trans-cis isomerization pathways in photoactive yellow protein visualized by picosecond X-ray crystallography," *Nature Chemistry*, vol. 5, no. 3, pp. 212–220, 2013.

[49] N. Purwar, J. M. McGarry, J. Kostera, A. A. Pacheco, and M. Schmidt, "Interaction of nitric oxide with catalase: structural

and kinetic analysis," *Biochemistry*, vol. 50, no. 21, pp. 4491–4503, 2011.

[50] J. R. Helliwell, Y. P. Nieh, J. Raftery et al., "Time-resolved structures of hydroxymethylbilane synthase (Lys59Gln mutant) as it is loaded with substrate in the crystal determined by Laue diffraction," *Journal of the Chemical Society*, vol. 94, no. 17, pp. 2615–2622, 1998.

[51] A. Cornish-Bowden, *Fundamentals of Enzyme Kinetics*, Wiley-VCH, 4th edition, 2012.

[52] M. Youngblut, E. T. Judd, V. Srajer et al., "Laue crystal structure of Shewanella oneidensis cytochrome c nitrite reductase from a high-yield expression system," *Journal of Biological Inorganic Chemistry*, vol. 17, no. 4, pp. 647–662, 2012.

[53] A. V. Cherepanov and S. De Vries, "Microsecond freeze-hyperquenching: development of a new ultrafast micro-mixing and sampling technology and application to enzyme catalysis," *Biochimica et Biophysica Acta*, vol. 1656, no. 1, pp. 1–31, 2004.

[54] D. P. Deponte, J. T. Mckeown, U. Weierstall, R. B. Doak, and J. C. H. Spence, "Towards ETEM serial crystallography: electron diffraction from liquid jets," *Ultramicroscopy*, vol. 111, no. 7, pp. 824–827, 2011.

[55] A. M. Gañán-Calvo, D. P. DePonte, M. A. Herrada, J. C. H. Spence, U. Weierstall, and R. B. Doak, "Liquid capillary micro/nanojets in free-jet expansion," *Small*, vol. 6, no. 7, pp. 822–824, 2010.

[56] R. G. Sierra, H. Laksmono, J. Kern et al. et al., "Nanoflow electrospinning serial femtosecond crystallography," *Acta Crystallographica D*, vol. 68, no. part 11, pp. 1584–1587, 2012.

[57] H. Y. Park, X. Qiu, E. Rhoades et al., "Achieving uniform mixing in a microfluidic device: hydrodynamic focusing prior to mixing," *Analytical Chemistry*, vol. 78, no. 13, pp. 4465–4473, 2006.

[58] H. S. Carslaw and J. C. Jaeger, *Conduction Heat in Solids*, Clarendon Press, Oxford, UK, 2nd edition, 1959.

[59] J. Hajdu, K. R. Acharya, D. I. Stuart et al., "Catalysis in the crystal: synchrotron radiation studies with glycogen phosphorylase b," *EMBO Journal*, vol. 6, no. 2, pp. 539–546, 1987.

[60] S. Geremia, M. Campagnolo, N. Demitri, and L. N. Johnson, "Simulation of diffusion time of small molecules in protein crystals," *Structure*, vol. 14, no. 3, pp. 393–400, 2006.

[61] A. Aquila, M. S. Hunter, R. B. Doak et al. et al., "Time-resolved protein nanocrystallography using an X-ray free-electron laser," *Optics Express*, vol. 20, no. 3, pp. 2706–2716, 2012.

[62] J. M. Holton and K. A. Frankel, "The minimum crystal size needed for a complete diffraction data set," *Acta Crystallographica Section D*, vol. 66, no. 4, pp. 393–408, 2010.

[63] L. Lomb, T. R. M. Barends, S. Kassemeyer et al., "Radiation damage in protein serial femtosecond crystallography using an X-ray free-electron laser," *Physical Review B*, vol. 84, no. 21, Article ID 214111, 2011.

[64] M. Schmidt, T. Graber, R. Henning, and V. Srajer, "Five-dimensional crystallography," *Acta Crystallographica A*, vol. 66, part 2, pp. 198–206, 2010.

[65] R. Neutze, R. Wouts, D. Van Der Spoel, E. Weckert, and J. Hajdu, "Potential for biomolecular imaging with femtosecond X-ray pulses," *Nature*, vol. 406, no. 6797, pp. 752–757, 2000.

[66] D. Giannakis, P. Schwander, and A. Ourmazd, "The symmetries of image formation by scattering. I. Theoretical framework," *Optics Express*, vol. 20, no. 12, pp. 12799–12826, 2012.

[67] P. Schwander, C. H. Yoon, A. Ourmazd, and D. Giannakis, "The symmetries of image formation by scattering. II. Applications," *Optics Express*, vol. 20, no. 12, pp. 12827–12849, 2012.

[68] R. Fung, V. Shneerson, D. K. Saldin, and A. Ourmazd, "Structure from fleeting illumination of faint spinning objects in flight," *Nature Physics*, vol. 5, no. 1, pp. 64–67, 2009.

[69] M. M. Seibert, T. Ekeberg, F. R. N. C. Maia et al., "Single mimivirus particles intercepted and imaged with an X-ray laser," *Nature*, vol. 470, no. 7332, pp. 78–82, 2011.

[70] D. K. Saldin, V. L. Shneerson, M. R. Howells et al., "Structure of a single particle from scattering by many particles randomly oriented about an axis: toward structure solution without crystallization?" *New Journal of Physics*, vol. 12, Article ID 035014, 2010.

[71] D. K. Saldin, H. C. Poon, P. Schwander, M. Uddin, and M. Schmidt, "Reconstructing an icosahedral virus from single-particle diffraction experiments," *Optics Express*, vol. 19, no. 18, pp. 17318–17335, 2011.

[72] R. A. Kirian, K. E. Schmidt, X. Wang, R. B. Doak, and J. C. H. Spence, "Signal, noise, and resolution in correlated fluctuations from snapshot small-angle X-ray scattering," *Physical Review E*, vol. 84, no. 1, part 1, Article ID 011921, 2011.

[73] D. K. Saldin, H. C. Poon, V. L. Shneerson et al., "Beyond small-angle X-ray scattering: exploiting angular correlations," *Physical Review B*, vol. 81, no. 17, Article ID 174105, 2010.

[74] H. C. Poon, M. Schmidt, and D. K. Saldin, "Extraction of fast changes in the structure of a disordered ensemble of photoexcited biomolecules," *Advances on Condensed Matter Physics*. In press.

5

A New Circuit Model for Spin-Torque Oscillator Including Perpendicular Torque of Magnetic Tunnel Junction

Hyein Lim, Sora Ahn, Miryeon Kim, Seungjun Lee, and Hyungsoon Shin

Department of Electronics Engineering, Ewha Womans University, 11-1 Daehyun-Dong, Seodaemoon-Gu, Seoul 120-750, Republic of Korea

Correspondence should be addressed to Hyungsoon Shin; hsshin@ewha.ac.kr

Academic Editor: Mohindar S. Seehra

Spin-torque oscillator (STO) is a promising new technology for the future RF oscillators, which is based on the spin-transfer torque (STT) effect in magnetic multilayered nanostructure. It is expected to provide a larger tunability, smaller size, lower power consumption, and higher level of integration than the semiconductor-based oscillators. In our previous work, a circuit-level model of the giant magnetoresistance (GMR) STO was proposed. In this paper, we present a physics-based circuit-level model of the magnetic tunnel junction (MTJ)-based STO. MTJ-STO model includes the effect of perpendicular torque that has been ignored in the GMR-STO model. The variations of three major characteristics, generation frequency, mean oscillation power, and generation linewidth of an MTJ-STO with respect to the amount of perpendicular torque, are investigated, and the results are applied to our model. The operation of the model was verified by HSPICE simulation, and the results show an excellent agreement with the experimental data. The results also prove that a full circuit-level simulation with MJT-STO devices can be made with our proposed model.

1. Introduction

Spin-torque oscillator (STO) is a promising new device which is based on the spin-transfer torque (STT) effect in magnetic multilayered nanostructure. According to the theory presented by Slonczewski [1] and Berger [2], spin-polarized current may produce precessing magnetization in the ferromagnetic material. Many experiments and researches have been carried out to support the theory about STO. Recently, interest in the magnetic tunnel junction (MTJ) which uses a thin dielectric spacer (e.g., MgO) as is shown in Figure 1(a) is rapidly growing because it may overcome the limitation of small power in giant magnetoresistance (GMR) devices.

The effect of STT in MTJ-STO can be explained by Landau-Lifshitz-Gilbert (LLG) equation that represents the movement of the magnetization. LLG equation has been proved very useful in describing the precessional motion of magnetization in the ferromagnetic material. In this equation, spin torque consists of two components: in-plane torque (a_J) and perpendicular torque (b_J). In a metallic GMR device, the perpendicular torque is very small compared with the in-plane torque such that it can be safely ignored. In MTJ devices, however, the perpendicular torque has a magnitude of about 30% of the in-plane torque. Thus, when an MTJ device is modeled, the perpendicular torque has to be considered.

Researches on STOs have been mostly focused on physical characteristics of the device with respect to the composing material and the structure whereas study on the application of STO devices to CMOS circuits is still very immature. One of the major obstacles to the integration of STO devices with CMOS circuits is the lack of circuit-level model of STO devices. Therefore, the model that represents accurate physical characteristics of STO has to be developed to accelerate the convergence of STO devices with CMOS technology.

In our previous work, physics-based GMR-STO circuit model was proposed [3, 4]. In this paper, a new circuit model for MTJ-STO is proposed including the effect of perpendicular torque. The model is fully compatible with circuit-level simulators such as SPICE. A current mirror circuit that contains an STO element and a multistage wideband amplifier was designed to verify the correctness and usability of our model in real-world circuit design.

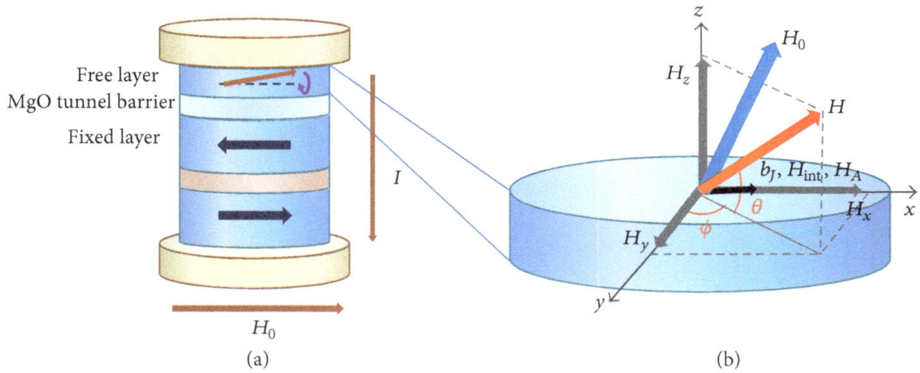

FIGURE 1: Simple schematic diagrams of (a) MTJ-STO device, (b) three axis components of the external and internal magnetic fields. The perpendicular torque b_J acts like a magnetic field and is added to the direction of the fixed layer magnetization.

2. Analytic Model of STO

LLG equation consists of three torques which are precessional torque, damping torque, and STT as shown in the following [5]:

$$\frac{d\widehat{M}_F}{dt} = -\gamma\left[\widehat{M}_F \times \widehat{H}_{\text{eff}}\right] + \frac{\alpha_G}{M_S}\left[\widehat{M}_F \times \frac{d\widehat{M}_F}{dt}\right] - \gamma\widehat{\tau}_{\text{STT}}, \quad (1a)$$

$$\widehat{\tau}_{\text{STT}} = a_J\widehat{M}_F \times \left(\widehat{M}_P \times \widehat{M}_F\right) + b_J\left(\widehat{M}_F \times \widehat{M}_p\right). \quad (1b)$$

In this equation, $\widehat{M}_F$ is the magnetization of the free layer, $\widehat{M}_P$ is the magnetization of the fixed layer, γ is the gyromagnetic ratio, $\widehat{H}_{\text{eff}}$ is the effective magnetic field, α_G is the Gilbert damping constant, M_S is the saturation magnetization, and $\widehat{\tau}_{\text{STT}}$ is STT.

As shown in (1b), STT consists of the in-plane torque component a_J and the perpendicular torque component b_J. In particular, the perpendicular torque is known as the field-like torque because of its characteristics inferred from the LLG equation. The perpendicular torque acts like a magnetic field that may be added to the effective field to the direction of the fixed layer magnetization. Therefore, the effective field in the MTJ devices can be increased or decreased depending on the magnitude and the direction of the perpendicular torque. If the magnitude and the direction of the effective field are changed, the precessional motion will be changed, and the characteristics of the STO (the generation frequency, the mean oscillation power, and full-width-at-half-maximum (FWHM) generation linewidth) are also changed, accordingly.

The objective of our model is to precisely replicate the three characteristics of MTJ-STOs. Governing equations for those three characteristics are primarily based on Slavin's analytic model [6] which states that mean oscillation power P_{exp}, generation frequency ω_g, and generation linewidth $2\Delta\omega$ are expressed as the following equations:

$$P_{\text{exp}} = R_{\text{rf}} I^2 \overline{p}, \quad (2a)$$

$$\omega_g = \omega_0 + N\overline{p}, \quad (2b)$$

$$2\Delta\omega = \left(1 + \nu^2\right)\Gamma_+\left(p_0\right)\frac{k_B T}{E\left(\overline{p}\right)}. \quad (2c)$$

In (2a), (2b), and (2c), R_{rf} is the magnetoresistance and defined as $R_{\text{rf}} = (R_L/2[R(\theta_0) + R_L]^2)(dR/d\theta)^2 \cdot 4(1+\varepsilon)$. R_L is load resistance, and ε is defined as $4\pi M_0/(H_0 + 4\pi M_0)$. $\overline{p} = (Q\eta/(Q+\zeta))[1+\exp(-(\zeta+Q)/Q^2\eta)/E_\beta((\zeta+Q)/Q^2\eta)] + ((\zeta-1)/(\zeta+Q))$ is the dimensionless power, Q is the nonlinear damping coefficient, η is the effective noise power, and ζ is the supercriticality parameter defined as (I/I_{th}). Moreover, $\beta = -(1+Q)\zeta/Q^2\eta$ and $E_n(x) = \int_1^\infty e^{-xt}/t^n dt$ are the exponential integral functions. ω_0 is the ferromagnetic resonance (FMR) frequency; N is the nonlinear frequency shift coefficient; ν is the normalized dimensionless nonlinear frequency shift; $\Gamma_+(p_0)$ is the positive damping rate, T is the temperature, $E(\overline{p}) = \lambda\omega_0\overline{p}$ is the oscillator energy, λ is $V_{\text{eff}}M_0/\gamma$, V_{eff} is the effective volume of the oscillator, and γ is the gyromagnetic ratio. A more detailed explanation of the parameters in (2a), (2b), and (2c) can be found in [6].

Three characteristics of an STO are determined by the bias current and external magnetic field. The field intensity, in-plane angle, and out-of-plane angle of the external magnetic field affect the behavior of the STO separately. In addition, anisotropy field and coupling field (interaction field) also affect the characteristics of STO. The strength and the direction of the internal magnetic field can be calculated from the magnitude and the direction of the external magnetic field and the anisotropy field using the following equations:

$$H\sin\theta = H_0\sin\theta_0 - 4\pi M_0\sin\theta, \quad (3a)$$

$$H\cos\theta\cos\phi = H_0\cos\theta_0\cos\phi_0 + H_A\cos\theta\cos\phi, \quad (3b)$$

$$H\cos\theta\sin\phi = H_0\cos\theta_0\sin\phi_0. \quad (3c)$$

In (3a), (3b), and (3c), H is the internal magnetic field magnitude. θ and ϕ are the internal out-of-plane angle and in-plane angle of the bias field, respectively. H_0 is the external magnetic field magnitude, and θ_0 and ϕ_0 are the external out-of-plane angle and in-plane angle of the external magnetic field, respectively. H_A is the anisotropy field and M_0 represents saturation magnetization. The internal field is thus obtained from the external magnetic field, and the anisotropy field is used in our previous model for GMR-STO. However, as was mentioned in [7], the perpendicular torque b_J and the coupling field H_{int} should be considered together

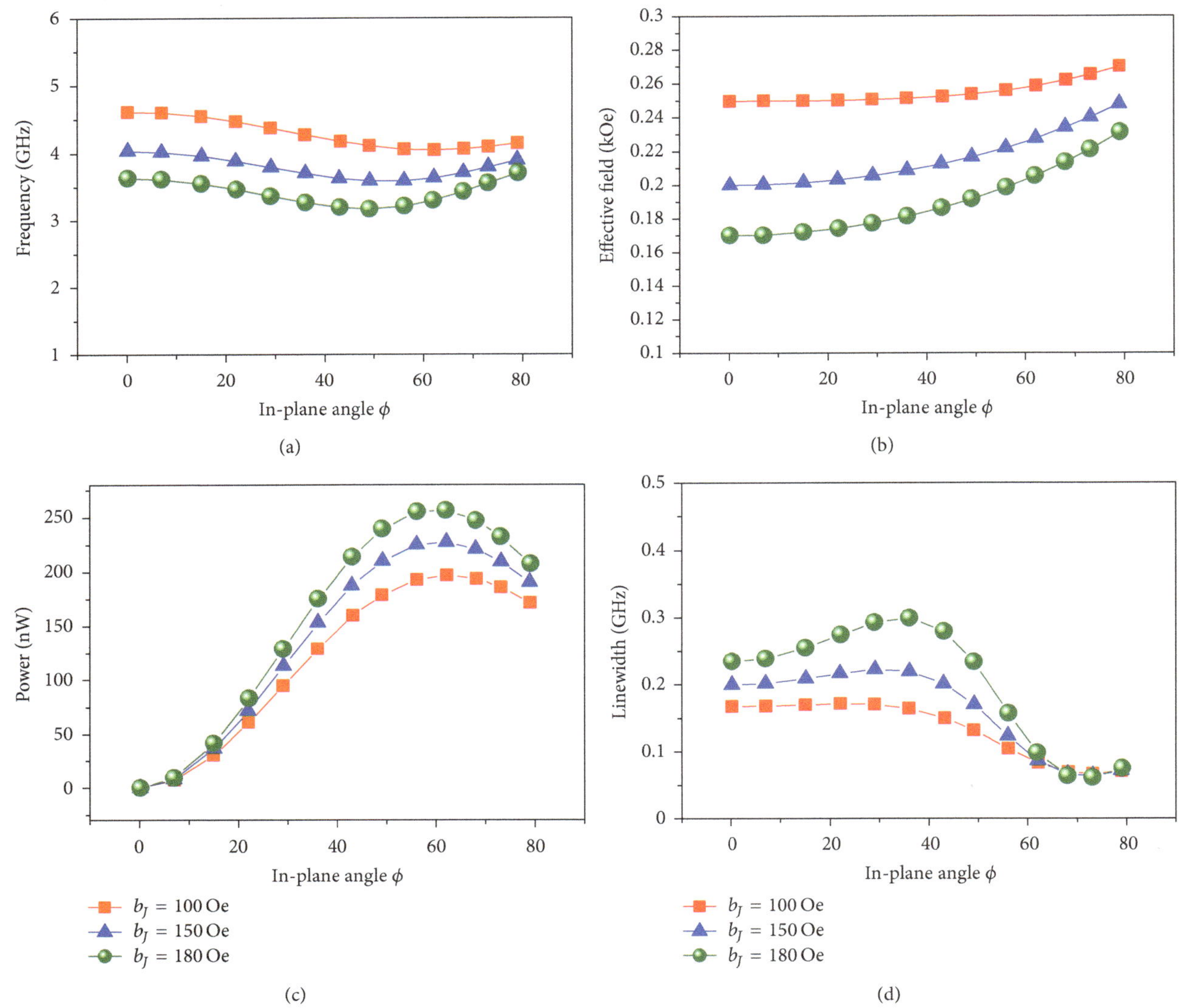

FIGURE 2: Characteristics of MTJ-STOs as functions of the perpendicular torque and the in-plane angle ϕ at $I = 1.5$ mA. (a) Generation frequency, (b) effective field magnitude, (c) mean oscillation power and (d) generation linewidth.

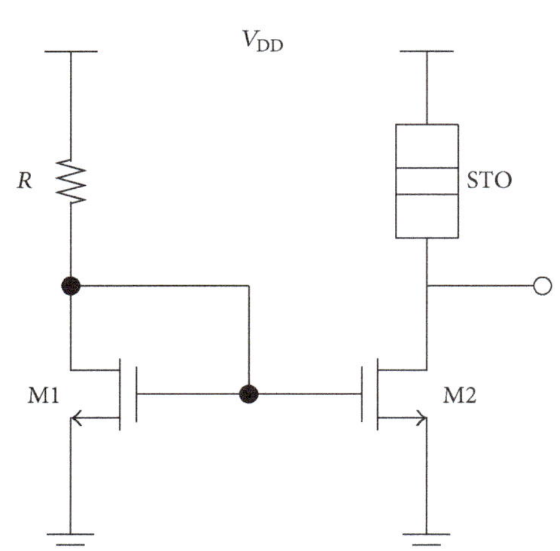

FIGURE 3: Current mirror circuit integrated with the MTJ-STO elements and the CMOS transistor.

to accurately model the behavior of MTJ-STO devices. In the proposed model, the perpendicular torque and coupling field are included as is shown in (4a), (4b), (4c), and (4d):

$$H_{x\_new} = H_x - H_{int} - b_J, \quad (4a)$$

$$H_{new} = \sqrt{H_{x\_new}^2 + H_y^2 + H_z^2}, \quad (4b)$$

$$\phi_{new} = \tan^{-1}\left(\frac{H_y}{H_{x\_new}}\right), \quad (4c)$$

$$\theta_{new} = \tan^{-1}\left(\frac{H_z}{\sqrt{H_{x\_new}^2 + H_y^2}}\right). \quad (4d)$$

The internal magnetic field H calculated in (3a), (3b), and (3c) is composed of three components, H_x, H_y, and H_z, as is shown in Figure 1(b). Both the coupling field and

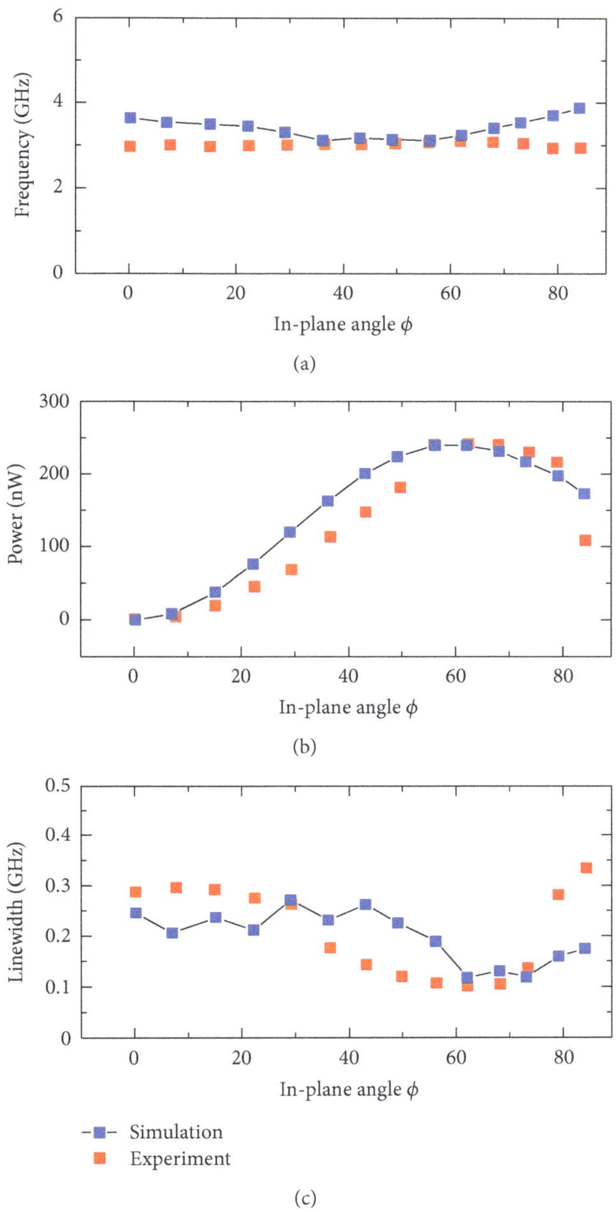

FIGURE 4: Comparison of simulation results for three characteristics of STOs with experimental data [7] as functions of the perpendicular torque and the inplane angle ϕ. (a) Frequency, (b) mean power, and (c) linewidth.

the perpendicular torque affect the easy axis component of the internal magnetic field. Therefore, the magnitude and the direction of the new internal magnetic field H_{new}, ϕ_{new}, and θ_{new} become different from the original internal magnetic field that has been calculated without considering the coupling field and the perpendicular torque. Oscillation power, generation frequency, and linewidth can now be expressed more accurately using the new internal magnetic field parameters.

The devices are assumed to be in antiparallel (AP) state as was assumed in [7]. The easy axis is set to positive x axis, and the magnetization of fixed layer is towards $-x$ direction

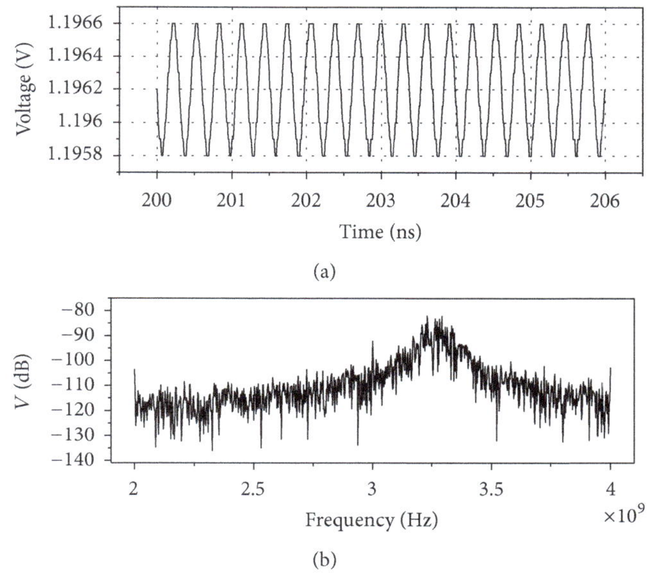

FIGURE 5: (a) The output voltage signal in the time domain, (b) the generation frequency in the frequency domain at 62° of the in-plane angle.

whereas the magnetization of the free layer is set to $+x$ direction.

The perpendicular torque is known to depend on the applied voltage, but the voltage dependence of the perpendicular torque has not been clearly defined. The perpendicular torque has quadratic voltage dependence according to Sankey et al. [8] and Kubota et al. [9]. Petit et al. [10] and Li et al. [11], however, reported that the perpendicular torque has linear voltage dependence. Recently, the voltage dependence of the perpendicular torque can be generally described by the lowest order Taylor expansion ($b_J = C_1V + C_2V^2$) that explain the main effects of asymmetric and symmetric bias dependence, simultaneously [5, 12]. Therefore, the perpendicular torque is assumed to have linear and quadratic voltage dependence in this paper.

3. Characteristics of STO with Perpendicular Torque

In this section, the characteristics of an MTJ-STO versus the in-plane angle of internal magnetic field are observed while varying the magnitude of the perpendicular torque. During the simulation, the geometrical parameter and layer composition of the MTJ device are assumed to be the same as [7]. The dimension of the device is 140 nm × 85 nm, and the free layer thickness is 1.8 nm. In this model, the value of the saturation magnetization and the anisotropy field are assumed to be 12 kOe and 100 Oe, respectively. Also, as the perpendicular torque cannot be measured directly from the given experimental data, its proper value can only be found by fitting the experimental data with our MTJ-STO model by adjusting the magnitude of the perpendicular torque using C_1 and C_2. Thus, the characteristics of MTJ-STO are investigated

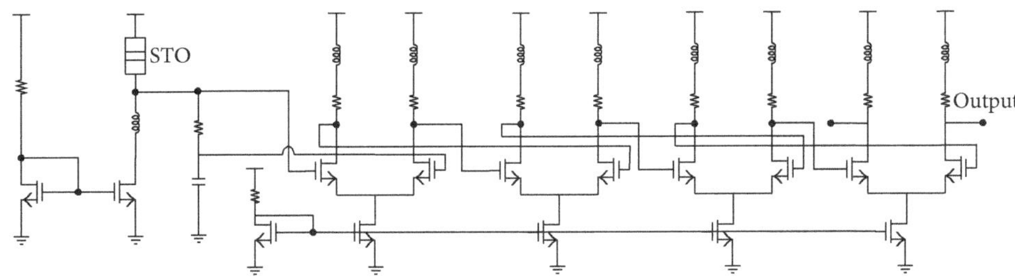

FIGURE 6: Circuit schematic of the current mirror circuit with a 4-stage common source amplifier.

by varying the magnitude of the perpendicular torque as is shown in Figure 2.

First, the generation frequency of STO as a function of the perpendicular torque is observed through the simulation. Figure 2(a) shows that the generation frequency decreases when the magnitude of the perpendicular torque increases. In particular, the frequency variation due to the perpendicular torque becomes large when the in-plane angle of the internal magnetic field is small. From the LLG equation, the perpendicular torque is added to the effective field with the direction of the magnetization of the fixed layer, so that the perpendicular torque affects the x axis component of the internal effective field. Also, the x axis component of the internal effective field is dominant when the in-plane angle is small. Consequently, the generation frequency decreases with increasing the perpendicular torque. In other words, the perpendicular torque tends to push the effective field to the hard axis so that the magnitude of the effective field becomes smaller at the small in-plane angle, and this phenomenon is illustrated in Figure 2(b).

Second, the mean power and the generation linewidth of the STO with perpendicular torque are observed and shown in Figures 2(c) and 2(d). Both the mean power and the generation linewidth increase with increasing the magnitude of the perpendicular torque. It is worth noting that the mean power gradually increases with increasing in-plane angle and has a maximum value at around 60° while the generation linewidth decreases and has a minimum value at the point where the power has the maximum. As previously explained, the reason that the maximum power and minimum linewidth occur at a certain in-plane angle rather than the hard axis direction is the perpendicular torque which affects the effective field.

4. Simulation Results

Verification of our model is performed using HSPICE circuit simulator. A current mirror CMOS circuit is integrated with an MTJ-STO element as shown in Figure 3.

Proposed STO model is written in Verilog-A, which is a hardware description language for mixed-mode circuit design such that it is fully compatible with circuit-level simulators such as HSPICE. The correctness of our model was verified by comparing the simulation results with the previously reported experimental data [7]. The simulation results at $I = 1.5\,\text{mA}$ are shown in Figure 4. As the magnitude of the perpendicular torque is not given with the experimental data, the different values of the perpendicular torque are tried in our simulation to find the best match. In this case, the simulation results show that the perpendicular torque of 180 Oe gives the best fit with the experimental data ($C_1 = 29\,\text{Oe/V}$, $C_2 = 160\,\text{Oe/V}^2$). The generation frequency is shown in Figure 4(a), and the simulation results show good agreement with the experimental data. In Figure 4(b), the mean power increases with increasing in-plane angle. It has the maximum value at 60° both in the simulation results and the experimental data. The experimental data of the generation linewidth in Figure 4(c) shows a rapid growth at large in-plane angle. That phenomenon is due to the thermal fluctuation which is not considered in our model. The output voltage signal in the time domain and the generation frequency in the frequency domain at 62° of the in-plane angle are shown in Figure 5. There are some researches that present the method for the measurement of the STT experimentally [8, 12]; however, our model is focused on the practicality of the circuit model that expresses the characteristics of the MTJ-STO. Simply adjust the coefficient of the perpendicular torque so that the accurate representation of the MJT-STO characteristics can be possible.

Even when an MTJ-STO gives out more power than a GMR-STO, the output signal needs more amplification to be used as a high frequency microwave source to CMOS circuits. A four-stage common source wideband amplifier is designed and added to the current mirror circuit as shown in Figure 6. The bandwidth of the amplifier is about 8 GHz, and the gain is 40 dB. Figure 7 shows the simulation results of the amplifier with our MTJ-STO model. The output spectrum of the amplifier has a same generation frequency, and the power gain is about 35 dB.

5. Conclusion

In this paper, we propose a circuit-level model of MTJ-STOs. The perpendicular torque and coupling field which have been ignored in GMR-STO model are included in our new model for MTJ-STO. As the perpendicular torque magnitude increases, the force that pushes the effective magnetic field toward the hard axis becomes stronger so that the magnitude of the effective magnetic field and the generation frequency decreases accordingly. In addition, power and the linewidth increases as the perpendicular torque increases. As the perpendicular torque cannot be measured directly from the

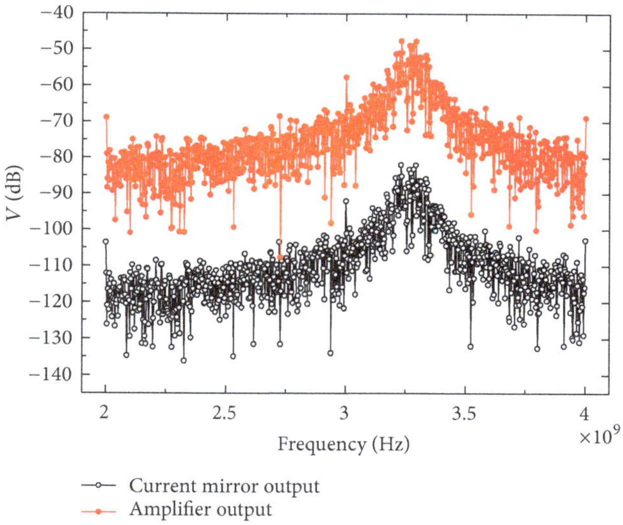

FIGURE 7: The frequency spectrum of the output signal from current mirror stage and the amplifier stage.

given experimental data, it needs to be derived from multiple simulation data with different values of the torque by best-fitting technique. In order to verify the MTJ-STO model, the current mirror CMOS circuit was composed with an MTJ-STO and simulated with HSPICE circuit simulator. The simulation results are compared with the experimental data, and they show good agreement with existing experimental data when the perpendicular torque is properly determined. Also, multistage wideband CMOS amplifier is designed and simulated with the current mirror circuit to prove the usability of our model in real circuit design environment.

As this model is based on the basic theory of microwave generation in magnetic nanostructures, it can be applied to various types of STO elements for broad application area.

Acknowledgments

This research was supported by the Basic Science Research Program through the National Research Foundation of Korea (NRF) funded by the Ministry of Education, Science, and Technology (no. 2011-0016277). And the authors would like to thank A. Slavin for the helpful discussions about the subject.

References

[1] J. Slonczewski, "Current-driven excitation of magnetic multilayers," *Journal of Magnetism and Magnetic Materials*, vol. 159, pp. L1–L7, 1996.

[2] L. Berger, "Emission of spin waves by a magnetic multilayer traversed by a current," *Physical Review B*, vol. 54, pp. 9353–9358, 1996.

[3] H. Lim, S. Ahn, S. Lee, and H. Shin, "Physics-based SPICE model of spin-torque oscillators," *Japanese Journal of Applied Physics*, vol. 51, Article ID 04DM03, 2012.

[4] S. Ahn, H. Lim, S. Lee, and H. Shin, "Analytic Model of Spin-Torque Oscillators (STO) for Circuit-Level Simulation," *Journal of Semiconductor Technology and Science*, vol. 13, article 028, 2013.

[5] S.-C. Oh, S.-Y. Park, A. Manchon et al., "Bias-voltage dependence of perpendicular spin-transfer torque in asymmetric MgO-based magnetic tunnel junctions," *Nature Physics*, vol. 5, no. 12, pp. 898–902, 2009.

[6] A. Slavin and V. Tiberkevich, "Nonlinear auto-oscillator theory of microwave generation by spin-polarized current," *IEEE Transactions on Magnetics*, vol. 45, no. 4, pp. 1875–1918, 2009.

[7] Z. M. Zeng, P. Upadhyaya, P. Khalili Amiri et al., "Enhancement of microwave emission in magnetic tunnel junction oscillators through in-plane field orientation," *Applied Physics Letters*, vol. 99, no. 3, Article ID 032503, 2011.

[8] J. C. Sankey, Y.-T. Cui, J. Z. Sun, J. C. Slonczewski, R. A. Buhrman, and D. C. Ralph, "Measurement of the spin-transfer-torque vector in magnetic tunnel junctions," *Nature Physics*, vol. 4, no. 1, pp. 67–71, 2008.

[9] H. Kubota, A. Fukushima, K. Yakushiji et al., "Quantitative measurement of voltage dependence of spin-transfer torque in MgO-based magnetic tunnel junctions," *Nature Physics*, vol. 4, no. 1, pp. 37–41, 2008.

[10] S. Petit, C. Baraduc, C. Thirion et al., "Spin-torque influence on the high-frequency magnetization fluctuations in magnetic tunnel junctions," *Physical Review Letters*, vol. 98, no. 7, Article ID 077203, 2007.

[11] Z. Li, S. Zhang, Z. Diao et al., "Perpendicular spin torques in magnetic tunnel junctions," *Physical Review Letters*, vol. 100, no. 24, Article ID 246602, 2008.

[12] C. Wang, Y.-T. Cui, J. A. Katine, R. A. Buhrman, and D. C. Ralph, "Time-resolved measurement of spin-transfer-driven ferromagnetic resonance and spin torque in magnetic tunnel junctions," *Nature Physics*, vol. 7, no. 6, pp. 496–501, 2011.

6

Temperature Dependence of the Tilt Angle for the Smectic A-Smectic C* Transition in a Mixture of C7-70PDOB Ferroelectric Liquid Crystals near the Tricritical Point

Mustafa Kurt[1] and Hamit Yurtseven[2]

[1] *Department of Physics, Canakkale Onsekiz Mart University, 17100 Canakkale, Turkey*
[2] *Department of Physics, Middle East Technical University, 06531 Ankara, Turkey*

Correspondence should be addressed to Hamit Yurtseven, hamit@metu.edu.tr

Academic Editor: Durga Ojha

The temperature dependence of the tilt angle θ is studied in the smectic C* phase near the smectic A-smectic C* tricritical point for a mixture of 70PD0B in the ferroelectric liquid crystal C7 ($X = 16.92$). The mean-field models with the biquadratic $P^2\theta^2$ (P is the spontaneous polarization) and $P^2\theta^2 + P\theta$ (bilinear) coupling terms in the free energy expansion are used to analyze the experimental data for the tilt angle in this binary mixture. From our analysis, the coefficients given in the free energy expansion of the mean-field models are determined. Our results show that the mean-field theory explains adequately the observed behaviour of the C7-70PD0B mixture near the AC* tricritical point.

1. Introduction

Smectic transitions in ferroelectric liquid crystals have been the subject of various experimental and theoretical studies. In particular, smectic A and C (or C*) transitions have been investigated and reported in the literature.

In the smectic A phase, the liquid molecules are oriented along the director (orientational order parameter) and they show a transitional order in layers, whereas in the smectic C (or C*) phase, additionally they make a tilt angle with the director which is perpendicular to the smectic layers. If the chiral molecules, which are optically active, exist in the liquid crystalline material, then the smectic C* phase occurs.

The smectic A-smectic C* (AC*) transition has been predicted by de Gennes [1] as the transition which belongs to the three-dimensional XY universality class ($n = 2$, $n = 3$). This has not been confirmed by some early experiments which are in agreement with a Landau mean-field theory of the AC* transition [2–7]. Some experimental [8–13] and theoretical [3, 14, 15] studies have shown that from the first-order to a second order AC* transition, there occurs a tricritical point (TCP). In our recent studies, tricritical behaviour of mixtures of C7+10.04 [16] and SCE9+SCE10 [17, 18] has been shown using the mean-field models. Mean-field models are applicable to many liquid crystalline systems and they predicted adequately the observed behaviour of those materials exhibiting phase transitions. The physical quantities measured to high accuracy can then be analyzed by a mean-field model and the type of transition (first order, second order, or tricritical) can be characterized. From this point of view, mean-field model is a more realistic theoretical model to be applicable to the experiments.

The mean-field models which describe the AC (or AC*) transitions have the free energy expanded in terms of the order parameters (tilt angle θ and the spontaneous polarization P) and their bilinear ($P\theta$) and/or biquadratic ($P^2\theta^2$) couplings [3, 14, 19, 20]. Dipolar interactions ($P\theta$ coupling) are of a chiral character which becomes importantly close to the AC* transition, whereas quadrupolar interactions ($P^2\theta^2$ coupling) are of a nonchiral one which covers a wide range of temperatures below the transition temperature in the smectic C or AC* phase. On the basis of our earlier mean-field models with the biquadratic coupling ($P^2\theta^2$), recently we have calculated tilt angle and the temperature shifts as a function of concentration for a mixture of 10.04+C7 [16] and the temperature dependence of the spontaneous polarization and the tilt angle for a mixture of SCE9+SCE10 [17, 18] for the AC* transition. Very recently, we have studied the dielectric constant as a function of temperature for the AC* transition in 4-(3-methyl-2-chlorobutanoyloxy)-4′-heptyloxybiphenly (A7) [21] and we have calculated a generalized smectic-hexatic phase diagram in various mixtures of liquids crystals [22].

In this study, we focus on the temperature dependence of the tilt angle for a mixture of C7-70PDOB ferroelectric liquid crystal close to the tricritical point (TCP). The ferroelectric material 4-(3-methyl-2-chlorobutanoyloxy)-4′-heptyloxy biphenly (C7) exhibits a large spontaneous polarization [11]. It undergoes a first-order AC* transition. When adding a second compound of 4-heptyloxy-4′-decloxybenzoate (7OPDOB), this transition is driven towards a second order by passing through a tricritical point (TCP) [9]. The transition temperatures of C7 for the relevant phases are as follows: SmC* (53.9°C) and SmA (54.06°C) [8]. The AC* transition temperatures of a binary mixture of C7+7OPDOB depending on the X values (mol percent of 7OPDOB) [9] are given in Table 1. Ferroelectric liquid crystals with large spontaneous polarizations such as C7 can be used in switching display devices and liquid crystals devices. Their carbon-13 nuclear magnetic resonance can be obtained experimentally to determine the temperature dependence of order parameter. In our previous study [21], we used our mean-field model with the $P^2\theta^2$ coupling and calculated the dielectric constant (dielectric susceptibility) at various temperatures under constant electric field for A7. Our calculation for the tilt angle (order parameter) as a function of temperature was performed [21] using the molecular field theory [23] for A7. The experimental data for the dielectric constant of A7 [24] was then analyzed according to the expression derived from the mean-field model [21]. Similarly, we have used our mean-field model with the quadratic couplings $\Psi^2\theta^2$ (Ψ is the orientational order parameter and θ is the tilt angle) for the smectic-hexatic phase transitions in various mixtures of liquid crystals in our recent study [22]. By expanding the free energy in terms of the order parameters (Ψ and θ), we derived the phase line equations and calculated a generalized phase diagram (T-X) on the basis of the experimental data [25].

In the present study, not only the quadric coupling $P^2\theta^2$ but also $P^2\theta^2 + P\theta$ coupling in the free energy of our mean-field model is considered and the temperature dependence of the tilt angle θ is derived in both cases. Instead of using the temperature dependence of the order parameter from the order parameter, from the molecular field theory [23], as we have also needed in our recent study [21], the θ versus T relation derived here is directly used to analyze the experimental data [9]. This is more general in the sense that the experimental data can be analyzed freely with the fitted parameters determined, whereas the molecular field theory [23] predicts the temperature dependence of the order parameter according to the critical exponent 1/2. This, however, can restrict the analysis of the experimental data for various mixtures of liquid crystals. In order to get good fits, our θ versus T expression is favorable. On this basis, we have fitted to the experimental data [9] the expression for the tilt angle as a function of temperature, which we derive from the mean-field models with the $P^2\theta^2$ and $P\theta + P^2\theta^2$ couplings for the AC* transition close to the TCP. Coefficients given in the mean-field energy are calculated from our fits and they are interpreted within the AC* transition in this mixture of C7-70PDOB.

In Section 2, we give an outline of our mean-field models, where the temperature dependence of the tilt angle and the spontaneous polarization are derived. In Section 3, the tilt angle expression is fitted to the experimental data and the results are presented. Sections 4 and 5 give our discussion and conclusions, respectively.

2. Theory

The free energy of the smectic C* phase can be expanded in terms of the tilt angle θ and the spontaneous polarization P for the AC* transition in ferroelectric liquid crystals. By considering the quadrupolar interaction between the θ and P (biquadratic coupling $P^2\theta^2$), the free energy can be expressed as

$$g = \frac{1}{2}a(T - T_0)\theta^2 + \frac{1}{4}b\theta^4 + \frac{1}{6}c\theta^6 + \frac{1}{2\chi_0\varepsilon_0}P^2 - DP^2\theta^2 + \frac{1}{4}eP^4 - EP \quad (1)$$

in the presence of the electric field E, where a, b, c, D, and e are constants. T_0 is the transition temperature, χ_0 is the static dielectric susceptibility, and ε_0 is the permittivity in free space. In the free energy expansion, the tilt angle θ is taken as the primary order parameter which does not exist in the smectic A phase. The spontaneous polarization P is the secondary order parameter which is defined in both the smectic A and C* phases. We have introduced this model in our previous study [19].

In order to describe the AC* transition, the energy is minimized with respect to the order parameters P and θ. By minimizing the free energy g with respect to the spontaneous polarization P, we get at the zero electric field ($E = 0$),

$$\frac{1}{\chi_0\varepsilon_0} - 2D\theta^2 + eP^2 = 0. \quad (2)$$

Also, by minimizing the free energy g with respect to the tilt angle θ, we find that

$$a(T - T_0) + b\theta^2 + c\theta^4 - 2DP^2 = 0. \quad (3)$$

Equations (2) and (3) give the temperature dependence of the spontaneous polarization and tilt angle, respectively. By substituting (2) into (3), the temperature dependence of the primary order parameter θ (tilt angle) can be written as

$$(T - T_C) = \frac{-c}{a}\theta^4 + \frac{1}{a}\left(\frac{4D^2}{e} - b\right)\theta^2 - \frac{3}{16 \cdot c \cdot a}\left(b - \frac{4D^2}{e}\right)^2. \quad (4)$$

TABLE 1: Transition temperature (T_{AC^*}) for the different mixtures of C7+7OPDOB system [9].

Mole percent	T_{AC^*} (°C)
16.92	44.668
15.49	45.120
14.02	45.738
13.60	46.095
13.30	46.271

In (4) the temperature shifts,

$$\Delta T = T_C - T_0 = \frac{3}{16ac}\left(b - \frac{4D^2}{e}\right)^2 - \frac{2D}{ea\chi_0\varepsilon_0}, \quad (5)$$

were used for a first-order AC* transition [19]. In (5) T_C denotes the experimentally measured AC* transition.

The free energy of the smectic C* phase can also be expanded in terms of the tilt angle θ and the spontaneous polarization P by considering both the dipolar interactions (bilinear coupling $P\theta$) and the quadrupolar interactions (biquadratic coupling $P^2\theta^2$) for the AC* transition in ferroelectric liquid crystals. The free energy can then be expressed as

$$g = \frac{1}{2}a(T - T_0)\theta^2 + \frac{1}{2}b\theta^4 + c\theta^6 + \frac{1}{2\chi_0\varepsilon_0}P^2 - DP^2\theta^2 - CP\theta. \quad (6)$$

By minimizing the free energy with respect to the spontaneous polarization P and the tilt angle θ as before, one gets

$$\frac{P}{\chi_0\varepsilon_0} - C\theta - 2DP\theta^2 = 0, \quad (7)$$

$$a(T - T_0)\theta + b\theta^3 + c\theta^5 - CP - 2DP^2\theta = 0, \quad (8)$$

respectively. By substituting (7) into (8), the temperature dependence of the tilt angle θ can be written as

$$(T - T_0) = \frac{-b}{a}\theta^2 - \frac{c}{a}\theta^4 + \frac{1}{a}\left(\frac{\chi_0\varepsilon_0 C^2}{1 - 2D\chi_0\varepsilon_0\theta^2}\right) + \frac{1}{a}\left(\frac{2D\chi_0^2\varepsilon_0^2 C^2\theta^2}{\left(1 - 2D\chi_0\varepsilon_0\theta^2\right)^2}\right). \quad (9)$$

Using (1) and (6) with the $P^2\theta^2$ and $P^2\theta^2 + P\theta$ coupling in the free energies, respectively, the temperature dependence of the tilt angle θ can be predicted for the AC* transition in ferroelectric liquid crystals.

3. Calculations and Results

The temperature dependence of the tilt angle θ was calculated for the AC* transition of a binary mixture of 70PDOB in C7 near the tricritical point ($X = 16.92$). The value X is the mol percent (mol %) of 70PDOB in C7, as given in Table 1 [9]. For this calculation, (4) was fitted to the experimental data [9] and the coefficients were determined. Table 2 gives values of the coefficients with the uncertainties from (4) according to the biquadratic coupling ($P^2\theta^2$) for this mixture of 70PDOB+C7 when the tricritical concentration of 70PDOB is $X = 16.92$ for the AC* transition. We plot θ versus $T - T_C$ in Figure 1. We also calculated the temperature dependence of the tilt angle θ by fitting (9) to the experimental data [9] in the case of both biquadratic ($P^2\theta^2$) and bilinear ($P\theta$) couplings. Coefficients with the uncertainties and plots of θ versus $T - T_0$ are given in Table 2 and Figure 2, respectively.

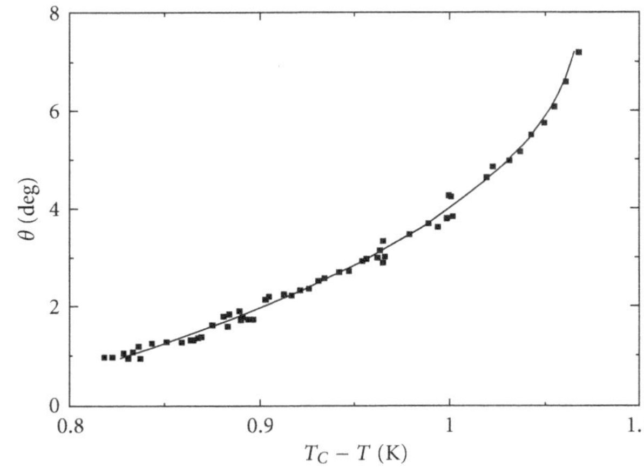

FIGURE 1: Tilt angle θ as a function of $T - T_C$ for the AC* transition in a mixture of 70PDOB+C7 near the tricritical point ($X = 16.92$) according to (4) with the $P^2\theta^2$ coupling.

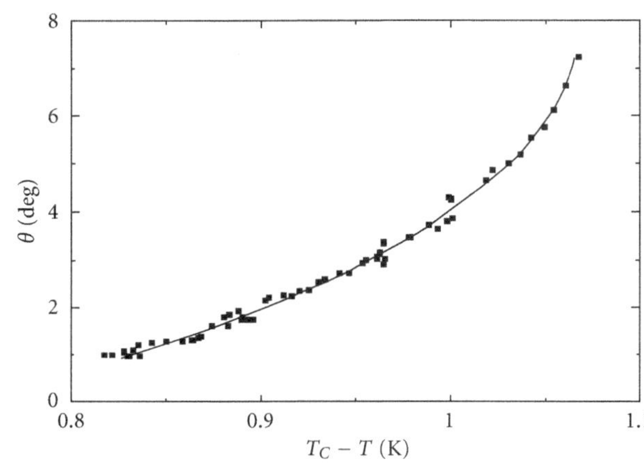

FIGURE 2: Tilt angle θ as a function of $T - T_C$ for the AC* transition in a mixture of 70PDOB+C7 near the tricritical point ($X = 16.92$) according to (9) with the $P\theta + P^2\theta^2$ coupling.

4. Discussion

The tilt angle θ was calculated as a function of temperature in the C* phase for the AC* transition near the tricritical concentration of 70PDOB ($X = 16.92$) in C7 using the mean-field models with the $P^2\theta^2$ (1) and $P\theta + P^2\theta^2$ (6)

TABLE 2: Values of the parameters given in the equations indicated for the AC* transition in a mixture of 70PDOB+C7 near the tricritical point ($X = 16.92$).

χ_0	T_0 (°C)	a (J/m³ K deg²)	b (J/m³ deg⁴)	c (J/m³ deg⁶)	C (J/m³ C deg)	D (J·m/C² deg²)	e (Jm⁵/C⁴)	Equation
4.59	44.668	1642.41 ± 1.30	15.41 ± 0.01	-5.32×10^{-5} $\pm 4.78 \times 10^{-8}$	—	2.546 ± 0.001	3.557 ± 0.002	(4)
4.59	44.668	1.77×10^{17} $\pm 1.7 \times 10^{14}$	8.31×10^{14} $\pm 4.6 \times 10^{12}$	3.18×10^{12} $\pm 1.13 \times 10^{11}$	6.83×10^{13} $\pm 2.6 \times 10^{7}$	4.05×10^{6} $\pm 1.4 \times 10^{5}$	—	(9)

couplings, as shown in Figures 1 and 2, respectively. Both equations, (4) with the $P^2\theta^2$ coupling and (9) with the $P\theta + P^2\theta^2$, when fitted to the experimental data [9], describe the observed behaviour of the tilt angle θ near the tricritical point (TCP) satisfactorily. We also calculated the standard deviations of the fitted parameters for both mean-field models, as given in Table 2. The uncertainties in the coefficients of the free energies ((1) and (6)) are comparatively small in most cases (Table 2). However, these uncertainties can be significant for the coefficients in the free energies for both models since the critical behaviour of the mesomorphic mixture depends on those coefficients, in particular the coefficients of the coupling terms ($DP^2\theta^2$ and $CP\theta$). For the mesomorphic mixture studied here, regarding the values of C and D (Table 2) for the second mean-field model (6), the bilinear $P\theta$ coupling is dominant in comparison with the biquadratic coupling $P^2\theta^2$. However, considering all the values of the fitted parameters in the free energy expansion (Table 1), the mean-field model with the $P^2\theta^2$ coupling (1) can still be preferred since (4) provides the values of the fitted parameters which are physically meaningful. This then indicates that the dominant mechanism of the smectic AC* transition near the tricritical point is due to quadratic interactions for a mixture of 70PDOP+C7. In (1) the biquadratic coupling terms ($P^2\theta^2$) characterizes the nonchiral properties and induces a transverse quadrupolar ordering. This is the dominant term in a wide temperature range far away from T_C. To stabilize the mesomorphic mixture for far away T_C, we included the P^4 term, which has also been considered in the generalized mean-field model for the smectic A-chiral-smectic C phase transition in the case of p-(n-decyloxybenzylidence)-p-amino-(2-methylbutly) cinnamate (DOBAMBC) in an earlier study [3]. Since the bilinear coupling term ($P\theta$) characterizes the chiral properties close to the T_C, the mesomorphic mixture is already stabilized with the biquadratic coupling term ($P^2\theta^2$) and, no additional term such as P^4 is required, in (6). So that the free energy of the smectic C* phase is given in terms of the spontaneous polarization up to P^2 with the bilinear coupling term $P\theta I$ as also given previously [3] and with this free energy the system is stable. Thus, the temperature dependence of θ is for both mean-field models studied here. In fact the case of DOBAMBC, the relative importance of biquadratic term $P^2\theta^2$ and P^4 term in comparison with terms involving $P\theta$ and P^2 has been demonstrated numerically using the generalized mean-field model for the AC* transition [3, Table III]. Thus, as explained previously [3], when relatively reliable experimental data is obtained in the temperature range studied, the biquadratic term $P^2\theta^2$ and P^4 are not negligible as we also demonstrate in our study here (Table 2). Thus, the terms θ^6, $P^2\theta^2$, θ^4, and P^4 (1), and $P^2\theta^2$, $P\theta$, and P^2 (6) are required to explain the major features of the AC* transition in C7-7OPDOB. We also note that from the data analysis point of view, if the P^4 term was added to the free energy of the C* phase (6), the fitting of the model to the experimental data would not have been a straightforward problem. This is due to the P^3 term appearing in (7) when the free energy was minimized with respect to the polarization P. As a cubic polarization equation, two solutions would correspond to a local minimum of the free energy, which may then incorporate the thermal fluctuations in the mean-field model, as also pointed out previously [14].

The temperature dependence of the tilt angle θ was calculated here near the tricritical point ($X = 16.92$) using the experimental data [9] where they obtained $\beta = 0.436$ according to a power-law formula

$$\phi = \phi_0 \left[\frac{T_{AC^*} - T}{T_{AC^*}} \right]^{\beta}. \tag{10}$$

In (10) ϕ is the tilt angle, ϕ_0 is the amplitude, T_{AC^*} is the AC* transition temperature, and β is the critical exponent for the order parameter [26]. The experimental tilt-angle data were analyzed within the temperature interval of 240 mK for $X = 16.92$ mixture of 70PDOB in C7 [9]. The experimental tilt-angle (ϕ) data were also the same as those obtained from the analyses within the temperature intervals of 103 mK ($\beta = 0.5$), 490 mK ($\beta = 0.38$), and 930 mK ($\beta = 0.343$) [9]. Similarly, the temperature dependence of the tilt angle can also be calculated for the AC* transition in a mixture of 70PDOB-C7 in the case of various concentrations of 70PDOB on the basis of our mean-field models given here. Since the tilt angle ϕ has been obtained from the layer spacing data as a function of $T_{AC^*} - T$ for C7 and its mixture with 70PDOB for concentrations of 0 (C7), 5.15, 9.68, 15.0, and 19.4 [8], (4) and (9) can be fitted to those data to describe the AC* transition in this mixture. Since $X = 0$ corresponds to a first-order AC* transition in C7, as the concentration of 70PDOB increases towards $X = 19.4$, the AC* transition then becomes a second order through $X = 16.92$ (tricritical transition) for this mixture of 70PDOB-C7. Thus, from a first order to a second order via the tricritical point (TCP) AC* transition, as the concentration (70PDOB) increases, can be described adequately by calculating the temperature dependence of the tilt angle using the experimental data [9] on the basis of our mean-field models studied. Mean-field to tricritical crossover behaviour near the AC* tricritical point can then be better understood for a mixture of 70PDOB-C7. This work is under progress.

5. Conclusions

The tilt angle was predicted as a function of temperature for the AC* transition of a mixture of 70PDOB-C7 close to the tricritical point. The mean-field models which consider the quadratic and dipolar interactions between the tilt angle and the spontaneous polarization were used for this calculation of the tilt angle. Expressions for tilt angle which we derived from our mean-field models were fitted to the experimental data and the coefficients were determined.

Our results show that quadratic interactions seem to be dominant in the mechanism of the AC* transition for the 70PDOB-C7 mixture near the tricritical point in regard to the experimental data.

References

[1] P. G. de Gennes, "Some remarks on the polymorphism of smectics," *Molecular Crystals and Liquid Crystals*, vol. 21, no. 1-2, pp. 49–76, 1973.

[2] C. C. Huang and J. M. Viner, "Nature of the smectic-A-smectic-C phase transition in liquid crystals," *Physical Review A*, vol. 25, no. 6, pp. 3385–3388, 1982.

[3] C. C. Huang and S. Dumrangrattana, "Generalized mean-field model for the smectic-A-chiral-smectic-C phase transition," *Physical Review A*, vol. 34, no. 6, pp. 5020–5026, 1986.

[4] T. Chan, Ch. Bahr, G. Heppke, and C. W. Garland, "Landau second order to tricritical crossover behaviour for smectic A-smectic C* transitions," *Liquid Crystals*, vol. 13, no. 5, pp. 667–675, 1993.

[5] F. Yang, G. W. Bradberry, and J. R. Sambles, "Optical confirmation of the extended mean-field theory for a smectic-C*-smectic-A transition," *Physical Review E*, vol. 50, no. 4, pp. 2834–2838, 1994.

[6] F. Mercuri, M. Marinelli, U. Zammit, C. C. Huang, and D. Finotello, "Critical behavior of thermal parameters at the smectic-A-hexatic-B and smectic-A-smectic-C phase transitions in liquid crystals," *Physical Review E*, vol. 68, Article ID 051705, 10 pages, 2003.

[7] K. Denolf, B. Van Roie, G. Pitsi, and J. Thoen, "Investigation of the smectic-A-smectic-C* transition in liquid crystals by adiabatic scanning calorimetry," *Molecular Crystals and Liquid Crystals*, vol. 449, no. 1, pp. 47–55, 2006.

[8] B. R. Ratna, R. Shashidhar, G. G. Nair, S. K. Prasad, Ch. Bahr, and G. Heppke, "Evidence of a first-order smectic-smectic-C* transition and its approach to tricritical behavior," *Physical Review A*, vol. 37, no. 5, pp. 1824–1826, 1988.

[9] R. Shashidhar, B. R. Ratna, G. G. Nair, S. K. Prasad, Ch. Bahr, and G. Heppke, "Mean-field to tricritical crossover behavior near the smectic-A-smectic-C* tricritical point," *Physical Review Letters*, vol. 61, no. 5, pp. 547–549, 1988.

[10] H. Y. Liu, C. C. Huang, Ch. Bahr, and G. Heppke, "Tricriticality near the smectic-A-smectic-C transition of a liquid-crystal compound," *Physical Review Letters*, vol. 61, no. 3, pp. 345–348, 1988.

[11] Ch. Bahr and G. Heppke, "Influence of electric field on a first-order smectic-A-ferroelectric-smectic-C liquid-crystal phase transition: a field-induced critical point," *Physical Review A*, vol. 41, no. 8, pp. 4335–4342, 1990.

[12] Ch. Bahr and G. Heppe, "Critical exponents of the electric-field-induced smectic-A-ferroelectric-smectic-C liquid-crystal critical point," *Physical Review A*, vol. 44, no. 6, pp. 3669–3672, 1991.

[13] J. Zubia, M. Castro, J. A. Puertolas, J. Etxebarria, M. A. P. Jubindo, and M. R. de la Fuante, "Character of the smectic-A-chiral-smectic-C phase transition near a chiral-nematic-smectic-A-chiral-smectic-C point," *Physical Review E*, vol. 48, no. 3, pp. 1970–1976, 1993.

[14] T. Carlsson, B. Zeks, A. Levstik, C. Filipic, I. Levstik, and R. Blinc, "Generalized Landau model of ferroelectric liquid crystals," *Physical Review A*, vol. 36, no. 3, pp. 1484–1487, 1987.

[15] P. K. Mukherjee, "Tricritical behavior of the smectic-A to smectic-C* transition," *Journal of Chemical Physics*, vol. 131, no. 7, Article ID 074902, 2009.

[16] H. Yurtseven and M. Kurt, "Tilt angle and the temperature shifts calculated as a function of concentration for the AC* phase transition in a binary mixture of liquid crystals," *International Journal of Modern Physics B*, vol. 25, no. 13, pp. 1791–1806, 2011.

[17] H. Yurtseven and M. Kurt, "Spontaneous polarization and the tilt angle as a function of temperature in chiral smectic C phase of SCE9 and SCE10," *Balkan Physics Letters*, vol. 19, no. 191039, Article ID 19 191039, pp. 335–340, 2011.

[18] H. Yurtseven and M. Kurt, "Temperature dependence of the order parameters (spontaneous polarization and the tilt angle) at various concentrations for a mixture of SCE9 + SCE10," *Liquid Crystals*, vol. 39, no. 5, pp. 539–544, 2012.

[19] S. Salihoglu, H. Yurtseven, A. Giz, D. Kayişoglu, and A. Konu, "The mean field model with $P^2\theta^2$ coupling for the smectic A-smectic C* phase transition in liquid crystals," *Phase Transitions*, vol. 66, no. 1–4, pp. 259–270, 1998.

[20] S. Salihoglu, H. Yurtseven, and B. Bumin, "Concentration dependence of polarization for the AC* phase transition in a binary mixture of liquid crystals," *International Journal of Modern Physics B*, vol. 12, no. 20, pp. 2083–2090, 1998.

[21] H. Yurtseven and E. Kilit, "Critical behaviour of the dielectric constant for the smectic AC* transition under the electric field in ferroelectric liquid crystals," *Physics and Chemistry of Liquids*, vol. 50, no. 3, pp. 367–377, 2012.

[22] H. Yurtseven and H. Karacali, "Calculation of a generalised smectic-hexatic phase diagram in liquid crystals," *Physics and Chemistry of Liquids*, vol. 50, no. 3, pp. 302–315, 2012.

[23] R. Brout, *Phase Transitions*, chapter 2, Bnjamin, New York, NY, USA, 1965.

[24] Ch. Bahr, G. Heppke, and N. K. Sharma, "Dielectric studies of the smectic C*-smectic A transition of a ferroelectric liquid crystal with high spontaneous polarization," *Ferroelectrics*, vol. 76, no. 1, pp. 151–157, 1987.

[25] Z. Kutnjak and C. W. Garland, "Generalized smectic-hexatic phase diagram," *Physical Review E*, vol. 57, no. 3, pp. 3015–3020, 1998.

[26] L. Martinez-Miranda, A. R. Kortan, and R. J. Birgeneau, "Phase diagram, fluctuations, and phase transitions near the liquid-crystal nematic-smectic-A-smectic-C multicritical point," *Physical Review A*, vol. 36, no. 5, pp. 2372–2383, 1987.

7

Dopant Effect and Cell-Configuration-Dependent Dielectric Properties of Nematic Liquid Crystals

Feng-Ching Lin,[1] Po-Chang Wu,[2] Bo-Ru Jian,[1] and Wei Lee[1,2,3]

[1] *Department of Physics, Chung Yuan Christian University, Chung-Li 32023, Taiwan*
[2] *College of Photonics, National Chiao Tung University, Tainan 71150, Taiwan*
[3] *Center for Nanotechnology, Chung Yuan Christian University, Chung-Li 32023, Taiwan*

Correspondence should be addressed to Wei Lee; wlee@cycu.edu.tw

Academic Editor: Hamit Yurtseven

The dielectric polarizations induced by impurity ions in both high- and low-resistivity nematic liquid crystals (NLCs) were investigated. Upon adding carbon nanotubes (CNTs) as a dopant, the former showed no distinct change in the low-frequency ($< 10^2$ Hz) dielectric spectrum, whereas the latter exhibited dramatic decrease in dielectric constant in that the CNTs remarkably trapped the impurity ions in the NLC of high ionic content. Consequently, the dopant raised the voltage holding ratio by 26% and prolonged the lifetime of the cell. Also investigated were ionic behaviors in the low-resistivity NLC confined in cells with three different configurations. The diffusion constant of the ions in homeotropic cells was found to be the greatest, and the dc conductivity, determined by the diffusion constant, was also higher in the homeotropic ones.

1. Introduction

Nematic liquid crystals (NLCs) have ubiquitously been exploited in photonic devices, especially in current flat panel displays. Regardless of their origins, impurity ions existing in liquid-crystal (LC) cells have long been a nuisance to the global LC display (LCD) industry. It is well known that the cumulative impurity ions produce an internal, counteracting electric field in an LC cell under voltage application, causing many problems such as the reduction of voltage holding ratio, increase in threshold voltage, image sticking, gray-level shift, image flickering, and the slow-down of switching response [1–4]. In order to solve these issues, some research groups have adopted a postsynthetic approach [5] and demonstrated that doping an adequate amount of carbon nanotubes (CNTs) into NLCs can help suppress the unwanted field-screening effect [6]. The fact that the impurity ions can be effectively trapped by CNTs via charge transfer substantially improves the electrooptical performance of an LC device [7, 8]. With a compatible shape to the rodlike molecules of NLCs, tubular CNTs are known as a one-dimensional nanomaterial, possessing unique material characteristics to be promisingly applied in modern photonic and semiconductor devices. As a matter of fact, recent laboratory studies have demonstrated their potential for flat panel display applications, permitting their roles in electrodes, thin-film transistors, alignment layers, and field-emission devices.

The quality of an LC material for photonic applications can be examined by low-frequency dielectric spectroscopy [9]. It has been established that the LC dielectric behavior at low frequencies is dictated by the charge carrier transport process and is susceptible to the material properties of the alignment layers if any [10]. Interestingly but not surprisingly, Costa et al. revealed that impurity ions in the NLC E7 travel to electrodes faster in homeotropic cells than in homogeneous cells due to the anisotropy in diffusion constant [11]. Thus, the underlying mechanisms depend upon the materials used and the properties of the cell.

In this comparative study we used a low-resistivity NLC (E7) and a high-resistivity NLC (CYLC-01) to observe their low-frequency polarizations contributed by mobile ions. The dielectric spectra of the two eutectic NLC materials were compared between 10 Hz and 10 kHz. The dopant effects of CNTs on the dielectric properties of the two NLCs were investigated. In addition, by expanding the frequency range to 10^{-1}–10^{4} Hz, the ionic transport behaviors were analyzed for three distinctive (homogeneous, twisted-nematic, and homeotropic) cell configurations for the ion-rich NLC E7.

2. Experimental

Both low- and high-resistivity NLC materials were employed for comparison; they were E7 and CYLC-01, respectively. With a typical bulk resistivity of ~10^{11} Ω-cm, E7 is a well-known commercial eutectic mixture (from Merck), consisting of primarily 4′-*n*-pentyl-4-cyanobiphenyl and three other cyano compounds. Owing to its high polarity, E7 exhibits a strong permanent dipole moment along the molecular axis, possesses large dielectric anisotropy ($\Delta\varepsilon' \sim 14$ at 1 kHz and 20°C [12]), and is notorious for its vulnerability to ionic contamination. On the other hand, the mixture CYLC-01 is an ultrapure CF_2O-based TFT-grade NLC. Its excellent material quality, including high bulk resistivity of ~10^{14} Ω-cm, allows its appearance in the modern active-matrix LC displays. Both the LC substances are of positive dielectric anisotropy (defined as the difference between the parallel component $\varepsilon_{\parallel}$ and perpendicular component $\varepsilon_{\perp}$ of the real dielectric constant). The dopant material used was ultrapure multiwalled carbon nanotubes (CNTs) with inner diameters of 2-3 nm and ~21 graphene layers (from SRS Research). According to our prior experience, the optimal CNT concentration of 0.05 wt% was adopted [13]. Four material samples were prepared, including pure E7, E7/CNT, pure CYLC-01, and CYLC-01/CNT, which were individually injected into empty homogeneous-alignment cells by capillary action in the isotropic phase. The cell gap was 11.3 μm between the conducting glass substrates, and the effective electrode area was 1 cm^2. For the second part of the experiment concerning the influence of cell configuration, we prepared five pristine E7 samples for each type of cell structure. The cell configurations investigated in this work included the homogeneous (i.e., planar alignment), 90° twisted-nematic, and homeotropic (i.e., vertical alignment) structures. The alignment layers for the 6.2 μm thick homogeneous and twisted cells were the same polyimide (Nissan SE-2170). The homeotropic cells of ~5 μm in thickness were purchased from EHC, Japan, and the aligning material is unknown.

The complex dielectric constant was acquired with an LCR meter (HIOKI 3522-50 LCR HiTESTER) having capability of measurement in the frequency range from 1 mHz to 100 kHz and basic measurement accuracy of ±0.08%. The probe voltage adopted in this study was 50 mV_{rms} in the sinusoidal waveform, which was much smaller than the LC threshold voltage. The voltage holding ratio (VHR) experiment [14] was carried out to simulate the charging and discharging properties of E7 cells, by which the cell performance of the doped one was examined. The temperatures for the first part and second part of the experiment were 303 K and 296 K, respectively.

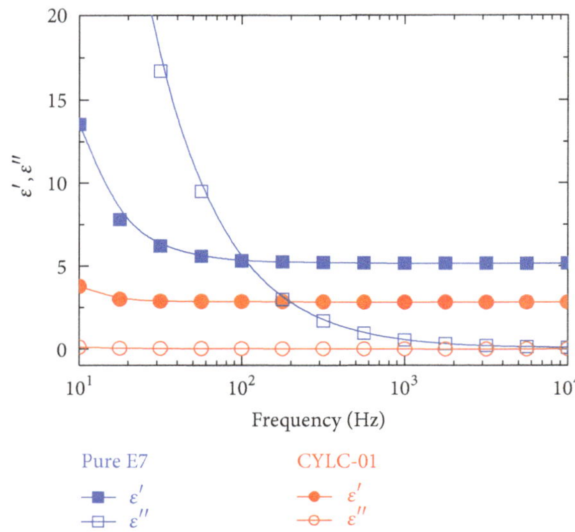

FIGURE 1: The dielectric spectra of the low-resistivity NLC E7 and high-resistivity NLC CYLC-01 confined in homogeneous cells at 303 K. ε' and ε'' are the real and imaginary parts of the complex dielectric function.

3. Results and Discussion

Figure 1 shows the frequency dependence of the real and imaginary parts of the complex dielectric function ($\varepsilon' - i\varepsilon''$) of pure E7 and pure CYLC-01. In the low-frequency region ($f < 10^2$ Hz), the dielectric behavior was attributable to the space-charge polarization. Both frequency-varying ε' and ε'', dominated by the ionic effect in this frequency region, are proportional to $f^{-3/2}$ and f^{-1}, respectively. Taking the diffusive motion of the impurity ions into account, the expressions for the ionic contributions to the dielectric dispersion ε' and dielectric loss ε'' in the appropriate frequency range in the nematic LC cell are [15]

$$\varepsilon'(f) = \frac{nq^2 D^{3/2}}{\pi^{3/2}\varepsilon_0 L k_B T} f^{-3/2} + \varepsilon'_b, \quad (1)$$

$$\varepsilon''(f) = \frac{nq^2 D}{\pi \varepsilon_0 k_B T} f^{-1}, \quad (2)$$

respectively, where n is the mobile ion concentration, q is the charge of an ion, D is the diffusion constant of the ion, L is the thickness of the cells, ε_0 is the permittivity in free space, k_B is the Boltzmann constant, T is the absolute temperature, and ε'_b is the intrinsic dielectric constant of the NLC bulk. The ionic contribution decreased with increasing frequency. At low frequencies the impurity ions had enough time to reach the electrodes to form dipole moments, while at high frequencies the space-charge polarization could not keep up with the polarity alteration of the electric field, which, in turn, led to the dielectric relaxation. The constant value of ε' between 10^3 and 10^4 Hz was mainly contributed by the intrinsic LC dielectric term ε'_b. It is clear from Figure 1 that the dielectric relaxation in the low-frequency region (10 Hz < f < 100 Hz) was not evident in the dielectric spectrum

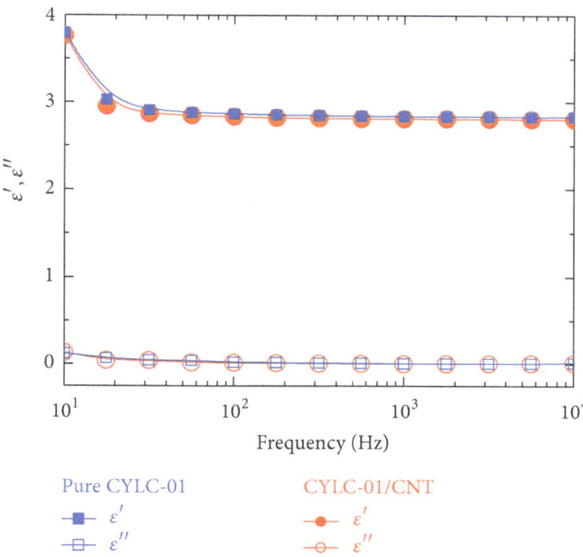

FIGURE 2: The dielectric spectra of pristine and 0.05 wt% CNT-doped CYLC-01 in homogeneously aligned cells at 303 K.

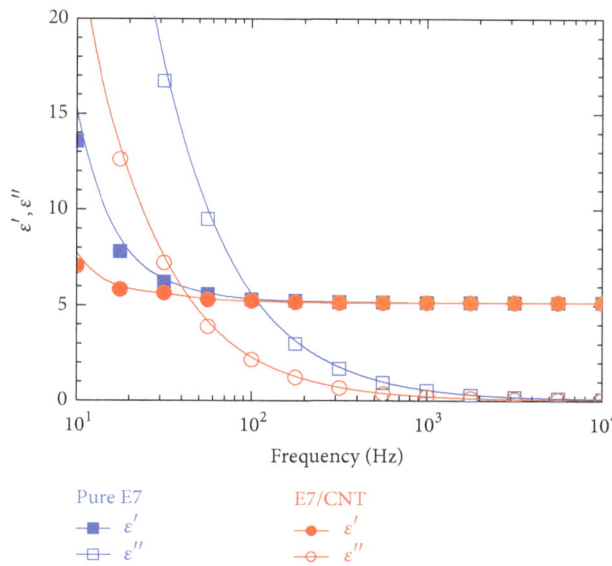

FIGURE 3: The dielectric spectra of pure and 0.05 wt% CNT-doped E7 in homogeneous cells at 303 K.

of the superfluorinated LC mixture CYLC-01. As mentioned above, CYLC-01 was a high-resistivity NLC, connoting that it contained a very limited amount of impurity ions. Consequently, the ionic contribution was not as significant at low frequencies compared with the E7 counterpart.

Figure 2 depicts the dielectric spectra of pure as well as CNT-doped CYLC-01 samples. Since their real parts almost coincide with each other and imaginary parts are virtually identical, the dopant effect on the dielectric behavior (in the frequency range studied) was trivial in the TFT-grade NLC. In contrast, the impact of the dopant on the low-frequency dielectric spectrum was very pronounced for the E7 host as shown in Figure 3. One can see from this figure that, after doping 0.05-wt% CNTs in E7, the dielectric permittivity (ε') decreased in the low-frequency region and the relaxation frequency shifted to the lower frequency. Again, E7 is a low-resistivity NLC, which contains a large amount of impurity ions. The CNTs as an additive in the ion-rich nematic host effectively trapped mobile ions [5] to give rise to the reduction in space-charge polarization.

The VHR is an important measure of the LCD performance. It is conventionally defined as

$$\mathrm{VHR}\,(\%) = \frac{V_t}{V_i} \times 100\%, \qquad (3)$$

where V_i is the initial (or input) pulse voltage (5 V in this study) and V_t is the terminal voltage of the LC cell during a frame time of 16.67 ms. One can see from Figure 4 that, by adding an adequate amount of CNTs in E7, the VHR was promoted from 48% to 61%, yielding a substantial increase by 26%. In order to monitor the degradation of the cell manifested by the decrease in VHR over time, a four-day observation was made. Figure 5 illustrates the time-evolved VHR curves for the doped and undoped cells. The monotonic fall of the VHR in both cells can be explained by the ascending

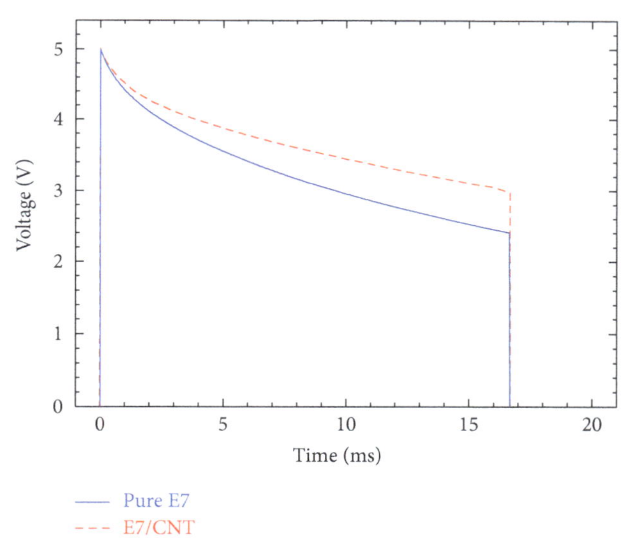

FIGURE 4: Voltage responses of doped and undoped E7 cells to a 5-V square-wave pulse of 15.6 μs.

ionic concentration after the cell fabrication [16]. Using the monoexponential decay expression $\mathrm{VHR}(t) = Ae^{-t/\tau}$, where A is the VHR at time $t = 0$ and τ is the characteristic lifetime, one can fit the experimental data and obtain a lifetime of 118 h for the doped cell compared with that of 59 h for the undoped counterpart. Obviously, the addition of CNTs in E7 prolonged the cell lifetime before degradation.

Figure 6 shows the dielectric spectra of E7 in various cells with homogeneous, twisted, and homeotropic alignment. In the 10^3–10^4 Hz frequency region, the dielectric constant ε' of E7 was the largest in the homeotropic configuration, because

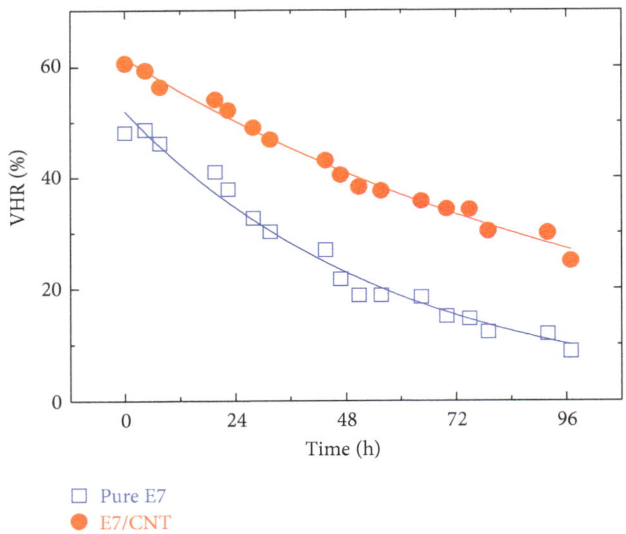

FIGURE 5: Time-evolved VHR curves in a period of 96 h.

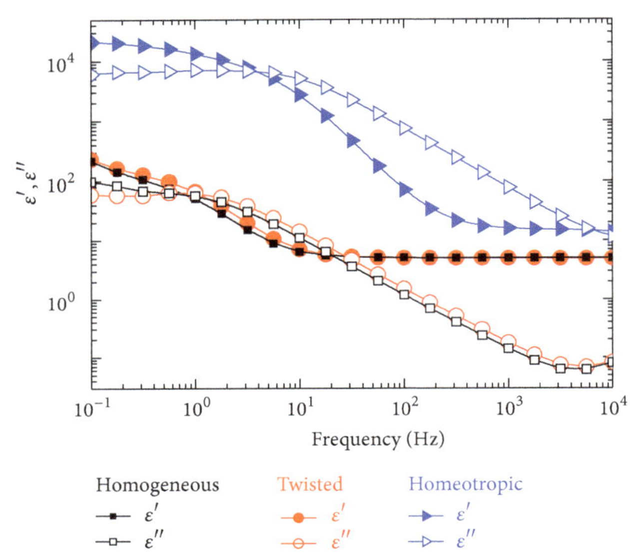

FIGURE 6: The dielectric spectra of E7 in homogeneous, twisted, and homeotropic cells at 296 K.

E7 possessed positive dielectric anisotropy. Note that the effective dielectric constant ε can be expressed by [17]

$$\varepsilon = \varepsilon_{\perp} + \Delta\varepsilon \sin^2\theta, \tag{4}$$

where θ is the tilt angle between the substrate plane and the LC director. Thus, the complex dielectric constant measured in a homeotropic cell is the sole contribution of the parallel component $\varepsilon_{\parallel}$. In the frequency region of $f < 10^3$ Hz, both ε' and ε'' functions can be individually fitted by (1) and (2), respectively, for calculation of their ion density n and diffusion constant D. The results are listed in Table 1.

The impurity ions in the LC bulk can be contributed by the dissociation of the LC material itself and ion injection from the alignment or electrode layer [18, 19]. Based on the data displayed in Table 1, the ion concentration in the homeotropic cells was much (two orders of magnitude) higher than that in the other cells with different configurations. Here, the most likely explanation is the occurrence of ion injection into the LC bulk from the homeotropic alignment layers because the electrodes are identical. The ionic concentrations in the homogeneous and twisted-nematic cells are quite the same due to the use of the same aligning material.

TABLE 1: Ion concentration n and diffusion constant D in E7 for homogeneous, twisted, and homeotropic cells at 296 K.

Cell type	n (cm^{-3})	D (cm^{2}s^{-1})
Homogeneous	1.80×10^{13}	2.67×10^{-7}
Twisted	1.74×10^{13}	3.55×10^{-7}
Homeotropic	5.12×10^{15}	5.72×10^{-7}

The diffusion constant D is anisotropic and related to the viscosity and electrical conductivity of the medium [7, 11]. Table 1 reveals that D in a homeotropic cell was larger in comparison with that in a homogeneous one. This result, arising from the fact that the diffusion constant along the molecular axis is greater (namely, $D_{\parallel} > D_{\perp}$), is in good agreement with an earlier study by Costa et al. [11]. We also observed that D in the twisted cells is somewhat larger than that in the homogeneous cells. Since the viscosity of the LC E7 in splay and twist molecular configurations is identical [20], we believe that the difference can be ignored due to the experimental error and uncertainties in the curve-fitting process. As a result, it can be summarized that the alignment layer and molecular orientation play an important role in ionic concentration and diffusion constant, respectively. The dc conductivity can be written as

$$\sigma_{dc} = qn\mu, \tag{5}$$

where μ is the ion mobility. According to the Einstein relation

$$\mu = \frac{q}{k_B T} D, \tag{6}$$

(5) becomes

$$\sigma_{dc} = \frac{nq^2 D}{k_B T}, \tag{7}$$

indicating that the ionic dc conductivity σ_{dc} is not only proportional to the ion concentration n, but also to the diffusion constant D. Thus, the conductivity of an LC cell changes with the variation of D in different LC alignment.

4. Conclusions

This paper presents the low-frequency ($f < 10^2$ Hz) dielectric behaviors induced by impurity ions in LC cells under different conditions. In the high-resistivity NLC (CYLC-01) which contained a low concentration of impurity ions, we did not observe an evident sign of the mobile-ion polarization and the dielectric constant did not seem to change after doping

CNTs. By contrast, the dielectric relaxation was obviously observed in the low-resistivity NLC (E7) rich in impurity ions. Our experimental results suggested that CNTs as a dopant can effectively suppress the ion effect in the low-resistivity NLC, thereby promoting the VHR substantially. In addition, the low-resistivity NLC confined in homeotropic cells, in comparison with cells in the homogeneous and twisted configurations, exhibited the highest impurity-ion concentration, implying significant ion injection from the corresponding alignment layers.

Acknowledgment

The authors are thankful for the financial support of the National Science Council of Taiwan under Grant nos. NSC 98-2112-M-033-004-MY3, NSC 101-2112-M-009-018-MY3, and NSC 101-2811-M-009-059.

References

[1] S. H. Perlmutter, D. Doroski, and G. Moddel, "Degradation of liquid crystal device performance due to selective adsorption of ions," *Applied Physics Letters*, vol. 69, no. 9, pp. 1182–1184, 1996.

[2] C. Colpaert, B. Maximus, and A. de Meyere, "Adequate measuring techniques for ions in liquid crystal layers," *Liquid Crystals*, vol. 21, no. 1, pp. 133–142, 1996.

[3] V. A. Tsevetkov and O. V. Tevetkov, "Ions influence on electrooptical characteristics of NLC," *Molecular Crystals and Liquid Crystals*, vol. 368, no. 1, pp. 625–632, 2001.

[4] L. O. Palomares, J. A. Reyes, and G. Barbero, "Optical response of a nematic sample submitted to a periodic external electric field: role of the ionic impurities," *Physics Letters A*, vol. 333, no. 1-2, pp. 157–163, 2004.

[5] M. Rahman and W. Lee, "Scientific duo of carbon nanotubes and nematic liquid crystals," *Journal of Physics D*, vol. 42, Article ID 063001, 12 pages, 2009.

[6] H. Y. Chen and W. Lee, "Suppression of field screening in nematic liquid crystals by carbon nanotubes," *Applied Physics Letters*, vol. 88, Article ID 222105, 3 pages, 2006.

[7] B. R. Jian, C. Y. Tang, and W. Lee, "Temperature-dependent electrical properties of dilute suspensions of carbon nanotubes in nematic liquid crystals," *Carbon*, vol. 49, no. 3, pp. 910–914, 2011.

[8] H. H. Liu and W. Lee, "Ionic properties of liquid crystals dispersed with carbon nanotubes and montmorillonite nanoplatelets," *Applied Physics Letters*, vol. 97, no. 17, Article ID 173501, 3 pages, 2010.

[9] C. Y. Tang, S. M. Huang, and W. Lee, "Electrical properties of nematic liquid crystals doped with anatase TiO_2 nanoparticles," *Journal of Physics D*, vol. 44, Article ID 355102, 5 pages, 2011.

[10] S. Murakami and H. Naito, "Electrode and interface polarizations in nematic liquid crystal cells," *Japanese Journal of Applied Physics*, vol. 36, no. 4, pp. 2222–2227, 1997.

[11] M. R. Costa, R. A. C. Altafim, and A. P. Mammana, "Ionic impurities in nematic liquid crystal displays," *Liquid Crystals*, vol. 28, no. 12, pp. 1779–1783, 2001.

[12] Merck data sheets.

[13] H. Y. Chen, W. Lee, and N. A. Clark, "Faster electro-optical response characteristics of a carbon-nanotubenematic suspension," *Applied Physics Letters*, vol. 90, Article ID 033510, 3 pages, 2007.

[14] C. Y. Tang, Y. C. Lin, and W. Lee, "Voltage holding ratio of a nematic liquid crystal cell under UV exposure," in *Proceedings of the Photonics Global Conference*, Suntec Singapore International Convention & Exhibition Centre, Singapore, December 2010.

[15] G. Barbero and A. L. Alexe-Ionescu, "Role of the diffuse layer of the ionic charge on the impedance spectroscopy of a cell of liquid," *Liquid Crystals*, vol. 32, no. 7, pp. 943–949, 2005.

[16] H. H. Liu and W. Lee, "Time-varying ionic properties of a liquid-crystal cell," *Applied Physics Letters*, vol. 97, Article ID 023510, 3 pages, 2010.

[17] W. Lee, J. S. Gau, and H. Y. Chen, "Electro-optical properties of planar nematic cells impregnated with carbon nanosolids," *Applied Physics B*, vol. 81, no. 2-3, pp. 171–175, 2005.

[18] S. Murakami and H. Naito, "Charge injection and generation in nematic liquid crystal cells," *Japanese Journal of Applied Physics*, vol. 36, no. 2, pp. 773–776, 1997.

[19] H. de Vleeschouwer, A. Verschueren, F. Bougrioua et al., "Dispersive ion generation in nematic liquid crystal displays," *Japanese Journal of Applied Physics A*, vol. 41, no. 3, pp. 1489–1494, 2002.

[20] H. J. Coles and M. S. Bancroft, "Viscosity coefficients and elastic constants of nematic solutions of a side-chain polymer," *Molecular Crystals and Liquid Crystals Science and Technology. Section A*, vol. 237, no. 1, pp. 97–110, 1993.

8

Thermodynamic, Electromagnetic, and Lattice Properties of Antiperovskite Mn_3SbN

Ying Sun,[1] Yan-Feng Guo,[2] Yoshihiro Tsujimoto,[3] Xia Wang,[2] Jun Li,[2] Clastin I. Sathish,[2,4] Cong Wang,[5] and Kazunari Yamaura[2,4]

[1] *International Center for Materials Nanoarchitectonics (WPI-MANA), National Institute for Materials Science, 1-1 Namiki, Ibaraki, Tsukuba 305-0044, Japan*
[2] *Superconducting Properties Unit, National Institute for Materials Science, 1-1 Namiki, Ibaraki, Tsukuba 305-0044, Japan*
[3] *Materials Processing Unit, National Institute for Materials Science, 1-1 Namiki, Ibaraki, Tsukuba 305-0044, Japan*
[4] *Graduate School of Chemical Sciences and Engineering, Hokkaido University, Hokkaido, Sapporo 060-0810, Japan*
[5] *Department of Physics, Center for Condensed Matter and Materials Physics, Beihang University, Beijing, Haidian 100191, China*

Correspondence should be addressed to Ying Sun; sun.ying@nims.go.jp

Academic Editor: Y. Sun

The physical properties of polycrystalline Mn_3SbN were investigated using measurements of the magnetic, calorimetric, and electronic transport properties. At room temperature, the phase crystallizes in a tetragonal structure with $P4/mmm$ symmetry. A remarkably sharp peak in the heat capacity versus temperature curve was found near 353 K. The peak reaches 723 J mol^{-1} K^{-1} at its highest, which corresponds to a transition entropy of 10.2 J mol^{-1} K^{-1}. The majority of the large entropy change appears to be due to lattice distortion from the high-temperature cubic structure to the room-temperature tetragonal structure and the accompanying Ferrimagnetic transition.

1. Introduction

Antiperovskite compounds with the formula Mn_3XN or Mn_3XC (X = Cu, Zn, Ga, Cu, In, or Sn) were discovered in the middle of the last century [1]. Recently, interest in these materials has intensively renewed owing to discoveries of new, potentially useful properties [2–4] such as the giant magnetoresistance of Mn_3GaC [5], negative thermal expansion (NTE) of $Mn_3Cu(Ge)N$ [6] and $Mn_3Zn(Ge)N$ [7], magnetostriction of Mn_3CuN [8] and Mn_3SbN [9], and near-zero temperature coefficient of the resistivity of Mn_3CuN [10] and Mn_3NiN [11]. Specifically, Takenaka and Takagi found that Ge-doped Mn_3CuN compound has a large NTE (NTE parameter = -25×10^{-6} K^{-1}) [12]; using neutron diffraction, the broad NTE was determined to be associated with the local T_4 structure [6]. Asano et al. discovered large magnetostriction in tetragonal Mn_3CuN; it expands 0.2% and shrinks 0.1% in the directions parallel and perpendicular to an external 90 kOe magnetic field, respectively [8]. In previous studies, we found a peculiar phase separation and accompanying anomaly in the electronic transport properties of Mn_3ZnN [13, 14], while further study indicated that the thermal expansion properties of Mn_3ZnN can be controlled by introducing Zn vacancies [15]. In addition, Song et al. observed a canonical spin-glass state in Mn_3GaN below the spin-freezing temperature of 133 K [16]. Lukashev et al. systematically studied the spin density of the spin-frustrated state of a Mn-based antiperovskite under mechanical stress [17].

The above-mentioned properties enable a variety of potential applications for this type of material. Although the prospective industrial markets are expected to be large and much effort has already been devoted to studying their structural, electromagnetic, and transport properties, further investigations on antiperovskite materials are still required. In this study, the thermodynamic, electromagnetic, and electronic transport properties of Mn_3SbN are investigated. In particular, we focused on the notable transition entropy that

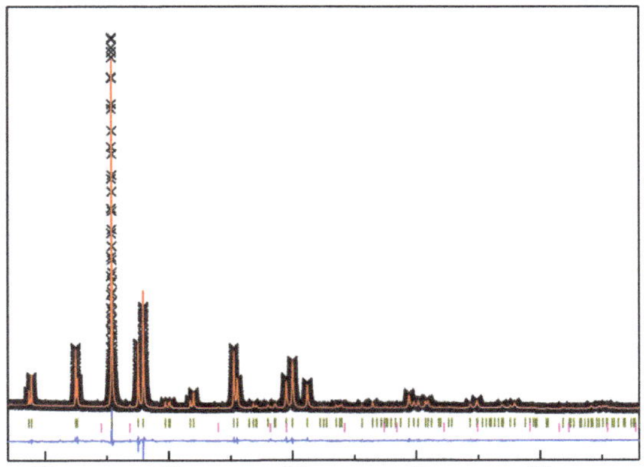

FIGURE 1: Rietveld refinement of the synchrotron XRD pattern of Mn_3SbN. The crosses and solid lines represent the observed and calculated profiles, respectively, and the difference between the curves is shown at the bottom. Bragg positions are marked by small ticks (upper: Mn3ZnN, lower: MnO).

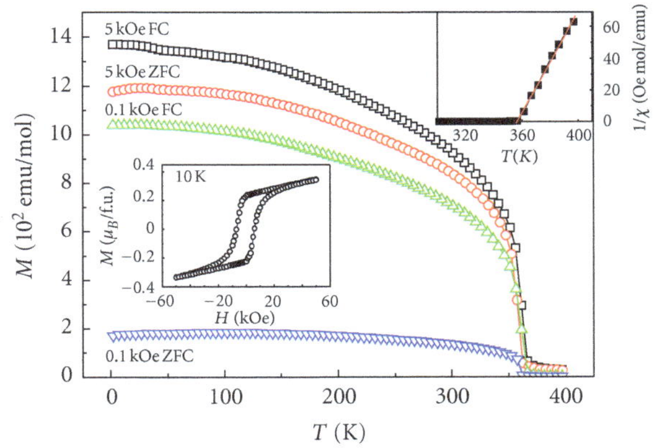

FIGURE 2: Temperature dependence of magnetization in magnetic fields of 0.1 and 5 kOe. The left and right insets show the isothermal magnetization curve at 10 K and χ^{-1} versus T plot of the FC data at 0.1 kOe, respectively.

accompanies the magnetic and crystal structure transition above room temperature.

2. Experimental Details

A polycrystalline Mn_3SbN sample was prepared via the solid-state reaction of fine powders of Sb (99.99%, Rare Metallic Co.) and Mn_2N, which was synthesized by firing Mn powder (99.99%, Sigma Aldrich Co.) in nitrogen at 800°C for 60 h. Stoichiometric amounts of the starting materials were thoroughly mixed, and the mixture was pressed into a pellet. The pellet was sealed in an evacuated quartz tube, heated in a box furnace at 800°C for three days, and then slowly cooled to room temperature in the furnace.

The crystal structure of Mn_3SbN was analyzed by synchrotron X-ray diffraction (SXRD) using a large Debye-Scherrer camera at the BL15XU NIMS beam line of the SPring-8 facility in Hyogo, Japan. The SXRD data were collected for 2θ ranging from 2° to 60° at intervals of 0.003°. The incident beam was monochromatized at $\lambda = 0.65297$ Å. The evolution of the Mn_3SbN crystal structure with temperature was also determined via the measurement of the SXRD patterns.

The temperature dependence of magnetization was measured between 2 and 400 K with applied magnetic fields of 0.1 and 5 kOe using a Magnetic Property Measurements System (Quantum Design). The measurements were conducted on loosely gathered powder under both zero-field cooling (ZFC) and field cooling (FC) conditions. The isothermal magnetization curve was recorded at 10 K between −50 and 50 kOe.

Specific heat (C_p) values were measured between 2 and 400 K with cooling using a Physical Properties Measurement System (Quantum Design). The sample was fixed on a stage using a small amount of grease; the heat capacity of the grease was measured first and subtracted from the total C_p.

The electrical resistivity (ρ) was measured between 2 and 400 K with cooling and heating using a conventional four-probe techniques with the same apparatus. The AC gauge current and frequency were 10 mA and 30 Hz, respectively. The electrical contacts were prepared on the surface of a bar-shaped piece of the pellet using silver paste and Pt wires.

3. Results and Discussion

As shown in Figure 1, the synchrotron XRD pattern at room temperature fit well with a model pattern of the proposed structure (space group: *P4/mmm*). The structural parameters of Mn_3SbN were refined by the Rietveld method using the RIETAN-FP program [18]. The occupancy factors of Sb, N, Mn1, and Mn2 were refined to be 1 (fixed), 1 (fixed), 0.97(1), and 0.99(1), respectively, while the isotropic atomic displacement parameters were 0.42(1), 0.84(5), 0.86(1), and 0.78(1) Å^2, respectively. The lattice constants were calculated to be $a = b = 4.17994(4)$ Å and $c = 4.27718(5)$ Å. The final R_{wp} and R_p reliability indexes were below 5.56% and 4.09%, respectively. The analysis revealed 1.91 mass% MnO in the sample as an impurity; as shown later, the magnetic, C_p, and ρ measurements suggest that the impurity does not significantly impact the measurements of Mn_3SbN in this study.

Figure 2 displays the temperature dependence of magnetization of polycrystalline Mn_3SbN. The magnetization steeply increases upon cooling to around 353 K, which suggests the establishment of long-range magnetic order at the magnetic transition temperature (T_c). In addition, a small hysteresis can be observed between the heating and cooling process, implying the first-order character of the magnetic transition. The remarkable bifurcation between the ZFC and FC curves may originate from the spontaneous alignment of random magnetic Mn moments in domain boundaries. It is worth noting that the hysteresis is less significant at a higher magnetic field of 5 kOe, which supports the domain picture.

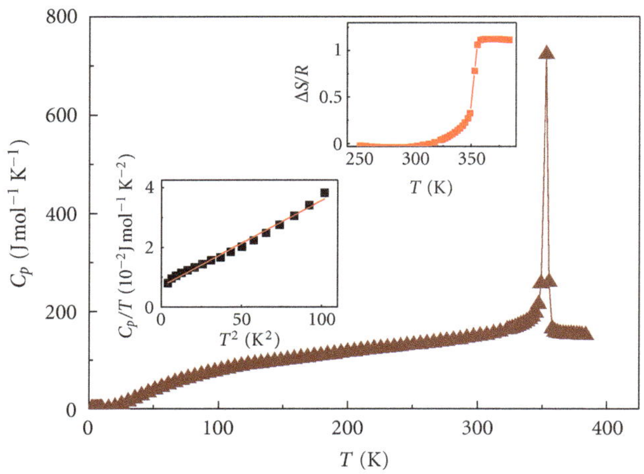

FIGURE 3: Temperature dependence of the C_p of Mn_3SbN. The left and right insets show a linear fit to the C_p/T versus T^2 curve and an estimation of ΔS, respectively.

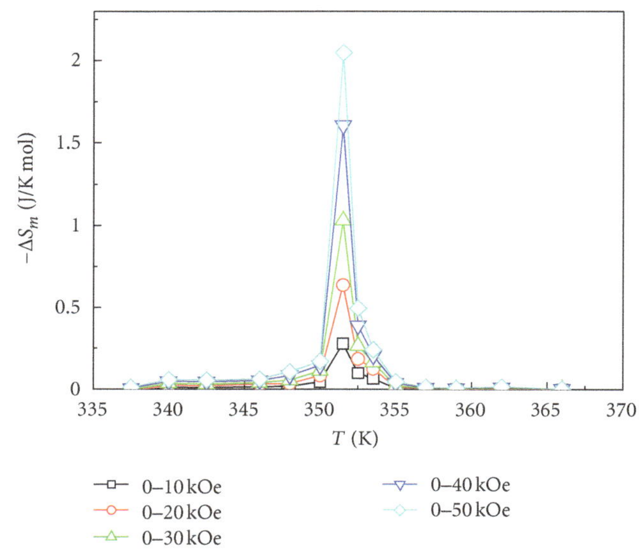

FIGURE 4: Temperature dependence of the magnetic entropy change when the magnetic field changes from 0 to 10, 20, 30, 40, and 50 kOe, respectively.

To further study the magnetic properties, we applied the Curie-Weiss law to the paramagnetic portion. As shown in the right inset of Figure 2, the χ^{-1}-T plot is well represented by the Curie-Weiss law, that is, the spin-only expression for magnetic susceptibility: $\chi(T) = C/(T - \Theta_W)$, where C is the Curie constant and Θ_W is the Weiss temperature. The value of Θ_W was determined to be 354 K, which suggests that ferromagnetic correlation is dominant in the spin system. The effective Bohr magneton (μ_{eff}) was estimated to be 1.28 μ_B/Mn from $\mu_{\text{eff}} = 2.83(C/\eta)^{0.5}\mu_B$, where η is the number of magnetic atoms in the molecular formula ($\eta = 3$ in the present case). The value of μ_{eff} is much lower than that of other antiperovskite manganese nitrides (e.g., 2.87 μ_B for Mn_3ZnN [14]) and even lower than the expected moment for localized $S = 1/2$ spins, suggesting an itinerant character of the 3d electrons in Mn_3SbN.

From the isothermal magnetization curve (see the inset of Figure 2), it was found that the magnetization at 50 kOe is ~0.35 μ_B/Mn, which is too small to be caused by full ferromagnetic order. The gap suggests that the spins of the Mn atoms are possibly Ferrimagnetically ordered. This Ferrimagnetic interaction is also suggested by the magnetization characteristics above 10 kOe, that is, the magnetization continuously increases with increasing magnetic field without approaching saturation. The Ferrimagnetic order of a related Mn-based antiperovskite compound was explained by a Γ^{4g} spin structure, where two of the three Mn magnetic moments are antiferromagnetically coupled and the third exhibits FM behavior [19]. It is possible that a similar magnetic structure is established in Mn_3SbN below 353 K.

To further characterize the magnetic transition, the specific heat was measured from 400 to 2 K. As shown in Figure 3, the temperature dependence of C_p features a sharp and narrow peak around T_c ($\Delta C_p/R = 87$ and $\Delta T = 3$ K, where R is the ideal gas constant). This is indicative of a first-order-like transition, as discussed in [20].

An estimation of entropy change is essential to understanding the nature of the transition of Mn_3SbN. The peak was roughly separated from the baseline using a polynomial function. Analysis indicates that the total transition entropy (ΔS) is ~1.23R (10.2 J/mol K). Since the total entropy change comprises all contributions, including the lattice, electronic, and magnetic changes [20], we evaluated each contribution independently.

For the present compound, the abrupt change of the magnetization at T_c may induce a large $\partial M/\partial T$; therefore, a large magnetic entropy change (ΔS_m) is expected. A series of magnetization curves with small temperature steps were measured; the data allow for a rough estimation of the magnetic entropy change via the thermodynamic Maxwell relation, as follows [21]:

$$\left(\frac{\partial S(T,H)}{\partial H}\right)_T = -\left(\frac{\partial M(T,H)}{\partial T}\right)_H. \quad (1)$$

The magnetic entropy change, $\Delta S_m(T,H)$, can be calculated by

$$\Delta S_m(T,H) = S_m(T,H) - S_m(T,0) = \int_0^H \left(\frac{\partial M(T,H)}{\partial T}\right)_H dH. \quad (2)$$

The temperature dependence of ΔS_m calculated from (2) with fields of 10, 20, 30, 40, and 50 kOe is shown in Figure 4. The ΔS_m is maximized around T_c, and the maximum is estimated to be ~2.1 J mol^{-1} K^{-1}, which implies that the lattice and electronic changes provide a fairly large contribution to the total entropy change.

To investigate the electronic contribution (i.e., the Sommerfeld coefficient or γ), the C_p/T versus T^2 plot below 10 K was analyzed by applying the approximate Debye model, as

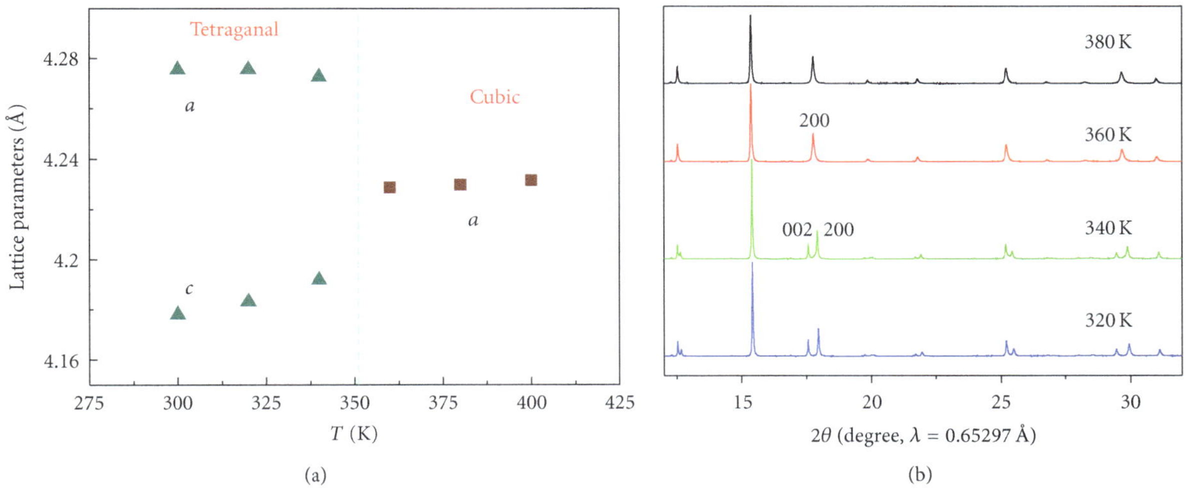

FIGURE 5: Variation in (a) lattice parameters and (b) synchrotron XRD patterns with temperature for Mn_3SbN.

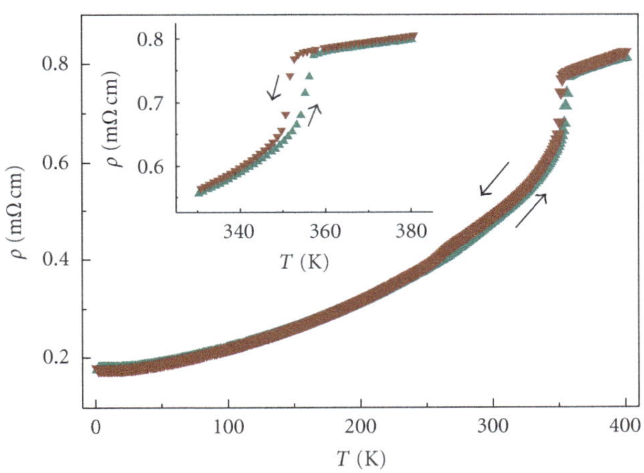

FIGURE 6: Variation of ρ with temperature for Mn_3SbN upon cooling and warming. The inset shows an enlarged view of the ρ variation around the magnetic transition.

follows: $C(T)/T = \gamma + 2.4\pi^4 n\ N_0 k_B (1/\Theta_D{}^3) T^2$ ($T \ll \Theta_D$), where n denotes the number of atoms per formula unit, k_B is the Boltzmann constant, N_0 is the Avogadro constant, and Θ_D is the Debye temperature. Fitting to the linear part of the C_p/T versus T^2 plot using the least-squares method yielded γ and Θ_D values of ~7.03(1) mJ mol^{-1} K^{-2} and 326(2) K, respectively. Compared with the parameters determined for other antiperovskite nitrides, Mn_3SbN has a much lower γ, which indicates that the electronic correlation is somewhat weakened [20]. Thus, the electronic contribution might not be a dominant contributor to the total transition entropy.

In addition to the magnetic and electronic contributions, a possible lattice change may need to be investigated to analyze the total transition entropy. The variation of the synchrotron XRD pattern with temperature was measured. As shown in Figure 5(b). It can be seen that some typical reflections disappear with temperature, for example, the two reflections (002) and (200) for the $P4/mmm$ lattice merge to one reflection. By the Rietveld analysis of the synchrotron XRD patterns, the structural change from tetragonal to cubic was defined, and the lattice constants were determined as a function of temperature, as shown in Figure 5(a). It is obvious that lattice parameter c increases slightly with increasing temperature, whereas a gradually decreases. When the temperature crosses T_c, the tetragonal structure completely transforms to an unidentified cubic structure. Hence, the lattice distortion must contribute to the total entropy change.

According to the thermodynamic relation, the magnetization (M) is equal to the first derivative of the magnetic free energy by the magnetic field, that is, $df(M)/dH$ [22]. Therefore, the sharp transition indicates that the energy barrier in the free energy that separates the paramagnetic and ferromagnetic states is large. Accordingly, T_c and the energy barrier height probably correlate with the electronic density of states, which exhibits a sharp peak near the Fermi level [23]; therefore, the large entropy change is possibly related to the reconstruction of the electronic structure, which could induce the magnetic and structural transition. Since such an electronic reconstruction is often sharply reflected in a ρ-T curve, the electronic transport properties of Mn_3SbN were carefully measured (shown in Figure 6). It is evident that an abnormal drop appears at T_c in the ρ-T curve, which is indicative of an electronic structure reconstruction. In addition, as shown in the inset of Figure 6, a small hysteresis was observed between the warming and cooling curves; this is in agreement with a first-order transition.

4. Conclusions

In conclusion, the thermodynamic, electromagnetic, and transport properties of antiperovskite Mn_3SbN were studied. The phase crystallizes in a tetragonal structure with $a = b = 4.17994(4)$ Å and $c = 4.27718(5)$ Å at room temperature. The C_p measurements revealed a sharp endothermic peak in the C_p-T curve at 353 K, which corresponds to a large entropy change (~10.2 J mol^{-1} K^{-1}). The present study clearly indicates that the entropy change is accompanied with a Ferrimagnetic transition and lattice distortion as well as a possible electronic structure reconstruction.

Conflict of Interests

The authors declare that they have no conflict of interests.

Acknowledgments

The authors would like to thank the staff members at BL15XU, the National Institute for Materials Science (NIMS), and SPring-8 for their help in the use of the beamline. The SXRD measurements were performed with the approval of the NIMS Beamline Station (Proposal no. 2011A4502). This study was supported in part by the World Premier International Research Center Initiative of the Ministry of Education, Culture, Sports, Science, and Technology (MEXT); a Grant-in-Aid for Scientific Research grant (no. 22246083) from the Japan Society for the Promotion of Science (JSPS); the Funding Program for World-Leading Innovative R&D on Science and Technology (FIRST Program) of the JSPS; the Advanced Low Carbon Technology Research and Development Program (ALCA) of the Japan Science and Technology Agency (JST).

References

[1] G. W. Wiener and J. A. Berger, "Structure and magnetic properties of some transition metal nitrides," *Transactions of the American Institute of Mining, Metallurgical, and Petroleum Engineers*, vol. 203, pp. 360–368, 1955.

[2] Y. Sun, Y. C. Wen, L. H. Chu, M. Nie, and C. Wang, "Abnormal thermal expansion in anti-perovskite manganese nitride," *Journal of the Chinese Ceramic Society*, vol. 37, no. 5, pp. 73–77, 2009.

[3] B. S. Wang, P. Tong, Y. P. Sun et al., "Large magnetic entropy change near room temperature in antiperovskite $SnCMn_3$," *Environment People Law*, vol. 85, no. 4, Article ID 47004, 2009.

[4] R. Huang, L. Li, F. Cai, X. Xu, and L. Qian, "Low-temperature negative thermal expansion of the antiperovskite manganese nitride Mn_3CuN codoped with Ge and Si," *Applied Physics Letters*, vol. 93, no. 8, Article ID 081902, 3 pages, 2008.

[5] K. Kamishima, T. Goto, H. Nakagawa et al., "Giant magnetoresistance in the intermetallic compound Mn_3GaC," *Physical Review B*, vol. 63, no. 2, Article ID 024426, 3 pages, 2001.

[6] S. Iikubo, K. Kodama, K. Takenaka, H. Takagi, M. Takigawa, and S. Shamoto, "Local lattice distortion in the giant negative thermal expansion material $Mn_3Cu_{1-x}Ge_xN$," *Physical Review Letters*, vol. 101, no. 20, Article ID 205901, 4 pages, 2008.

[7] Y. Sun, C. Wang, Y. C. Wen, K. G. Zhu, and J. T. Zhao, "Lattice contraction and magnetic and electronic transport properties of $Mn_3Zn_{1-x}Ge_xN$," *Applied Physics Letters*, vol. 91, Article ID 231913, 3 pages, 2007.

[8] K. Asano, K. Koyama, and K. Takenaka, "Magnetostriction in Mn_3CuN," *Applied Physics Letters*, vol. 92, no. 16, Article ID 161909, 3 pages, 2008.

[9] T. Shimizu, T. Shibayama, K. Asano, and K. Takenaka, "Giant magnetostriction in tetragonally distorted antiperovskite manganese nitrides," *Journal of Applied Physics*, vol. 111, Article ID 07A903, 3 pages, 2012.

[10] E. O. Chi, W. S. Kim, and N. H. Hur, "Nearly zero temperature coefficient of resistivity in antiperovskite compound $CuNMn_3$," *Solid State Communications*, vol. 120, no. 7-8, pp. 307–310, 2001.

[11] Y. Sun, C. Wang, L. Chu, Y. Wen, M. Nie, and F. Liu, "Low temperature coefficient of resistivity induced by magnetic transition and lattice contraction in Mn_3NiN compound," *Scripta Materialia*, vol. 62, no. 9, pp. 686–689, 2010.

[12] K. Takenaka and H. Takagi, "Giant negative thermal expansion in Ge-doped anti-perovskite manganese nitrides," *Applied Physics Letters*, vol. 87, no. 26, Article ID 261902, 3 pages, 2005.

[13] Y. Sun, C. Wang, Q. Z. Huang et al., "Neutron diffraction study of unusual phase separation in the antiperovskite nitride Mn_3ZnN," *Inorganic Chemistry*, vol. 51, no. 13, pp. 7232–7236, 2012.

[14] Y. S. Sun, Y. F. Guo, X. X. Wang et al., "Resistive switching phenomenon driven by antiferromagnetic phase separation in an antiperovskite nitride Mn_3ZnN," *Applied Physics Letters*, vol. 100, Article ID 161907, 4 pages, 2012.

[15] C. Wang, L. H. Chu, Q. R. Yao et al., "Tuning the range, magnitude, and sign of the thermal expansion in intermetallic $Mn_3(Zn, M)_xN$ (M = Ag, Ge)," *Physical Review B*, vol. 85, Article ID 220103, 5 pages, 2012.

[16] B. Song, J. Jian, H. Bao et al., "Observation of spin-glass behavior in antiperovskite Mn_3GaN," *Applied Physics Letters*, vol. 92, no. 19, Article ID 192511, 3 pages, 2008.

[17] P. Lukashev, R. F. Sabirianov, and K. Belashchenko, "Theory of the piezomagnetic effect in Mn-based antiperovskites," *Physical Review B*, vol. 78, no. 18, Article ID 184414, 5 pages, 2008.

[18] F. Izumi and K. Momma, "Three-dimensional visualization in powder diffraction," *Solid State Phenomena*, vol. 130, pp. 15–20, 2007.

[19] D. Fruchart and E. F. Bertaut, "Magnetic studies of the metallic perovskite-type compounds of manganese," *Journal of the Physical Society of Japan*, vol. 44, no. 3, pp. 781–791, 1978.

[20] J. García, J. Bartolomé, D. González, R. Navarro, and D. Fruchart, "Thermophysical properties of the intermetallic Mn_3MN perovskites II. Heat capacity of manganese zinc nitride: Mn_3ZnN and manganese gallium nitride: Mn_3GaN," *The Journal of Chemical Thermodynamics*, vol. 15, no. 11, pp. 1041–1057, 1983.

[21] M. H. Phan and S. C. Yu, "Review of the magnetocaloric effect in manganite materials," *Journal of Magnetism and Magnetic Materials*, vol. 308, no. 2, pp. 325–340, 2007.

[22] A. Fujita, S. Fujieda, Y. Hasegawa, and K. Fukamichi, "Itinerant-electron metamagnetic transition and large magnetocaloric effects in $La(Fe_xSi_{1-x})_{13}$ compounds and their hydrides," *Physical Review B*, vol. 67, no. 10, Article ID 104416, 2003.

[23] J. P. Jardin and J. Labbe, "Phase transitions and band structure in metallic perovskites (carbides and nitrides)," *Journal of Solid State Chemistry*, vol. 46, pp. 275–293, 1983.

Ellipsometric Characterization of Thin Films from Multicomponent Chalcogenide Glasses for Application in Modern Optical Devices

R. Todorov,[1] J. Tasseva,[1] V. Lozanova,[1] A. Lalova,[1] Tz. Iliev,[2] and A. Paneva[3]

[1] *Institute of Optical Materials and Technologies "Acad. J. Malinowski," Bulgarian Academy of Sciences, Acad. G. Bonchev Street, Building 109, 1113 Sofia, Bulgaria*
[2] *Geological Institute, Bulgarian Academy of Sciences, Acad. G. Bonchev Street, Building 24, 1113 Sofia, Bulgaria*
[3] *Faculty of Physics, Sofia University "St. Kliment Ohridski," 5 James Boucher Avenue, 1164 Sofia, Bulgaria*

Correspondence should be addressed to R. Todorov; rossen@iomt.bas.bg

Academic Editor: R. N. P. Choudhary

A review is given on the application of the reflectance ellipsometry for optical characterization of bulk materials and thin films with thickness between $\lambda/20$ and 2λ (at $\lambda = 632.8$ nm). The knowledge of the optical constants (refractive index, n, and extinction coefficient, k) of thin films is of a great importance from the point of view of modelling and controlling the manufacture of various optical elements, such as waveguides, diffraction gratings, and microlenses. The presented results concern the optical properties of thin films from multicomponent chalcogenide glasses on the base of As_2S_3 and GeS_2 determined by multiple-angle-of-incidence ellipsometry and regarded as a function of the composition and thickness. The homogeneity of the films is verified by applying single-angle calculations at different angles. Due to decomposition of the bulk glass during thermal evaporation, an optical inhomogeneity of the thin As (Ge)-S-Bi(Tl) films is observed. The profile of n in depth of thin As-S-Tl (Bi) films was investigated by evaporation of discrete layers. It is demonstrated that homogenous layers from the previous compounds with controlled composition can be deposited by coevaporation of As_2S_3 and metals or their compounds (Bi, Tl, In_2S_3).

1. Introduction

The optical constants of the thin films are commonly determined by methods such as spectrophotometry, prism-coupling technique, and ellipsometry. The ellipsometric methods are fast, nondestructive, highly sensitive to the presence of surface inclusions and provide higher accuracy for determination of the refractive index and thickness of thin layers in comparison with the spectrophotometric methods [1–4]. Currently various forms of ellipsometry exist and are applied for investigation of the physical properties of thin films [3]. The most common among these methods are the spectral ellipsometry and multiple-angle-of-incidence ellipsometry [5]. Spectral ellipsometry includes measurement and analysis of the complex reflection ratio in wide range of wavelengths at a fixed angle of light incidence. The method provides information for the optical response functions (complex refractive index, $\hat{n}$, or complex permittivity, $\hat{\varepsilon}$) [6]. The main problem of the spectral forms of ellipsometry is the selection of an appropriate dispersion model for description of the optical constants [6]. Single-wavelength ellipsometry offers an advantage in this respect since it does not require a dispersion law and enables the *in situ* control of the process of thin film deposition [5].

Chalcogenide glasses exhibit transmittance in wide spectral range (1–20 μm), low phonon energy and bandgap, high linear and nonlinear refractive index, and high photosensitivity when illuminated by near bandgap light [7–9]. Mixing chalcogenide glasses with a relevant oxide allows extended variation of the refractive index, for example, in GeO_2-GeS_2 system [10]. The chalcogenide glasses can be deposited as thin films by a variety of techniques as thermal and flash evaporation [11, 12], spin-coating [13] and pulsed laser deposition [14]. As a result, they have found practical applications in manufacturing of various elements for optics in visible and infrared spectral region [15], photonics [16], and all optical processing [17] such as waveguides [18], optics amplifiers [19], diffraction gratings [20], microlenses [21], and phase-change recording media [22]. To realize all these devices and applications, one should know and be able to precisely control the optical parameters of the materials. Ellipsometric measurements have been applied for investigation of the optical properties of chalcogenide films [23–35] and of different photo-induced effects as photocrystaleization [29, 31], volume changes [26, 34], and photodissolution of metals [35]. The comparison of dispersion models for analysis of the measured spectra of ellipsometric angles (Ψ and Δ) of thin As_2S_3 was made in [36]. The authors [37] found that the model of Urbach-Cody-Lorentz and parameterization of the density of the electronic states model give the best fit of spectrophoptometric (transmittance and reflectance) and/or ellipsometric experimental data for thin As-S films.

This is a review paper on the optical properties and their changes after illumination of thin chalcogenide films from As-S(Se), As-S-Ge(Se), and As(Ge)-S-Tl(Bi, In) as studied by means of multiple-angle-of-incidence ellipsometry. We discuss the applicability of multiple-angle-of-incidence ellipsometry (MAI-ellipsometry) for characterization of chalcogenide layers from the above systems exclusively when irregular gradient of the refractive index is involved, resulting from compositional variation in the depth of the layers. The possibility to deposit homogeneous multicomponent chalcogenide films by coevaporation is demonstrated.

2. Experimental Details

Bulk glasses from the system As_xS_{100-x} ($28 \le x \le 45$), As_xSe_{100-x} ($28 \le x \le 50$), Ge_xS_{100-x} ($20 \le x \le 40$), $As_{40}S_{60-x}Se_x$ ($0 \le x \le 60$), and $(As_2S_3)_{100-x}Bi(Tl)_x$ ($0 \le x \le 10$) were synthesized in a quartz ampoule from elements of purity 99,999% by the method of melt quenching [25, 26, 38]. Thin films with thickness between 30 and 1000 nm were deposited from a Mo or Ta boat with a preliminarily calculated amount of bulk materials for each particular thickness. The layers were deposited by thermal evaporation at deposition rates of 0.4–0.7 nm/s on optical glass BK-7 substrates. The films were exposed to white light (Mercury lamp, 20 mW/cm^2). The composition of the bulk samples and thin films deposited on graphite substrates was determined by scanning electron microscopy with an X-ray microanalyzer (Jeol Superprobe 733, Japan) [39]. The composition of the films from binary and ternary As-S-Se and As-S-Ge systems [25, 26] is close to that of the bulk samples, while it is observed that the amount of Bi or Tl in the thin As-S-Bi(Tl) layers is smaller compared to that in the bulk samples [39].

The ellipsometric measurements were carried out at three angles of light incidence: 45, 50, and 55° using LEM-3M1 ellipsometer equipped with low-intensity He-Ne laser (λ = 632.8 nm). The measurements were performed at angles of incidence below the Brewster's angle in order to remove reflection from the back side of the substrate. The accuracy of determination of Ψ and Δ was 0.01° and 0.02°, respectively. The systematic errors were eliminated through averaging of the four zones measurements. The duration of each ellipsometric measurement was about 5 min, and all measurements were carried out within 24 hours after the films deposition. Thus, the fast relaxation in selenium rich films [29] and oxidation of germanium containing films [40] would have not produced an artefact.

3. Theory and Modelling

Reflective ellipsometry is applied for quantitative estimation of an optical system by the so-called relative coefficient of reflectance, ρ [1]. The basic equation of the ellipsometry is

$$\rho = \frac{R_p}{R_s} = \tan\Psi \cdot e^{i\Delta}, \quad (1)$$

where R_p and R_s are complex reflection coefficients for p- and s-polarization, respectively. $\tan\Psi$ is the ratio amplitude of the reflection coefficients R_p and R_s and Δ is the difference between their phases:

$$\tan\Psi = \frac{r_p}{r_s}, \qquad \Delta = \delta_p - \delta_s, \quad (2)$$

where r_p and r_s are the modulus of R_p and R_s and δ_p and δ_s are their phases. In the case of a three-phase system (ambient, thin film, substrate) (Figure 1(a)), (1) can be written as:

$$\rho = \tan\Psi \cdot e^{i\Delta} = \frac{r_{01p} + r_{12p}e^{-i2\beta}}{1 + r_{01p}r_{12p}e^{-i2\beta}} \times \frac{1 + r_{01s}r_{12s}e^{-i2\beta}}{r_{01s} + r_{12s}e^{-i2\beta}}, \quad (3)$$

where r_{01p}, r_{01s}, r_{12p}, and r_{12s} are the Fresnel coefficients for ambient, thin layer and thin layer, substrate boundaries, respectively, and β is the phase thickness:

$$\beta = 2\pi\left(\frac{d_1}{\lambda}\right)\widehat{n}_1\cos\phi_1, \quad (4)$$

where $\widehat{n}_1 = n_1 - ik_1$ is the complex refractive index of the thin layer with thickness d. ϕ and λ are the angle of incidence and wavelength of the light, respectively.

A schematic model of an inhomogeneous layer is presented in Figure 1(b). In this case the Fresnel coefficients in (3) for ambient, thin layer and thin layer, substrate boundaries will include different refractive indices for the thin film, n_1' and n_2'', respectively [41].

In (4) for the phase thickness, β, the refractive index and angle of incidence are generalized by their averaged values, the product of which satisfies the following integral:

$$n_{av}\cos\phi_{av} = \frac{1}{d}\int_0^d n(z)\cos[\phi(z)]\,dz. \quad (5)$$

Two-layer models are usually used for analysis of roughness and surface phenomena as oxidation. Rough-surface models involving ridges, pyramids, or other similar geometric forms have been used to simulate surface roughness [42]. For the selection of the most appropriate model, where no data

TABLE 1: Optical parameters (refractive index, n, and thickness, d) of thin films from As-S-Bi (Tl, In) systems, determined by ellipsometric and Swanepoel's spectrophotometric methods at $\lambda = 632.8$ nm. The composition of the thin films determined by X-ray microanalysis is given in parentheses.

Composition of thin film	Spectrophotometric		Ellipsometric				
			Multiple-angle calculation		Single-angle calculation		
	d_{unexp} [nm]	n_{unexp}	d_{unexp} [nm]	n_{unexp}	n_{unexp} (45°)	n_{unexp} (50°)	n_{unexp} (55°)
$As_{28}S_{72}$ ($As_{28.5}S_{71.5}$)	977	2.35	978	2.347	2.353	2.352	2.351
$As_{40}S_{60}$ ($As_{40.7}S_{59.3}$)	943	2.46	937	2.457	2.452	2.457	2.454
$As_{45}S_{55}$ ($As_{43.3}S_{56.7}$)	898	2.45	897	2.449	2.451	2.448	2454
GeS_2 ($Ge_{35.1}S_{64.9}$)	987	2.16	970	2.172	2.167	2.163	2.168
$As_{10}Ge_{30}S_{60}$ ($As_{10.7}Ge_{32.9}S_{56.4}$)	1217	2.34	1208	2.360	2.362	2.353	2.346
$As_{40}S_{30}Se_{30}$ ($As_{39}S_{32}Se_{29}$)	996	2.72	997	2.722	2.719	2.719	2.722
$Ge_{32}S_{64}Ga_4$ ($Ge_{39.6}S_{58.6}Ga_{1.6}$)	905	2.25	871	2.236	2.215	2.321	2.346
$Ge_{32}S_{64}In_4$ ($Ge_{38.5}S_{60.1}In_{1.4}$)	1270	2.27	970	2.220	2,086	2,215	2,299
$Ge_{32}S_{64}Tl_4$ ($Ge_{33.3}S_{65.0}Tl_{1.7}$)	1388	2.26	1408	2.204	2.235	2.389	2.447
$Ge_{28}S_{66}Bi_6$ ($Ge_{31.4}S_{65.9}Bi_{2.7}$)	1390	2.19	1291	2.256	2.161	2.264	2.255
$As_{39}S_{58}Bi_3$ ($As_{46.9}S_{52.0}Bi_{1.1}$)	1014	2.29	1122	2.367	2.493	2.477	2.467
$As_{38}S_{56}Bi_6$ ($As_{45.0}S_{51.7}Bi_{2.3}$)	1201	2.49	996	2.789	2.805	2.751	2.694
$As_{38}S_{56}Tl_6$ ($As_{43.6}S_{54.0}$ $Tl_{2.4}$)	1117	2.64	930	2.712	2.690	2.718	2.721
As_2S_3/Bi ($As_{41.3}S_{53.5}Bi_{5.2}$)	1065	2.64	1067	2.632	2.639	2.632	2.627
As_2S_3/Tl ($As_{43.5}S_{54.5}Tl_{2.0}$)	1341	2.499	1342	2.492	2.493	2.495	2.496
As_2S_3/In_2S_3 ($As_{31.1}S_{60.7}In_{8.2}$)	805	2.53	800	2.528	2.524	2.528	2.527

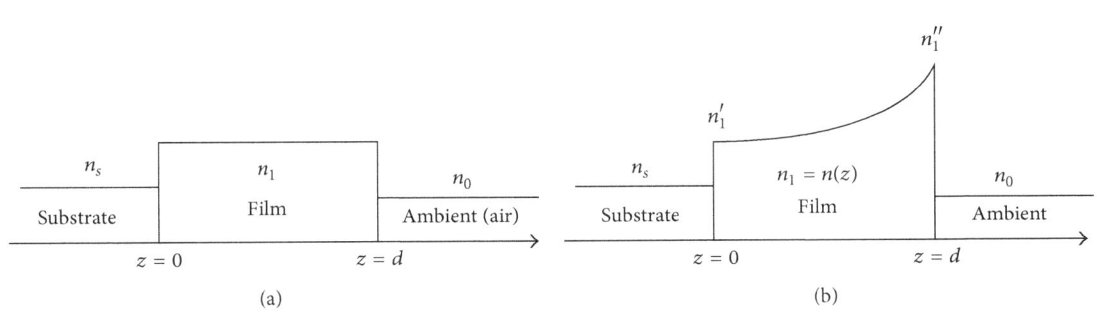

FIGURE 1: Schematic presentation of homogenous thin film with refractive index, n_1, and thickness, d, (a) and inhomogeneous thin film (b).

is available for the profile, best-fit result out of the experimental set of values of Ψ and Δ should be used.

The dependence of Ψ and Δ on the system parameters in (3) for a three-phase system can be expressed by the following analytical function:

$$\Psi = \tan^{-1}\left[\rho\left(n_0, n_1, k_1, n_s, k_s, d, \phi, \lambda\right)\right],$$
$$\Delta = \arg\left[\rho\left(n_0, n_1, k_1, n_s, k_s, d, \phi, \lambda\right)\right], \quad (6)$$

where n_0 is the refractive index of ambient (in case of air, $n_0 = 1$), and n_s and k_s are the refractive index and extinction coefficient of the substrate, respectively. An essential problem of all forms of ellipsometry is the impossibility of inverting (6) in order to obtain the complex refractive index parameters. The methods applied for solving the system of (6) using multiangle measurements were described in [1]. In many cases these parameters are varied in order to minimize some function. One of the most often used estimators is

$$G = \sum_{i=1}^{M}\left(\frac{\left(\Delta_i^c - \Delta_i^m\right)^2}{\delta\Delta^2} + \frac{\left(\Psi_i^c - \Psi_i^m\right)^2}{\delta\Psi^2}\right), \quad (7)$$

where M is the number of the experimental points (angles of incidence in our case), Δ_i^c and Ψ_i^c are the theoretical values of the ellipsometric angles at a determined group of optical parameters for the applied model, and Δ_i^m and Ψ_i^m are the measured values. The fitting process includes a choice of suitable initial values for the unknown parameters and the minimization of the function G using the fitting procedure based on the Levenberg-Marquardt algorithm. The initial estimates of the parameters were simultaneously varied until a fit was obtained between the experimental data and the values calculated from the thin film formula (3). The good coincidence between the measured and theoretical values of the ellipsometric angles within the defined accuracy is an evidence of felicitous choice of a model of the investigated system [5, 43]. The error is calculated from the elements of the correlation matrix which the Levenberg-Marquardt algorithm produces during the minimization procedure. The calculation procedure is described in [1, 5].

4. Results and Discussion

Knowledge of the optical parameters of the single layers is important for successful engineering of multilayer structures. Firstly, we determined the optical parameters of the thin films using the single homogenous absorbing layer model and applying multiangle measurements (45, 50, and 55°). In this case the refractive index, n, extinction coefficient, k, and thickness, d, are simultaneously calculated (three-parameter calculation). We found that the maximal experimental error of n was $\Delta n = \pm 0.005$–0.007 [26]. The basic advantage of the multiangle measurements consists in their independence on the half-wave point of light incidence angle [44]. Further, we used the calculated optical parameters from multiangle measurement as an initial estimate of the unknown parameters for single-angle calculations in order to determine the degree of inhomogeneity of the thin layers. We searched for two optical parameters of the thin film and the third one has been iterated with small steps around its approximate value until minimum value of G was found [26]. The single-angle calculation allows the film's thickness to be calculated with an accuracy better than 0.3%. We showed that in case of homogenous thin films, both single- and multiple-angle methods gave the same results.

In Table 1 a comparison of the thickness and refractive index values of thin films calculated from ellipsometric measurements and/or from their transmission spectra using Swanepoel's method [45] is presented. The measurements were performed on one spot of each sample. The areas of the incident probe beams of the spectrophotometer and the ellipsometer were 50 and 1 mm^2, respectively.

For thin As-S films the values for the refractive index within the limits of the experimental error obtained through ellipsometry are in a good agreement with those calculated from spectrophotometric measurements. The single-angle calculations at different angles show that thin films from As-S(Se), Ge-S, As-Ge-S systems are homogenous or their inhomogeneity cannot be revealed since the discrepancies are within the experimental error. It is seen that the optical constants and thickness for layers from Ge-S-Ga(In, Tl, Bi) and As-S-Bi(Tl) systems as derived by ellipsometry differ from those calculated from spectrophotometric measurements. At the same time the difference in the values of n independently determined at 45°, 50°, and 55° is higher than the experimental error $\Delta n = \pm 0.005$. It is seen from (4) and (5) that in case of inhomogeneous coatings, due to different paths of the laser beam of the ellipsometer, different values for n would be observed. These considerations suggest inhomogeneity of the thin films deposited form bulk glasses from As-S-Bi(Tl) systems. The derived values of n, thus, can be regarded just as effective values. We used the method of thermal coevaporation to deposit thin films from ternary As_2S_3-In_2S_3 system [46] and thin films from multicomponent $GeSe_2$-Sb_2Se_3-AgI system [47]. In Table 1 the refractive index of thin As_2S_3/Bi, As_2S_3/Tl, and As_2S_3/In_2S_3 films deposited by coevaporation is given. It is seen from single-angle calculations that the co-evaporation from two sources makes possible the deposition of homogenous coatings from the above multicomponent glasses.

Our results for the optical constants and thickness of thin films from As-S-Ge, As_xSe_{100-x}, and Ge_xS_{100-x} systems and their changes after illumination by halogen lamp are presented in Table 2. It is seen that the refractive index of the chalcogenide films varies in wide diapason, that is, from 2.05 to 3.15 for GeS_2 and As_2Se_3, respectively. The results for n and k of the films from As-S=(Se) systems are in agreement with the data reported in the literature [29, 48–50], With the substitution of Se for S in thin As-S-Se layers, the refractive index of the layers increases and the highest value is reached for As_2Se_3 film. The changes in the optical parameters of thin As-Se films under illumination are more pronounced compared to those in As-S layers. Maximum changes in the refractive index of the As-Se films were observed for the thin film with composition $As_{50}Se_{50}$ ($\Delta n = +0.23$). The refractive index of the thin films from Ge-S system decreases after

TABLE 2: Optical constants (refractive index, n, extinction coefficient, k) at λ = 632.8 nm and thickness, d of thin chalcogenide films before and after light exposure.

Composition of thin film	n_{unexp}	n_{exp}	k_{unexp}	k_{exp}	d_{unexp} [nm]	d_{exp} [nm]
$As_{40}S_{60}$	2.458	2.531	0.001	0.003	1032	996
$As_{10}Ge_{30}S_{60}$	2.320	2.226	0	0.001	1208	1279
$As_{20}Ge_{20}S_{60}$	2.393	2.300	0	0	1172	1202
$As_{30}Ge_{10}S_{60}$	2.438	2.415	0.003	0.003	1128	1136
$Ge_{20}S_{80}$	2.138	2.079	0	0	741	775
$Ge_{33}S_{67}$	2.172	2.056	0.005	0	970	990
$Ge_{40}S_{60}$	2.577	2.524	0.001	0	1187	1218
$As_{28}Se_{72}$	2.979	3.047	0.039	0.037	893	929
$As_{40}Se_{60}$	3.031	3.155	0.036	0.054	822	801
$As_{50}Se_{50}$	2.874	3.101	0.014	0.041	803	795
$As_{40}S_{45}Se_{15}$	2.567	2.687	<0.001	<0.001	1045	1035
$As_{40}S_{30}Se_{30}$	2.763	2.820	0.007	0.013	997	985
$As_{40}S_{15}Se_{45}$	2.845	2.991	0.013	0.043	879	865

TABLE 3: Refractive index, n, and thickness, d, of thin films, determined by ellipsometric, triple and double spectrophotometric methods. T is the transmission, R_f, R_b, and R_m are the reflections from the front and back side of the BK-7 substrate and from the film deposited on Si-wafer substrate, respectively.

Composition	(T, R, R_m) method		Ellipsometric		(T, R_f) method		Ellipsometric	
	d_{unexp} [nm]	d_{exp} [nm]	d_{unexp} [nm]	d_{exp} [nm]	n_{unexp}	n_{exp}	n_{unexp}	n_{exp}
$As_{40}S_{60}$	74	72	74.0	72.5	2.46	2.58	2.457	2.622
GeS_2	48	50	47.0	49.7	2.15	2.11	2.131	2.096
$As_{20}Ge_{20}S_{60}$	80	85	81.8	86.2	2.35	2.30	2.369	2.280
$As_{10}Ge_{30}S_{60}$	95	91	93.6	91.5	2.32	2.23	2.336	2.258
$As_{40}S_{30}Se_{30}$	73	72	74.2	73.7	2.74	2.94	2.722	2.949
$As_{40}Se_{60}$	70	65	69.3	65.6	3.05	3.31	3.056	3.316

exposure. The greatest changes were observed for thin film with stoichiometric composition GeS_2.

The photo-induced volume effects in thin chalcogenide films are connected with changes of their thickness. From practical point of view, volume changes make chalcogenide layers perspective materials for fabrication of microlenses [21]. An effect of photocontraction was found after exposure to light of thin films from As-S(Se) (see Table 2). In Figure 2 the relative changes of the thickness, $\Delta d/d_{unexp}$ as a function of the composition for As-S, Ge-S, $As_{40}S_{60-x}Se_x$, and As-S-Ge films are presented. It is seen that the layer from As-S and $As_{40}S_{60-x}Se_x$ systems exhibited effect of photocontraction. The changes in the thickness increase with sulphur contain in layers from binary As-S system. A pure effect of photoexpansion is observed for thin Ge-S layers after illumination, while for As-S-Ge system, both effects of photocontraction and photoexpansion are presented. We have obtained the maximal value of photoexpansion for thin film with composition $As_{10}Ge_{30}S_{60}$ ($\Delta d/d = +7.6\%$). We have obtained the maximal value of photoexpansion for thin film with composition $As_{10}Ge_{30}S_{60}$ ($\Delta d/d = +7.6\%$). This result is in agreement with data published by other authors [51].

To understand the origin of inhomogeneity of the thin films, we investigated the thickness dependence of the refractive index. For the most photonic devices, it is necessary that the thin layers are with an exact thickness. The accumulated knowledge of the refractive index and its photo-induced changes would support the modelling and manufacturing of 1D photonic crystal [52]. The results from our investigations on the thickness dependence of the optical properties of thin films from As-Se, Ge-S, As-Ge-S, and As-S-Se systems [25–27] showed that the refractive index is independent of d for $d > 50$ nm. For acquiring of information about the possible changes of the optical constants during the process of measurement with He-Ne laser, we compared the values of n, k, and d for thin films obtained by MAI-ellipsometry with those determined by two and three spectrophotometric measurements [53] (Table 3).

In Figure 3(a) a theoretical, as well as, experimental Ψ-Δ dependence at room temperature is given for thin As_2S_3 films (d = 25–100 nm) with refractive index n = 2.45 at λ = 632.8 nm and angle of light incidence $\varphi = 50^\circ$. It is seen that the experimental results follow the theoretical curve for $d > 50$ nm. The thickness dependence of the ellipsometric angles Ψ-Δ of thin films with compositions $As_{40}S_{30}Se_{30}$ and $As_{40}Se_{60}$ is presented in Figures 3(b) and 3(c), respectively. The parameters of theoretical curves are n = 2.72 and k = 0.001 or n = 3.05 and k = 0.03 at λ = 632.8 nm and angle of light incidence $\varphi = 50^\circ$ for thin films with composition $As_{40}S_{30}Se_{30}$ and $As_{40}Se_{60}$, respectively.

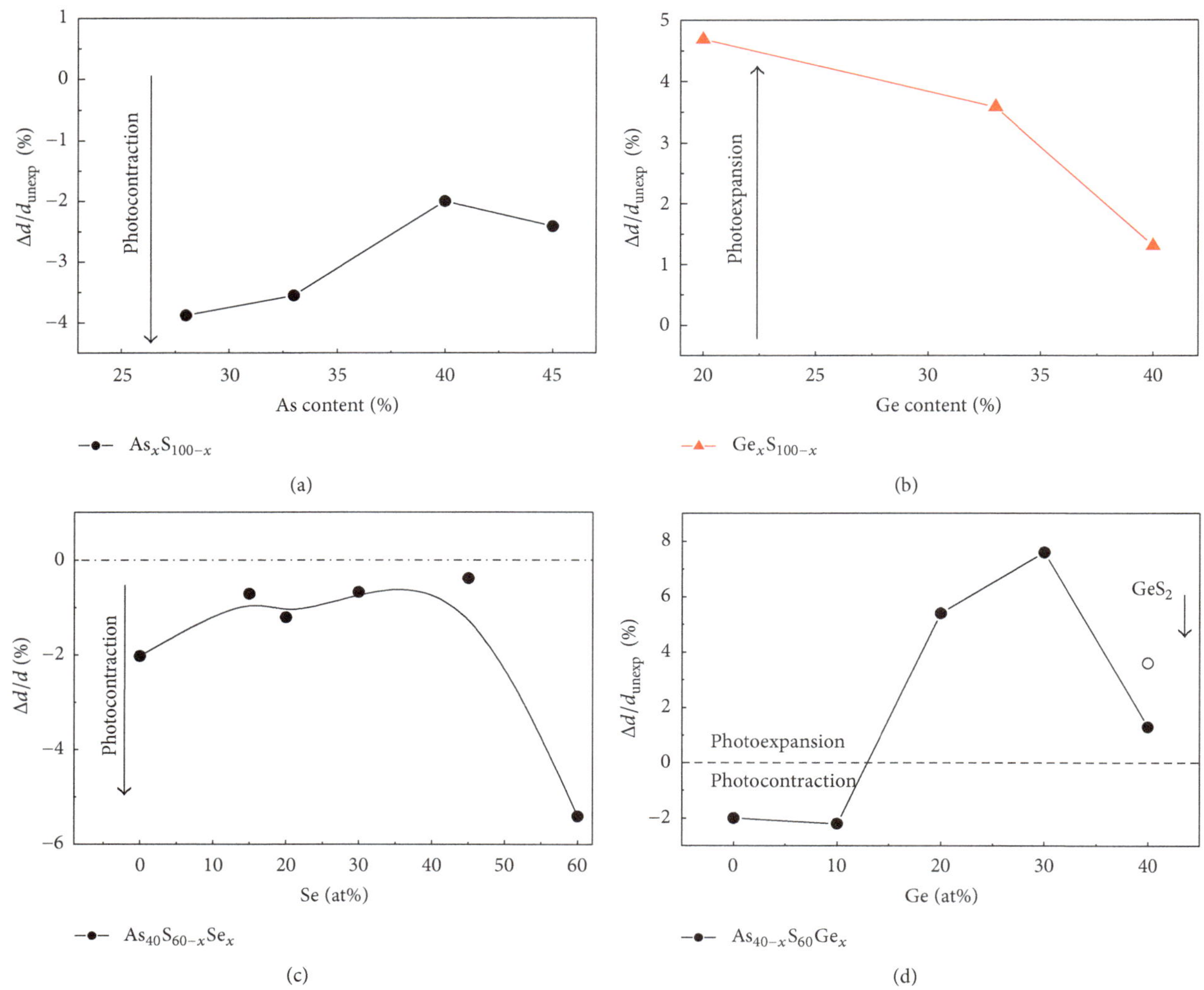

FIGURE 2: Relative changes in the thickness, $\Delta d/d_{unexp}$, of thin films from the As-S (a), Ge-S (b), As-S-Se (c), and As-S-Ge (d) systems after exposure to white light as a function of Se and Ge content, respectively, (the lines connecting the experimental points are only drawn as a guide to the eye).

It is observed that n is independent of the thickness for $d > 50$ nm. We believe that the optical properties of the films thinner than 50 nm are strongly influenced by the substrate and the process of the thin films growth. The single homogenous absorbing layer model is only an approximation, since it does not consider either the existence of the substrate/film interface, the surface roughness, or oxidation. We can expect that the error of determination of d increases extremely for $d < 30$ nm [54, 55]. The relative error in the thickness determination is independent of d for layers with $d > 70$ nm (±0.3%), and downwards it increases up to 1% for layers with d~30 nm. The error of the refractive index determination is weakly affected (±0.2-0.3%) by the layer's thickness [24].

It is seen from Table 1 that the composition of thin films from As-S-Bi is different from that of the respective bulk sample. We made the composition analysis of the residue in the boat that was found to be $As_{0.5}S_{59.3}Bi_{40.2}$ after evaporation of thin film from bulk $As_{38}S_{56}Bi_6$ glass. That is why firstly we analyzed the thin As-S-Bi films using the two-layer model considering an absorbing sublayer on the upper part of the coating (closer to air) when the whole quantity in the boat was evaporated. This model gives the best fit of the experimental results (Figure 4). The results show a higher extinction coefficient, k, of the upper sublayer. The thicknesses of the layers for thin films with compositions $As_{39}S_{58}Bi_3$ and $As_{34}S_{54}Bi_{10}$ system were 39 and 70 nm, respectively. The X-ray microanalyses at different acceleration voltage [39] showed that concentration of Bi or Tl increases in direction from the substrate to the interface of the layer with the ambient. In [56] it was found by mass-spectroscopy analysis that the vapour of As_2S_3, formed in the process of thermal evaporation, consists of different sulphur- or arsenic-rich fractions. These fractions polymerize on the substrate forming the glass network of the thin film. This finding suggests that the melt in the boat should become poorer of light fractions during the thermal

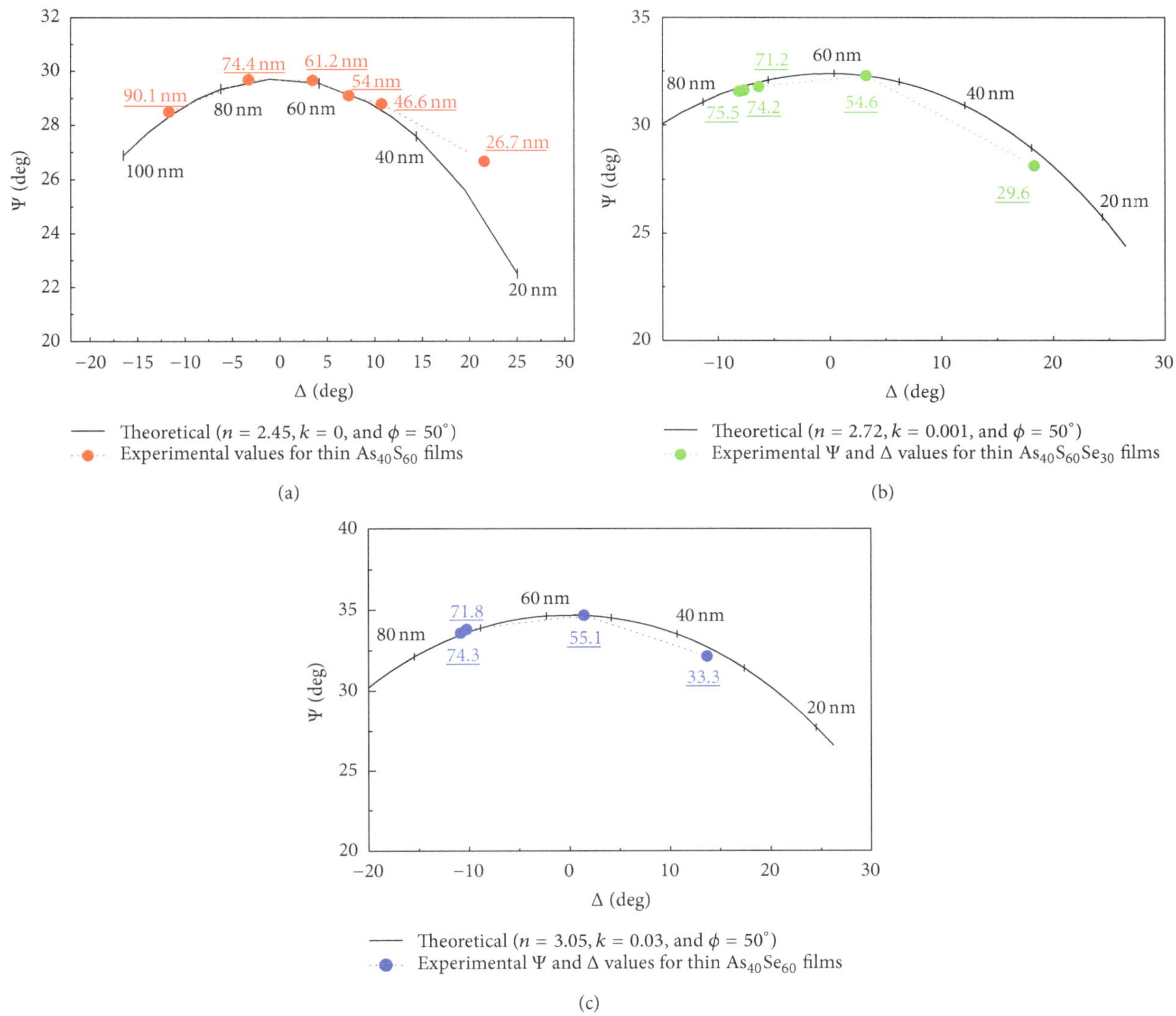

FIGURE 3: Theoretical Ψ-Δ dependence for thin $As_{40}S_{60-x}Se_x$ films and the experimental data for thin (a) $As_{40}S_{60}$, (b) $As_{40}S_{30}Se_{30}$, and (c) $As_{40}Se_{60}$ films. The parameters of the theoretical curves are given in the graphs. The calculated film's thickness is underlined (the dotted lines connecting the experimental points are only drawn as a guide to the eye).

evaporation, and therefore the upper part of the thin films should be enriched with structural units formed from heavier fractions. Most probably this is the reason why in some cases inhomogeneous layers are produced by thermal evaporation. This phenomenon is more pronounced for thin films from the ternary As-S-Tl(Bi) systems [39, 57].

Usually inhomogeneous thin films are presented as multilayer coatings where the refractive index is assumed to be variable under some mathematical function—linear, exponential and Gaussian [58]. In many cases, it is not possible to determine the manner in which the refractive index of the film varies. For estimation of the optical homogeneity in the volume of the thin layers, thin films (about 80–100 nm thick) were deposited consecutively on different substrates until the whole quantity in the boat was evaporated. We expect that the dependence of the refractive index of the discrete layers on the order of their deposition would be the same as the profile of n in the one micron integral coating. Figure 5 represents the distribution of the refractive index, n, on the total thickness of the sublayers for unexposed $As_{40}S_{60}$ and $As_{38}S_{56}Tl_6$ layers. It is seen that the refractive index for as-deposited $As_{40}S_{60}$ films does not change with the stack order, while n considerably increases for the last two numbers of the system $As_{38}S_{56}Tl_6$. It can be concluded therefore that the As_2S_3 thin films are homogeneous in composition in depth of the films, while during the process of vacuum deposition of As-S-Tl coatings, a process of decomposition takes place. This fact suggests that we can consider $As_{38}S_{56}Tl_6$ film to be composed of two parts: homogenous sublayer with low refractive index and inhomogeneous sublayer on the upper surface of the coating. The X-ray microanalysis of the composition of the sublayers revealed variation between

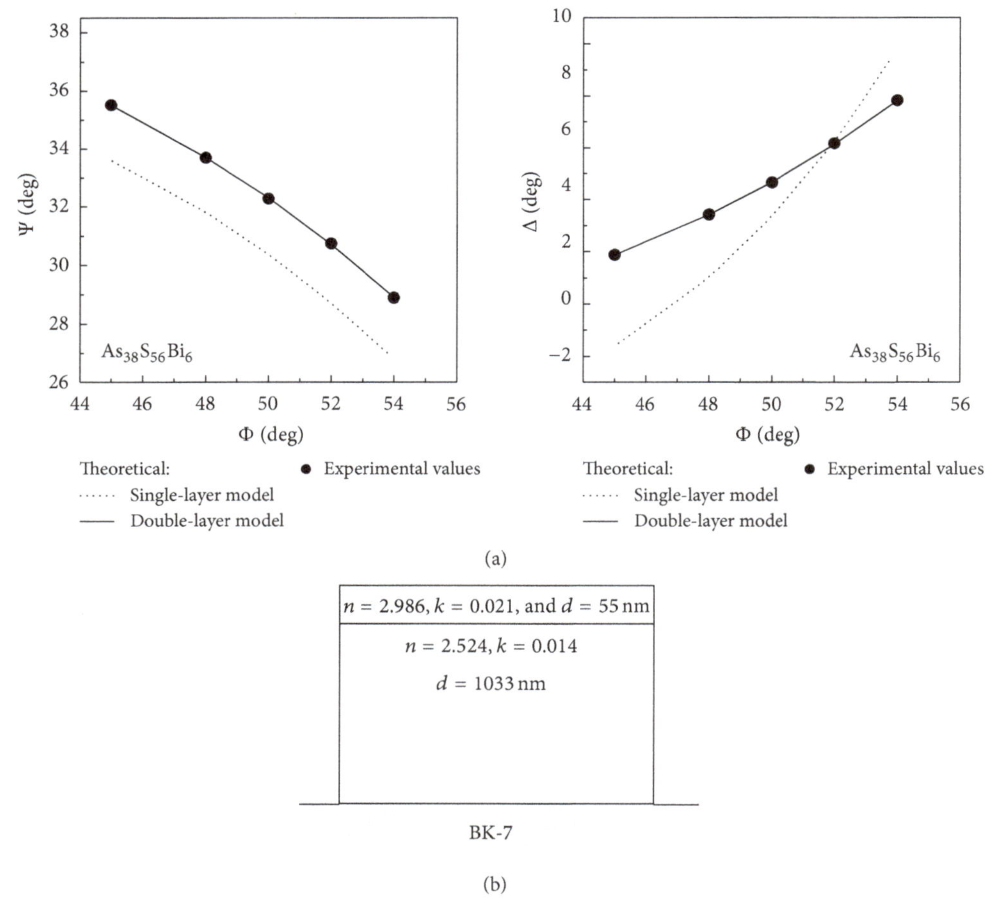

FIGURE 4: Comparison of theoretical (dotted and solid lines for single and double layer models, resp.) and experimental values (symbols) of Ψ and Δ for thin $As_{38}S_{56}Bi_6$ film depending on the angle of light incidence (a) and schematic representation of the two-layer model with calculated values for optical parameters (b).

0.1–0.6 at % of the thallium content in the samples from 1 to 6, while the compositions for the sublayers 7 and 8 were $As_{32.6}S_{62.3}Tl_{8.1}$ and $As_{5.6}S_{59.5}Tl_{34.9}$, respectively. It is known that in case linear profile of the average values for refractive index, n_{av}, is equal to the refractive index at the thickness level 0.5 d [59]. The dependence presented in Figure 5 can be extrapolated by exponential function of the type $n = n_0 + A \cdot \exp(x/B)$, where n_0 is the value of the refractive index of the film close to substrate. Having determined the optical parameters and thickness of the single layers, we calculated the theoretical ellipsometric angles Ψ and Δ of the complex structure [60]. Further to finding the effective value of n we used these values to calculate refractive index of such coating as the integral film. We found that the effective value of n was 2.679 that is close to the one determined for one-micron layer evaporated from bulk glass (see Table 1).

The thin films deposited from bulk $As_{38}S_{56}Bi_6$ glass showed similar dependence of n on their deposition number (Figure 6). This dependence suggests that metal content increases extremely in the last two samples. The calculated values for n for $i = 8$ is lower than those calculated from

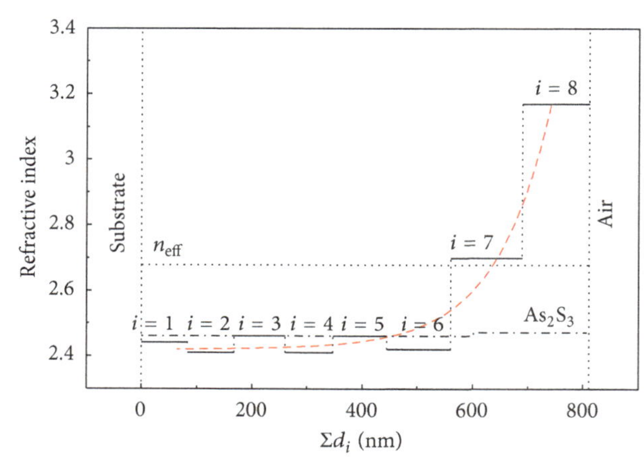

FIGURE 5: Variation of the refractive index depending on the order of deposition of discrete films ($i = 1 - 8$) from bulk $As_{38}S_{56}Tl_6$ glass. The effective value of the refractive index of the coating is given as n_{eff} (dotted line). The exponential extrapolation of the variation of n of sublayers is given by dashed line. The dashed-dotted line represents the same dependence for thin $As_{40}S_{60}$ layers.

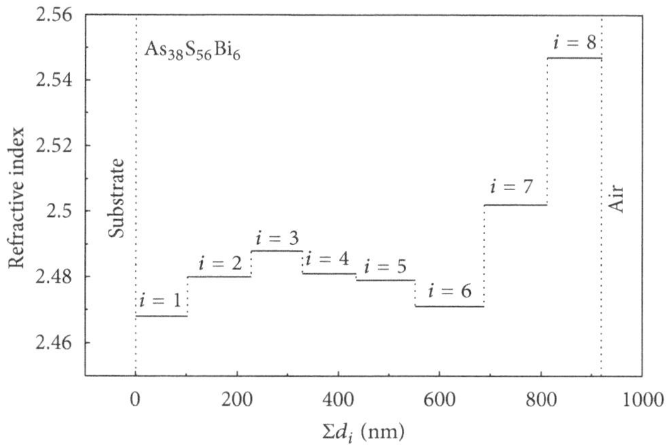

FIGURE 6: Variation of the refractive index depending on the order of deposition of discrete films ($i = 1 - 8$) from bulk $As_{38}S_{56}Bi_6$ glass.

two-layer model (see Figure 4(b)) since in the former case it is averaged over higher thickness ($d = 107$ nm).

In the existing methods the profile is determined using virtual slicing of the thin film into sublayers, suggestion of some function of the dependence of the refractive index in the depth of the coating (linear, exponential, or polynomial) and minimization of the discrepancy between theoretically calculated and measured values of Ψ and Δ [61]. In contrast, the previously suggested method requires no mathematical function and provides reliable information for the refractive index profile in the depth of thin films. The gradient of n due to variation of composition in the depth of the layer could replace the design of column structures in the construction of multilayer structure [62, 63].

5. Conclusion

The surface roughness plays a key role in the formation of the interface between layers in multilayer structures. The effects of surface oxidation observed in thin films [40] could be determining in the formation of this interface and their presence indirectly demonstrated. In [24] it was shown that the introducing of a surface layer with $n = 1.46$ considerably improves the accuracy of the fit of the experimental data for Ψ and Δ for As_2S_3 film. The thickness of the surface overlayer is in the range of 2–15 nm [26, 28]. Our investigation on surface roughness by White Light Interferometric Profiler [64] shows that the root mean square (rms) is about 1.0–8.6 nm for As-S-Se/Ag layer with thickness about 700 nm. In spite of the great number of papers dealing with surface phenomena in chalcogenide thin films, the investigations by ellipsometric measurements are still in the beginning.

In the present paper the optical properties of thin films on the base of As_2S_3 and GeS_2 as a function of the composition and thickness were investigated by MAI ellipsometry. The optical parameters of the thin films from the previous systems were determined with an accuracy of 0.2-0.3% for thickness in the range from $\lambda/20$ to 2λ. A good agreement was demonstrated between the values of the optical constants as obtained by ellipsometric and spectrophotometric measurements in the case of homogeneous thin films from As-S(Se) and As-S-Ge systems. The possibility is shown for the refractive index to be varied in wide range from 2.05 to 3.15. The inhomogeneity of thin As-S-Tl(Bi) films leads to discrepancy of the results derived from both types of measurements. The refractive index profile of thin $As_{38}S_{56}Tl_6$ and $As_{38}S_{56}Bi_6$ films is investigated. It was found that the variation of the refractive index with the distance from the substrate to the surface was about 0.8 and 0.1, respectively. This result presents an opportunity for utilization of inhomogeneous layers from As-S-Tl(Bi) system in photonic multilayer structures. The possibility for deposition of homogenous As-S-Tl(Bi, In) layers by coevaporation of As_2S_3 and metal compounds is demonstrated.

Acknowledgment

The authors would like to acknowledge the financial support from Project BG05PO001-3.3-005/0001 "Science & Business" branch, operational program "Human Resource Development."

References

[1] R. V. A. Azzam and N. M. Bashara, *Ellipsometry and Polarized Light*, chapter 4, North-Holland, Amsterdam, The Netherlands, 1977.

[2] L. Ward, "The accuracy of some mixed photometric and polarimetric functions in the determination of the optical constants of thin films," *Journal of Physics D*, vol. 17, no. 9, pp. 1781–1790, 1984.

[3] K. Vedam, "Spectroscopic ellipsometry: a historical overview," *Thin Solid Films*, vol. 313-314, pp. 1–9, 1998.

[4] D. E. Aspnes, "Expanding horizons: new developments in ellipsometry and polarimetry," *Thin Solid Films*, vol. 455-456, pp. 3–13, 2004.

[5] T. E. Jenkins, "Multiple-angle-of-incidence ellipsometry," *Journal of Physics D*, vol. 32, no. 9, pp. R45–R56, 1999.

[6] D. E. Aspnes and A. A. Studna, "Dielectric functions and optical parameters of Si, Ge, GaP, GaAs, GaSb, InP, InAs, and InSb from 1.5 to 6.0 eV," *Physical Review B*, vol. 27, no. 2, pp. 985–1009, 1983.

[7] A. Zakery and S. R. Elliott, "Optical properties and applications of chalcogenide glasses: a review," *Journal of Non-Crystalline Solids*, vol. 330, no. 1–3, pp. 1–12, 2003.

[8] B. F. Bowden and J. A. Harrington, "Fabrication and characterization of chalcogenide glass for hollow Bragg fibers," *Applied Optics*, vol. 48, no. 16, pp. 3050–3054, 2009.

[9] E. Márquez, J. M. González-Leal, A. M. Bernal-Oliva, T. Wagner, and R. Jiménez-Garay, "Preparation and optical dispersion and absorption of Ag-photodoped $Ge_xSb_{40-x}S_{60}$ ($x \leq 10$, 20 and 30) chalcogenide glass thin films," *Journal of Physics D*, vol. 40, no. 17, pp. 5351–5357, 2007.

[10] N. Terakado and K. Tanaka, "Photo-induced phenomena in GeO_2-GeS_2 glasses," *Japanese Journal of Applied Physics B*, vol. 46, no. 12–16, pp. L265–L267, 2007.

[11] K. Petkov and B. Dinev, "Photo-induced changes in the optical properties of amorphous As–Ge–S thin films," *Journal of Materials Science*, vol. 29, no. 2, pp. 468–472, 1994.

[12] J. Tasseva, K. Petkov, D. Kozhuharova, and T. Iliev, "Light-induced changes in the physico-chemical and optical properties of thin Ge–S–Se–As films," *Journal of Optoelectronics and Advanced Materials*, vol. 7, no. 3, pp. 1287–1292, 2005.

[13] C. Tsay, Y. Zha, and C. B. Arnold, "Solution-processed chalcogenide glass for integrated single-mode mid-infrared waveguides," *Optics Express*, vol. 18, no. 25, pp. 26744–26753, 2010.

[14] A. V. Rode, A. Zakery, M. Samoc, R. B. Charters, E. G. Gamaly, and B. Luther-Davies, "Laser-deposited As_2S_3 chalcogenide films for waveguide applications," *Applied Surface Science*, vol. 197-198, pp. 481–485, 2002.

[15] X. H. Zhang, L. Calvez, V. Seznec et al., "Infrared transmitting glasses and glass-ceramics," *Journal of Non-Crystalline Solids*, vol. 352, no. 23-25, pp. 2411–2415, 2006.

[16] K. Suzuki, Y. Hamachi, and T. Baba, "Fabrication and characterization of chalcogenide glass photonic crystal waveguides," *Optics Express*, vol. 17, no. 25, pp. 22393–22400, 2009.

[17] J. M. Laniel, N. Hô, R. Vallée, and A. Villeneuve, "Nonlinear-refractive-index measurement in As_2S_3 channel waveguides by asymmetric self-phase modulation," *Journal of the Optical Society of America B*, vol. 22, no. 2, pp. 437–445, 2005.

[18] K. Ogusu, J. Yamasaki, S. Maeda, M. Kitao, and M. Minakata, "Linear and nonlinear optical properties of Ag–As–Se chalcogenide glasses for all-optical switching," *Optics Letters*, vol. 29, no. 3, pp. 265–267, 2004.

[19] A. Andriesh, "Chalcogenide glasses as multifunctional photonic MATERIALS," *Journal of Optoelectronics and Advanced Materials*, vol. 7, no. 6, pp. 2931–2939, 2005.

[20] T. Wágner and P. J. S. Ewen, "Photo-induced dissolution effect in $Ag/As_{33}S_{67}$ multilayer structures and its potential application," *Journal of Non-Crystalline Solids*, vol. 266–269, pp. 979–984, 2000.

[21] G. Beadie, W. S. Rabinovich, J. Sanghera, and I. Aggarwal, "Fabrication of microlenses in bulk chalcogenide glass," *Optics Communications*, vol. 152, no. 4–6, pp. 215–220, 1998.

[22] T. Kawaguchi and S. Maruno, "Kinetic study of metallic silver photoinduced surface deposition phenomenon," *Japanese Journal of Applied Physics A*, vol. 33, no. 6, pp. 3417–3418, 1994.

[23] R. A. Synowicki and T. E. Tiwald, "Optical properties of bulk c-ZrO2, c-MgO and a-As_2S_3 determined by variable angle spectroscopic ellipsometry," *Thin Solid Films*, vol. 455-456, pp. 248–255, 2004.

[24] D. Franta, I. Ohlídal, M. Frumar, and J. Jedelský, "Optical characterization of chalcogenide thin films," *Applied Surface Science*, vol. 175-176, pp. 555–561, 2001.

[25] R. Todorov and K. Petkov, "Light induced changes in the optical properties of thin As–S–Ge(Bi, Tl) films," *Journal of Optoelectronics and Advanced Materials*, vol. 3, no. 2, pp. 311–317, 2001.

[26] R. Todorov, A. Paneva, and K. Petkov, "Optical characterization of thin chalcogenide films by multiple-angle-of-incidence ellipsometry," *Thin Solid Films*, vol. 518, no. 12, pp. 3280–3288, 2010.

[27] R. Todorov, T. Iliev, and K. Petkov, "Light-induced changes in the optical properties of thin films of Ge–S–Bi(Tl, In) chalcogenides," *Journal of Non-Crystalline Solids*, vol. 326-327, pp. 263–267, 2003.

[28] L. F. Santos, A. Ganjoo, H. Jain, and R. M. Almeida, "Optical and spectroscopic characterization of germanium selenide glass films," *Journal of Non-Crystalline Solids*, vol. 355, no. 37–42, pp. 1984–1988, 2009.

[29] M. V. Sopinskyy, P. E. Shepeliavyi, A. V. Stronski, and E. F. Venger, "Ellipsometry and AFM study of post-deposition transformations in vacuum-evaporated As–S–Se films," *Journal of Optoelectronics and Advanced Materials*, vol. 7, no. 5, pp. 2255–2266, 2005.

[30] V. Pamukchieva and A. Szekeres, "Optical properties of $Ge_xSb_{20-x}Te_{80}$ thin films and their changes by light illumination," *Optical Materials*, vol. 30, no. 7, pp. 1088–1092, 2008.

[31] T. Ide, M. Suzuki, and M. Okada, "Spectroscopic ellipsometry studies on optical constants of $Ge_2Sb_2Te_5$ used for phase change optical disks," *Japanese Journal of Applied Physics B*, vol. 34, no. 4, pp. L529–L532, 1995.

[32] M. Yaman, H. E. Kondakci, and M. Bayindir, "Large and dynamical tuning of a chalcogenide fabry-perot cavity mode by temperature modulation," *Optics Express*, vol. 18, no. 3, pp. 3168–3173, 2010.

[33] S. Y. Kim, S. J. Kim, H. Lim et al., "Spectro-ehipsometry investigation of cascaded crystallization behavior of phase-change Ge–Sb–Te alloy," *Japanese Journal of Applied Physics B*, vol. 40, no. 3, pp. 1575–1577, 2001.

[34] S. H. Messaddeq, V. K. Tikhomirov, Y. Messaddeq, D. Lezal, and M. Siu Li, "Light-induced relief gratings and a mechanism of metastable light-induced expansion in chalcogenide glasses," *Physical Review B*, vol. 63, no. 22, Article ID 224203, 5 pages, 2001.

[35] Y. Murakami, T. Ogawa, M. Wakaki, and S. Kawabata, "In-situ ellipsometric observations of thickness change in the layers of Ag/a-As_2S_3 film system with progression of photodoping," *Japanese Journal of Applied Physics A*, vol. 39, no. 2, pp. 509–510, 2000.

[36] V. Pamukchieva, A. Szekeres, and D. Arsova, "Spectroscopic ellipsometry study of the effect of illumination and thermal annealing on the optical constants of thin Ge–As–S films," *Physica Scripta*, vol. 83, no. 2, Article ID 025405, 2011.

[37] I. Ohlídal, D. Franta, M. Šiler et al., "Comparison of dispersion models in the optical characterization of As–S chalcogenide thin films," *Journal of Non-Crystalline Solids*, vol. 352, no. 52–54, pp. 5633–5641, 2006.

[38] J. Tasseva, R. Todorov, D. Tsankov, and K. Petkov, "Optical properties of multi-component arsenic-containing chalcogenide thin films," *Journal of Optoelectronics and Advanced Materials*, vol. 9, no. 2, pp. 344–347, 2007.

[39] K. Petkov, T. Iliev, R. Todorov, and D. Tzvetkov, "X-ray microanalysis and optical properties of thin As–S–Bi (Tl) films," *Vacuum*, vol. 58, no. 2, pp. 321–326, 2000.

[40] L. Tichý, H. Tichá, P. Nagels, and E. Sleeckx, "A review of the specific role of oxygen in irreversible photo- and thermally induced changes of the optical properties of thin film amorphous chalcogenides," *Optical Materials*, vol. 4, no. 6, pp. 771–779, 1995.

[41] C. K. Carniglia, "Ellipsometric calculations for nonabsorbing thin films with linear refractive-index gradients," *Journal of Optical Society of America A*, vol. 7, no. 5, pp. 848–856, 1990.

[42] D. E. Aspnes, J. B. Theeten, and F. Hottier, "Investigation of effective-medium models of microscopic surface roughness by spectroscopic ellipsometry," *Physical Review B*, vol. 20, no. 8, pp. 3292–3302, 1979.

[43] G. E. Jellison Jr., "Spectroscopic ellipsometry data analysis: measured versus calculated quantities," *Thin Solid Films*, vol. 313-314, pp. 33–39, 1998.

[44] M. K. Smit and J. W. Verhoof, "Accuracy analysis in multiple angle of incidence ellipsometry," *Thin Solid Films*, vol. 189, no. 2, pp. 193–203, 1990.

[45] R. Swanepoel, "Determination of the thickness and optical constants of amorphous silicon," *Journal of Physics E*, vol. 16, no. 12, pp. 1214–1222, 1983.

[46] R. Todorov, D. Tsankov, J. Pirov, and K. Petkov, "Structure and optical properties of thin As_2S_3-In_2S_3 films," *Journal of Physics D*, vol. 44, no. 30, Article ID 305401, 2011.

[47] K. Petkov, G. Vassilev, R. Todorov, J. Tasseva, and V. Vassilev, "Optical properties and structure of thin films from the system $GeSe_2$-Sb_2Se_3-AgI," *Journal of Non-Crystalline Solids*, vol. 357, no. 14, pp. 2669–2674, 2011.

[48] K. Tanaka, "Optical properties and photo-induced changes in amorphous As–S films," *Thin Solid Films*, vol. 66, no. 3, pp. 271–279, 1980.

[49] M. I. Kozak, V. N. Zhikharev, V. Y. Loya, I. P. Studenyak, I. I. Shpak, and I. I. Turok, "An ellipsometric study of relaxation-induced changes in the optical characteristics and structural inhomogeneity of As_2S_3 glassy thin films," *Technical Physics Letters*, vol. 32, no. 5, pp. 456–458, 2006.

[50] J. Tasseva, R. Todorov, and K. Petkov, "Linear and non-linear optical properties of thin films from the system As–S–Se," *Journal of Optoelectronics and Advanced Materials*, vol. 11, no. 9, pp. 1257–1260, 2009.

[51] E. Vateva, D. Arsova, E. Skordeva, and V. Pamukchieva, "Irreversible and reversible changes in band gap and volume of chalcogenide films," *Journal of Non-Crystalline Solids*, vol. 326-327, pp. 243–247, 2003.

[52] R. Todorov, J. Tasseva, T. Babeva, and K. Petkov, "Multilayer As_2Se_3/GeS_2 quarter wave structures for photonic applications," *Journal of Physics D*, vol. 43, no. 50, Article ID 505103, 8 pages, 2010.

[53] F. Abelès and M. L. Thèye, "Méthode de calcul des constantes optiques des couches minces absorbantes à partir de mesures de réflexion et de transmission," *Surface Science*, vol. 5, no. 3, pp. 325–331, 1966.

[54] I. Chambouleyron, S. D. Ventura, E. G. Birgin, and J. M. Martínez, "Optical constants and thickness determination of very thin amorphous semiconductor films," *Journal of Applied Physics*, vol. 92, no. 6, pp. 3093–3102, 2002.

[55] R. Andrade, E. G. Birgin, I. Chambouleyron, J. M. Martinez, and S. D. Ventura, "Estimation of the thickness and the optical parameters of several stacked thin films using optimization," *Applied Optics*, vol. 47, no. 28, pp. 5208–5220, 2008.

[56] T. P. Martin, "Stability of As–S clusters," *The Journal of Chemical Physics*, vol. 80, no. 1, pp. 170–175, 1984.

[57] K. Petkov, R. Todorov, D. Kozhuharova, L. Tichy, E. Cernoskova, and P. J. S. Ewen, "Changes in the physicochemical and optical properties of chalcogenide thin films from the systems As–S and As–S–Tl," *Journal of Materials Science*, vol. 39, no. 3, pp. 961–968, 2004.

[58] F. L. McCrackin and J. P. Colson, "Computational techniques for the use of the exact drude equations in reflection problems," in *Ellipsometry in the Measurement of Surfaces and Thin Films*, E. Passaglia, R. R. Stromberg, and J. Kruger, Eds., National Bureau of Standards Miscellaneous Publication 256, pp. 61–82, National Bureau of Standards, Washigton, DC, USA, 1964.

[59] R. Swanepoel, "Determination of surface roughness and optical constants of inhomogeneous amorphous silicon films," *Journal of Physics E*, vol. 17, no. 10, pp. 896–903, 1984.

[60] O. S. Heavens, *Optical Properties of Thin Solid Films*, Butterworth Scientific, London, UK, 1955.

[61] J. P. Borgogno, B. Lazarides, and E. Pelletier, "Automatic determination of the optical constants of inhomogeneous thin films," *Applied Optics*, vol. 21, no. 22, pp. 4020–4029, 1982.

[62] S. Chattopadhyay, Y. F. Huang, Y. J. Jen, A. Ganguly, K. H. Chen, and L. C. Chen, "Anti-reflecting and photonic nanostructures," *Materials Science and Engineering R*, vol. 69, no. 1–3, pp. 1–35, 2010.

[63] A. Lalova and R. Todorov, "Optical properties of porous chalcogenide films for sensor application," *Journal of Physics: Conference Series*, vol. 398, no. 1, Article ID 012023, 2012.

[64] J. Tasseva, R. Todorov, T. Babeva, and K. Petkov, "Structural and optical characterization of Ag photo-doped thin $As_{40}S_{60-x}Se_x$ films for non-linear applications," *Journal of Optics*, vol. 12, no. 6, Article ID 065601, 2010.

10

Progress in Dual (Piezoelectric-Magnetostrictive) Phase Magnetoelectric Sintered Composites

Rashed Adnan Islam[1] and Shashank Priya[2]

[1] *Philips Lumileds Lighting Co., 370 W. Trimble Road, San Jose, CA 95131, USA*
[2] *Materials Science and Engineering, Virginia Tech, Blacksburg, VA 24061, USA*

Correspondence should be addressed to Rashed Adnan Islam, rashed_ms@yahoo.com

Academic Editor: Vladimir Petrov

The primary aims of this review article are (a) to develop the fundamental understanding of ME behavior in perovskite piezoelectric-spinel magnetostrictive composite systems, (b) to identify the role of composition, microstructural variables, phase transformations, composite geometry, and postsintering heat treatment on ME coefficient, and (c) to synthesize, characterize, and utilize the high ME coefficient composite. The desired range of ME coefficient in the sintered composite is 0.5–1 V/cm·Oe. The studies showed that the soft piezoelectric phase quantified by smaller elastic modulus, large grain size of piezoelectric phase (~1 μm), and layered structures yields higher magnitude of ME coefficient. It is also found that postsintering thermal treatment such as annealing and aging alters the magnitude of magnetization providing an increase in the magnitude of ME coefficient. A trilayer composite was synthesized using pressure-assisted sintering with soft phase [0.9 PZT–0.1 PZN] having grain size larger than 1 μm and soft ferromagnetic phase of composition $Ni_{0.8}Cu_{0.2}Zn_{0.2}Fe_2O_4$ [NCZF]. The composite showed a high ME coefficient of 412 and 494 mV/cm·Oe after sintering and annealing, respectively. Optimized ferrite to PZT thickness ratio was found to be 5.33, providing ME coefficient of 525 mV/cm·Oe. The ME coefficient exhibited orientation dependence with respect to applied magnetic field. Multilayering the PZT layer increased the magnitude of ME coefficient to 782 mV/cm·Oe. Piezoelectric grain texturing and nanoparticulate assembly techniques were incorporated with the layered geometry. It was found that with moderate texturing, d_{33} and ME coefficient reached up to 325 pC/N and 878 mV/cm·Oe, respectively. Nanoparticulate core shell assembly shows the promise for achieving large ME coefficient in the sintered composites. A systematic relationship between composition, microstructure, geometry, and properties is presented which will lead to development of high-performance magnetoelectric materials.

1. Magnetoelectric Effect

The coexistence of coupled electrical and magnetic properties in the "magnetoelectric" material has led to the possibility of developing smarter and smaller electronic components. In order to make this possibility a reality, significant efforts are required to understand the science of magnetoelectric (ME) behavior and apply this understanding to develop higher sensitivity material. The primary aim of this review is to identify the role of composition, microstructural variables, composite geometry, texturing, postsintering heat treatment, and nanoscale assembly on ME coefficient. The overall objective is to analyze the synthesis process and properties and provide new applications for ME composite.

Magnetoelectric effect is the combination of two types of materials property such as magnetostriction and piezoelectricity [1–3]. Magnetoelectric effect can be described as the induced electrical polarization under magnetic field or induced magnetization under electric field. Under applied alternating magnetic field, the magnetostrictive phase produces strain which is transferred on to the piezoelectric phase that converts strain into electric charge. On the other hand, under applied electric field, piezoelectric phase produces strain which is transferred on to the magnetostrictive phase that converts it into magnetic field. The former is called the direct magnetoelectric effect and the later is called the converse magnetoelectric effect [4, 5]. These materials are extremely promising for applications such as sensors, actuator, transducer, storage devices, and various other applications [6–8]. The interrelation between ferroelectricity and magnetism allows the magnetic control of ferroelectric properties and vice versa [9–12]. Figure 1 shows the interrelation between electric field, magnetic field, and stress. A ferromagnetic and ferroelectric phase can be combined in several different ways, including naturally occurring composites, artificially designed composites, and *in-situ* sintered composites [13, 14].

2. Thermodynamic Considerations of Magnetoelectricity

The thermodynamic consideration of magnetoelectric effect is obtained from the expansion of free energy of the system in terms of magnetic and electric field, such as

$$G(\vec{E},\vec{H}) = G_o - P_s\,\bar{E} - M_s\,\bar{H} - \frac{1}{2}\chi_{ij}(E)E_iE_j - \frac{1}{2}\chi_{ij}(H)\bar{H}_i\bar{H}_j - \alpha_{ij}\bar{E}_i\bar{H}_j, \quad (1)$$

where $\bar{E}$ and $\bar{H}$ are the electric field and magnetic field, respectively [15]. Differentiation of (1) gives us polarization and magnetization as follows:

$$P_i = -\left(\frac{\delta G}{\delta E}\right) = P_s + \chi_{ij}E_j + \alpha_{ij}H_j, \quad (2)$$

$$M_i = -\left(\frac{\delta G}{\delta H}\right) = M_s + \chi_{ij}H_j + \alpha_{ij}. \quad (3)$$

Here, α_{ij} is the magnetoelectric tensor. Magnetoelectric effect combines two important materials property, permittivity, and permeability, and for a single phase material, they define the upper limit of α_{ij} as follows [15]:

$$\alpha_{ij} < \sqrt{\varepsilon_{ij}\mu_{ij}}. \quad (4)$$

Many of the single-phase materials posses either low permittivity or low permeability or both, as a consequence of which the magnetoelectric coupling is small. For magnetoelectric composite materials, an indirect coupling between ferroelectric and ferromagnetic phase can be established via strain. Piezoelectricity defines the materials property that converts applied stress into proportional electric charge. The linear constitutive equations for a piezoelectric material are given as

$$D_3 = \varepsilon_{33}^{T}E_3 + d_{33}T_3, \quad (5)$$

$$S_3 = d_{33}E_3 + s_{33}^{E}T_3, \quad (6)$$

where D, E, T, and S are the dielectric displacement, electric field, stress, and strain, ε is the permittivity, s is elastic compliance, and d is piezoelectric charge constant. The constitutive equations for magnetostriction are given as [16]

$$S = s^{H}T + qH, \quad (7)$$

$$B = qT + \mu^{T}H, \quad (8)$$

where B, q, μ, and H are magnetic induction, piezo magnetic coefficient, permeability, and magnetic field. Magnetoelectric coefficient of a composite can be described in direct notation of tensors as

$$T = cS - e^{T}E - \alpha S^{ms}, \quad (9)$$

$$D = eS + \varepsilon E + \alpha H, \quad (10)$$

$$B = \mu(\varepsilon, E, H)H, \quad (11)$$

where σ, ε, D, E, B, H, c, and K are the stress, strain, electric displacement, electric field, magnetic induction, magnetic field, stiffness constant at constant field, and dielectric constant at constant strain and magnetic field, respectively [17]. The models proposed in literature show variation of ME coefficient with piezoelectric coefficient, piezomagnetic coefficient, and elastic compliances of piezoelectric and magnetostrictive phases. Srinivasan et al. have proposed the ME coefficient as

$$\frac{\delta E_3}{\delta H_1} = \frac{-2d_{31}^{p}q_{11}^{m}v^{m}}{(s_{11}^{m}+s_{12}^{m})\varepsilon_{33}^{T,P}v^{p} + \left(s_{11}^{p}+s_{12}^{p}\right)\varepsilon_{33}^{T,P}v^{m} - 2\left(d_{31}^{p}\right)^{2}v_{m}}, \quad (12)$$

where d_{31}^{p} is the piezoelectric coefficient, v^m and v^p are the volume of magnetic and piezoelectric phase, t^m and t^p are the thickness of magnetic and piezoelectric phase, s_{11}^{p}, s_{12}^{p} are the elastic compliances for piezoelectric phase, s_{11}^{m}, s_{12}^{m} are the elastic compliances for magnetostrictive phase, q_{11} is the piezomagnetic coefficient of the magnetic phase, and $\varepsilon_{33}^{T,P}$ is the permittivity of the piezoelectric phase. Dong et al. have derived the expression for the magnetoelectric coefficient in T-T mode as

$$\left|\frac{dV}{dH}\right|_{T-T} = \beta\frac{n(1-n)Ad_{33,m}d_{31,p}g_{31,p}}{S_{11}^{E}\left[nS_{11}^{E}(1-k_{31}^{2}) + (1-n)S_{11}^{H}\right]}, \quad (13)$$

where β is a constant related to DC magnetic field with a maximum value of 1, n is the thickness ratio of magnetostrictive layer to composite thickness, d is the piezoelectric strain constant, s is the elastic constant, g is the piezoelectric voltage constant, A is the cross-sectional area of the laminate, and k is the electromechanical coupling factor. The relevance of the constant β was not explained in this model. Clearly, the ME coefficient is directly related to piezoelectric constant which is related to dielectric permittivity [$d_{31}^{2} = k_{31}^{2}(s_{11}^{E}\varepsilon_{33}^{T})$] and piezomagnetic coefficient which is related to permeability [$q_{11,m} = \mu_{33}.s_{33}.\lambda_{33}$].

3. Magnetoelectric Single Phase Materials

The history of the magnetoelectric effect dates back to 1894 when Pierre Curie observed that a body with asymmetric molecules gets electrically polarized under magnetic field and vice versa. In 1958, Landau and Lifshitz conducted the feasibility study on the presence of magnetoelectric effect in crystals of specific symmetry. Later Dzyaloshinskii showed the violation of time-reversal symmetry explicitly for a single-phase antiferromagnetic system (Cr_2O_3). In 1960, Astrov presented the experimental evidence of ME effect in Cr_2O_3 in a wide temperature range of 80–330 K [18, 19]. Smolensky and Ioffe synthesized the antiferromagnetic ferroelectric perovskite ceramic $Pb(Fe_{1/2}Nb_{1/2})O_3$(PFN). Later, single crystals of PFN were grown and the presence of weak spontaneous magnetic moment in the ferroelectric phase below 9 K was confirmed [20]. Recently, single-phase thin-film multiferroic systems, such as perovskite type $BiFeO_3$ or $BiMnO_3$, the boracite family, $BaMF_4$ families where M is the divalent transition metal, hexagonal $RMnO_3$ where R is rare earth metals, are of interest to understand the magnetoelectric coupling between two types of dipoles [21–24]. These single-phase materials show two transitions, one corresponding to paraelectric to ferroelectric phase and another corresponding to ferro/ferri/antiferromagnetic to paramagnetic phase. The mechanism for the observance of magnetoelectric effect in single-phase materials is specific to a family. For example—stereochemical activity of Bi lone pair for $BiFeO_3$, frustrated spin resulting from magnetoelastic lattice modulations for $TbMnO_3$, and charge ordering in doped perovskite [14]. Single-phase magnetoelectric (ME) materials suffer from the drawback that the ME effect is considerably weak. They can be used only at very low temperature and involve expensive materials and processing technique, and they have disadvantage of degradation under cyclic conditions. The reason behind this is that the single-phase materials have two order parameters: polarization and magnetization. One of the order parameters is generally large and the other is small, which results in minor ME exchange between the two subsystems.

4. Magnetoelectric Composite Materials

Better alternatives to single-phase materials are ME composites that have large magnitudes of ME voltage coefficient. The ME effect can be realized by using composites of piezomagnetic and piezoelectric phases or magnetostrictive and piezoelectric phases. Moreover, these composites are easy to fabricate compared to the single-phase materials, cost-effective, and they have higher working temperature range [2, 25, 26].

5. Properties of Composite

Composite properties can be divided into three categories; sum properties, product properties, and combination properties. A sum property is average of the contribution from individual phases present in the composite such as density, resistivity, and so forth. Product property can be explained as follows: If one phase shows the property A → B and the other shows B → C, then the composite will show new property A → C. Magnetoelectric effect is product property of the composite. Other examples of product property are piezoresistance (piezoelectric and magnetoresistance), magnetoresistance (magnetostriction and piezoresistance), photostriction (photoconductivity and electrostriction), and piezoluminescence (piezoelectricity and electroluminescence) [27]. If the average magnitude of the property in composite exceeds the value of the end components, then it is an example of combination properties. If two phases show convex and concave type of sum properties, then the combination of these two will produce a maximum at an intermediate phase composition.

6. Connectivity of Composite

Composite geometry can be designed based upon the connectivity. For two-phase system, there are ten types of connectivity. For three- and four-phase system, there are 20 and 35 types of connectivity. The connectivity in general can be given by the ratio $((n + 3)!)/3!n!$, where n is number of phases. Figure 1 shows the geometry for different connectivities (1-0, 2-0, 3-0, 1-1, 2-1, 3-1, 2-2, 2-3, and 3-3) in two, phase composite. A 3-0 connectivity implies particulate type of composite where one phase will be the matrix and the other phase will be dispersed in the matrix. A 2-2 connectivity represents the layered geometry which can be bilayer, trilayer, or multilayer. The phases are connected in plane but not out of plane. This connectivity is most commonly used for multilayer capacitors and laminated magnetoelectric composite [28].

7. Types of Magnetoelectric Composites

Among various possibilities for magnetoelectric composites, *in-situ*, sintered particulate and nanoparticulate composites are of 0-3 connectivity whereas laminate and multilayer composites are mainly of 2-2 type and the nanopillars are 1-3 composites. In the following subsection, each of the composite systems will be discussed separately with a list of different reported values for ME coefficient.

7.1. In-Situ Composite Using Unidirectional Solidification. The original work on ME composites was done at Philips Laboratories [30–34]. The ME composites were synthesized by unidirectional solidification of eutectic composition in quinary system Fe-Co-Ti-Ba-O. The unidirectional solidification results in decomposition of eutectic liquid into alternate layers of the constituent phases; a piezoelectric perovskite phase (P) and a piezomagnetic spinel phase (S) (L → P + S). The results showed that presence of excess TiO_2(1.5 wt%) gives a high ME voltage coefficient of dE/dH = 50 mV/cm·Oe [31]. However, other compositions showed a lower dE/dH in the range of 1–4 mV/cm·Oe. In subsequent work, a high ME coefficient of 130 mV/cm·Oe was obtained for eutectic composition of $BaTiO_3$-$CoFe_2O_4$ by unidirectional solidification [33]. This value is about an order of magnitude higher than that of single phase Cr_2O_3 (dE/dH = 20 mV/cm·Oe). However, unidirectional solidification has several complications for implementing in practical applications such as long duration, and it requires critical control over the composition [32]. This type of composite allows forming spinel dendrites in perovskite matrix, which are detrimental for obtaining large piezoelectric and magnetoelectric effect. The resistivity of these *in-situ*

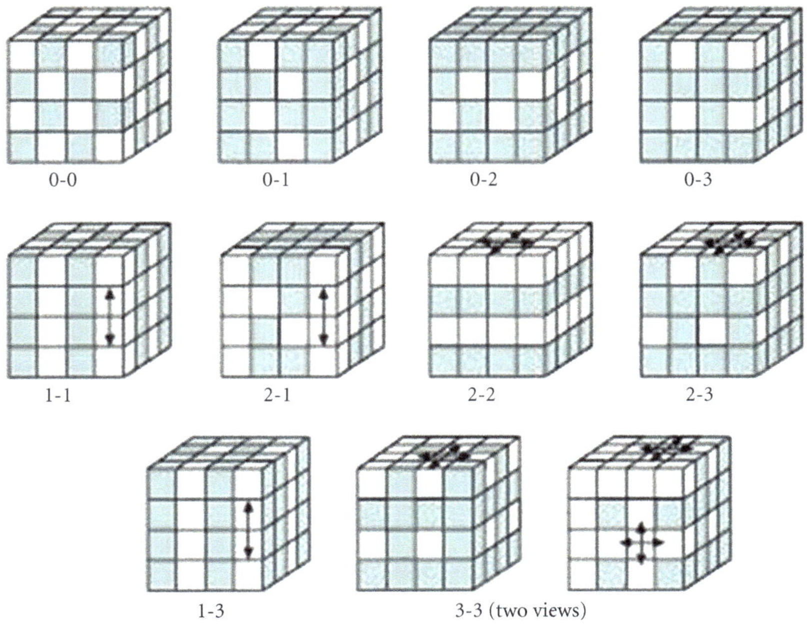

FIGURE 1: Different types of connectivity of the two-phase composite system [28, 29].

composites is very low of the order of 10^8–10^9′Ω-cm, which presents difficulty in electrical poling.

7.2. Laminated Composite. Recently, giant ME response has been reported for laminate composites, fabricated by bonding piezoelectric plate/disk/fibers between two layers of magnetostrictive plates/disks/foils [34–37]. Figures 2(a) and 2(b) show the piezoelectric plates or fiber sandwiched between two magnetostrictive plates or foils, respectively. In these designs, the external magnetic field creates strain in the magnetostrictive layers which is transferred on to the piezoelectric material through bonding layer thereby producing electrical charge across the electrodes. The electrical charge is amplified and converted into voltage using a capacitive circuit. A magnetic DC bias of varying magnitude is essential to obtain the magnetostrictive strain. The magnetoelectric laminate composite, embedding piezoelectric PMN-PT single crystal between magnetostrictive Terfenol-D single crystal, exhibited the ME coefficient of 10.30 V/cm·Oe, which is ~80 times higher than that previously reported in either naturally occurring magnetoelectrics or artificially designed composites (ADCs). Srinivasan et al. have investigated the bilayer and multilayer structure of PZT and ferromagnetic ferrites. It was shown that multilayer structures can provide ME coefficient as high as 1500 mV/cm·Oe [38, 39]. Laminated structures of Ni-PZT-Ni were found to exhibit ME coefficient of 400–450 mV/cm·Oe [40]. Three-phase composites of PZT/epoxy/Terfenol-D or PZT/PVDF/Terfenol-D have been reported to posses the ME coefficient in the range of 50–150 mV/cm·Oe [41]. Dong et al. have reported composite structures consisting of piezofiber laminated between two three-phase composites of PZT/epoxy/Terfenol-D or PZT/PVDF/Terfenol-D have been reported and the magnitude of ME coefficient was found to be in the range of 50–150 mV/cm·Oe [41]. Dong et al. have reported composite structure consisting of piezofiber laminated between two sheets of high-permeability magnetostrictive FeBSiC alloys [42]. This structure exhibited a giant response of 22 V/cm·Oe at 1 Hz.

7.3. ME Micro/Nanostructures. Recently, magnetoelectric composite-based nanostructures have been synthesized using thin-film deposition techniques such as sol-gel deposition, PLD, and RF magnetron sputtering. The commonly experimented geometries are bilayers, hetrostructures, self-assembled structures, and nanowires. Zheng et al. have synthesized the self-assembled nanostructure in $BaTiO_3$-$CoFe_2O_4$ [BTO-CFO] composite system consisting of CFO nanopillars embedded in BTO matrix [8]. TEM investigation of self-assembled nanopillars show 20–30 nm CFO pillars embedded in BTO matrix. Compared to bulk ME composites, nanostructures offer the possibility of much higher interlayer interaction. The disadvantage of nanopillars is high leakage current due to continuous path of low-resistance ferrite pillars. The multilayer geometry has a disadvantage of clamping effect of the substrate. ME nanoscale heterostructures of PZT-CFO double-layer thin films, PZT-NFO nanoparticulate composites, and PZT-CFO nanowires have also been reported [44]. Figure 3 shows the HRTEM and cross-sectional TEM of 4 different systems. Figure 3(a) shows the HRTEM image of BTO-NFO heterostructure with sharp SAED pattern in the inset. The misfit strain is around −0.8% which is smaller than that for bulk ceramics. The lattice constant calculated from SAED pattern was 3.943 and 4.023z Å for BTO and 8.244 Å for NFO [45]. Figure 3(b) shows the cross-sectional HRTEM for PZT-CZFO composite thin film. Clear interface between PZT and CZFO was observed as well as the interface between the CZFO and

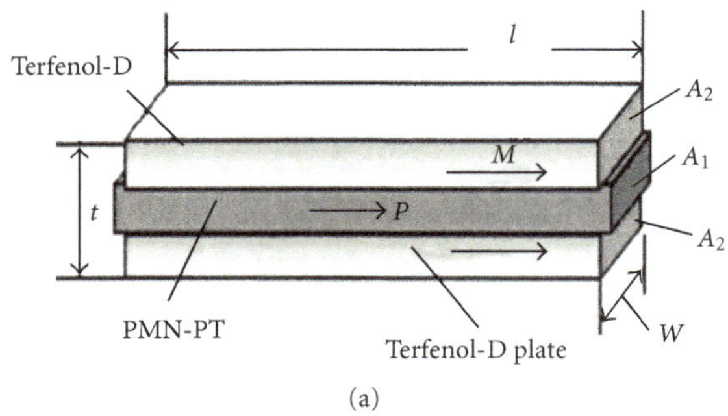

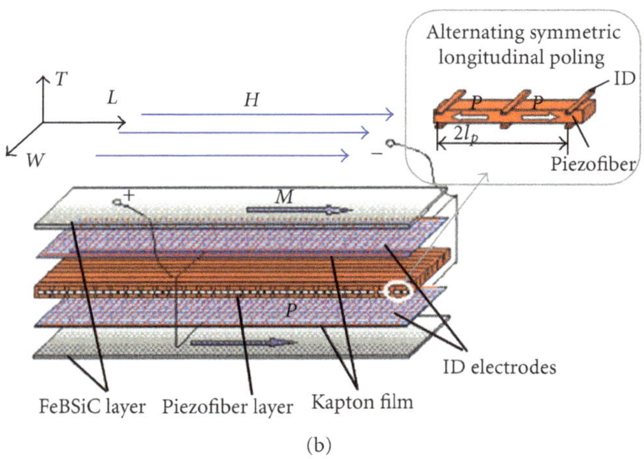

FIGURE 2: Schematic diagram of laminated composite (a) Terfenol-D sandwich structure with PMN-PT in the middle and (b) Magnetic FeBSiC top and Bottom layer with intermediate layer as piezoelectric fiber [42, 43].

the substrate. No evidence of second phase was found in these layered films by HRTEM. In the same literature, it was found that if the PZT was deposited earlier, then it has a preferential growth along the (100) crystallographic orientation [46]. Figure 3(c) shows the HRTEM image of PZT-NFO nanoparticulate composite thin film. Clearly a well-defined boundary with different atomic arrangements was observed in this kind of composite which indicates good dispersion of NFO particles in PZT matrix [47]. The cross-sectional TEM image shown in Figure 3(d) corresponds to PZT (440 nm) LSMO (330 nm) bilayer deposited on top of (001) $LaAlO_3$ substrate. There were large amounts of void grain boundaries observed in LSMO phase, which may help to release the magnetostrictive strain along the normal of the plain direction [48].

7.4. Sintered Composites. Magnetoelectric (ME) effect in particulate sintered composites has generally been obtained by combining magnetostrictive and piezoelectric phases. Sintered composites fabricated by mixing of piezoelectric and ferromagnetic phases have been widely studied because of its simplicity in synthesis using well-established conventional ceramic processing technique. Various particulate composites of piezoelectric ($BaTiO_3$ or $Pb(Zr,Ti)O_3$) and magnetostrictive ($LiFe_5O_8$, $Ni_{1-x}Zn_xFe_2O_4$, $CoFe_2O_4$, or $CuFe_2O_4$) phases with different dimensional connectivity's have been reported in literature. Although the ME effect is known in various binary systems, little information is available on structure-property relationship. Sintered particulate composites have inferior properties compared to laminated ones because of drawbacks of low resistivity, interface defects, interface diffusion, and mismatch of elastic compliances. To enhance the ME coefficient of sintered composite, it is necessary to optimize the composition, microstructural features (grain size, grain orientation), and sintering parameters. In particulate composites, higher ME coefficient implies higher elastic coupling between the magnetic and piezoelectric phases. The elastic coupling can be maximized by having coherent response from the magnetostrictive phase under dc bias, so that the stress on the piezoelectric lattice across the grains is in phase with each other. This is only possible if there is uniform distribution of the magnetostrictive phase with soft piezoelectric phase. If the particles are crystallographically aligned with the matrix, more benefits can be obtained because of the built-in directionality. However, this is quite difficult to obtain in practice. An attempt has been made to achieve this by synthesizing textured composites. In addition to conventional ceramic processing and sol-gel route, one pot [49] is a process where the two phases are simultaneously synthesized in the same pot in order to get better homogenization and dispersion of magnetic phase. Heterocoagulation has been demonstrated to provide the core-shell structures of ferroelectric and ferromagnetic composite particles. Heterocoagulation process consists of coagulation of particles with opposite surface charges dispersed in a solution. At some specific pH value, one species has a positive surface charge density whereas the other one has a negative charge.

8. Problems and Shortcoming of Sintered Composite

The trend of the ME research since 2002 is to use a high magnetostriction material (Terfenol-D) or a high permeability material (Metglas) and use epoxy to bond with a PZT plate, rod, disk, fiber, and so forth. Different piezoelectric vibration modes such as bending, longitudinal, radial, and push-pull and so forth. have been utilized in laminated geometries. However, these reported structures have so far failed to identify the synthesis process that can eliminate the need for epxoy. Practically, industry would like to have a ceramic composition with relatively simple processing technique that can provide higher sensitivity. In order to get higher ME coefficient, the following problems needs to be addressed.

(1) How can the ME coefficient be enhanced in the bulk sintered composites?

(2) What are the microstructural variables that make significant contribution towards ME coefficient?

(3) Define a composition space that allows identification of specific elements providing enhanced ME coefficient.

(4) Define the geometry of the sintered composite that will lead to better coupling in dual phase structures.

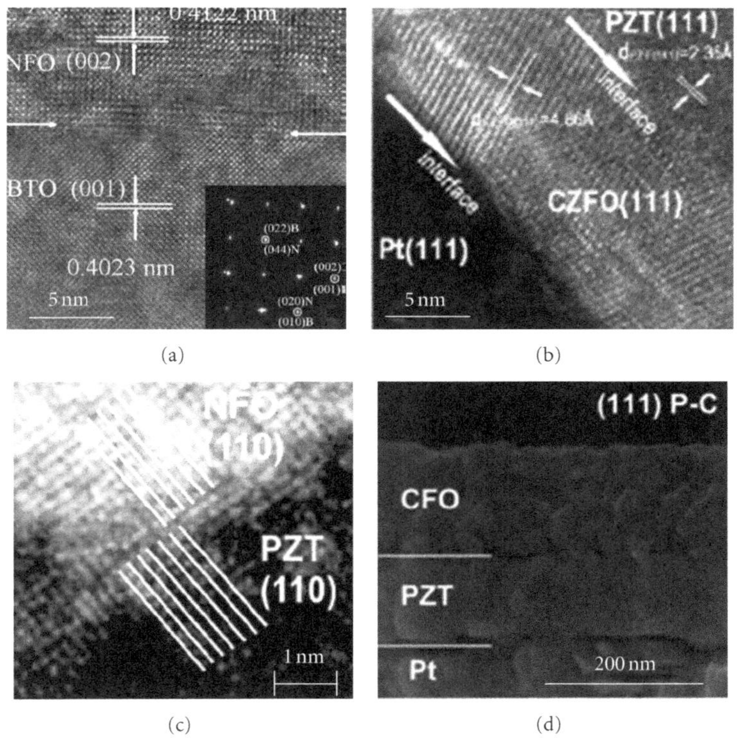

Figure 3: (a) HRTEM of BTO-NFO bilayer thin-film structure, (b) HRTEM of PZT-CZFO bilayer thin-film structure showing interface, (c) HRTEM of PZT-NFO nanoparticulate composite showing orientation relationship of NFO (110), and PZT (110), and (d) cross-sectional TEM image of PZT-CFO bilayer film [45–48].

(5) Establish the composition-microstructure-property relationships for dual phase sintered composites.

In order to solve these problems, several factors need to be investigated [1–31].

(1) Effect of the piezoelectric and magnetostrictive composition.

(2) New microstructures that can provide enhanced resistivity.

(3) Processing steps that can provide equilibrium between the two phases.

(4) Removal of defects at the piezoelectric—magnetostrictive interface formed due to mismatch in shrinkage rate and sintering temperatures of individual phases.

(5) Control of grain size and orientation.

(6) Periodic distribution of two phases.

(7) Geometry of the phase distribution.

A significant effort is required to address the above-mentioned factors. Realization of sintered composite with large ME coefficient will provide an adequate energy conversion device from magnetic to electric. In addition to magnetic field sensing and current probes, various other commercial application are possible such as gyrator, read-only memory device, transducers, filters, recording heads, and sensors. Magnetoelectricity in two-phase (piezoelectric-magnetostrictive) composite is a strain mediated effect where magnetostrictive strain produced by applied magnetic field is transferred on to the piezoelectric phase. In order to achieve efficient strain transfer, piezoelectric materials with matching elastic properties and higher piezoelectric voltage constant (g) are suitable. The mechanical properties of PZT-based ceramics are nearly similar for various compositional modifications. However, the electrical properties can be changed to a large extent by tailoring the microstructure, defect concentration, and domain states. Figure 4 shows the design concept for a high-performance magnetoelectric composite. The prime requirement for this composite is high piezoelectric voltage constant, low loss material. This goal correlates with the design of high-energy density material as it will be shown here that the coefficients for energy are dependent on "g" constant.

9. Scope of this Review

The scope of improvement is in meeting with the challenges of sintering which lead to low ME coefficient as identified in previous sections. First of all, chemical reaction, low resistivity, and interface defects such as porosity, micro cracks make very weak coupling between the piezoelectric and magnetostrictive phase. Secondly, if the resistivity of the composite is low, electrical poling becomes very difficult due to high leakage current. Lastly, the connectivity of

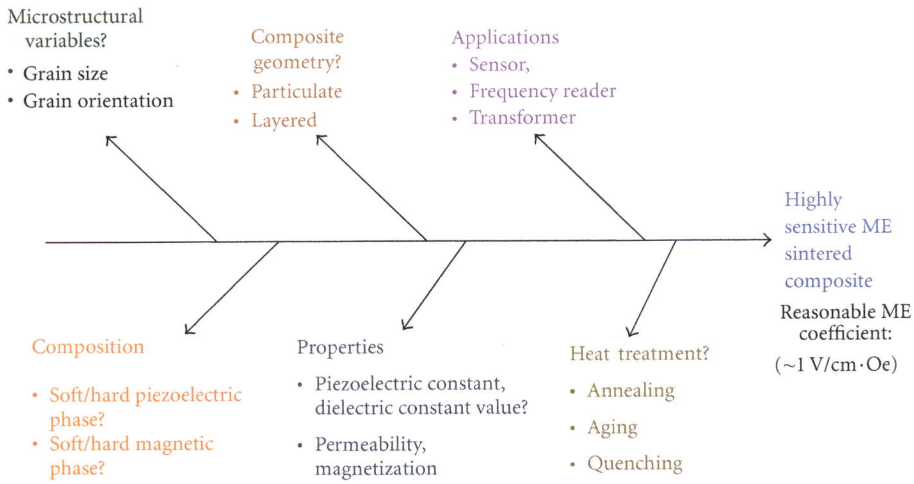

FIGURE 4: Fish-bone diagram for solving ME sintered composite issues.

the ferrite grains plays an important role. Good dispersion of ferrite grains inside the piezoelectric matrix is essential for sustaining sufficient electrical resistivity in the matrix. To further improve the property of the composite, comprehensive understanding of each parameter that controls the magnetoelectric coefficient is necessary and efforts have to be made to optimizing these controlling variables. Also the fabrication process for the composites should be designed to provide a cost-effective ME composites. ME composites can lead to miniaturization of several electronic components and thus cost-effective methodology will facilitate their deployment. Magnetoelectric MEMS devices can be designed to enhance the functionality of the system.

To understand the science of magnetoelectric materials and hence to synthesize the composite with high ME response an optimum system with best possible composite geometry needs to be realized. To start with, the selection of appropriate composition with suitable doping is the important step. The effect of microstructural features, sintering parameters, and heat treatment on the ME response should be analyzed. The specific factors that need to be considered are as following.

9.1. System Selection. The figure of merit for large magnetoelectric coefficient in a ferromagnetic-ferroelectric composite is given by the (14) as

$$\text{Figure or merit} = \sqrt{\mu\varepsilon}, \tag{14}$$

where μ is the magnetic permeability and ε is the dielectric permittivity. Thus, the primary criterion for the selection of the individual phases in the composite is to identify materials having similar crystallographic symmetry and possessing large magnetic permeability and dielectric permittivity. As is well known that compared to $BaTiO_3$ (BT), Pb (Zr Ti) O_3 [PZT] has stronger piezoelectric and dielectric properties, higher Curie temperature, higher resistivity, and lower sintering temperature. Also there is no high-temperature phase transition in the PZT system. Investigations have shown that the nickel-based ferrite particles are very stable in the doped-PZT matrix and do not react at high sintering temperatures of 1250°C [50]. Ni-based ferrites are soft compared to Co-ferrite and have higher permeability. Also an addition of Zn in the ferrite structure improves the permeability, magnetization and reduces the Curie temperature. Thus, PZT and Ni ferrite satisfies the phase equilibrium criterion for large ME effect.

9.2. Composition of Multiphase System. Composition of the piezoelectric and magnetostrictive phase needs to be optimized. For piezoelectric phase using PZT as the base composition, the composition can be varied using acceptor or donor dopant. Acceptor dopants (Mn, Zn, etc.) result in decreased piezoelectric, dielectric constant, and dielectric loss. On the other hand, donor dopants (Nb) are higher in valance and create cation vacancies, resulting in increased piezoelectric and dielectric constant with increased loss. The composition for ferromagnetic phase can be varied to change the magnetic properties such as magnetization, permeability, and coercivity. Ni- and Co-based ferrites are the base composition because ferrites have higher resistivity than metallic ferromagnets. To improve the resistivity, Mn can be doped, and to increase the magnetization, Zn can be doped.

9.3. Microstructural Features: Grain Size. Effect of grain size on piezoelectric, dielectric, and ferroelectric properties has been widely studied in literature [51–55]. It is well known that piezoelectric and dielectric properties drop rapidly below critical grain size (~100 nm) [56, 57]. SEM analysis combined with the surface tension measurement indicate that the surface bond contraction due to small size induces a compressive stress on the inner part of a grain and this effect plays an important role in ferroelectric materials in the nanometer size range. The induced stress causes decrease of Curie temperature and spontaneous polarization with decreasing grain size. The domain wall contribution has an opposite effect as compared with the surface bond contraction-induced effect. When the grain size decreases to a value comparable to the width of domain walls, pinning points develop inside the grains and the domain wall motion is inhibited. The reduced wall mobility causes a decrease in the relative permittivity. The measured value is a competition between the increase of relative permittivity by the surface bond contraction effect and its decrease by the domain wall pinning effect.

Although the effect of grain size on piezoelectric, dielectric, and ferroelectric properties has been studied, the effect of microstructure on magnetoelectric property has not been studied in detail. As it is well known that the magnetoelectric property depends upon the piezoelectric and dielectric property, which in turn depends on grain size, magnetoelectric property should also depend upon grain size. Optimization of the grainsize will lead to the increased ME coefficient of the composite.

9.4. Microstructural Features: Texturing. Crystallographic texturing of piezoelectric grains in certain orientation can improve the piezoelectric, ferroelectric response because better spontaneous polarization can be achieved than in randomly oriented grain. Many researchers have reported the improvement of piezoelectric responses by incorporating the textured grain growth in the ceramic structure [58–60]. The most common fabrication processes for texturing are hot forging, hot pressing, gradient sintering, and templated grain growth. Templated grain growth is the most popular grain orientation technique, where a seed crystal is added to the matrix containing much finer equiaxed powder. During sintering, the matrix densifies around the templates with the same orientation as that of the templates.

9.5. Composite Geometry. In bulk composite, the resistivity of the composite drops rapidly as the connectivity of the ferrite particles forms a conductive path along the sample. This allows charge leakage through the ferrite. Also, the flux lines cancel each other resulting in a low stress generation. Layered structures are much more attractive as the stress generation is much higher and each phase preserves its physical properties. Also in applications like magnetic field sensing devices, two factors should be clearly addressed: (i) variation of the magnetoelectric (ME) response as a function of the magnetic field direction, and (ii) variation of the ME response as a function of the DC bias. Ideally, the ME response should be a high number without any DC bias. In order to fabricate miniatured sensors, the approach based on cofiring is more attractive similar to that of multilayer ceramic capacitors.

9.6. Postsinter Heat Treatment. For high magnetoelectric coefficient, homogenized piezoelectric grain, even-distribution of magnetostrictive particle, stress-free interface, and most importantly an equilibrium structure are necessary. A postsintering treatment is necessary to promote the equilibrium condition and to reduce the stress arising from the misfit strain [61–63]. Optimized annealing and aging condition improve the individual piezoelectric and ferromagnetic properties.

10. Specific Results

As mentioned earlier, it is important to understand the factors controlling the magnetoelectric coefficient. Experimental investigation was conducted to identify the correlation between composition, microstructure, and property. The optimization was conducted in terms of magnetoelectric coefficient (mV/cm·Oe) with supporting structural, electrical, and magnetic characterization.

11. High-Energy Density Piezoelectric Phase for ME Composites

The magnitude of the transduction is governed by the effective piezoelectric strain constant, d, and the effective *piezoelectric voltage constant,* g. Using the linear constitutive piezoelectric equations, a relation between the energy density of the piezoelectric material and the transduction coefficient (d.g) under an applied stress X can be derived Under an applied force ($F = X \cdot A$, where A is the area), the open circuit output voltage (V) of the ceramic is given as

$$V = E \cdot t = -g \cdot X \cdot t = -\frac{g \cdot F \cdot t}{A}, \quad (15)$$

where t is the thickness of the ceramic, E is the electric field, and g is the piezoelectric voltage coefficient given as

$$g = \frac{d}{\varepsilon_0 \varepsilon^X}, \quad (16)$$

where ε^X is the dielectric constant under constant stress condition. The charge (Q) generated on the piezoelectric ceramic is given by the relation:

$$D = \frac{Q}{A} = \frac{E}{\beta^X} = \frac{V \cdot \varepsilon_0 \varepsilon^X}{t}, \quad (17)$$

or

$$\frac{Q}{V} = \frac{\varepsilon^X \varepsilon_0 A}{t} = C, \quad (18)$$

where D is the dielectric displacement, C is the capacitance, and ${}^X\beta$ is the dielectric susceptibility of the material under constant stress condition. Dielectric susceptibility is equal to the inverse of dielectric permittivity tensor component and can be defined from the constitutive equation for a linear piezoelectric material as

$$E = -g X + \beta X D. \quad (19)$$

Equation (18) shows that at low frequencies, a piezoelectric plate can be assumed to behave like a parallel plate capacitor. Hence, electric energy available under ac stress excitation is given as $U = 1/2CV^2$ or energy per unit volume

$$u = \frac{1}{2}(\mathrm{d.g}) \cdot \left(\frac{F}{A}\right)^2. \quad (20)$$

Equations (15) and (20) show that for a given material of fixed area and thickness, a material with high (d.g) product and high g constant will generate high voltage and power when the piezoelectric ceramic is directly employed for mechanical to electrical energy conversion. Since, in the case of magnetoelectric composites, the magnetostrictive phase applies the stress on the piezoelectric phase through elastic coupling the high-energy density material will lead to higher response.

12. Design of High "g" Materials

Enhancement in the magnitude of the piezoelectric voltage constant requires an optimization in the magnitude of the piezoelectric stress constant and dielectric constant. In general, the variation in the magnitude of piezoelectric stress constant and dielectric constant is similar and they increase or decrease simultaneously upon modification with dopants or processing technique resulting in minor changes in the magnitude of their ratio. A novel processing strategy is required to obtain large g constants [64]. The criterion for maximization of the product (d.g) with respect to the microstructure variable x (such as grain size) can be determined as follows:

$$\begin{aligned}\frac{\partial}{\partial x}(\mathrm{d.g}) &= d\cdot\left(\frac{\partial g}{\partial x}\right)+g\cdot\left(\frac{\partial d}{\partial x}\right)\\ &= 2\cdot\left(\frac{d}{\varepsilon}\right)\cdot\left(\frac{\partial d}{\partial x}\right)-\left(\frac{d}{\varepsilon}\right)^2\left(\frac{\partial \varepsilon}{\partial x}\right)=0\end{aligned} \tag{21}$$

or

$$\frac{\partial(\ln d^2)}{\partial x}=\frac{\partial(\ln\varepsilon)}{\partial x}\Longrightarrow d=\varepsilon^{0.5},\ \text{where } \varepsilon=\varepsilon^X\varepsilon_o. \tag{22}$$

Once again, differentiating (21) with respect to x and substituting the condition given by (22), it can be shown that

$$\left(\frac{\partial^2}{\partial x^2}\right)(\mathrm{d.g})=\left(\frac{\partial\varepsilon}{\partial x}\right)^2\frac{1}{\varepsilon^{1.5}}\left(1-\frac{1}{\varepsilon^{0.5}}\right)<0. \tag{23}$$

Thus, the condition given by (22) represents the maximum magnitude of the product (d.g). In general, the condition given by (22) can be expressed as [64]

$$|d|=\varepsilon^n. \tag{24}$$

Using (24), the maximum magnitude of the product (d.g) is given as

$$\mathrm{Max}(\mathrm{d.g})=\varepsilon^{2n-1}, \tag{25}$$

where n is the material constant which is fixed by the magnitude of the piezoelectric and dielectric constants. It has been shown that for all the commercial polycrystalline ceramic materials, the magnitude of n lies in the range of 1.1 to 1.3, (24) represents a deterministic rule for realizing the high-energy density piezoelectric materials. According to this equation, a giant enhancement in the magnitude of the g constant will be obtained by reducing the magnitude of the n. As the magnitude of n decreases and approaches the theoretically minimum possible value of 0.5, the magnitude of product (d.g) reaches the maximum possible magnitude of unity. In order to obtain large magnitude of the product (d.g), the piezoelectric compositions can be tailored by modifying with the dopant or processing technique such that the change in piezoelectric and dielectric constant leads to a decrement in the magnitude of constant n.

The piezoelectric constant, d, can be written in terms of the electromechanical coupling factor, k, and the elastic compliance, s, as follows:

$$d=k\sqrt{\varepsilon.s}. \tag{26}$$

Substituting (24) into (26), it can be shown that

$$n-\frac{1}{2}=\frac{\ln(k)}{\ln(\varepsilon)}+\frac{1}{2}\frac{\ln(s)}{\ln(\varepsilon)}. \tag{27}$$

Eqution (27) indicates that a material with low magnitude of n will require low magnitude of coupling coefficient and elastic compliance. However, large coupling factor and low losses are important for efficient energy conversion.

13. Defect Chemistry

Defects and substituents are known to affect the electromechanical properties of the normal ferroelectrics in both poled and unpoled conditions [65–70] Generally, acceptor doping (Fe, Mn, Ni, Co) in the PZT-based ceramics results in (i) decreased dielectric constant and loss, (ii) lower elastic compliance, (iii) lower electromechanical coupling factor, and (iv) lower electromechanical losses. In contrast, donor doping (La, Sb, Bi, W) results in (i) increased dielectric constant and loss, (ii) increased elastic compliance, (iii) increased electromechanical coupling factor, and (iv) increased electromechanical losses. Thus, an acceptor modified material seems to be appropriate for obtaining low value of n. In acceptor modified materials, oxygen vacancies are created for charge compensation, which are frozen into the ceramic during quenching from high temperature. The electroneutrality reaction representing the equilibrium between acceptor ions (A) and oxygen vacancies is expressed as

$$n+2[A']=p+[V_o^{\bullet\bullet}]. \tag{28}$$

Assuming a constant lead activity, the formation of oxygen vacancy can be expressed as

$$O_o\longrightarrow\frac{1}{2}O_2+V_o^{\bullet\bullet}+2e'. \tag{29}$$

The rate constant for this reaction can be expressed as

$$K=[V_o^{\bullet\bullet}]PO_2^{1/2}e^2, \tag{30}$$

where PO_2 is the partial pressure of the oxygen. If PO_2 is constant, then the formation of $V_o^{\bullet\bullet}$ can be expressed as:

$$\mathrm{Pb}(g)\longrightarrow\mathrm{Pb_{Pb}}+V_o^{\bullet\bullet}+2e'. \tag{31}$$

These oxygen vacancies can reside in the material in many possible ways: (i) associated defect dipole structure with acceptor, (ii) diffuse to domain walls, or (iii) compensate the local charge fluctuation at domain walls. The mechanism for introduction of "hard" properties depends on the way in which oxygen vacancies distribute themselves. In Mn-modified PZT ceramics, it is believed that hardening occurs due to the pinning of domain boundaries by Mn—oxygen vacancy defect structure [67]. Further, it is believed that Mn^{+3} attaches strongly to oxygen vacancies due to Jahn Teller effect. This prevents conduction and reduces the domain wall mobility resulting in reduced losses [65]. The MPB (Zr : Ti = 52 : 48) composition in PZT shows the highest piezoelectric properties such as high piezoelectric constant d_{33}, coupling factor k, and dielectric constant. These properties drop on either side of the MPB. The compositions on tetragonal side (Zr : Ti ratio of 56 : 44) have less variation in the dielectric and piezoelectric properties as compared to composition on the rhombohedral side with reference to MPB [71]. Thus, by selecting a composition on the tetragonal side and doping with suitable donor (Nb^{+5}, etc.) and acceptor (Zn^{+2}, Ni^{+2}, etc.) ions, it is possible to formulate a composition, which has relatively higher piezoelectric constant d_{33} but lower dielectric constant.

14. Effect of Doping: Piezoelectric Phase

Different piezoelectric composition can be fabricated using hard and soft characteristics. Two different compositions such as 0.9 $Pb(Zr_{052}Ti_{0.48})O_3$-0.1 Pb $(Zn_{1/3}Nb_{2/3})O_3$ [PZT(soft)] and 0.9 $Pb(Zr_{0.56}Ti_{0.44})O_3$-0.1 $Pb[(Zn_{0.8/3}Ni_{0.2/3})Nb_{2/3}]$ O_3 + 2 (mol %) MnO_2 [PZT (hard)] were used to see the effect of soft and hard nature of PZT on magnetoelectric coefficient. The magnetostrictive phase was fixed at 20 mole% $Ni_{0.8}Zn_{0.2}Fe_2O_4$. It was found that as the composition becomes electrically and mechanically soft, the polarization increases. Figure 5(a) shows that the polarizations of PZT, PZT (hard), and PZT (soft) were recorded as 14.61, 23.54, and 31.65 $\mu C/cm^2$, respectively. As we dope the PZT with Nb, the structure becomes softer, and therefore, piezoelectric, dielectric property increases. This also affects the piezoelectric, dielectric, and magnetoelectric properties of the composite. Piezoelectric constant (d_{33}) increases from 72 to 125 pC/N and dielectric constant increases from 635 to 915. From Figure 5(b), it is clearly observed that the PZT (soft)-20 NZF shows the maximum magnetoelectric response of 186.49 mV/cm·Oe, whereas pure PZT shows the value of 127.96 mV/cm·Oe, an increase of about ~50%. Corresponding elastic compliances (S_{11}) of these two compositions are 1.11 x 10^{-11} and 1.734 x 10^{-11} m^2/N—this data indicates a simple metric that softer the materials better is the magnetoelectric response.

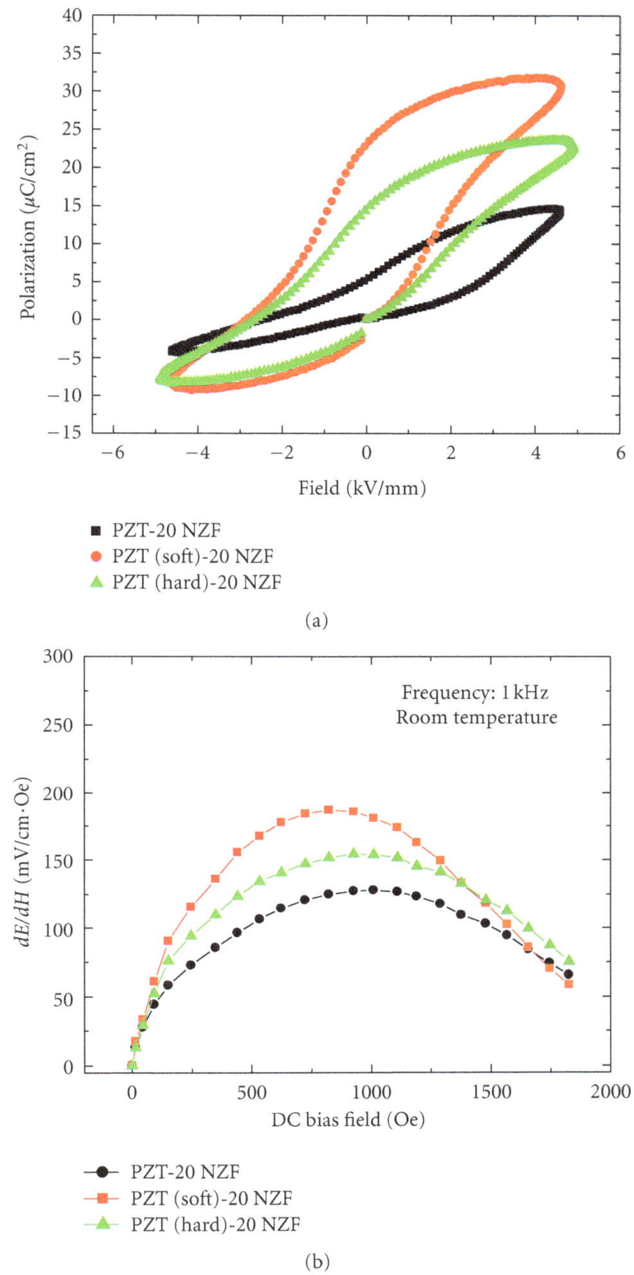

FIGURE 5: Effect of piezoelectric phase on the saturation polarization and magnetoelectric coefficient of ME composites.

15. Effect of Doping: Magnetostrictive Phase

Magnetic property, such as saturation magnetization and coercivity, controls the magnetoelectric property of the composite. A soft high initial permeability magnetic phase with magnetostriction will lead to larger domain rotation. Saturation magnetization is directly related to the permeability and the magnetostriction. The higher the magnetization, the higher is the permeability and the magnetostriction. Several compositions such as $NiFe_{1.9}Mn_{0.1}O_4$(NFM), $Ni_{0.8}Zn_{0.2}Fe_2O_4$ (NZF), $CoFe_2O_4$ (CFO), and $Co_{0.6}Zn_{0.4}Fe_2O_4$ (CZF) have been studied and it was found that the $Ni_{0.8}Zn_{0.2}Fe_2O_4$ has the smallest coercivity (52 Oe) and also it has the largest magnetoelectric coefficient. CFO is magnetically hard (coercivity 318.34 Oe) compared

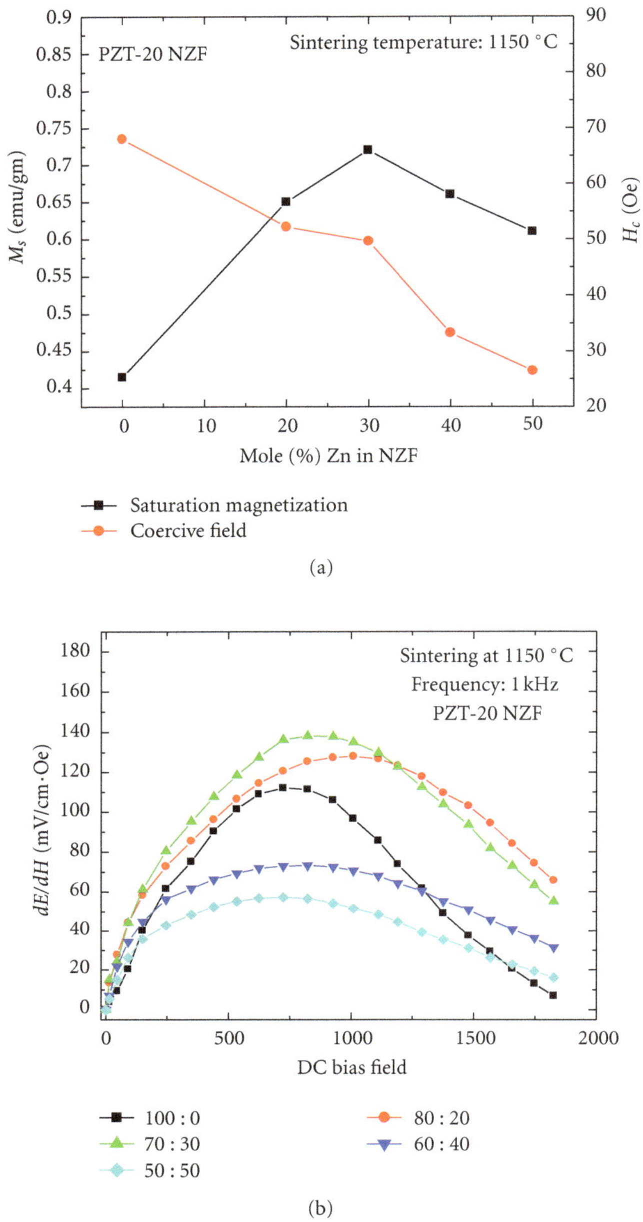

FIGURE 6: Effect of Zn doping on magnetic and magnetoelectric properties of PZT-20 NZF.

to NZF and has very small magnetoelectric response due to limits in domain wall rotation. Also, the control of Zn doping has a major effect of the magnetic property. It was found from Figure 6(a) that with the increase in Zn concentration, the coercivity drops. But the magnetization shows maxima at 30% Zn doping. The optimum concentration of Zn doping was found to be 30 mole%, at which the highest saturation magnetization was observed with moderate coercivity. At this composition, high ME coefficient of 144 mV/cm·Oe was recorded as observed from Figure 6(b)—an increase of about 25%.

16. Effect of Grain Size

Figures 7(a)–7(d) show the SEM microstructure of ME composite with 830, 758, 111, and 97 nm average grain size, respectively. All the microstructures were found to be dense and well sintered. The density data shows that as the grain size of the composite decreases the density increases, and at grain size of 97 nm the densification reaches about 98%. The densification decreases to 93% as the grain size increases to 750 nm. Figure 8(a) shows the variation of longitudinal piezoelectric strain constant (d_{33}) and dielectric constant ($\varepsilon_r/\varepsilon_o$) with grain size. It can be noticed in this figure that there is a critical grain size below which the piezoelectric properties drop rapidly. The critical grain size is in the range of 100–150 nm. Above 200 nm, the piezoelectric properties increase slowly and saturate at 600 nm. The magnitude of d_{33} and $\varepsilon_r/\varepsilon_o$ for ME composite with grain size of 97 nm were 44 pC/N and 659. This magnitude increases rapidly to 65 pC/N, and 746 at grain size of 215 nm. Above 600 nm, the piezoelectric and dielectric properties saturate at the magnitude of 72 pC/N and 781. The increase in the piezoelectric and dielectric properties can be explained in terms of domain wall motion. For larger grains, where the size of the domain is smaller than the size of the grains, the movement of the domain walls is easier. On the other hand, in small grains, the movement of domain walls is restricted by the grain boundary. As a result during poling, the domain switching will be difficult which reduces the piezoelectric properties [52–72]. Figure 8(b) shows the variation of the ME coefficient as a function of the grain size. As the grain size decreases, ME coefficient drops and shows similar trend as that of piezoelectric property. Above 200 nm grain size, the ME coefficient increases and it saturates above 600 nm as shown in Figure 8(b). For the grain size of 97 and 830 nm, the ME coefficient was found to be as 54.4 and 157.5 mV/cm·Oe, respectively. So, it can be noticed that larger grain size of around 1 μm or more will enhance the magnetoelectric property.

17. Effect of Geometry

In particulate composite, as the piezoelectric and magnetostrictive particles are randomly mixed, a large number of interfaces are created. It was found in our previous studies that at the interfaces of eutectic-based $BaTiO_3$-$CoFe_2O_4$ (BTO-CFO) composite, elemental interdiffusion occurs. In the BTO-30 CFO composite, where BTO rich islands were observed in BTO-CFO matrix, considerable amount of interdiffusion occurs. Elemental mapping of Co and Fe show that Fe and Co have diffused through $BaTiO_3$. Therefore, it is advantageous to create artificial interfaces as bilayer, trilayer, and so forth. The advantages of these types of artificial interfaces are interdiffusion is reduced; resistivity is high, less chance of chemical reaction, and piezoelectric/magnetostrictive properties are higher. Figure 9(a) shows the high magnification interface microstructure of PZT (hard)-$Ni_{0.6}Zn_{0.2}Cu_{0.2}Fe_2O_4$ (NCZF) cofired bilayer composite observed using Zeiss Leo Smart SEM on the polished and thermally etched samples. The

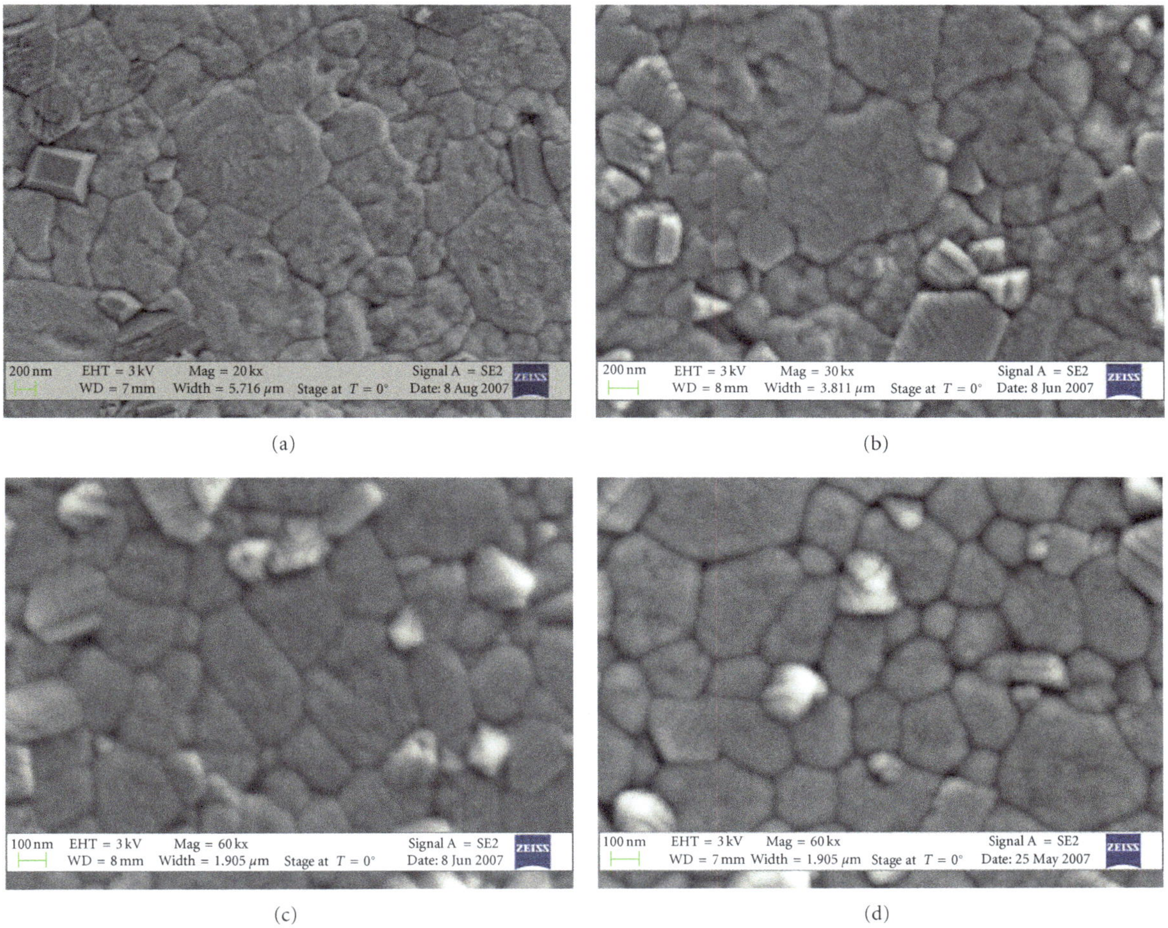

FIGURE 7: Microstructures of ME composites with varying grain size (a) 830 nm, (b) 758 nm, (c) 111 nm, and (d) 97 nm.

PZT grain size in the vicinity of the interface was larger than the grains far away from the interface. Liquid-phase sintering was found to occur in the PZT phase near the interface which may be associated to the diffusion of Cu at the interface and formation of low melting temperature phase. A low magnification microstructure of the interface showed the presence of porosity along the interface. An SEM elemental mapping of Cu shows that more concentration of Cu was found in the PZT phase close to interface which may be responsible for the liquid phase sintering [73–75]. Figure 9(b) shows the ferroelectric property of the cofired bilayer composite. Polarization of 60 $\mu C/cm^2$ and 0.1% strain was recorded at 4.5 kV/mm.

A square P-E loop was measured for the bilayer which signifies the high resistivity. The shape of the strain versus electric field loop was a normal butterfly loop observed for PZT. Figure 10(a) compares the magnetization-magnetic field hysteresis curve of bilayer and bulk composites measured with an alternating gradient magnetometer (AGM) up to 1.4T. It was found that the saturation magnetization (M_s) is higher in bilayer composite than the bulk, and the coercivity (H_c) drops from 43 Oe to a very low magnitude of 2.8 Oe. Figure 10(b) shows the temperature dependence of magnetization for bilayer and bulk composites measured in a high-temperature oven installed with Quantum design physical property measurement system (PPMS). The magnetic Curie temperature (T_c) was determined from the intersection of extrapolations of the greatest slope and flat region above T_c. It was found that T_c of the bilayer composite was higher than that of bulk composite. It can be noticed from Figure 10(b) that the magnetization-temperature behavior of the bilayers shows a change in slope at 530 K (~257 °C). This temperature range is similar to where peak in the dielectric behavior was observed. The results indicate the presence of secondary phase in the bilayer composites which is reflected both in dielectric and magnetic properties. This transition could be related to the presence of Cu_2O/CuO or a second ferromagnetic phase evolving from $CuFe_2O_4$ which forms a gradient composite structure with the PZT. Shrotri et al. have shown that doping above 20 mole % Cu in $(Ni,Zn)Fe_2O_4$ makes it unstable and at first CuO starts to form followed by a secondary-phase corresponding to the formulation $CuFe_2O_4$ [76, 77]. The free CuO if present on the piezoelectric side will react with the PZT-PZN to form low-temperature melting composition which is commonly observed in the low-temperature sintering of

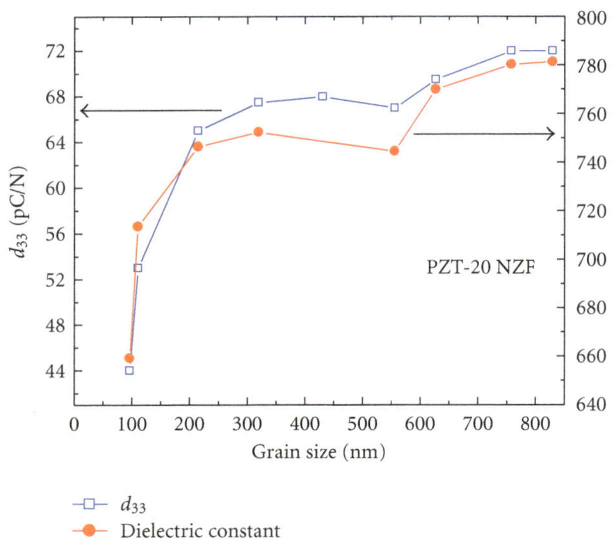

(a)

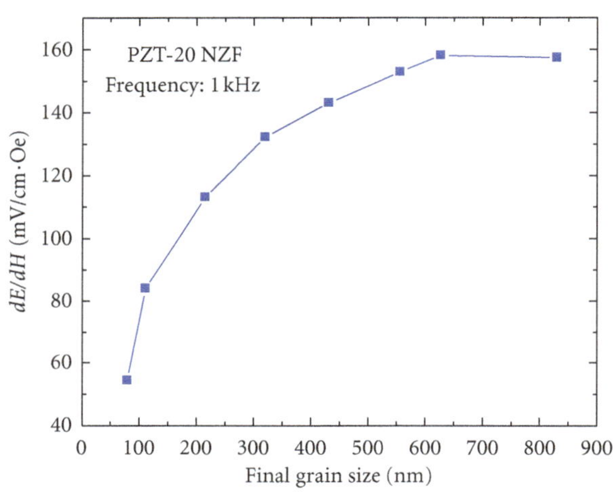

(b)

Figure 8: Effect of ferroelectric grain size on the piezoelectric, dielectric and magnetoelectric property of PZT-20 NZF composite.

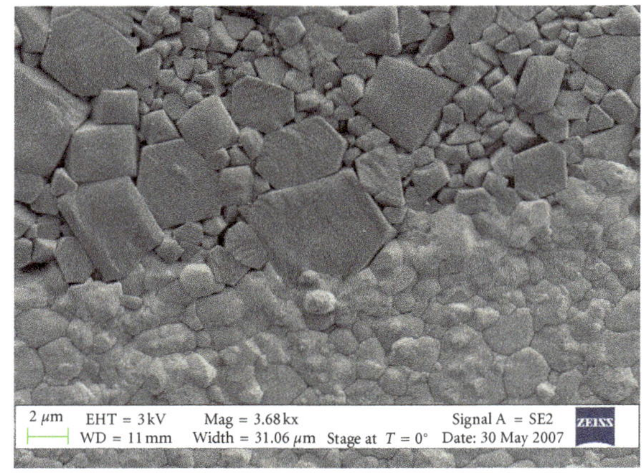

(a)

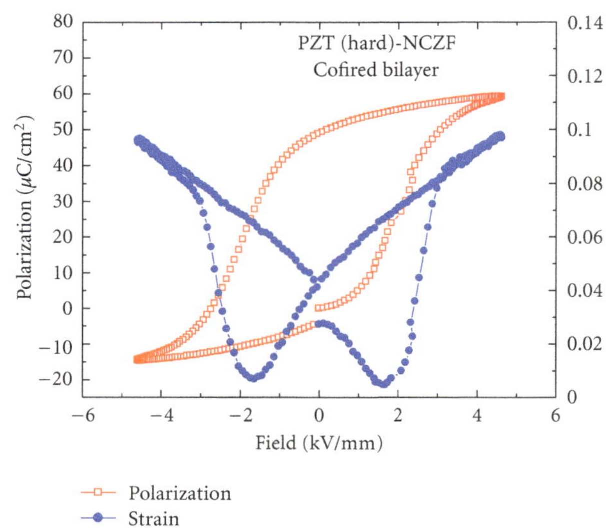

(b)

Figure 9: SEM images of PZT (hard)-NCZF bilayer composite and its magnetoelectric response.

ceramics. CuO is a common liquid-phase sintering agent in that case. The composite of this low temperature melting phase and $CuFe_2O_4$ phase may be associated with the transitions observed in the PZT-NCZF composites though more careful studies using high-temperature X-ray diffraction are required to confirm this hypothesis. The magnetization versus temperature behavior for bilayer composites can be resolved as summation of two curves, shown by different colored dashed line in Figure 10(b). $CuFe_2O_4$ has a Curie temperature of 743 K [78] and the dashed line (dark blue) indicates the Curie temperature of around 720–730 K. The second line (light blue) corresponds to the $(Ni,Zn)Fe_2O_4$, which has a Curie temperature of about ~640K. This value approximately corresponds to the curie temperature of $Ni_{0.65}Zn_{0.35}Fe_2O_4$ [79]. Thus, the overall magnetic response is average of the individual contributions from $CuFe_2O_4$ and $Ni_{0.65}Zn_{0.35}Fe_2O_4$ proportional to the volume fraction. It was found that in T-T mode maximum magnetoelectric coefficient of around 200 mV/cm·Oe was found in bilayer composite, more than 60% enhancement in the sensitivity. The enhancement can be explained due to the improved ferroelectric and piezoelectric property of the composite.

18. Effect of Annealing and Aging

Optimized post sinter thermal treatments can be done by annealing the composites at 800°C for 10hrs, followed by quenching in air and subsequent aging at 300°C for 5 hours. It was found on annealing and aging that (i) size and density of NFM phase reduced; (ii) PZT lattice constant expanded from (a = 3.87 Å, c = 4.07 Å) to (a = 4.07 Å, c = 4.09 Å); (iii) the ferroelectric and ferromagnetic Curie temperatures decreased by 8°C and 33°C, respectively; (iv) the magnetoelectric coefficient increased by ~50%. During annealing as the structure finds more time to reduce the defects, sharp grain boundaries were observed.

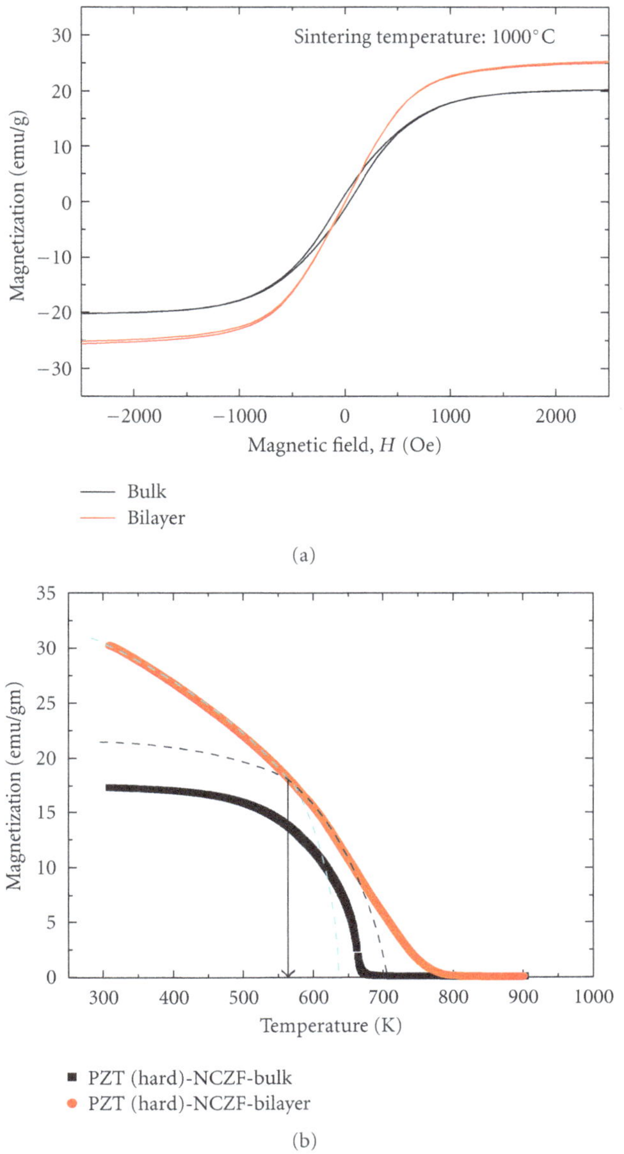

Figure 10: Ferromagnetic property of two types of composite. (a) Moment versus Field (hysteresis loop). (b) Magnetization as a function of temperature.

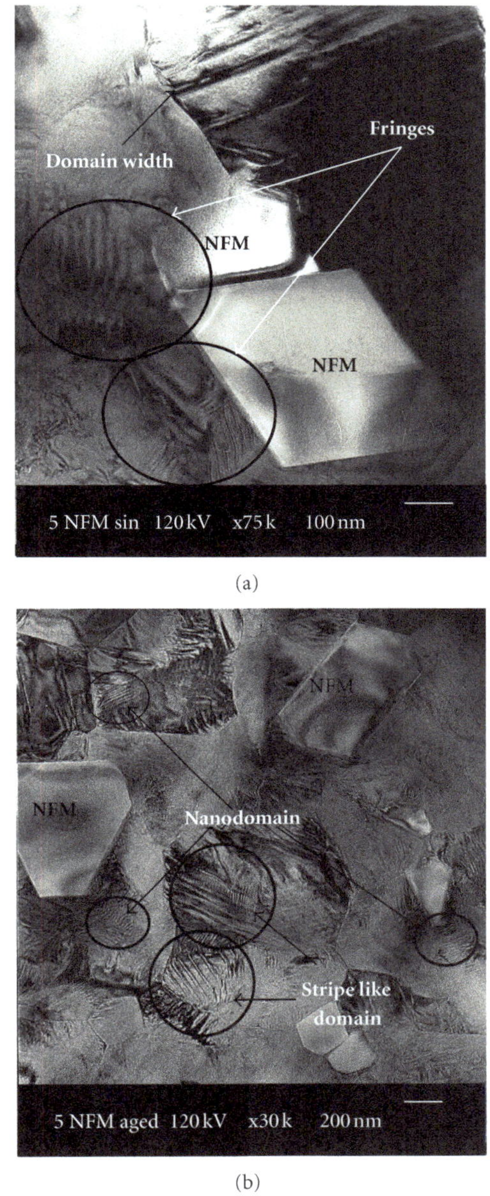

Figure 11: Bright-field TEM image of PZT-5 NFM composite. (a) After sintering, (b) after annealing and aging.

The sintered samples were found to consist of high misfit strain fields near the PZT/NFM interface which develop to accommodate the mismatch in the PZT and NFM lattice as shown by the bright field TEM images in Figure 11(a). The domain patterns had larger width characteristic of 90° domains and there is intergranular heterogeneity in domain width. The image (Figure 11(b)) after the aging process shows significantly reduced strains. The domain structure has stripe-like morphology and extend from grain boundary to grain boundary. A finer scale domain structure is also observed to exist within larger domain patterns. This finer domain pattern has striation-like morphology and is periodically spaced. Figures 12(a) and 12(b) show the dielectric constant and magnetization as a function of temperature for a PZT-5 at% NFM composite. Data are shown in each figure for the three different thermal conditions. The ferroelectric Curie temperature, as identified by the temperature of the maximum in the dielectric constant, was only decreased slightly (~8°C) by annealing and aging. However, the ferromagnetic one was decreased notably more: from 828 K in the as-sintered condition, to 788 K after annealing. In addition, it should be noted that the saturation magnetization at room temperature was increased from about 0.58 emu/gm to 0.64 emu/gm after annealing. This modest change in magnetization after annealing at 800°C may be related to an Mn valence change in the ferrite phase. Previously, it has been shown that $Mn^{+3} \rightarrow Mn^{+2}$

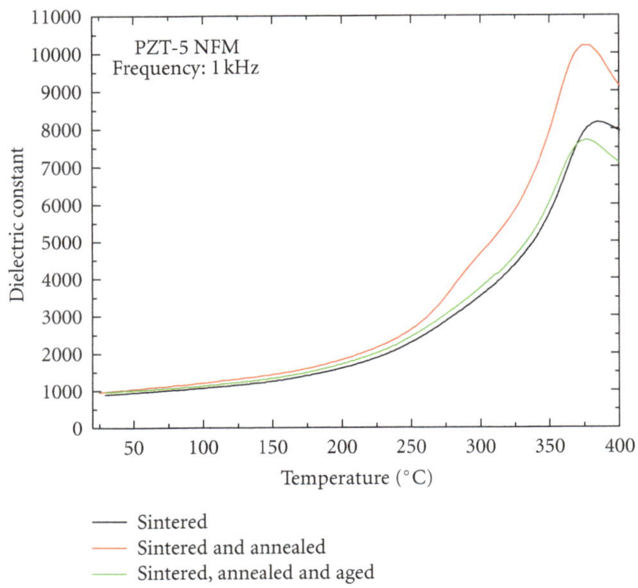

(a)

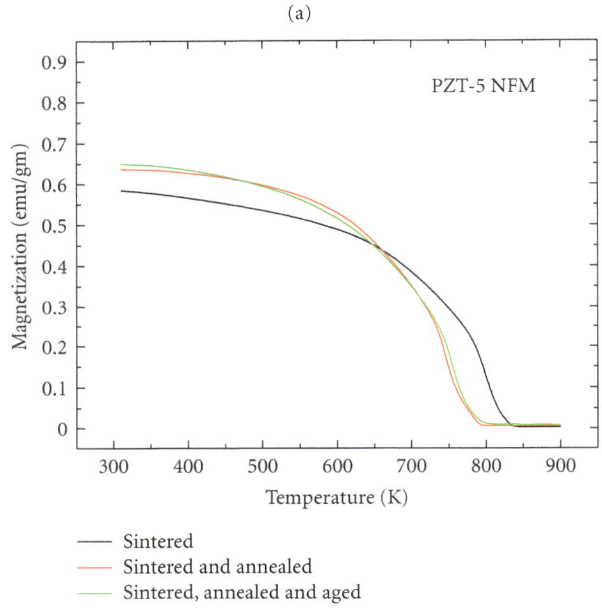

(b)

FIGURE 12: Temperature-dependent ferroelectric and ferromagnetic property of PZT-5 NFM composite. Temperature dependent, (a) dielectric constant, (b) saturation magnetization.

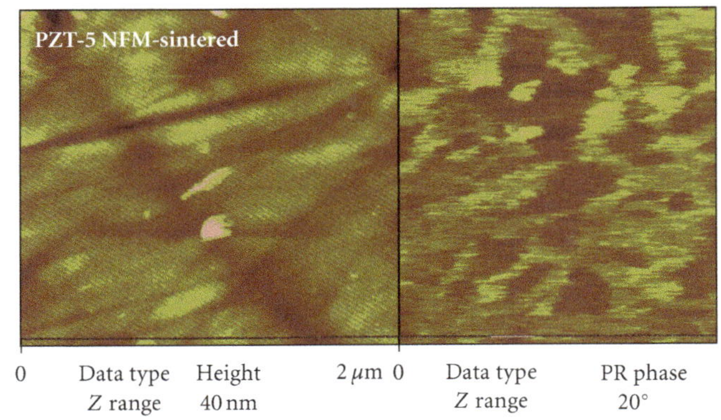

(a)

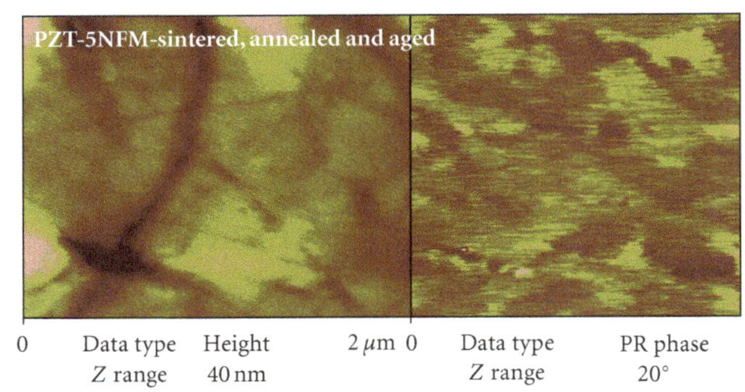

(b)

FIGURE 13: PFM images of PZT-5 NFM after (a) sintering, (b) annealing/aging.

begins to occur for temperatures above 700°C [80], resulting in an adjustment in the stoichiometry of the ferrite phase to $[Ni(Mn)_y]Fe_{1.9}Mn_{(0.1-y)}O_4$, where y is the concentration of Mn^{+2}. On quenching, this adjusted stoichiometry will be metastably frozen-in to room temperature. Since Mn^{+2} has a higher magnetic moment than Mn^{+3}, it is reasonable to expect a slight increase in saturation magnetization after annealing. ME coefficient of PZT-10 NFM was increased from 60mV/cm·Oe to 88mV/cm·Oe by annealing and aging: nearly a 50% increase. We attribute this increase is due to the expansion of lattice constant in PZT phase and reduction of defects at that interphase interfaces after annealing. Figures 13(a) and 13(b) compare the Piezo Force Microscopy images of the sintered and aged PZT-5NFM samples. It was found that in the sintered condition NFM phases are not homogenized as they are concentrated in one part. On the other hand in the aged condition, they are homogenized over the entire microstructure. Assuming that spinel phase is single domain, it can be hypothesized that the annealing and aging treatment results in redistribution of the magnetic phase.

The research on sintered composites has shown that cofiring multiple layers can provide higher ME coefficients. However, the drawback is that tape-casting process for synthesizing multilayers of heterogeneous materials is complex. Further, in order to improve the property of the sintered ME composites, the other variables such as composition, microstructure, geometry, texture, and postsinter heat treatment need to be optimized. As mentioned earlier, soft piezoelectric phase (high dielectric and piezoelectric constant), soft magnetic phase (high permeability and low coercivity) [81], large piezoelectric grain size (>1 μm) [82], layered structure (bilayer/trilayer) [83], and postsintering thermal treatment (annealing and aging) [84] improve the magnetoelectric property. In order to combine all the parameters together, the challenge is to develop a unique fabrication process, in which layers of piezoelectric and magnetostrictive can be cofired together. Moreover, the poling process of the piezoelectric phase requires high resistivity and thus electrodes have to be preserved during sintering.

19. Composite Synthesis

It was found that pressure-assisted sintering can produce trilayer composites with desired dimensions. Further, we designed the compositions such that sintering can be done at low temperatures of 900°C which resulted in stable electrodes. Powders of 0.9 Pb $(Zr_{0.52}Ti_{0.48})O_3$-0.1 Pb $(Zn_{1/3}Nb_{2/3})O_3$ [0.9 PZT-0.1 PZN] and $Ni_{0.6}Cu_{0.2}Zn_{0.2}Fe_2O_4$ [NCZF] were synthesized using conventional mixed oxide method. To fabricate a trilayer sample, at first a layer of NCZF powder (no binder was used) was pressed uniaxially at 52 MPa pressure inside a hardened steel die of diameter 12.5 mm and then Ag-Pd (DuPont 6160 conductor paste) electrode was printed. The second layer of 0.9 PZT-0.1 PZN was pressed on top of Ag-Pd under 5 MPa pressure inside the same die and then the electrode was again applied. The third and the final layer of NCZF layer was pressed on top of the second electrode layer. The five-layer structure NCZF/Ag-Pd/PZT-PZN/Ag-Pd/NCZF was then pressed under 20 MPa followed by using cold isostatic press (CIP) under 207 MPa. Trilayer composite was synthesized by sandwiching piezoelectric phase between magnetostrictive phases. The pressed pellet was dried at 100°C for 10 minutes and then placed inside a Lyndberg BlueM furnace by keeping 450 gm of load on top of it. The sintering was done at 900°C under a pressure of 30 kPa with very slow heating and cooling rate (~1°C/min). The sintered trilayer sample had dimensions of 3.25 mm in thickness and 11 mm in diameter. Annealing was done after sintering at 800°C for 10 hours under load. In order to polarize the samples, Ag-Pd electrode (DuPont 7714 conductor paste) was applied on top and bottom of the sample. The top electrode layer of the composite was then connected to the top of the PZT. Similarly, bottom electrode of the composite was connected to the bottom electrode of the PZT using the same Ag–Pd electrode. The poling was done at 110°C for 20 minutes under 2.2 kV/mm electric field. Microstructural analysis of the sintered samples was done by using Zeiss Leo Smart SEM on polished and thermally etched samples. Dielectric constant measurements were performed using an HP 4284 LCR meter; and the ME voltage coefficient by lock-in amplifier method. Figure 14(a) shows the low magnification SEM image of the sintered sample. The interface electrode layers are 10 μm thick, center layer is 0.9PZT-0.1 PZN, whereas outside layers are NCZF. There were some micro- and macrocracks noticed during the sintering which could be related to the thermal mismatch between these three layers and insufficient homogenization. Figure 14(b) shows the magnified view of the interface. The microstructure is dense in both the phases with excellent adherence with the Ag-Pd electrode. Figures 14(c) and 14(d) shows the low magnification image and interface microstructure of the annealed samples. The annealing treatment resulted in the homogenization at the interface reducing the porosity and microcrack density. The macrocracks were still present in the structure. Figures 14(e) and 14(f) show the high magnification grain structure of the sintered 0.9 PZT-0.1 PZN and NCZF. The grain size observed in 0.9 PZT-0.1 PZN was more than 1 μm and that in NCZF in the range of 3–5 μm.

A significant increase in piezoelectric and dielectric properties was observed after annealing which is consistent with our previous results. The magnitude of the electromechanical coefficients d_{33}, g_{33}, $\varepsilon_r/\varepsilon_o$ increased from 204 pC/N, 20.4×10^{-3} V-m/N, and 1132 to 250 pC/N, 22.4×10^{-3} V-m/N, and 1257. The magnitude of the loss factor (tanδ) decreased from 5.08% to 3.45% at room temperature. The loss factor increased sharply at high temperature which could be related to space charge polarization. At 100 kHz, the Curie temperature observed was around 330°C. From the resonance spectrum, it was found that the spectrum for annealed sample had less spurious and lower impedance at the resonance indicating high mechanical quality factor.

Figure 15 shows the magnetoelectric behavior of trilayer composite. The magnitude of the ME coefficient for sintered sample was of the order of 412 mV/cm·Oe with saturation field near 400 Oe. Annealed samples showed improved ME coefficient of 494 mV/cm.Oe about 20% higher than sintered ones. This magnitude is quite high from that previously reported for sintered composites. Srinivasan et. al. have derived the transverse ME coefficient for the layered composite as follows [25]:

$$\frac{\delta E_3}{\delta H_1} = \frac{-2d_{31}^{p}q_{11}^{m}v^{m}}{(s_{11}^{m}+s_{12}^{m})\varepsilon_{33}^{T,P}v^{p} + \left(s_{11}^{p}+s_{12}^{p}\right)\varepsilon_{33}^{T,P}v^{m} - 2\left(d_{31}^{p}\right)^2 v_m}, \tag{32}$$

$$\frac{\delta E_3}{\delta H_1} = \frac{-2d_{31}^{p}q_{11}^{m}\left(t_m/\left(t_m+t_p\right)\right)}{(s_{11}^{m}+s_{12}^{m})\varepsilon_{33}^{T,P}\left(t_p/t_m+t_p\right) + \left(s_{11}^{p}+s_{12}^{p}\right)\varepsilon_{33}^{T,P}\left(t_m/t_m+t_p\right) - 2\left(d_{31}^{p}\right)^2\left(t_m/t_m+t_p\right)}, \tag{33}$$

where d_{31}^{p} is the piezoelectric coefficient, v^m and v^p are the volume of magnetic and piezoelectric phase, t^m and t^p are the thickness of magnetic and piezoelectric phase, s_{11}^{p}, s_{12}^{p} are the elastic compliances for piezoelectric phase, s_{11}^{m}, s_{12}^{m} are the elastic compliances for magnetostrictive phase, q_{11} is the piezomagnetic coefficient of the magnetic phase, and $\varepsilon_{33}^{T,P}$ is the permittivity of the piezoelectric phase. For a radial mode disk, s_{11}^{p} can be calculated as follows:

$$s_{11}^{p} = \frac{\eta_1^2}{\rho(2\pi f_s a)^2\left[1-(\sigma^p)^2\right]}, \tag{34}$$

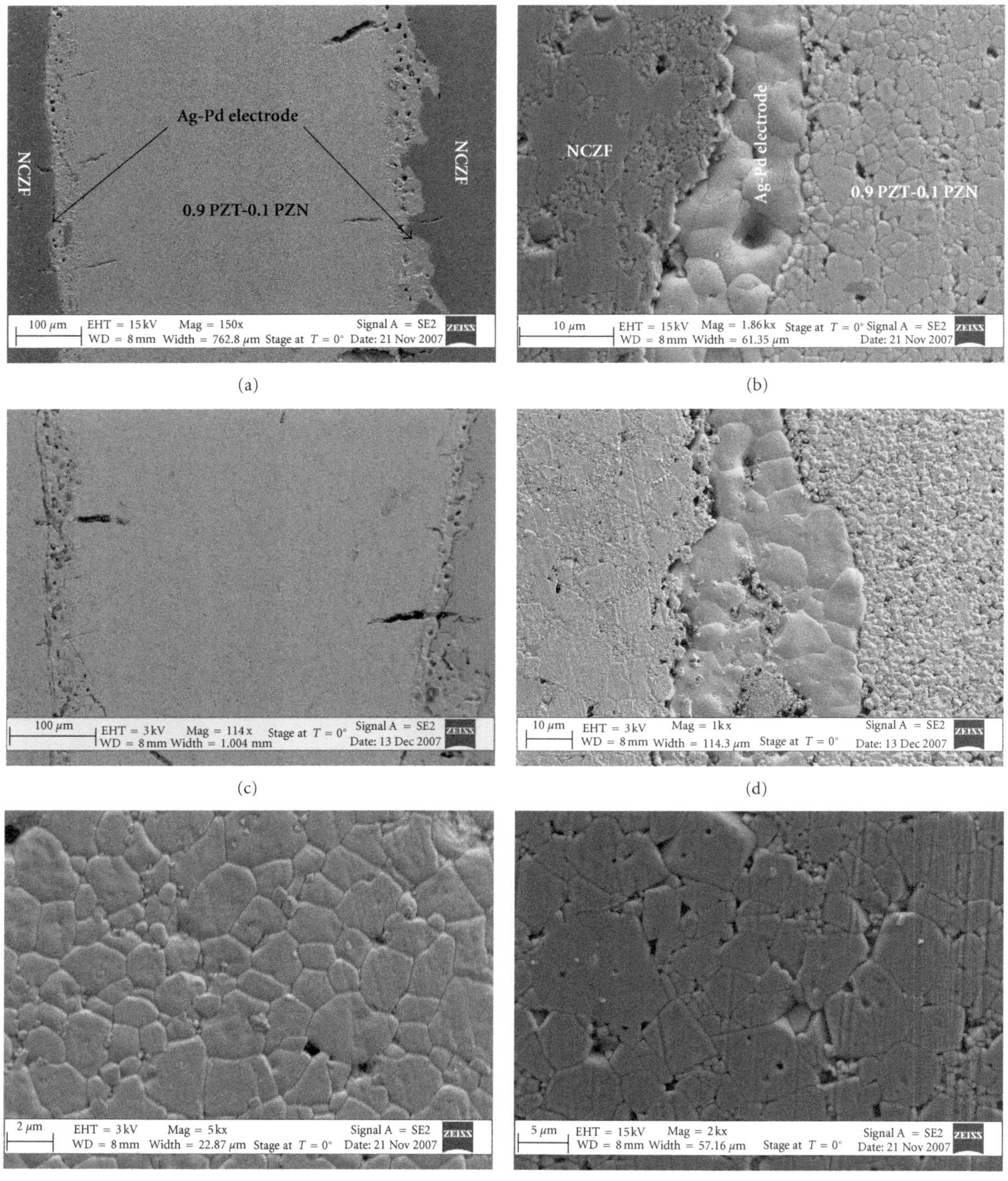

FIGURE 14: SEM micrographs. (a) low magnification image of sintered trilayer sample, (b) interface microstructure, (c) low magnification image of sintered and annealed sample, (d) interface microstructure of sintered and annealed trilayer sample, (e) grain structure of PZT side and (f) grain structure of NZF side.

where η_1 is the frequency constant, σ^p is the planar Poison's ratio, f_s is the resonance frequency, a is the radius, and ρ is the density (~7.52 gm/cc). The coefficients η_1 and σ^p can be found by measuring ratio of first overtone (~670 kHz for sintered samples) to fundamental resonance frequency (258 kHz) [26]. For the ratio of 2.602 η_1 and σ^p values were found to be 2.074 and 0.341, respectively, from IEEE standards. Using these values s_{11}^p can be calculated as 8.66×10^{-12} m^2/N. The transverse mode coupling constant k_{31} can be calculated by using the following expression [26]:

$$k_{31}^2 = k_p^2 \frac{\left(1 - \sigma_p\right)}{2}. \tag{35}$$

TABLE 1: Comparison between bulk, bilayer and trilayer ME composite

Composite	Geometry	Lattice constant	ME coefficient	DC bias
Bulk	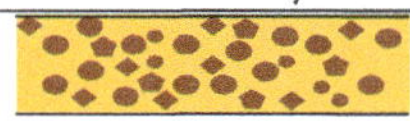	PZT (3.87 Å, $c = 4.07$ Å) and NFM = 8.42 Å	110 mV/cm·Oe for PZT-20 NZF	900 Oe (T-T) mode
Bilayer			200 mV/cm·Oe	800-900 Oe (T-T) mode
Trilayer	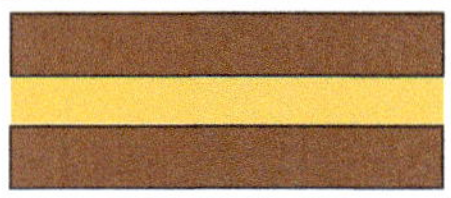	PZT (4.04 Å, $c = 4.11$ Å) and NCZF = 8.41 Å	526 mV/cm·Oe (single layer PZT) And 787 mV/cm·Oe (multilayer PZT)	400 Oe(L-T) mode

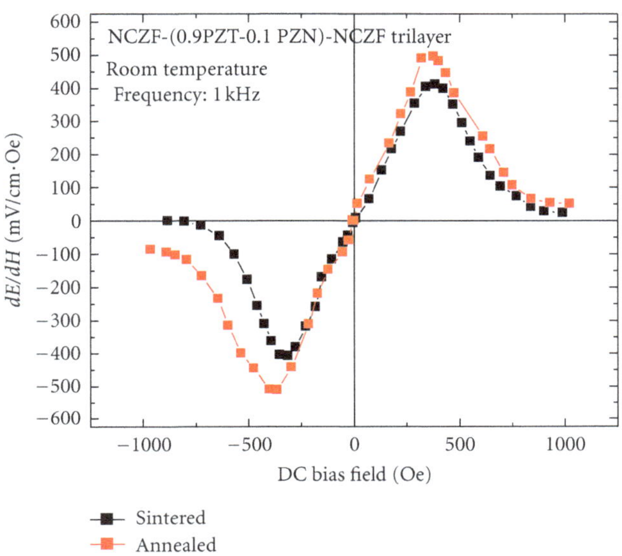

FIGURE 15: Magnetoelectric coefficient as a function of dc bias field.

The values of planar coupling constant k_p are listed in Table 1. Using these values, the magnitude of k_{31} is found to be 8.03%. If k_{31} and s_{11}^E are known,then d_{31} can be calculated as follows:

$$k_{31}^2 = \frac{d_{31}^2}{\left(s_{11}^E \varepsilon_{33}^T\right)}. \tag{36}$$

The magnitude of d_{31} was found to be as 23.66×10^{-12} C/N. The magnitude of s_{12}^E was found to be -2.96×10^{-12} m^2/N using the following expression [2]:

$$k_p^2 = \frac{2d_{31}^2}{\left[\left(s_{11}^E + s_{12}^E\right)\varepsilon_{33}^T\right]}. \tag{37}$$

Using the values of s_{11}^m, s_{12}^m as 6.5×10^{-12} m^2/N and -2.4×10^{-12} m^2/N [25], q_{11} as -65×10^{-9}/Oe (at 400 Oe field), $t_m/(t_m + t_p)$ and $t_p/(t_m + t_p)$ as 0.7 and 0.3, respectively, the magnitude of the ME coefficient computed from (33) was found to be 437.1 mV/cm.Oe. This calculated value is close to the experimentally measured value of 412 mV/cm.Oe. The calculated magnitude of the ME coefficient for the annealed samples was of the order of 468 mV/cm.Oe.

20. Effect of Thickness

The thickness ratio plays an important role in the trilayer geometry. From the expression for magnetoelectric coefficient for layered structure, it can be found that the magnetoelectric coefficient depends on piezoelectric coefficient (d_{31}), permittivity (ε_{33}), elastic compliance (s_{11} and s_{12}), thickness (t_P) of the piezoelectric phase, piezomagnetic coefficient (q_{11}), elastic compliance (s_{11} and s_{12}), and thickness (t_m) of the magnetic phase. Thus (33) can be rewritten as

$$\frac{\delta E_3}{\delta H_1} = \frac{-2d_{31}^P q_{11}^m \left(t_m/t_p\right)}{(s_{11}^m + s_{12}^m)\varepsilon_{33}^{T,P} + \left(s_{12}^P + s_{12}^P\right)\varepsilon_{33}^{T,P}\left(t_m/t_p\right) - 2\left(d_{31}^P\right)^2\left(t_m/t_p\right)} \tag{38}$$

In order to experimentally investigate the effect of thickness ratio and achieve a higher magnetoelectric coefficient, composites of different [0.9 PZT-0.1 PZN] thicknesses were synthesized. NCZF composition was used as the magnetostrictive layer and 0.9 PZT-0.1 PZN composition was used as piezoelectric layer. The amount of NCZF was fixed at 0.7 gm for top and bottom layer whereas the weight of 0.9PZT-0.1 PZN was varied from 0.8 to 0.3 gm. The interface electrode used was Dupont 6160 Ag-Pd conductor paste. The sintering was done at 900°C for 3 hrs using a load of 450 gm which is equivalent to 50 kPa. After sintering, each composite were cross-sectioned and polished for scanning electron microscopy. Figure 16 shows the cross sections of composites of PZT thickness of 0.45 and 1.2 mm, respectively. The thickness of interface electrode observed in these composites was in the range of

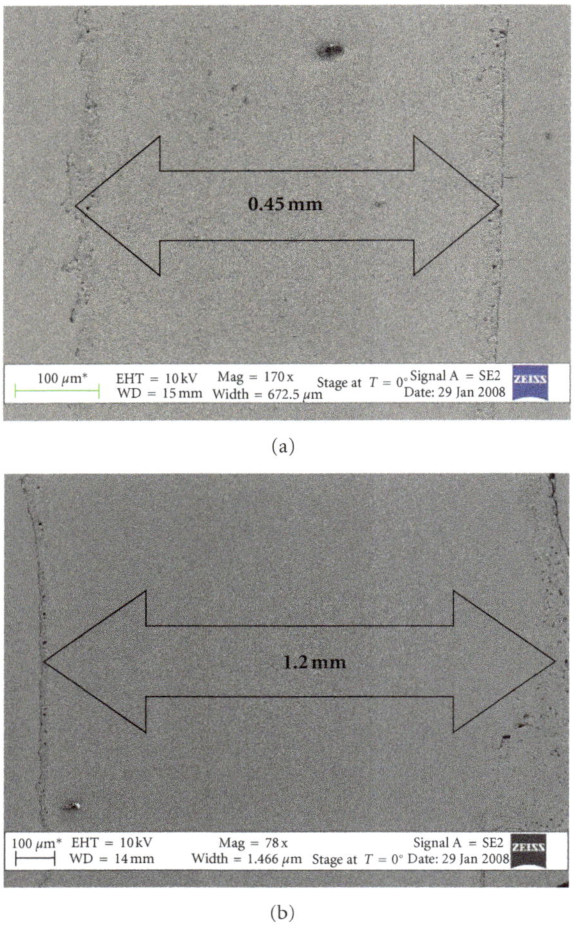

FIGURE 16: Composite of different thicknesses, (a) 0.45 mm, and (b) 1.2 mm.

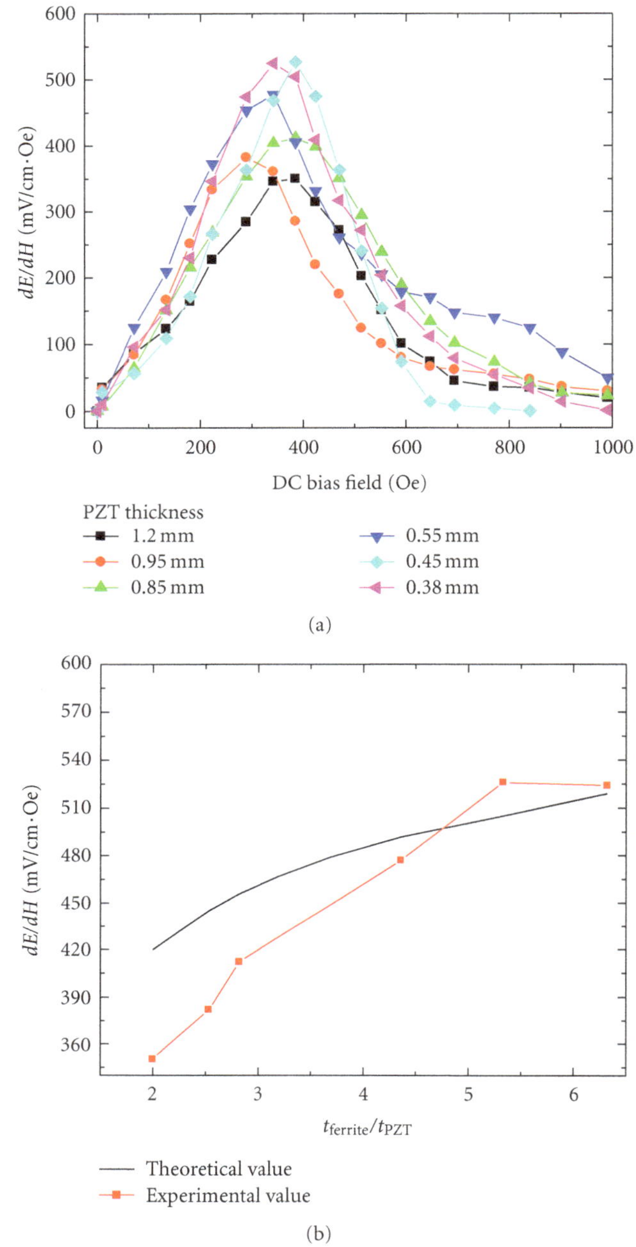

FIGURE 17: Magnetoelectric coefficient as a function of DC bias field for different thickness of [0.9 PZT-0.1 PZN].

5–10 μm. The adherence of the interface electrode with [0.9 PZ-0.1 PZN] and NCZF was found to be good. The [0.9 PZT-0.1 PZN] grain size observed in all the composites was above 1 μm. All the composites show piezoelectric charge constant above 225 pC/N, dielectric constant above 1150, and dielectric loss of 3–5%. X-ray elemental analysis was performed using the scanning electron microscope in order to identify any elemental diffusion through the electroded interface. The elemental analysis of Pb, Fe, and Ag shows a strong concentration of Pb was found on the [0.9 PZT-0.1 PZN] side and Fe on the NCZF side. Ag was found to be concentrated in the electroded area. By adding this interface electrode, the piezoelectric property was improved from 80 pC/N for cofired bilayer to 225 pC/N for trilayer composite. An improved dielectric constant and remarkable decrease in dielectric loss (from 29% to 5%) were recorded.

21. Effect of [0.9 PZT-0.1 PZN] Thickness and Orientation on ME Coefficient

Equation (38) shows the theoretical dependency of magnetoelectric coefficient on the magnetostrictive to piezoelectric layer thickness ratio. Figure 17 shows the magnetoelectric coefficient as a function of DC bias field for different thickness of [0.9 PZT-0.1 PZN]. It was found that as the thickness ratio of magnetic layer to piezoelectric layer increases (thickness of [0.9 PZT-0.1 PZN] decreases as the magnetic layer thickness was fixed), magnetoelectric coefficient increases from 354 to 526 mV/cm.Oe. The maximum magnetoelectric:

$$\sigma_{31f}^{E} = \frac{E_f E_p t_p \Delta\varepsilon_o}{(1-\gamma)\left(2E_f t_f + E_p t_p\right)}, \tag{39}$$

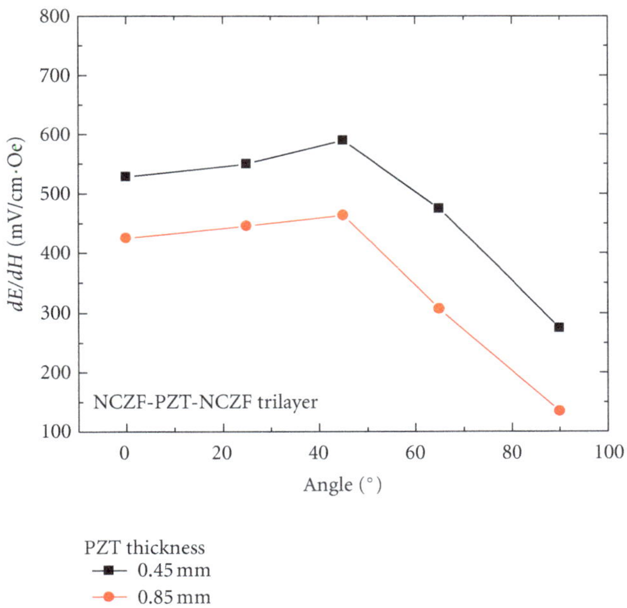

Figure 18: Effect of orientation on ME coefficient of NCZF-[PZT-0.1 PZN]-NCZF trilayer composite with two different [0.9 PZT-0.1 PZN] thickness.

$$\sigma_{31p}^{E} = -\frac{2E_f E_p t_p \Delta\varepsilon_o}{(1-\gamma)\left(2E_f t_f + E_p t_p\right)}. \quad (40)$$

As the thickness of the ferrite layer remains the same (1.2 mm on top and bottom), the compressive stress on [0.9 PZT-0.1 PZN] layer can be increased if the thickness of the [0.9 PZT-0.1 PZN] layer is decreased. The compressive stress is related to magnetoelectric coefficient as

$$\frac{dE}{dH} = \frac{2 \cdot g_{31} \cdot \sigma_{31P}^{E}}{H_{\text{ac}}}. \quad (41)$$

As the compressive stress in [0.9 PZT-0.1 PZN] is increased, magnetoelectric coefficient increases.

Figure 18 shows the magnetic field (AC and DC) orientation dependence of ME coefficient for trilayer composite ([0.9 PZT-0.1 PZN] thickness = 0.45 and 0.85 mm). It was found that composite shows an increase in ME coefficient from zero degree (when magnetic field direction is parallel to the sample surface) to 45°. Maximum magnetoelectric coefficient of 589 and 463 mV/cm·Oe was measured at 45°. As the angle is increased beyond 45°, the ME coefficients start to drop rapidly and when the sample surface is perpendicular to the magnetic field direction (T-T mode), a low value of 274 and 134 mV/cm·Oe was observed. It has been shown that with a real angle change, strain of the magnetic phase changes and is maximum around 51°. As the strain increases, compressive stress on [0.9 PZT-0.1 PZN] also increases which contributes to the high magnetoelectric coefficient as expressed by (41). At 90° angle, the maximum magnetoelectric coefficient was found to have low magnitude.

Figure 19: Schematic diagram for NCZF-Stack [0.9 PZT-0.1 PZN]-NCZF trilayer ME composite.

22. Effect of [0.9 PZT-0.1 PZN] Multilayering

In trilayer magnetoelectric composite based on NCZF-[PZT-PZN]-NCZF, composition optimum d_{33} and dielectric constant were found to be 225 pC/N and 1150. In order to achieve higher piezoelectric and dielectric constant, it is necessary to change the electrode pattern of the [0.9 PZT-0.1 PZN] layer. In this section, the introduction of stacked [0.9 PZT-0.1 PZN] configuration in trilayer geometry will be discussed. [0.9 PZT-0.1 PZN] stacking is most common design in actuator application. This type of stack actuator design provides large coupling, fast response time, and strong force. In our trilayer ME composite, we incorporated the stacked PZT–PZN as intermediate layer between two NCZF layers.

Figure 19 shows the schematic of new trilayer design with stacked PZT-PZN. Same fabrication process was used for the synthesis of new design, but instead of single PZT-PZN layer, there were three PZT-PZN layer and interdigital electrode layer along thickness direction. The yellow layer and the black lines are identified as PZT-PZN and electrodes, respectively. To improve the sinterability of PZT–PZN, 0.5 wt% of CuO was doped. The sintering temperature and time was 900°C for 3 hours with very slow heating and cooling rate (~1°C/min). Pressure applied during sintering was 50 KPa to avoid any delamination. After sintering, the trilayer was cross-sectioned and the optical images were taken in order measure the apparent thickness of the [0.9 PZT-0.1 PZN]. It was found that each layer of the [0.9 PZT-0.1 PZN] is around 300 μm with a combined thickness of 0.9 mm. NCZF layers were found to have thickness of 1.2 mm. Figure 20 shows the optical image of the cross-section of NCZF-stack PZT-PZN-NCZF trilayer.

The piezoelectric constant (d_{33}) increased from 204 to 535 pC/N and dielectric constant from 1132 to 5500. However, the dielectric loss of stack [0.9 PZT-0.1 PZN] increased from 5 to 13.4%, which may be due to the diffusion across the Ag-Pd electrode, which is a conductive layer. The increase in electromechanical coupling constant from 0.14 to 0.21 was observed. The impedance spectrum of the trilayer ME composite shows the resonance frequency at 523 kHz which is higher than that of single layer [0.9 PZT-0.1 PZN] which shows resonance at 258 kHz for the same dimensions. Also, the bandwidth (difference between resonance and antiresonance) was found to be higher (17 kHz) in case of trilayer composite with stack actuator configuration. Figure 21 shows the magnetoelectric coefficient of stack PZT-based trilayer composite. It shows a peak coefficient of

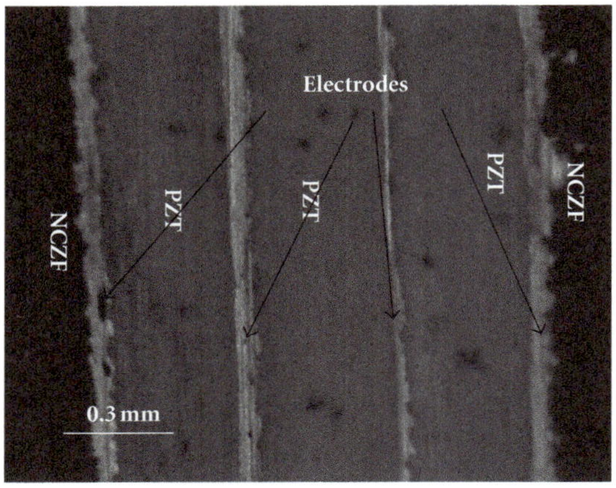

FIGURE 20: Cross sectional optical image of NCZF-Stack [0.9 PZT-0.1 PZN]-NCZF trilayer.

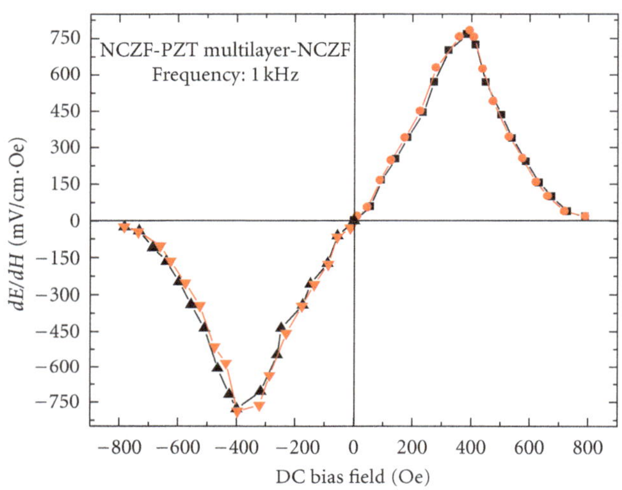

FIGURE 21: Magnetoelectric coefficient of multilayer [0.9 PZT-0.1 PZN] based trilayer composite.

782 mV/cm·Oe at 400 Oe magnetic DC bias. In order to check whether there is any hysteresis behavior shown by this sample, the ME coefficient was measured from negative to positive high field and vice versa. It was found that there is very small hysteresis in the response. Compared to the single-layer PZT-PZN in trilayer configuration (412 mV/cm·Oe), stack PZT-PZN shows much higher ME coefficient.

Figures 22(a) and 22(b) show the TEM micrographs of PZT-PZN and NCZF phases close to the interface. Inset of each image shows the SAED pattern. From the diffraction pattern, pure piezoelectric and magnetic phases can be identified as there is no superlattice structure. The grain size observed was around 1 μm for piezoelectric phase and ~500 nm for NCZF phases. The lattice parameter was calculated to be $a = 4.04$ and $c = 4.11$ Å with tetragonality (c/a ratio) of 1.017. Lattice constant for NCZF was calculated as 8.41 Å. The micrographs were taken within 10 μm from the interface on both sides. Table 1 compares the geometry,

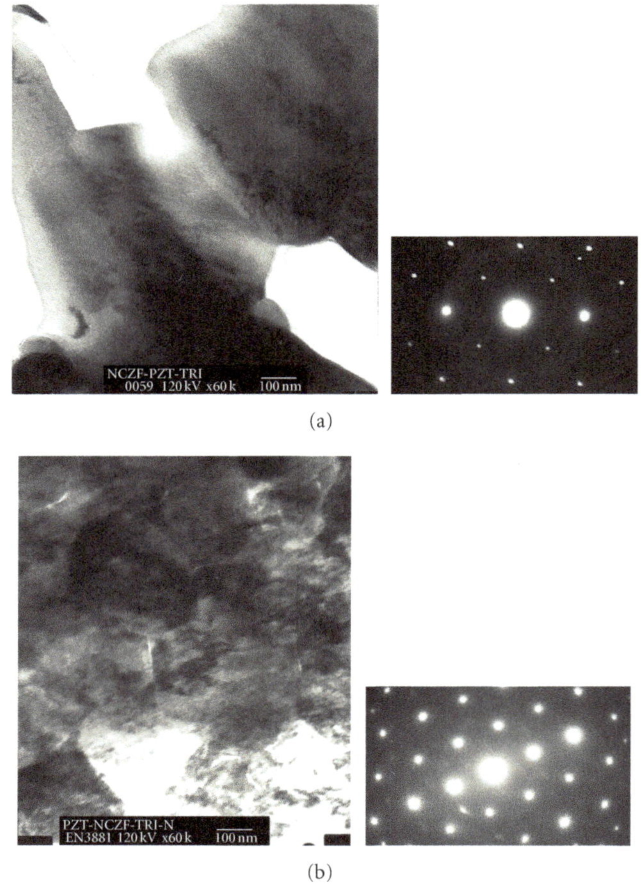

FIGURE 22: TEM micrograph and SAED pattern for (a) PZT-PZN and (b) NCZF near the interface.

lattice constant, ME coefficient, and DC bias field of bulk, bilayer, and trilayer composite. For bulk composite PZT, lattice constant was analyzed as strained by NFM lattice whereas for trilayer, the lattice constant was close to the pure phase. This clearly signifies that the interface diffusion didnot take place in the trilayer which was evident in bulk and bilayer composites. The interface diffusion observed in the bilayer composite was around 30 μm. Trilayer geometry with interface electrode was able to reduce the interface diffusion.

23. Grain Orientation

Crystallographic texturing of piezoelectric phase can improve the piezoelectric and ferroelectric response by exploiting the anisotropy of electrical properties. Results have been reported in literature on the effect of texturing in both lead-based and lead-free materials including $BaTiO_3$ (BTO), $Pb(Mg_{1/3}Nb_{2/3})O_3$-$PbTiO_3$ (PMN-PT), $(Na_{1/2}Bi_{1/2})TiO_3$-$BaTiO_3$ (NBT-BT), and (K,Na)NbO_3 (KNN) [85–88]. The goal of this study was to implement the known texturing process in heterogeneous system consisting of varying crystal structure. This will allow developing novel hybrid materials that exhibit maximum response. As a simple rule, rhombohedral phase oriented along the ⟨100⟩ direction provides higher piezoelectric response.

Texturing can be accomplished through various processes including, hot pressing (HP), templated grain growth (TGG) [89], and reactive template grain growth (RTGG). The advantage of TGG over HP is that the process is conducted using steps similar to that of conventional ceramic processing except with added seed material. Hence, in this study, TGG technique was adopted to synthesize the textured ME composite. The texturing process consists of two important steps: (i) seed fabrication, and (ii) seed alignment in the matrix [90]. To fabricate the seed (template), molten salt synthesis technique was utilized. It is desirable to have the seed crystal of same crystal structure as that of matrix. In this respect, BTO was chosen as the seed material since it can be grown in large size using Remeika process [91]. This synthesis process consists of heating the raw materials (BTO covered with KF salt in Pt crucible) in the temperature range of 1150–1225°C (±5°) for 8–12 hrs and then slowly cooling down to room temperature. The crystals were separated from KF salt by hot water. The size of synthesized BTO seed was found to be in the range of 75–200 μm. Some seeds were found to be as large as 2–3 mm plates. The optical images of BTO seed fabricated using Remeika method reveals facets at the surface which could be related to the twins. A high magnification optical image (inset) shows these facets of single crystal cover the whole surface. X-ray diffraction pattern of seed indicates strong intensity peak at 22.2° which was indexed as (001) peak of perovskite BTO.

The seed crystals were used as templates for texturing piezoelectric phase in the trilayer ME composite. For this purpose, small amount (1–5 wt.%) of seed was mixed with the piezoelectric powder 0.85 $Pb(Zr_{0.52}Ti_{0.48})O_3$–0.15 $Pb(Zn_{1/3}Nb_{2/3})O_3$ (PZT-PZN). The seeded powder was mixed thoroughly and dried. Trilayer was pressed with $Ni_{0.6}Cu_{0.2}Zn_{0.2}Fe_2O_4$(NCZF) top and bottom layer and seeded PZT-15% PZN in the middle. The sample was kept for sintering in an air atmosphere for 50 hr soaking time so that the grains can grow along the preferred orientation of seed crystal. A bilayer sample was also synthesized in order to study the structural characteristics of piezoelectric layer.

Figure 23 shows the X-ray diffraction pattern of the synthesized sample which shows considerable increase in (001)/(100), (111), (002)/(200) peak intensities compared to the (110)/(101) peak. Also, the peak splitting was clearly observed in (001)/(100) and (002)/(200) peaks, which indicates the increase in tetragonality. From this pattern, the lattice parameter can be calculated to be a = 4.008 Å, and c = 4.098 Å and the c/a ratio can be calculated as 1.022. Figure 23 also compares the XRD pattern of textured PZT-PZN with randomly oriented PZT-PZN. From X-ray diffraction patterns, the degree of texturing was calculated by Lotgering method [20] given as

$$f_{00l} = \frac{P_{00l} - P_O}{1 - P_O}, \qquad (42)$$

where $P_{00l} = (\sum I_{001})/(\sum I_{hkl})$ and $P_0 = (\sum I^0{}_{001})/(\sum I^0{}_{hkl})$ where I_{hkl} and $I^0{}_{hkl}$ are the intensities of hkl plane

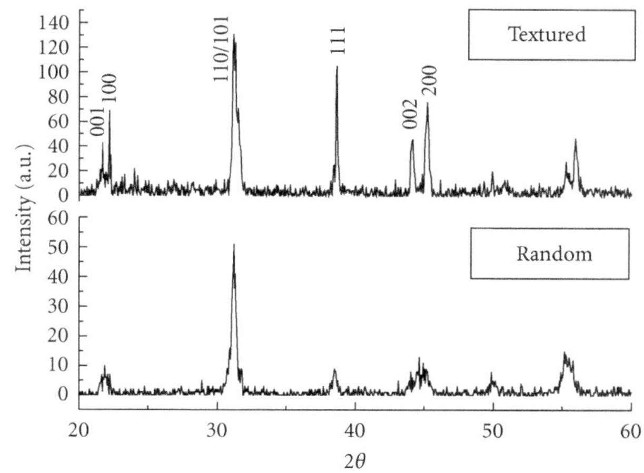

Figure 23: X-ray diffraction pattern of textured PZT-PZN compared to randomly oriented PZT-PZN.

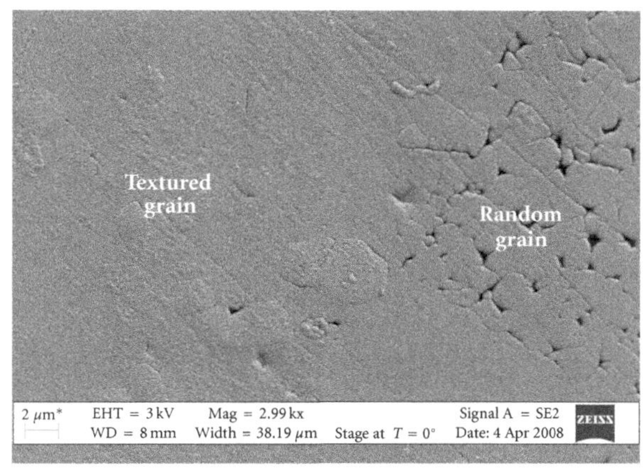

Figure 24: High-magnification SEM micrograph of PZT-PZN showing textured grain and randomly oriented grain interface.

for the textured and random sample and $\sum I_{001}$ is the summation of the intensities of all 001 planes. Using this formula, the Lotgering factor was calculated to be f = 0.35 for samples textured along ⟨001⟩ direction. Figure 24 shows the presence of very large grains which were considered to be as textured grains with average size in the range of 200–350 μm as compared to 1–2 μm grain size for surrounding grains. The grain size observed in Figure 24 is very high due to large soaking period of 50 hrs.

These are preliminary results related to our texturing are promising and more in-depth study is required to confirm the orientation and degree of texturing. The cross-section revealed by optical microscopy shows crack free interfaces with intact electrode layer. PZT-PZN layer was found to be around 0.5 ± 0.02 mm in thickness and NCZF thickness was of the order of 1.2 ±.05 mm. The samples were poled at 120°C for 20 minutes under a field of 2.5 kV/mm in a silicone oil bath. The longitudinal piezoelectric constant (d_{33}) was measured to be 325 ± 10 pC/N with dielectric constant of 1865 ± 30 and loss of 7% at room temperature at 1 kHz. Compared to the randomly oriented PZT-PZN

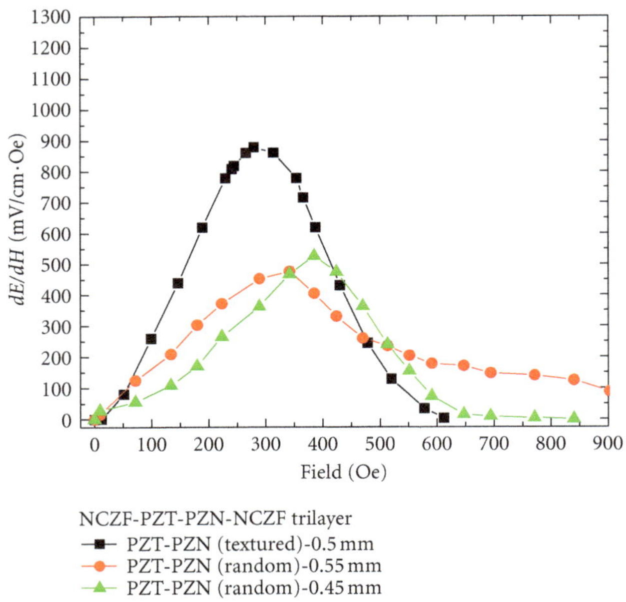

FIGURE 25: ME coefficient of partially textured trilayer composite compared with randomly oriented composite.

intermediate layer, d_{33} increases by 44% and dielectric constant increases by 50% by partial texturing. Figure 25 shows the ME coefficient of trilayer with textured PZT-PZN as function of DC bias. Three samples from each batch were measured and the average value was reported. The deviation observed from sample to sample within each batch was around 5%. The peak ME coefficient observed was around 878 ± 20 mV/cm·Oe DC bias field. For randomly oriented PZT-PZN layer, ME coefficient observed was around 526 ± 25 mV/cm·Oe for layer thickness of 0.45 mm and around 500 ± 15 mV/cm·Oe for thickness of 0.55 mm. This increase in piezoelectric and magnetoelectric coefficient in partially textured trilayer can be explained in terms of domain engineering.

24. Future of Sintered Composite

Laminate composites involve multiple step assembly process and are sensitive to the changes in configuration. Thus, our research emphasis has been on enhancing the ME coefficient in bulk sintered composites. The challenges associated with sintered composites are (i) reducing the diffusion across interface, (ii) reducing elastic softening at the interface, and (iii) reducing connectivity of the magnetostrictive phase without compromising the contact area with piezoelectric phase.

All these challenges can be addressed by synthesizing sintered composites from core-shell particles where piezoelectric phase is the core and magnetostrictive phase is the shell [2, 92]. The elastic coupling can be maximized by having coherent response from the magnetostrictive phase under dc bias such that the stress on the piezoelectric lattice across the grains is in phase with each other. This is only possible if there is uniform distribution of the magnetostrictive phase with respect to the piezoelectric matrix. This core-shell composite has the possibility to provide effective inter-phase exchange coupling originating from the maximized interphase contact. Thus, we synthesized a core-shell particulate composite of $Pb(Zr,Ti)O_3$ (PZT) and $NiFe_2O_4$ (NF), and characterized its magnetoelectric property.

PZT particles corresponding to formulation $Pb(Zr_{0.52}Ti_{0.48})O_3$ were synthesized using conventional ceramic processing route. Nano-NF particles were synthesized by chemical synthesis technique using the following processing steps. A mixture of 256 mg of $Ni(acac)_2$ and 500 mg of 1,2-hexadecanediol was added to a 125 mL European flask containing a magnetic stir bar. Benzyl ether 20 mL was then transferred into the flask and the contents were stirred while purging with N_2 for 20 min at room temperature. The flask was then heated to 120°C and the temperature was held for 20 min. During this time, 0.2 mmol of $Fe(CO)_5$ was injected into the flask while the N_2 purging continued. After 1 minute, 1 mL of oleic acid and 1 mL of oleylamine were injected and the mixture was maintained under N_2 blanket and heated to 160°C at a rate of 5°C per minute where it was held for 10 minutes. The flask was maintained at the refluxing temperature of 295°C for 30 min before cooling down to room temperature under the N_2 blanket. Size of $NiFe_2O_4$ was controlled by varying the amount of surfactants and heating rates. When faster heating rate and more surfactants were used, bigger NF particles were obtained. Coating of $NiFe_2O_4$ nanoparticles on PZT was done using similar recipe as that for synthesizing the NF nanoparticles. PZT particles were added in the beginning of synthesis keeping the other parameters constant. Amount of PZT particles was varied from 500 mg to 2 gm in order to vary the coating thickness and obtain different ratio of NF to PZT.

Figure 26(a) shows the TEM image of nanosize NF particles. The size of the particles was found to be in the range of 20–30 nm with hexagonal shape. Figure 26(b) shows the X-ray diffraction pattern of NF nanoparticles (red) compared with that of large size particles (black), both exhibiting spinel structure. The peak broadening occurred as the particle size became smaller. Figure 26(c) shows the magnetic hysteresis loop of the NF nanoparticles. Coercive field and magnetization were found to be lower in comparison with the large size particles, as we will discuss later. From Figures 26(b) and 26(c), it can be seen that excellent structural and magnetic properties in NF nanoparticles can be achieved using the chemical synthesis method.

Figure 27(a) shows the SEM images of PZT-NF core shell structure. EDS spot analysis on these particles showed presence of both PZT and NF. Figure 27(b) shows the TEM image of PZT particle coated with NF nanoparticles. The difference in contrast was clearly observed for two different sizes of particles. TEM investigations showed that the NF nanoparticle shell over the PZT core was 2-3 layers in thickness. PZT-NF core-shell particles were sintered at low temperature of 950°C and long holding time of 5 hrs. In order to increase the resistivity, a thin layer of pure PZT was pressed as top layer in a 0.25 inch die under hydraulic

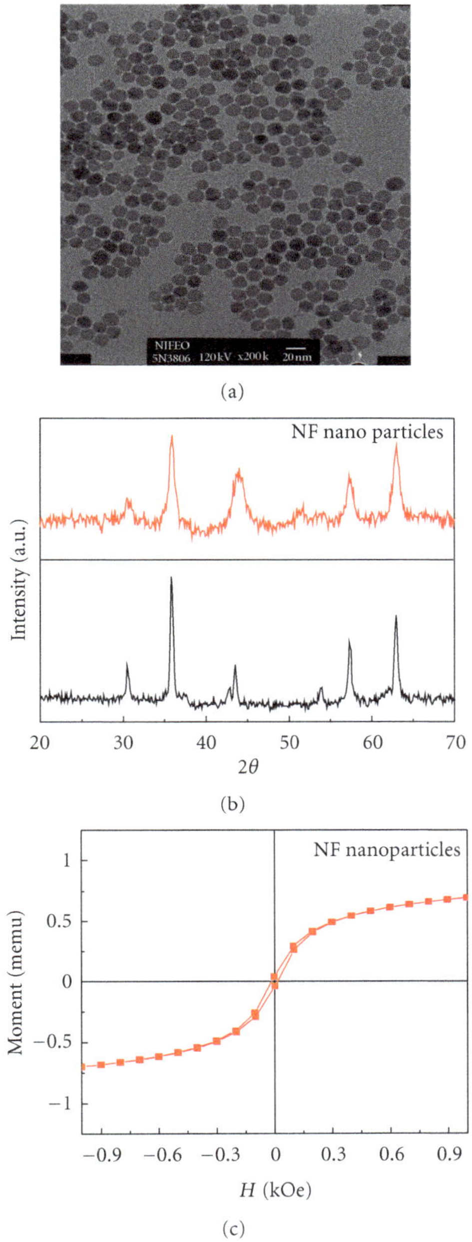

FIGURE 26: (a) TEM image, (b) X-ray diffraction pattern, and (c) magnetic properties of NF nano particles.

pressure of 2 ksi followed by cold isostatic pressure of 30 ksi. Figure 27(c) show the low magnification SEM image of PZT-NF core-shell structure. The microstructure of the sintered composite was dense with very small fraction of porosity. The grain size of the composite was found to be in the range of 500–800 nm. Inset of Figure 27(c) shows a magnified view of core-shell grains. TEM analysis revealed that the shell was present at the grain boundary of large fraction of grains. The coating thickness was equivalent to 2-3 layers, which varies from 40–70 nm. The size of nanoparticles in the shell was larger than the as-synthesized nanoparticles due to high-temperature sintering process.

Figure 27(d) shows the magnetic properties of as synthesized NF nanoparticles and PZT-NF core-shell structure. One major difference can be immediately noted in this figure related to increase in coercive field as compared to the nanoparticles. For PZT-NF core-shell particles, the coercive field was found to be 216 Oe whereas for as, synthesized nanoparticles, this magnitude was significantly lower. This may be attributed to the increase in size of the NFO shell during sintering process. As particle size decreases, coercivity decreases and for very low value of particle size ferromagnetic to superparamagnetic transition occurs [93, 94]. It has been shown in literature that with increasing particle size from 1 nm and above, the magnitude of coercive field increases and reaches maxima before dropping again with further increase in size. The particle size where maximum in coercive field occurs corresponds to the size of single-domain particle. The size of the shell in the sintered composite may be close to the single-domain particle size and hence the coercive field observed was high.

Figure 28(a) shows the TEM image of grain structure for sintered PZT-NF composite. Figure 28(b) shows the variation of magnetoelectric coefficient as a function of DC bias for the sintered composite. The composite was electroded and poled at 2.5 kV/mm for 30 minutes in silicone oil bath. The temperature was fixed at 120°C. The piezoelectric (d_{33}) and dielectric properties were measured on poled and aged sample. The composite had d_{33} of 60 pC/N, dielectric constant of 865, and dielectric loss of 5.45%. The peak ME coefficient of 195 mV/cm.Oe was observed at 454 Oe. This magnitude is much higher than that obtained for random polycrystalline-sintered composites [95]. Further, it was found that in the DC bias range of 265 to 617 Oe, the ME coefficient remained above 175 mV/cm·Oe, which is almost 90% of the peak value. This data is quite different as compared to that obtained for bulk particulates which exhibit sharp peaks. This wide bandwidth offers significant advantages for sensing applications. This may be due to the hard nature (larger coercive field) of the magnetic phase. The coercive field was of the order of 216 Oe for the core-shell structure. Above this field, the magnetization rotation contributes towards magnetostrictive strain. Above 617 Oe, the ME coefficient drops as the piezomagnetic coefficient starts to drop rapidly.

25. Conclusion

The improvement of the magnetoelectric coefficient of the sintered composite was discussed in this review along with composition-microstructure-property relationship. A new material constant n was derived from the condition for maximization of g constant ($|d| = \varepsilon^n$). According to the developed model, as the magnitude of n decreases towards unity a giant enhancement in the magnitude of the g constant can be obtained. The experimental data was found to be consistent with this prediction. Effect of matrix grain size on the magnetoelectric coefficient of PZT-20 NZF particulate composites was studied. It was found that grain size has significant effect on the piezoelectric, ferroelectric, and dielectric properties of the composite and

FIGURE 27: Micrographs for PZT-NF core-shell structure, (a) SEM and (b) TEM, (c) SEM microstructure of sintered PZT-NF core shell structure, and (d) comparison of magnetic hysteresis property of NF nanoparticles and PZT-NF core-shell structure.

hence influences the magnetoelectric property. There is a critical grain size below which the piezoelectric charge constant (d_{33}), dielectric constant, and piezoelectric voltage constant (g_{33}) drops rapidly. This critical grain size is in the range of 100–150 nm. Above 200 nm, the piezoelectric properties increases slowly and saturates at 600 nm. Samples having grain size of around 100 nm show small magnitude of ME coefficient (54.4 mV/cm·Oe) while those with grain size of 600 nm exhibit a value of 155 mV/cm·Oe.

The effect of postsintering thermal treatment of PZT-NFM composites show that there is a correlation between structural deformation in PZT matrix, presence of defects between component phases in piezoelectric-magnetostrictive composites, and change in magnetization with changes in the ME coefficient. It was found that lattice expansion of PZT [from (a = 3.87 Å, c = 4.07 Å) to (a = 4.07 Å, c = 4.09 Å)], removal of defects, and increase in magnetization contribute to the enhancement in ME coefficient after annealing and aging. There is a decrease in ferroelectric and ferromagnetic Curie temperatures by 8°C and 33°C and at least 50% enhancement was observed in magnetoelectric coefficient after annealing and aging. A sintering process was introduced to fabricate NCZF-(0.9 PZT-0.1 PZN)-NCZF trilayer composites. Soft piezoelectric phase with grain size of more than 1 μm, and low coercivity magnetostrictive phase with grain size of 3–5 μm were obtained with excellent adherence to Ag-Pd electrode. Sintered and annealed samples showed ME coefficient of ~412 and 494 mV/cm·Oe, respectively, which are quite large compared to the previously reported values. Optimization of magnetostrictive to piezoelectric thickness ratio showed a magnetoelectric coefficient of 525 mV/cm·Oe when the angle between the sample and the magnetic field direction is zero. A slightly higher value (595 mV/cm·Oe) was observed when the angle was changed to 45°. It was also found that when the intermediate piezoelectric layer is changed from single layer to trilayer stack geometry then piezoelectric and dielectric property of the composite improved remarkably

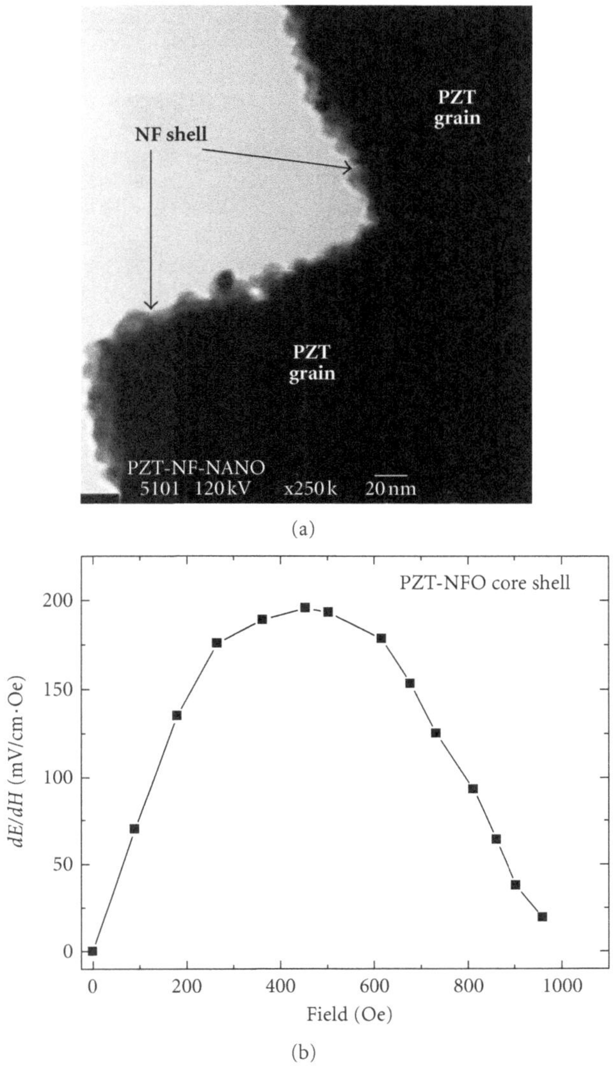

Figure 28: TEM image of sintered core-shell structure, (b) magnetoelectric coefficient as a function of DC bias field for sintered PZT-NF core shell structure.

from 204 pC/N and 1132 to 535 pC/N and 5500, respectively. This enhancement contributes to the high magnetoelectric response of 782 mV/cm·Oe.

Texturing the piezoelectric phase inside the trilayer ME composite improved the ME coefficient. X-ray diffraction and scanning electron microscopy showed that the texturing occurred inside the piezoelectric grains by using $BaTiO_3$ seeds. The degree of texturing was found to be 0.35 and 0.22 for (001) and (111) direction, respectively. For this moderate texturing, improvement in piezoelectric and dielectric constant was by 40–50%. It improved the ME coefficient by 67%, from 526 mV/cm·Oe to 878 mV/cm·Oe. Core shell particles of PZT and NFO were synthesized using chemical synthesis route with PZT as core and NFO as shell. The core-shell particles were sintered using compaction and annealed at 1025°C for 5 hours. The magnetic property (coercivity) of core shell particles was remarkably different than the as

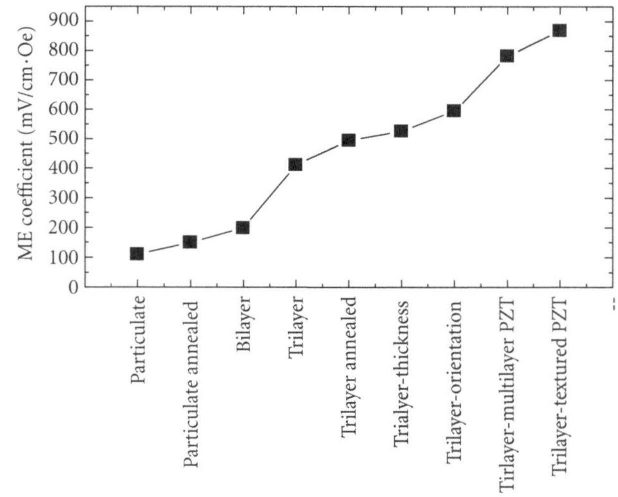

Figure 29: Progress in magnetoelectric sintered composite.

synthesized nanoparticles. Coercive field increased to 210 Oe, which is quite high for NFO nanoparticles. Due to this hard nature of NFO nanoparticles, high ME coefficient (>175 mV/cm·Oe) over the wide range of field was observed with peak ME coefficient of 195 mV/cm·Oe.

Figure 29 shows the progress in ME-sintered composite research starting from 110 mV/cm·Oe to 878 mV/cm·Oe ~ a 700% improvement—which is a significant contribution. Also at the end it was predicted that the nano-particulate core shell assembly has the promise for achieving large ME coefficient in the sintered composites.

Acknowledgment

S. Priya would like to acknowledge the financial support from NSF INAMM Program.

References

[1] J. Ryu, A. V. Carazo, K. Uchino, and H. E. Kim, "Piezoelectric and magnetoelectric properties of lead zirconate titanate/Ni-ferrite particulate composites," *Journal of Electroceramics*, vol. 7, no. 1, pp. 17–24, 2001.

[2] V. C. Flores, D. B. Baques, D. C. Flores, and J. A. M. Aquino, "Enhanced magnetoelectric effect in core-shell particulate composites," *Journal of Applied Physics*, vol. 99, no. 8, Article ID 08J503, 3 pages, 2006.

[3] T. G. Lupeiko, I. V. Lisnevskaya, M. D. Chkheidze, and B. I. Zvyagintsev, "Laminated magnetoelectric composites based on nickel ferrite and pzt materials," *Inorganic Materials*, vol. 31, no. 9, pp. 1139–1142, 1995.

[4] M. Feibig, "Revival of the magnetoelectric effect," *Journal of Physics D*, vol. 38, pp. R123–R152, 2005.

[5] C. Ederer and N. A. Spaldin, "Recent progress in first-principles studies of magnetoelectric multiferroics," *Current Opinion in Solid State & Materials Science*, vol. 9, pp. 128–139, 2005.

[6] J. van Suchetelene, "Product properties: a new application of composite materials," *Philips Research Report*, vol. 27, no. 3, pp. 28–37, 1972.

[7] G. Smolenskii and V. A. Ioffe, Colloque International du Magnetisme, Communication no. 71, 1958.

[8] J. Wang, J. B. Neaton, H. Zheng et al., "Epitaxial BiFeO," *Science*, vol. 299, no. 5613, pp. 1719–1722, 2003.

[9] J. Wang, H. Zheng, S. E. Lofland et al., "Multiferroic $BaTiO_3$–$CoFe_2O_4$ Nanostructures ," *Science*, vol. 303, no. 5658, pp. 661–663, 2004.

[10] C. Ederer and N. A. Spaldin, "A new route to magnetic ferroelectrics," *Nature Materials*, vol. 3, no. 12, pp. 849–851, 2004.

[11] W. Eerenstein, N. D. Mathur, and J. F. Scott, "Multiferroic and magnetoelectric materials," *Nature*, vol. 442, no. 7104, pp. 759–765, 2006.

[12] N. Hur, S. Park, P. A. Sharma, J. S. Ahn, S. Guha, and S. W. Cheong, "Electric polarization reversal and memory in a multiferroic material induced by magnetic fields," *Nature*, vol. 429, no. 6990, pp. 392–395, 2004.

[13] C. W. Nan, M. I. Bichurin, S. Dong, D. Viehland, and G. Srinivasan, "Multiferroic magnetoelectric composites: historical perspective, status, and future directions," *Journal of Applied Physics*, vol. 103, no. 3, Article ID 031101, 35 pages, 2008.

[14] S. Priya, R. Islam, S. Dong, and D. Viehland, "Recent advancements in magnetoelectric particulate and laminate composites," *Journal of Electroceramics*, vol. 19, no. 1, pp. 147–164, 2007.

[15] G. R. Harshe, *Magnetoelectric effect in piezoelectric—magnetostrictive composite*, Ph.D. dissertation, Pennsylvania State University, University Park, Pa, USA, 1991.

[16] J. F. Herbst, T. W. Capehart, and F. E. Pinkerton, "Estimating the effective magnetostriction of a composite: a simple model," *Applied Physics Letters*, vol. 70, no. 22, Article ID 3041, 3 pages, 1997.

[17] C. W. Nan, M. Liu, X. Feng, and S. Yu, "Possible giant magnetoelectric effect of ferromagnetic rare-earth-iron-alloys-filled ferroelectric polymers," *Applied Physics Letters*, vol. 78, no. 17, Article ID 2527, 3 pages, 2001.

[18] D. N. Astrov, "The magnetoelectric effect in antiferromagnetics," *Soviet Physics, JETP*, vol. 11, pp. 708–709, 1960.

[19] D. N. Astrov, B. I. Al'shin, R. V. Zhorin, and L. A. Drobyshev, *Soviet Physics, JETP*, vol. 28, p. 1123, 1968.

[20] G. Smolenskii and V. A. Ioffe, Colloque International du Magnetisme, Communication, no. 71, 1958.

[21] T. Kimura, T. Goto, H. Shintani, K. Ishizaka, T. Arima, and Y. Tokura, "Magnetic control of ferroelectric polarization," *Nature*, vol. 426, no. 6962, pp. 55–58, 2003.

[22] T. Lottermoser, T. Lonkai, U. Amann, D. Hohlwein, J. Ihringer, and M. Fiebig, "Magnetic phase control by an electric field," *Nature*, vol. 430, no. 6999, pp. 541–544, 2004.

[23] B. B. Van Aken, T. T. A. Palstra, A. Filippetti, and N. A. Spaldin, "The origin of ferroelectricity in magnetoelectric $YMnO_3$," *Nature Materials*, vol. 3, no. 3, pp. 164–170, 2004.

[24] M. Fiebig, T. Lottermoser, D. Fröhlich, A. V. Goltsev, and R. V. Pisarev, "Observation of coupled magnetic and electric domains," *Nature*, vol. 419, no. 6909, pp. 818–820, 2002.

[25] M. Zeng, J. G. Wan, Y. Wang et al., "Resonance magnetoelectric effect in bulk composites of lead zirconate titanate and nickel ferrite," *Journal of Applied Physics*, vol. 95, no. 12, Article ID 8069, 5 pages, 2004.

[26] J. Zhai, Z. Xing, S. Dong, J. Li, and D. Viehland, "Magnetoelectric laminate composites: an overview," *Journal of the American Ceramic Society*, vol. 91, no. 2, pp. 351–358, 2008.

[27] R. E. Newnham, "Composite Electro ceramics," *Ferroelectrics* , vol. 68, no. 1–4, pp. 1–32, 1986.

[28] T. H. Odell, "Magnetoelectrics—a new class of materials," *Electronics and Power*, vol. 11, no. 8, pp. 266–267, 1965.

[29] R. E. Newnham, "Composite piezoelectric transducers," *Materials in Engineering*, vol. 2, no. 2, pp. 93–106, 1980.

[30] J. V. D. Boomgaard, A. M. J. G. Van Run, and J. Van Suchtelen, "Magnetoelectricity in piezoelectric-magnetostrictive composite," *Ferroelectrics*, vol. 10, pp. 295–298, 1976.

[31] J. V. D. Boomgaard and R. A. J. Born, "A sintered magnetoelectric composite material $BaTiO_3$-Ni(Co, Mn) Fe_2O_4," *Journal of Materials Science*, vol. 13, no. 7, pp. 1538–1548, 1978.

[32] J. V. D. Van Den Boomgaard, D. R. Terrell, R. A. J. Born, and H. F. J. I. Giller, "An in situ grown eutectic magnetoelectric composite material—part I Composition and unidirectional solidification," *Journal of Materials Science*, vol. 9, no. 10, pp. 1705–1709, 1974.

[33] A. M. J. G. Van Run, D. R. Terrell, and J. H. Scholing, "An in situ grown eutectic magnetoelectric composite material—part 2 Physical properties," *Journal of Materials Science*, vol. 9, no. 10, pp. 1710–1714, 1974.

[34] J. Ryu, S. Priya, K. Uchino, and H.-E. Kim, "High Magnetoelectric Properties in $0.68Pb(Mg_{1/3}Nb_{2/3})O_3$-$0.32PbTiO_3$ Single Crystal 28 and Terfenol-D Laminate Composites," *Journal of the Korean Ceramic Society*, vol. 39, pp. 813–817, 2002.

[35] J. Ryu, S. Priya, K. Uchino, and H. E. Kim, "Magnetoelectric effect in composites of magnetostrictive and piezoelectric materials," *Journal of Electroceramics*, vol. 8, no. 2, pp. 107–119, 2002.

[36] S. Dong, J. F. Li, and D. Viehland, "Ultrahigh magnetic field sensitivity in laminates of TERFENOL-D and Pb $(Mg_{1/3}Nb_{2/3})O_3$–$PbTiO_3$ crystals," *Applied Physics Letters*, vol. 83, no. 11, Article ID 2265, 3 pages, 2003.

[37] S. Dong, J. F. Li, and D. Viehland, "Giant magneto-electric effect in laminate composites," *IEEE Transactions on Ultrasonics, Ferroelectrics, and Frequency Control*, vol. 50, no. 10, pp. 1236–1239, 2003.

[38] G. Srinivasan, E. T. Rasmussen, J. Gallegos, R. Srinivasan, Y. I. Bokhan, and V. M. Laletin, "Magnetoelectric bilayer and multilayer structures of magnetostrictive and piezoelectric oxides," *Physical Review B*, vol. 64, no. 21, Article ID 214408, 6 pages, 2001.

[39] G. Srinivasan, E. T. Rasmussen, and R. Hayes, "Magnetoelectric effects in ferrite-lead zirconate titanate layered composites: the influence of zinc substitution in ferrites," *Physical Review B*, vol. 67, no. 1, Article ID 014418, 11 pages, 2003.

[40] V. M. Laletin, N. Paddubnaya, G. Srinivasan et al., "Frequency and field dependence of magnetoelectric interactions in layered ferromagnetic transition metal-piezoelectric lead zirconate titanate," *Applied Physics Letters*, vol. 87, no. 22, Article ID 222507, 3 pages, 2005.

[41] N. Cai, C.-W. Nan, J. Zhai, and Y. Lin, "Large high-frequency magnetoelectric response in laminated composites of piezoelectric ceramics, rare-earth iron alloys and polymer," *Applied Physics Letters*, vol. 84, no. 18, Article ID 3516, 3 pages, 2004.

[42] S. Dong, J. Zhai, J. Li, and D. Viehland, "Near-ideal magnetoelectricity in high-permeability magnetostrictive/ piezofiber laminates with a (2-1) connectivity," *Applied Physics Letters*, vol. 89, no. 25, Article ID 252904, 3 pages, 2006.

[43] S. Dong, J. F. Li, and D. Viehland, "A longitudinal-longitudinal mode TERFENOL-*D*/$Pb(Mg_{1/3}Nb_{2/3})O_3$–$PbTiO_3$ laminate composite ," *Applied Physics Letters*, vol. 85, no. 22, Article ID 5305, 2 pages, 2004.

[44] J.-P. Zhou, H. He, Z. Shi, and C.-W. Nan, "Magnetoelectric $CoFe_2O_4/Pb(Zr_{0.52}Ti_{0.48})O_3$ double-layer thin film prepared

by pulsed-laser deposition," *Applied Physics Letters*, vol. 88, no. 1, Article ID 013111, 3 pages, 2006.

[45] C. Deng, Y. Zhang, J. Ma, Y. Lin, and C.-W. Nan, "Magnetic-electric properties of epitaxial multiferroic Ni Fe_2O_4–BaTi O_3 heterostructure," *Journal of Applied Physics*, vol. 102, no. 7, Article ID 074114, 5 pages, 2007.

[46] H.-C. He, J.-P. Zhou, J. Wang, and C.-W. Nan, "Multiferroic $Pb(Zr_{0.52}Ti_{0.48})O_3$–$Co_{0.9}Zn_{0.1}Fe_2O_4$ bilayer thin films via a solution processing," *Applied Physics Letters*, vol. 89, no. 5, Article ID 052904, 3 pages, 2006.

[47] H. Ryu, P. Murugavel, J. H. Lee et al., "Magnetoelectric effects of nanoparticulate $Pb(Zr_{0.52}Ti_{0.48})O_3$–$NiFe_2O_4$ composite films," *Applied Physics Letters*, vol. 89, no. 10, Article ID 102907, 3 pages, 2006.

[48] Y. G. Ma, W. N. Cheng, M. Ning, and C. K. Ong, "Magnetoelectric effect in epitaxial $Pb(Zr_{0.52}Ti_{0.48})O_3/La_{0.7}Sr_{0.3}MnO_3$ composite thin film," *Applied Physics Letters*, vol. 90, no. 15, Article ID 152911, 3 pages, 2007.

[49] S. Q. Ren, L. Q. Weng, S.-H. Song, F. Li, J. G. Wan, and M. Zeng, "$BaTiO_3/CoFe_2O_4$ particulate composites with large high frequency magnetoelectric response," *Journal of Materials Science*, vol. 40, no. 16, pp. 4375–4378, 2005.

[50] W. E. Kramer, R. H. Hopkins, and M. R. Daniel, "Growth of oxide in situ composites: the systems lithium ferrite-lithium niobate, lithium ferrite-lithium tantalate, and nickel ferrite-barium titanate," *Journal of Materials Science*, vol. 12, no. 2, pp. 409–414, 1977.

[51] B. S. Kang, D. G. Choi, and S. K. Choi, "Effects of grain size on pyroelectric and dielectric properties of $Pb_{0.9}La_{0.1}TiO_3$ ceramic," *Journal of the Korean Physical Society*, vol. 32, supplement 1, pp. S232–S234, 1998.

[52] H. T. Martirena and J. C. Burfoot, "Grain-size effects on properties of some ferroelectric ceramics," *Journal of Physics C*, vol. 7, no. 17, Article ID 3182, 1974.

[53] S. Choudhury, Y. L. Li, C. Krill, and L. Q. Chen, "Effect of grain orientation and grain size on ferroelectric domain switching and evolution: phase field simulations," *Acta Materialia*, vol. 55, no. 4, pp. 1415–1426, 2007.

[54] H. M. Duiker and P. D. Beale, "Grain-size effects in ferroelectric switching," *Physical Review B*, vol. 41, no. 1, pp. 490–495, 1990.

[55] V. Sundar, N. Kim, C. A. Randall, R. Yimnirun, and R. E. Newnham, "The effect of doping and grain size on electrostriction in $PbZr_{0.52}TiO_{0.48}O_3$," in *Proceedings of the 10th IEEE International Symposium on Applications of Ferroelectrics (ISAF '96)*, vol. 2, pp. 935–938, August 1996.

[56] C. J. Lu, S. B. Ren, H. M. Shen, J. S. Liu, and Y. N. Wang, "The effect of grain size on domain structure in unsupported $PbTiO_3$ thin films ," *Journal of Physics Condensed Matter*, vol. 8, no. 42, pp. 8011–8016, 1996.

[57] C. Sakaki, B. L. Newalkar, S. Komarneni, and K. Uchino, "Grain size dependence of high power piezoelectric characteristics in Nb doped lead zirconate titanate oxide ceramics," *Japanese Journal of Applied Physics 1*, vol. 40, no. 12, pp. 6907–6910, 2001.

[58] T. Takeuchi, T. Tani, and Y. Saito, "Piezoelectric properties of bismuth layer-structured ferroelectric ceramics with a preferred orientation processed by the reactive templated grain growth method," *Japanese Journal of Applied Physics 1*, vol. 38, no. 9, pp. 5553–5556, 1999.

[59] S. Kwon, E. M. Sabolsky, G. L. Messing, and S. Trolier-McKinstry, "High strain,⟨001⟩ textured $0.675Pb(Mg_{1/3}Nb_{2/3})O_3$–$0.325PbTiO_3$ ceramics:templated grain growth and piezoelectric properties," *Journal of the American Ceramic Society*, vol. 88, no. 2, pp. 312–317, 2005.

[60] J. L. Jones, S. C. Vogel, E. B. Slamovich, and K. J. Bowman, "Quantifying texture in ferroelectric bismuth titanate ceramics," *Scripta Materialia*, vol. 51, no. 12, pp. 1123–1127, 2004.

[61] R. A. Islam and S. Priya, "Enhanced magnetoelectric effect in $(1 - x)$ $Pb(Zr_{0.52}Ti_{0.48})O_3$–$xNiFe_{1.9}Mn_{0.1}O_4$ particulate composites—electrical and magnetic properties," *Integrated Ferroelectrics*, vol. 82, no. 1, pp. 1–24, 2006.

[62] R. A. Islam and S. Priya, "Annealing and aging effect in 0.95 $Pb(Zr_{0.52}Ti_{0.48})O_3$–0.05 $NiFe_{1.9}Mn_{0.1}O_4$ particulate magnetoelectric composites," *Japanese Journal of Applied Physics 2*, vol. 45, no. 4–7, pp. L128–L131, 2006.

[63] R. A. Islam and S. Priya, "Synthesis of high magnetoelectric coefficient composites using annealing and aging route," *International Journal of Applied Ceramic Technology*, vol. 3, no. 5, pp. 353–363, 2006.

[64] H. W. Kim, S. Priya, K. Uchino, and R. E. Newnham, "Piezoelectric energy harvesting under high pre-stressed cyclic vibrations," *Journal of Electroceramics*, vol. 15, no. 1, pp. 27–34, 2005.

[65] R. A. Islam and S. Priya, "Realization of high-energy density polycrystalline piezoelectric ceramics," *Applied Physics Letters*, vol. 88, no. 3, Article ID 032903, 3 pages, 2006.

[66] S. Priya, Ph.D. dissertation, Pennsylvania State University, University Park, Pa, USA, 2003.

[67] S. Priya and K. Uchino, "Dielectric and piezoelectric properties of the Mn-substituted $Pb(Zn_{1/3}Nb_{2/3})O_3$–$PbTiO_3$ single crystal ," *Journal of Applied Physics*, vol. 91, no. 7, Article ID 4515, 6 pages, 2002.

[68] S. Priya, K. Uchino, and D. Viehland, "Crystal growth and piezoelectric properties of Mn-substituted $Pb(Zn_1/3Nb_2/3)O_3$ single crystal," *Japanese Journal of Applied Physics 2*, vol. 40, no. 10, pp. L1044–L1047, 2001.

[69] S. Priya, K. Uchino, and D. Viehland, "Fe-substituted 0.92 $Pb(Zn_{1/3}Nb_{2/3})O_3$–$0.08PbTiO_3$ single crystals: a "hard" piezocrystal ," *Applied Physics Letters*, vol. 81, no. 13, Article ID 2430, 3 pages, 2002.

[70] S. Priya and K. Uchino, "High power resonance characteristics and dielectric properties of co-substituted $0.92Pb(Zn_{1/3}Nb_{2/3})O_3$–$0.08PbTiO_3$ single crystal," *Japanese Journal of Applied Physics 1*, vol. 42, no. 2, pp. 531–534, 2003.

[71] S. Priya, J. Ryu, K. Uchino, C. W. Ahn, and S. Nahm, "Induction of combinatory characteristics by relaxor modification of $Pb(Zr_{0.5}Ti_{0.5})O_3$," *Applied Physics Letters*, vol. 83, no. 24, Article ID 5020, 3 pages, 2003.

[72] C. A. Randall, N. Kim, J. P. Kucera, W. Cao, and T. R. Shrout, "Intrinsic and extrinsic size effects in fine-grained morphotropic-phase-boundary lead zirconate titanate ceramics," *Journal of the American Ceramic Society*, vol. 81, no. 3, pp. 677–688, 1998.

[73] K. I. Kobayashi, H. Morinaga, T. Araki, Y. Naka, and T. Oomura, "Low-loss Ni-Zn-Cu ferrite for deflection yoke," *Journal of Magnetism and Magnetic Materials*, vol. 104–107, part 1, pp. 413–414, 1992.

[74] I. Z. Rahman, T. T. Ahmed, and L. Powell, "Magnetic and physical characterization of nano granular Ni-Zn-Cu based ferrite powders," *Journal of Metastable and Nanocrystalline Materials*, vol. 17, pp. 9–16, 2003.

[75] W. C. Kim and S. J. Kim, "Magnetic properties of NiZnCu ferrite powders and thin films prepared by a sol-gel method," *IEEE Transactions on Magnetics*, vol. 37, no. 4, pp. 2362–2365, 2001.

[76] J. J. Shrotri, S. D. Kulkarni, C. E. Deshpande et al., "Effect of Cu substitution on the magnetic and electrical properties of Ni-Zn ferrite synthesized by soft chemical method," *Materials Chemistry and Physics*, vol. 59, no. 1, pp. 1–5, 1999.

[77] T. T. Ahmed, I. Z. Rahman, and M. A. Rahman, "Study on the properties of the copper substituted NiZn ferrites," *Journal of Materials Processing Technology*, vol. 153-154, no. 1–3, pp. 797–803, 2004.

[78] T. Okamura and Y. Kojima, "Ferromagnetic resonance in copper ferrite," *Physical Review*, vol. 86, no. 6, pp. 1040–1041, 1952.

[79] R. E. Newnham, "Electroceramics," *Reports on Progress in Physics*, vol. 52, no. 2, pp. 123–156, 1989.

[80] Z. J. Zhang, Z. L. Wang, B. C. Chakoumakos, and J. S. Yin, "Temperature dependence of cation distribution and oxidation. State in magnetic Mn-Fe ferrite nanocrystals," *Journal of the American Chemical Society*, vol. 120, no. 8, pp. 1800–1804, 1998.

[81] R. A. Islam, D. Viehland, and S. Priya, *Journal of Materials Science Letters*, 2007.

[82] R. A. Islam and S. Priya, *Journal of Materials Science Letters*, 2007.

[83] R. A. Islam and S. Priya, "Magnetoelectric properties of the lead-free cofired $BaTiO_3$–$(Ni_{0.8}Zn_{0.2})Fe_2O_4$ bilayer composite," *Applied Physics Letters*, vol. 89, no. 15, Article ID 152911, 3 pages, 2006.

[84] R. A. Islam, J. Jiang, S. Priya, F. Bai, and D. Viehland, "Correlation between structural deformation and magnetoelectric response in $(1 - x)$ Pb $(Zr_{0.52}\ Ti_{0.48})O_3$–$x$Ni $Fe_{1.9}\ Mn_{0.1}O_4$ particulate composites," *Applied Physics Letters*, vol. 91, no. 16, Article ID 162905, 3 pages, 2007.

[85] T. Takeuchi, T. Tani, and Y. Saito, "Piezoelectric properties of bismuth layer-structured ferroelectric ceramics with a preferred orientation processed by the reactive templated grain growth method," *Japanese Journal of Applied Physics 1*, vol. 38, no. 9, pp. 5553–5556, 1999.

[86] S. H. Hong, S. T. Trolier-McKinstry, and G. L. Messing, "Dielectric and electromechanical properties of textured niobium-doped bismuth titanate ceramics," *Journal of the American Ceramic Society*, vol. 83, no. 1, pp. 113–118, 2000.

[87] A. Moure, L. Pardo, C. Alemany, P. Millán, and A. Castro, "Piezoelectric ceramics based on Bi_3TiNbO_9 from mechanochemically activated precursors," *Journal of the European Ceramic Society*, vol. 21, no. 10-11, pp. 1399–1402, 2001.

[88] M. Allahverdi, A. Hall, R. Brennan, M. E. Ebrahimi, N. M. Hagh, and A. Safari, "An overview of rapidly prototyped piezoelectric actuators and grain-oriented ceramics," *Journal of Electroceramics*, vol. 8, no. 2, pp. 129–137, 2002.

[89] K. Watari, B. Brahmaroutu, G. L. Messing, S. Trolier-McKinstry, and S.-C. Cheng, "Epitaxial growth of anisotropically shaped, single-crystal particles of cubic $SrTiO_3$," *Journal of Materials Research*, vol. 15, no. 4, pp. 846–849, 2000.

[90] E. M. Sabolsky, S. Trolier-McKinstry, and G. L. Messing, "Dielectric and piezoelectric properties of ⟨001⟩ fiber-textured $0.675Pb(Mg_{1/3}Nb_{2/3})O_3$–$0.325PbTiO_3$ ceramics ," *Journal of Applied Physics*, vol. 93, no. 7, Article ID 4072, 9 pages, 2003.

[91] J. P. Remeika, *Notes*, vol. 76, p. 940, 1953.

[92] Y. S. Koo, T. Bonaedy, K. D. Sung et al., "Magnetodielectric coupling in core/shell $BaTiO_3$ /γ–Fe_2O_3 nanoparticles," *Applied Physics Letters*, vol. 91, no. 21, Article ID 212903, 3 pages, 2007.

[93] G. Rowlands, "The variation of coercivity with particle size ," *Journal of Physics D*, vol. 9, no. 8, Article ID 1267, 1976.

[94] K. Uestuener, M. Katter, and W. Rodewald, "Dependence of the mean grain size and coercitivity of sintered Nd-Fe-B magnets on the initial powder particle size," *IEEE Transactions on Magnetics*, vol. 42, no. 10, Article ID 2897, 2006.

[95] S. Priya, R. Islam, S. Dong, and D. Viehland, "Recent advancements in magnetoelectric particulate and laminate composites," *Journal of Electroceramics*, vol. 19, no. 1, pp. 147–164, 2007.

11

Broadband/Wideband Magnetoelectric Response

Chee-Sung Park and Shashank Priya

Center for Energy Harvesting Materials and Systems (CEHMS), Department of Materials Science and Engineering, Virginia Polytechnic Institute and State University, Blacksburg, VA 24061, USA

Correspondence should be addressed to Shashank Priya, spriya@vt.edu

Academic Editor: Amar Bhalla

A broadband/wideband magnetoelectric (ME) composite offers new opportunities for sensing wide ranges of both DC and AC magnetic fields. The broadband/wideband behavior is characterized by flat ME response over a given AC frequency range and DC magnetic bias. The structure proposed in this study operates in the longitudinal-transversal (L-T) mode. In this paper, we provide information on (i) how to design broadband/wideband ME sensors and (ii) how to control the magnitude of ME response over a desired frequency and DC bias regime. A systematic study was conducted to identify the factors affecting the broadband/wideband behavior by developing experimental models and validating them against the predictions made through finite element modeling. A working prototype of the sensor with flat bands for both DC and AC magnetic field conditions was successfully obtained. These results are quite promising for practical applications such as current probe, low-frequency magnetic field sensing, and ME energy harvester.

1. Introduction

Magnetoelectric (ME) materials have been investigated to find applications in sensors, transducers, actuators, energy harvesters, and servomechanism. Direct ME effect implies induction of electric polarization by applying an external magnetic field. It has been widely investigated in composite structures consisting of magnetostrictive and piezoelectric materials. In these composites, ME coefficient is dependent upon the elastic coupling occurring at the interface between piezoelectric and magnetostrictive phases [1–4]. Functional composites are defined by connectivity of materials such as 2-2 composite which is 2 dimensionally laminated, 3-0 composite which has dispersed particles in 3-dimensional matrix, and 3-1 composite which has 1-dmensional rods embedded in 3-dimensional matrix structures. Magnetostrictive-piezoelectric laminate composites with 2-2 connectivity have shown much higher ME coefficients than that of single-phase materials or particulate composites [4–8].

Experimental and analytical results on ME laminates have shown relatively large ME output voltage, but the peak in ME coefficient occurring at the optimum DC bias is generally sharp. This limits their ability to be utilized for AC magnetic field sensing when the optimized DC condition becomes slightly moved off by external noises. Moreover, high ME coefficient generally occurs near the electromechanical resonance (EMR) frequency. This narrow bandwidth also poses problem for stable sensing in a limited frequency range. The common modes utilized for ME laminates are longitudinal transversal (LT) and longitudinal longitudinal (LL). In these two modes, the ME composites exhibit maxima at a specific DC bias magnitude dependent upon the material properties. This maxima position is in the vicinity of ~5 Oe magnetic DC bias for 2-1 composites consisting of Metglas sheets and piezoelectric fibers which is probably the lowest reported value [9]. For sintered composites consisting of PZT and ferrite phases, the maximum occurs at the applied DC bias of ~400 Oe [10].

In order to overcome the problems of limited DC and AC magnetic field ranges associated with laminates, we attempted to design structures with flat ME responses for sensing wide ranges of both DC magnetic field and AC magnetic frequency as illustrated in Figure 1 [11, 12]. Metglas and lead zirconate titanate-lead zinc niobate (PZNT)-based composites were utilized for all the experimentation.

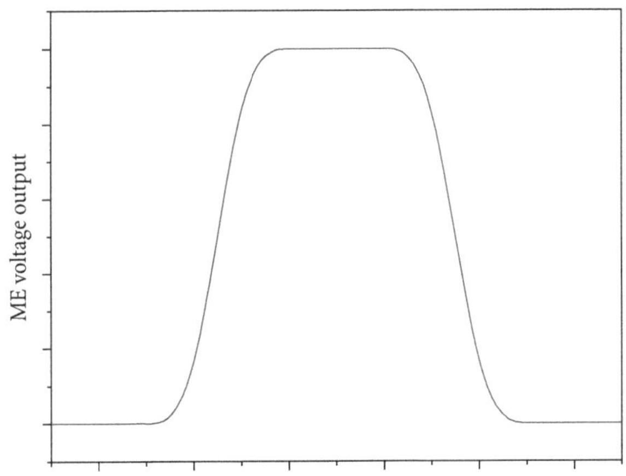

FIGURE 1: Schematic diagram of an ideal broadband/wideband ME sensor.

2. Broadband Magnetoelectric Composite

Metglas-PZT-based laminate composite structure was selected to understand ME phenomena and to overcome the problem of limited DC magnetic bias operating range. The selection of materials is based upon the fact that high magnetostriction in Metglas occurs at low magnetic DC bias magnitude while the piezoelectric voltage constant (g_{33}) in PZNT is high in the range of 23.41×10^{-3} Vm/N. Dong et al. have shown that the voltage generated through magnetoelectric composite structure is given as [13–15]

$$V = \frac{N d_{33,m} g_{33,p}}{N s_{33}^{E}(1 - k_{33}^{2}) + (1 - N) s_{33}^{H}} H \cdot t \quad (1)$$

(open circuit voltage),

where s_{33}^E and s_{33}^H are the elastic compliances for the piezoelectric and magnetostrictive layers, k_{33} the electromechanical coupling coefficient of the piezoelectric layer, $d_{33,m}$ and $g_{33,p}$ the longitudinal piezomagnetic and piezoelectric voltage coefficients, N is the thickness fraction of magnetostrictive layers, and t is the thickness of piezoelectric layer. This expression is an approximation and requires an additional correction factor to match the magnitude of voltage coefficient measured experimentally. However, it provides an insight into material selection, and based upon this expression it can be seen that piezoelectric voltage constants, elastic compliances (mechanical impedance matching), and piezomagnetic constants ($q_{ij} = \partial\lambda_{ij}/\partial H$) are the key factors controlling the magnitude of ME coefficients.

2.1. Effect of Shape and Laminate Configuration. Piezoelectric square plate and disks with the composition $Pb(Zn_{1/3}Nb_{2/3})_{0.2}(Zr_{0.5}Ti_{0.5})_{0.8}O_3$ [PZNT] were used in this work as schematically depicted in Figure 2(a). The piezoelectric constant of poled PZNT specimens was found to be 500 pC/N, and dielectric constant was found to be 2219 at 1 kHz. On these piezoelectric specimens, 25 μm-thick Metglas (2605SA1, Metglas Inc, USA) sheets of desired dimensions were attached. Twenty layers of Metglas were stacked on one side of square PZNT plate ($15.5 \times 15.5\,mm^2$) to achieve the unimorph structure [Type I]. For the PZNT disk with diameter of 10.2 mm, twenty layers of Metglas were stacked on both sides to form the sandwich structure [Type II]. Figure 2(b) shows the ME responses of both composite structures measured under $H_{ac} = 1$ Oe at 1 kHz. Type I exhibited maximum output voltage of 62 mV/cm.Oe at 215 Oe while Type II showed ME coefficient of 73 mV/cm·Oe at 570 Oe. It is interesting to note that by changing the structure from "unimorph" to "sandwich" and from square shape to disk, the maximum peak position in ME coefficient was shifted by more than 2 times (355 Oe shift). These results bring several interesting questions: (i) what is the effect of stacking configurations on the maximum position of ME coefficient?, (ii) how can the two configurations be combined such that ME coefficient has maximum magnitude over the DC bias range of 215 Oe to 570 Oe?, and (iii) what is the effect of interface area on ME coefficient? These questions will be answered in the next few sections.

2.2. Effect of Number of Metglas Layers. In order to control ME behavior, it is important to understand the role and nature of magnetostrictive layer. The role of number of Metglas layers on ME response was investigated by using the unimorph structure where 2 layers of Metglas with dimension of $7 \times 9\,mm^2$ were incrementally stacked on PZNT plate of $7 \times 15\,mm^2$. Figure 3(a) shows the variation of ME coefficient depending upon the number of Metglas layers and applied DC magnetic field. The composite with 4 layers of Metglas exhibited maximum magnitude of ME coefficient. With further increase in the number of Metglas layers, the ME output steadily decreases. The optimum magnitude of magnetic DC bias was found to increase with increase in the number of Metglas layers. Figure 3(b) summarizes the changes in maximum ME coefficient and optimum DC magnetic field as a function of number of Metglas layers. These results indicate that as the number of Metglas layers increases, higher DC bias is required to reach the maximum ME coefficient. This implies that it is possible to control the ME output value and saturation point by changing the number of Metglas layers.

2.3. Effect of Elastic Coupling on Electrode Structure. In order to gain more insight into elastic coupling process of laminate configuration, a sectioned electrode pattern was designed on PZNT plate of dimension $15.5 \times 15.5\,mm^2$. As schematically shown in Figure 4(a), the electrode pattern on top of the PZNT plate was divided into two parts, a stripe in between the "U" structure. The gap between the two sections was of the order of 1 mm. Ten layers of Metglas was attached on PZNT covering the whole surface area. In the center, on top of the stripe electrode, the number of Metglas layers was increased to thirty layers as shown in Figure 4(a). ME coefficient was first measured separately from the two sections, and then combined response was measured from two sections as shown in Figure 4(b). The ME responses in both scenario were the same regardless

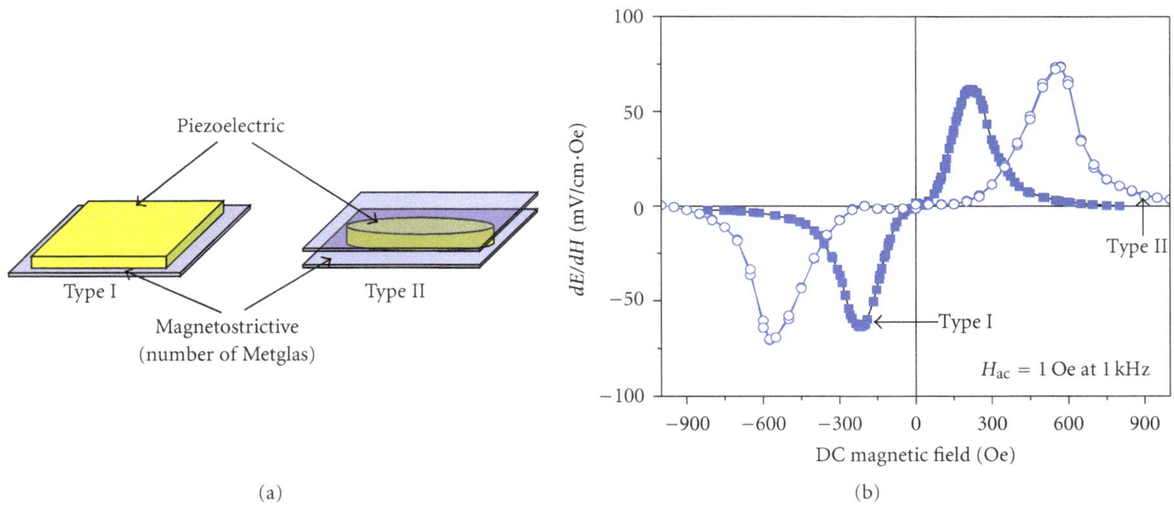

FIGURE 2: (a) Schematic diagram of fabricated samples: type I—15.5 × 15.5 mm^2 PZNT plate with Metglas attached on one side only, and Type II—10.2 mm diameter PZNT disk with Metglas attached on both sides. (b) Magnetoelectric output voltage as a function of DC magnetic field [11].

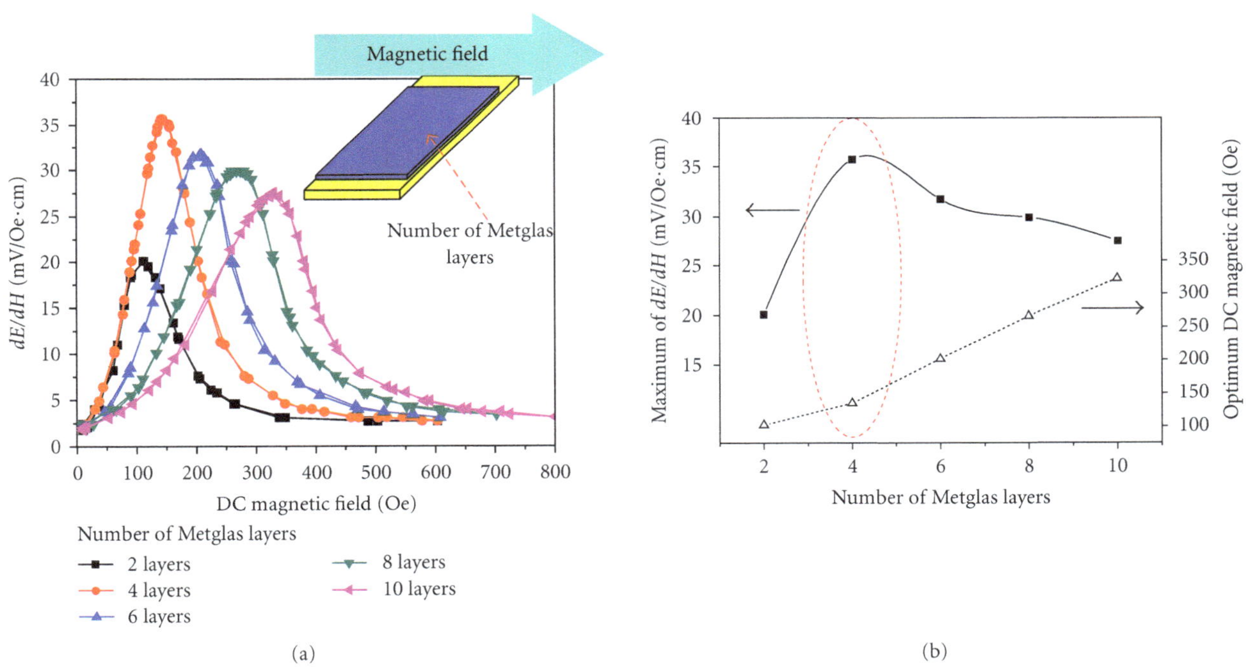

FIGURE 3: (a) ME coefficient as a function of the number of Metglas layers and DC magnetic field. (b) Maximum ME coefficient and optimum DC magnetic field as a function of the number of Metglas layers on PZNT. All measurements were conducted under the constant condition of H_{ac} = 1 Oe at f = 1 kHz [11].

of electrical connections. This phenomenon gives us an important insight. Although locally different ME output behavior was expected from each section because of the difference in number of Metglas layers, the magnitudes of ME coefficient and optimum DC magnetic field were found to be identical from both sections. This indicates that average strain distribution is homogeneous even though locally different strains are generated. Therefore, a larger physical separation will be required to achieve differences between the peak positions of ME coefficient from two separated sections.

2.4. Effect of Composite Dimensions on ME Behaviors. PZNT plates with different planar dimensions (15.5 × 15.5, 12.5 × 12.5, and 8 × 8 mm^2), but the same thickness of 1 mm was used in this experiment. Twenty-layers of Metglas were attached on each plate as shown in Figure 5 (unimorph

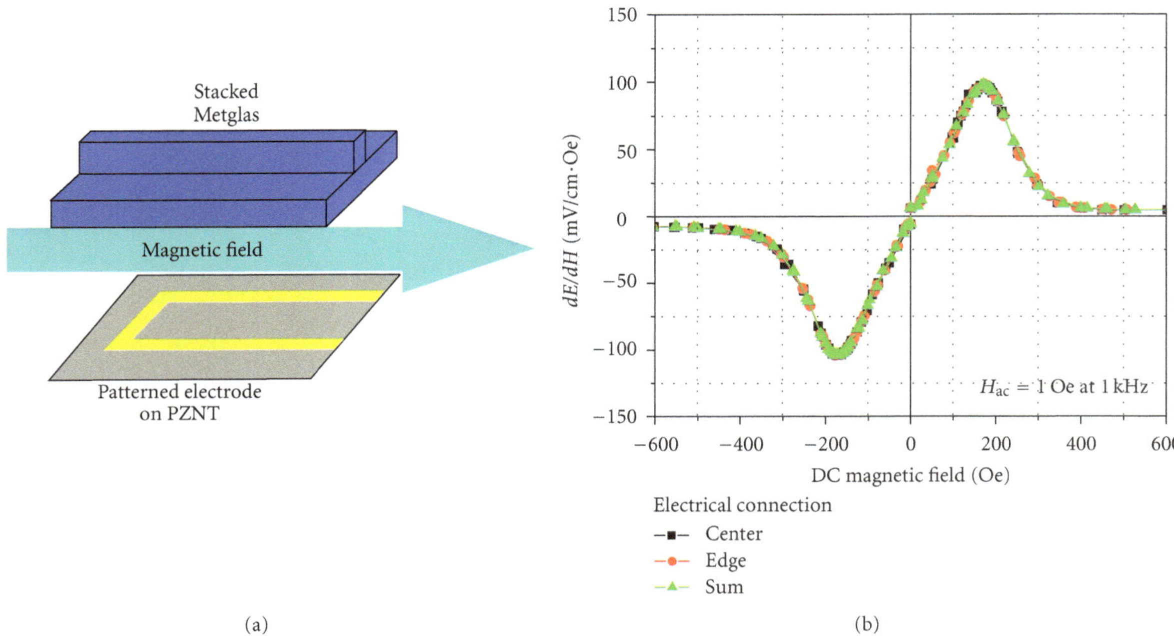

FIGURE 4: (a) Schematic representation of sample preparation with pyramid-shaped Metglas structure on separately electroded PZNT plate. (b) Magnetoelectric coefficients depending on electrical connections [11].

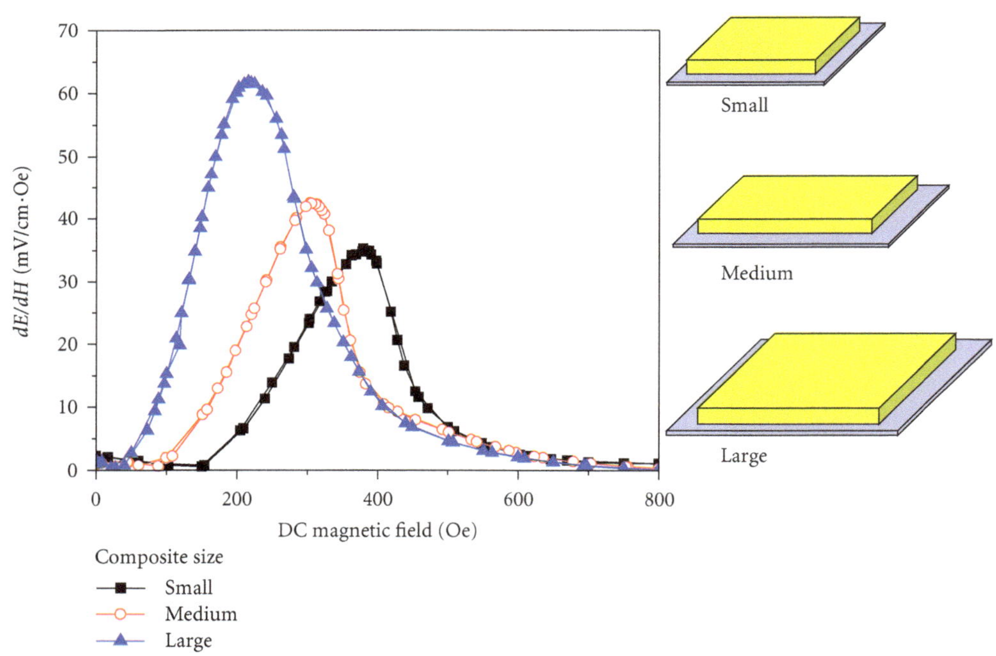

FIGURE 5: ME coefficient variation as a function of DC magnetic field for various dimensions. 20 layers of Metglas were attached on all PZNT plates. All ME measurements were conducted under $H_{ac} = 1$ Oe at 1 kHz [11].

configuration). Higher number of Metglas layers was used to avoid variation as a function of number of Metglas layers (saturated response in Figure 3(b)). Thus, the ME coefficient is dependent only upon the planar dimensions of the composite. Figure 5 shows the ME behavior as a function of DC magnetic field for three different samples. As the planar area of composite decreases, the ME voltage coefficient decreases, while optimum DC magnetic field increases. As a result, the ME coefficient and optimum DC magnetic field exhibit inverse behaviors with increasing planar area. In other words, when the planar geometry is larger, the ME coefficient is larger with smaller optimum

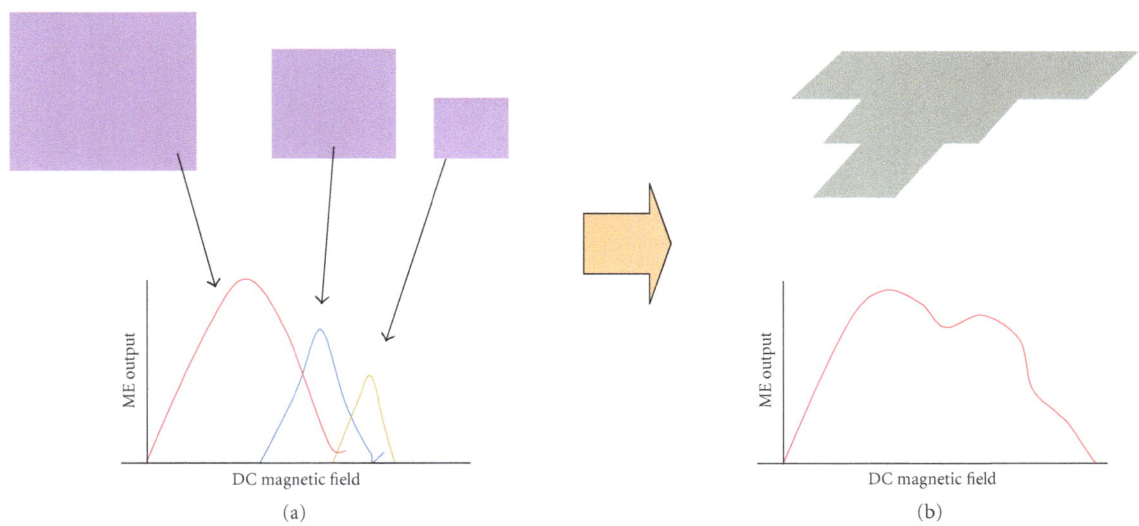

FIGURE 6: Schematic diagram of modified sensor structure. Based on the results of ME dependency on the size (a), dimensionally gradient structure was designed as shown in (b).

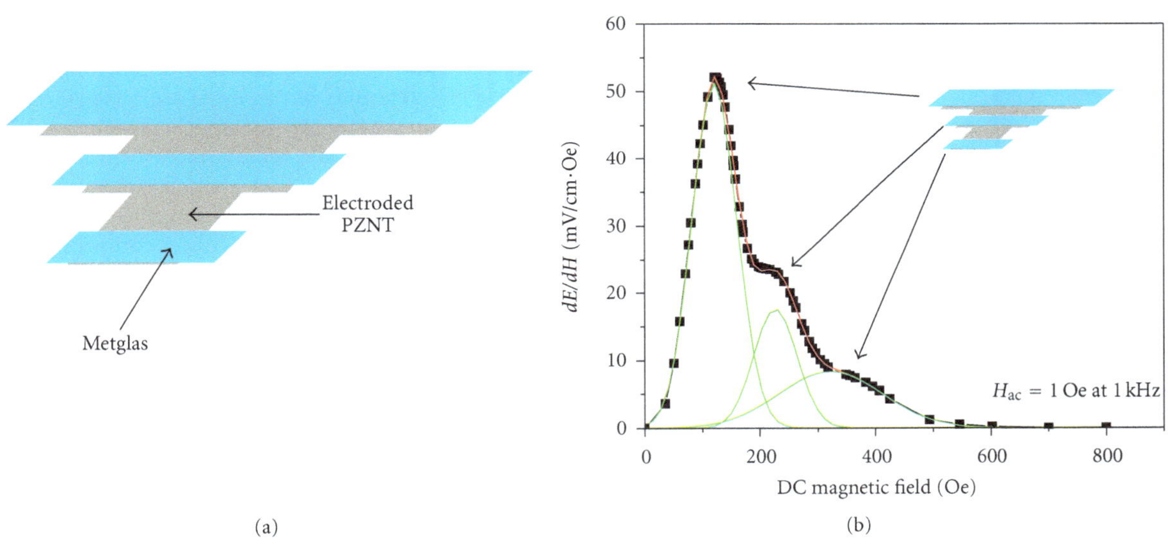

FIGURE 7: (a) Schematic structure of the dimensionally gradient laminate composite. 20 layers of Metglas were attached on each rectangular section, (b) ME coefficient of the dimensionally gradient sample under H_{ac} = 1 Oe at 1 kHz [11].

magnetic DC bias. The results of Figure 5 can be explained on the basis of edge effect called "shear lagging" [16]. As dimensions of ME composites become smaller, the relative surface and edge areas increase leading to shear lagging. This results in the change of magnetization behavior and piezomagnetic coefficient ($q_{ij} = \partial\lambda_{ij}/\partial H$); subsequently, both optimum magnetic bias and ME coefficient are changed with a trade-off relationship between them.

This result is important to realize different peak positions of ME coefficient from different sections. In combination with the result of Figure 3, it is quite interesting to note that optimum number of layers of Metglas on piezoelectric phase with large planar dimensions can provide higher ME coefficient at smaller magnetic DC bias. These results (Figures 3 and 5) lead us to the design rule that maximum peak of the ME coefficient can be shifted by altering the geometry of laminate composites and by designing structure with gradient dimensions as illustrated in Figure 6.

2.5. Effect of Dimensionally Gradient Laminate Composite. In order to achieve the individual response from separated sections of composite, a new geometrical structure was conceived as schematically shown in Figure 6. The composite design was guided by the results of previous sections that indicated (i) different dimensions resulted in different peak positions with respect to magnetic DC bias, and (ii) physical

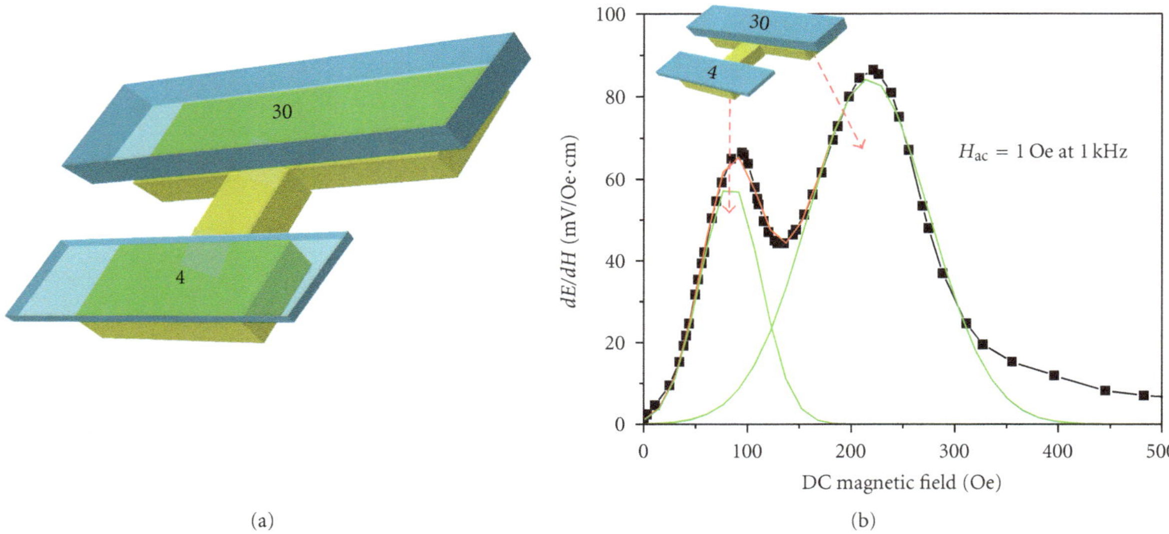

FIGURE 8: (a) Schematic diagram of the asymmetric H-shaped composite structure. (b) Broadband ME behavior of the asymmetric H-shaped laminate [11].

separation between the Metglas layers was necessary to create difference in the peak positions of ME coefficient. We incorporated these two factors by designing a dimensionally gradient composite with the separation of Metglas sections, as shown in Figure 7(a). The dimensions of the PZNT plate in each section were as following: 15.5 × 5, 10.6 × 5.5, and 6.2 × 5 mm^2 (width of about 5 mm was reduced per 5 mm length). The thickness of the plate was 1 mm. On this structure, 20 layers of Metglas with proportional surface area were attached where each Metglas section was separated from each other by 1.5 mm. The stack of twenty-layers of Metglas was intentionally selected so that the observed effect is solely from the variation of the dimension across the composite. Figure 5(b) shows the measured ME response of this composite structure. Interestingly, total ME response of the structure was in combination of the three individual effects as shown by the peak fitting; three-typical components were found in the same specimen. As the planar dimension of composite decreases, the saturation magnetic field increases in providing three separate peaks from three different sections. The larger composite section exhibits higher ME coefficient at smaller magnetic DC bias in agreement with our result in Figure 5 and expectation in Figure 6.

The results of Figures 3–7 lead us to an important conclusion that high magnitude and saturation point of ME coefficient can be controlled by composite structure having separated sections, dimensions of composite, and number of Metglas layers on the piezoelectric planar area.

2.6. Observation of Broadband Behavior. In order to enhance the physical separation, an asymmetric H-shaped PZNT plate was fabricated with the dimensions shown in Figure 8(a) (15.2 × 5.3 and 8.5 × 5 mm^2 areas were connected with the bridge of 4 × 5 mm^2). In this design, the bridge connecting two different dimensions of laminate composites will lead to two separate responses. On this PZNT plate, 4 layers of Metglas were attached on the smaller area, and 30 layers of Metglas were attached on the larger area. It should be noted here that a smaller dimensions lead to smaller ME coefficient, but smaller number of Metglas layers lead to larger ME coefficient. Thus, a compromise between the two opposing effects was calculated to be in the range equivalent to that of the large composite with 30 layers of Metglas. Figure 8(b) shows the measured ME response from this composite structure. The first peak of ME coefficient was found at 94 Oe which was associated with that of 4-layered Metglas structure on smaller piezoelectric surface area. The second peak in ME coefficient was found at 220 Oe resulting from the 30-layered Metglas structure on the larger piezoelectric area. There is a slight drop in the magnitude of ME coefficient between the two peaks, which reflects a mismatch in tailoring the dimensions of two sections. By further adjusting the dimensions, it is possible to bring the two peaks closer to each other and achieve almost a square wave response. However, the results of this figure clearly demonstrate the idea in designing the broadband magnetoelectric sensor.

2.7. Frequency Dependence of Broadband ME Sensor. The frequency dependence of the ME coefficient for broadband composite in the range of 40 to 10^5 Hz is shown in Figure 9(a). The measurement was conducted under the AC magnetic field of H_{ac} = 1 Oe and under applied DC magnetic fields of 94 and 220 Oe. The peaks in this figure correspond to electromechanical resonances occurring at 20, 29, 49, and 70 kHz. It is interesting to note that regardless of the applied DC magnetic field in the range of 94–220 Oe, the broadband ME sensor shows similar range of ME output voltage, ~4 V/cm·Oe, at 49 kHz. The maximum

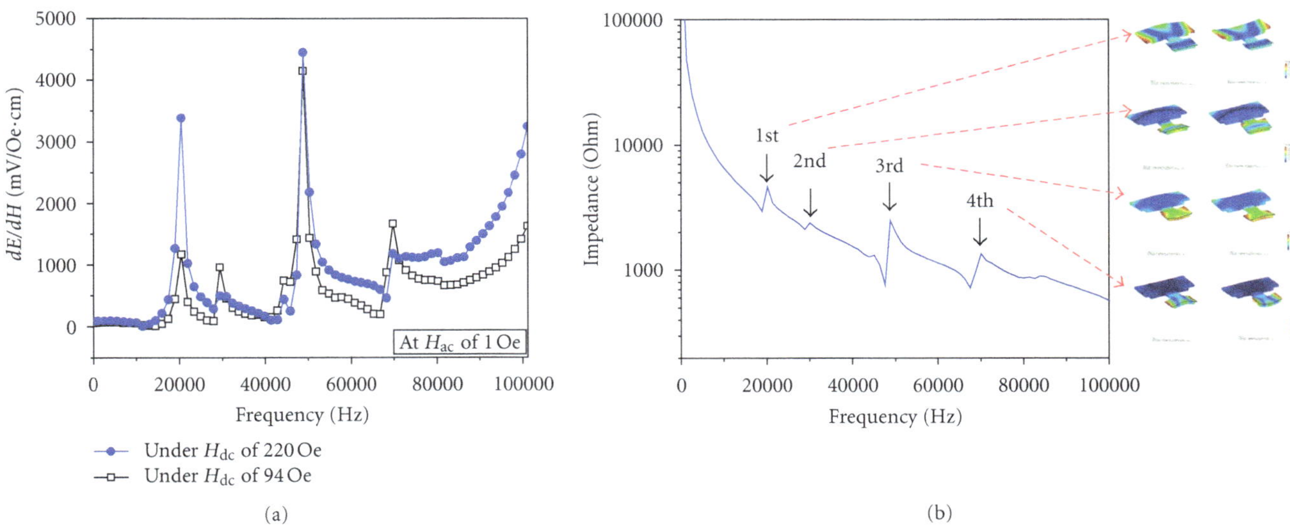

Figure 9: (a) Frequency dependence of the ME coefficient for broadband laminate composite. Measurements were conducted under H_{ac} = 1 Oe and H_{dc} = 94 and 220 Oe. (b) Impedance spectrum and FEM analysis of bending resonance modes corresponding to the peak ME responses [11].

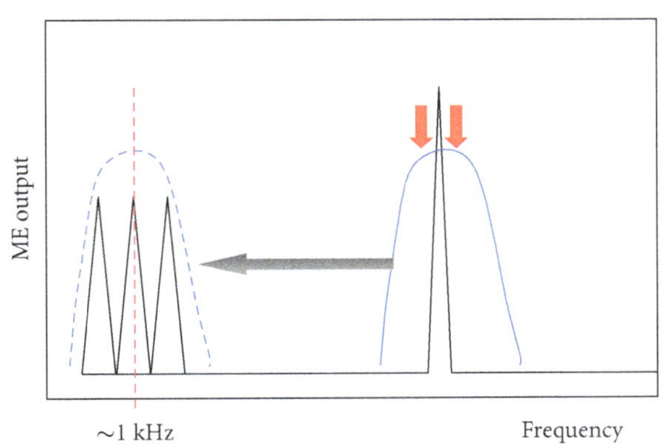

Figure 10: Schematic diagram of the goal to achieve wideband ME behavior @ ~1 kHz by generating multiresonances and merging them.

ME output was 4.5 V/cm·Oe under the conditions of H_{ac} = 1 Oe at 49 kHz and H_{DC} = 220 Oe. Figure 9(b) shows the impedance spectrum for H-shaped broadband laminate. The impedance spectrum was found to exhibit resonance peaks at the same position as that observed in ME measurements as a function of frequency. Further, this indicates that all the peaks were correlated to electromechanical resonance modes. An FEM analysis was conducted using ATILA to find the electromechanical resonance modes of the laminate composite. The bending oscillation modes for the four resonances observed in the impedance spectrum are shown in Figure 9(b). The first mode at 20 kHz was found to be from bending of the large rectangular section in the H laminate (Figure 9(b)-1st). The second mode at 29 kHz was found to be from the bending of bridge structure joining the two rectangular sections (Figure 9(b)-2nd). The output voltages from 2nd mode were relatively low which could be associated with the small deformation occurring in the bridge. The third mode at 49 kHz was related to the combined bending response from bridge and small rectangular section of the composite (Figure 9(b)-3rd). This mode generated the highest output voltages as shown in Figure 7. The fourth mode at 70 kHz was found to be related to the bending of small rectangular section of composite (Figure 9(b)-4th). The results indicate that third resonance mode at 49 kHz is combination of second mode occurring at 29 kHz and fourth mode occurring at 70 kHz and leads to larger displacements in the structure resulting in higher ME coefficient. These results further provide the insight in designing a composite structure that could provide large ME coefficient over a wide frequency range.

3. Wideband Magnetoelectric Composite

In previous sections, a near-flat ME response over a magnetic DC bias range of 90–220 Oe at f = 1 kHz was demonstrated by using dimensionally gradient composite structure. Next, we attempt to design composite structure with flat ME behavior as a function of frequency. The electromechanical resonance (EMR) frequency of structure in previous section was still high in the range of 20 kHz, and the ME coefficient exhibited a sharp peak as a function of frequency. As shown in Figure 9(a), the composites generally show enhanced ME coefficient at EMR frequency, but high ME coefficient near EMR is characterized by a sharp peak with very narrow range of magnetic DC bias [3, 14, 17, 18]. In conjunction with the fact that EMR in laminate composites is dependent upon the dimensions, this last observation has limited the application

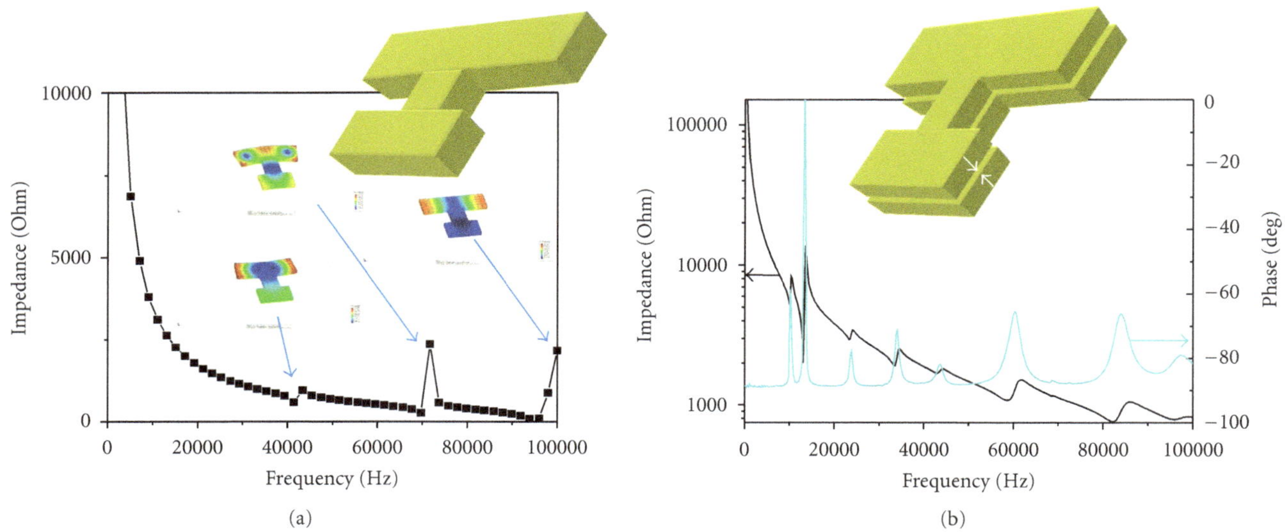

FIGURE 11: Resonance analysis: (a) unimorph and (b) bimorph piezoelectric asymmetric H-shaped structure [12].

of ME composites in magnetic field sensing; that is, large dimensions are required to achieve low EMR frequency.

A study for wideband ME response by combining several Terfenol-D/epoxy-Pb(Zr,Ti)O_3 bilayers through parallel and series electrical connections was conducted by Yu et al. [19]. However, this approach has obvious problems as follows: (i) several laminate composites are required with varying EMR frequency which increases the overall dimensions, (ii) the overall magnitude of ME coefficient at EMR frequency decreases significantly because of mismatch in electrical output, (iii) the magnitude of ME coefficient fluctuates depending upon the gradient in external magnetic field, and (iv) operational frequency is still quite high.

It is well known that, by combining the piezoelectric element having capacitance, C_p, in parallel with shunt having capacitance C_s, the total capacitance of the system changes to $C = C_p + C_s$, which affects the natural resonance frequency (ω) given as [20]

$$\omega = \sqrt{\frac{K_{\text{eff}} + C^{-1}d^2}{m_{\text{eff}}}}, \tag{2}$$

where K_{eff} is equivalent stiffness of the cantilever beam, m_{eff} is the effective mass, and d is the electromechanical coupling. Charnegie has shown that, for a three-layer structure consisting of two piezoelectric beams bonded on to a substrate which is nonpiezoelectric (three layer laminate), the EMR frequency of one piezoelectric layer with capacitance C_p can be shifted by changing the capacitance of other piezoelectric layer C_s through the expression [21]

$$\omega = \sqrt{\frac{3\left(s_{11} - d_{31}^2 A/t\left(C_p + C_s\right)\right)^{-1} I}{L^3 m_{\text{eff}}}}, \tag{3}$$

where A is area, t is the thickness, L is the length, s_{11} is the mechanical compliance of piezoelectric material, d_{31} is the electromechanical coupling coefficient, and I is the moment of inertia. Equation (3) also predicts that by combining several laminates in series and parallel connection the EMR frequency can be modified. This technique has been recently utilized for energy harvesters to modify the operating frequency range [22].

In order to achieve the wideband behavior for the magnetic frequency regime, multiple resonance modes in close vicinity to each other should be the key factor, as schematically illustrated in Figure 10. With reduction in the resonance frequency closer to ~1 kHz, wideband behaviors both in terms of magnetic DC bias and frequency will be described in the next few sections.

3.1. Bimorph-typed Dimensionally Gradient Laminate Composite. Figure 11 shows the difference in resonant frequencies between the asymmetric H-shaped piezoelectric unimorph and bimorph configurations. The resonance spectrum of the asymmetric H-shaped single plate was analyzed using the FEM method (ATILA), as shown in Figure 11(a). The FEM results demonstrate electromechanical resonances occurring at 40, 70, and 94 kHz as shown in Figure 11(a). The first mode at 40 kHz was associated with bending of the larger rectangular section. The second mode at 70 kHz was found to be from bending of two rectangular sections. The third mode at 94 kHz was related to combined lateral response from large rectangular section.

On the other hand, the asymmetric H-shaped piezoelectric bimorph structure was found to generate additional resonant modes and reduce the resonant frequency. Two piezoelectric plates with the thickness of 500 μm were machined to have the asymmetric H shapes (15.5 × 5 and 9.5 × 5 mm^2 areas were connected with the bridge of 4 × 5 mm^2). Next, two piezoelectric plates with opposite poling directions were bonded together as shown in the inset of Figure 11(b). The asymmetric H-shaped bimorph exhibited multiple resonances below 100 kHz as shown in Figure 11(b). The impedance and phase spectrums showed

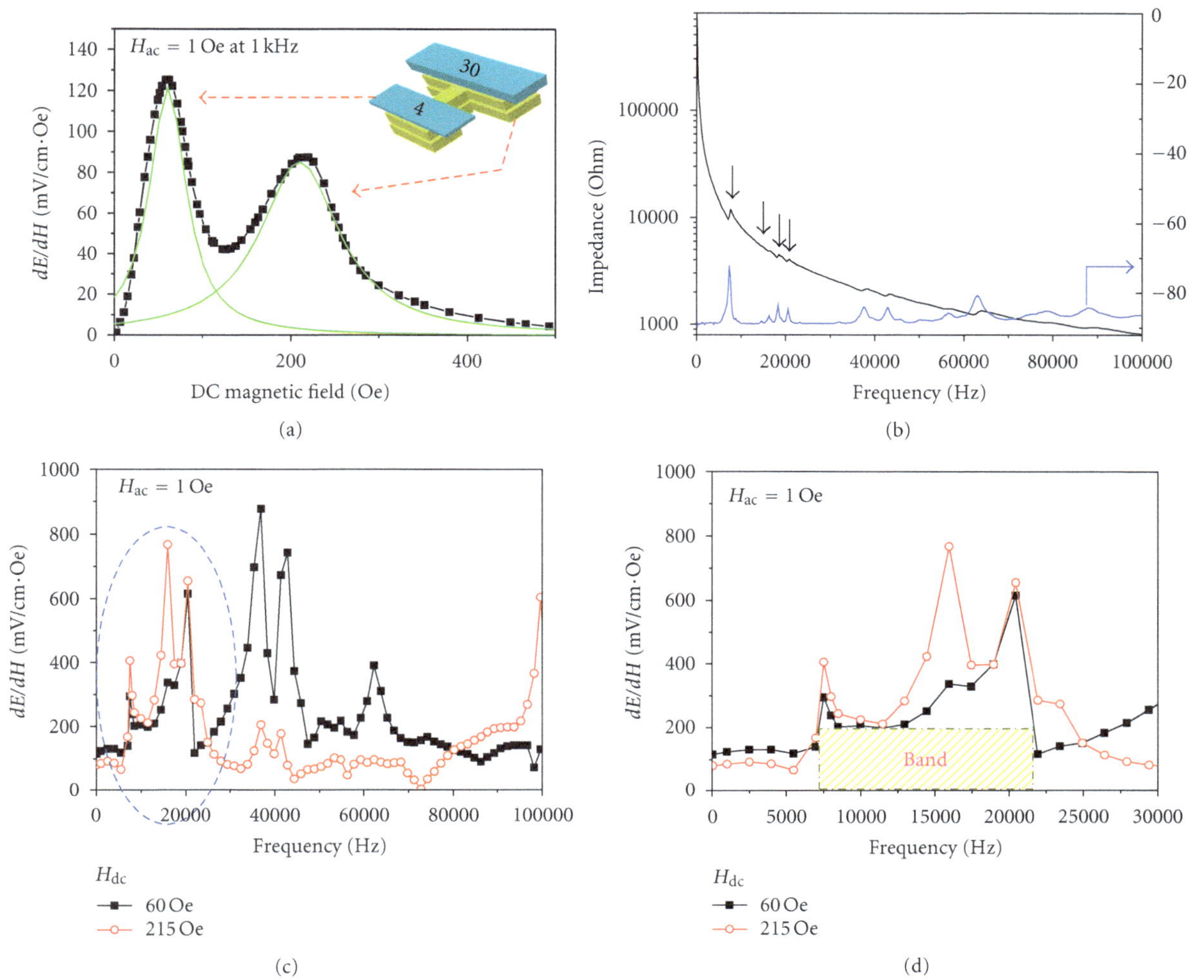

FIGURE 12: (a) ME coefficient as a function of magnetic DC bias at frequency of 1 kHz and applied AC magnetic field of H_{ac} = 1 Oe. Inset: schematic diagram of Metglas attached to the bimorph. Four layers of Metglas of area 15 × 7 mm^2 were attached on smaller rectangular section while thirty layers of Metglas of area 20 × 7 mm^2 were attached on the larger rectangular section, (b) impedance and phase spectrums as a function of frequency, (c) ME coefficient as a function of frequency until 100 kHz, and (d) wideband ME behavior as a function of frequency until 30 kHz [12].

the resonance peaks at 10.09, 13.08, 23.32, 33.31, 43.05, 58.54, and 82.26 kHz. These peaks are resultant of bimorph configuration which indicates coupling between the two piezoelectric plates. The main peak for bimorph was observed at 13.08 kHz. The presence of these additional resonant peaks allows us to merge them in the desired operating range resulting in wide band response.

3.2. Observation of Wideband Behavior. On the asymmetric H-shaped piezoelectric bimorph, four layers of Metglas with area of 15 × 7 mm^2 were attached at the smaller section, and thirty layers of Metglas with area of 20 × 7 mm^2 were attached at the larger section to have configuration similar to that of previous approach (Figure 8(a)). There are two variables here which can be adjusted to achieve averaging of the ME response from two sections of the asymmetric H laminate. First, if the area of two sections is same, then one with smaller number of Metglas layers will show higher ME coefficient. Second, if a number of Metglas layers are same, then one with smaller area will show smaller ME coefficient [11, 12]. Thus, by adjusting the ratio of Metglas layers to area of the piezoelectric rectangular sections, an average response can be obtained from the asymmetric H-shaped structure. Figure 12(a) shows the measured ME response from bimorph composite structure as a function of magnetic DC bias at frequency of 1 kHz with H_{ac} = 1 Oe. The peak at 60 Oe was associated with four-layered Metglas section on the smaller piezoelectric surface area. The second peak of ME coefficient at 215 Oe was associated with thirty-layered Metglas section with the larger piezoelectric area.

Figure 12(b) shows the impedance and phase angle spectrum of this laminate. After Metglas was attached on the PZNT plate, the intensity of resonances became smaller which can be explained by dampening effect and the position

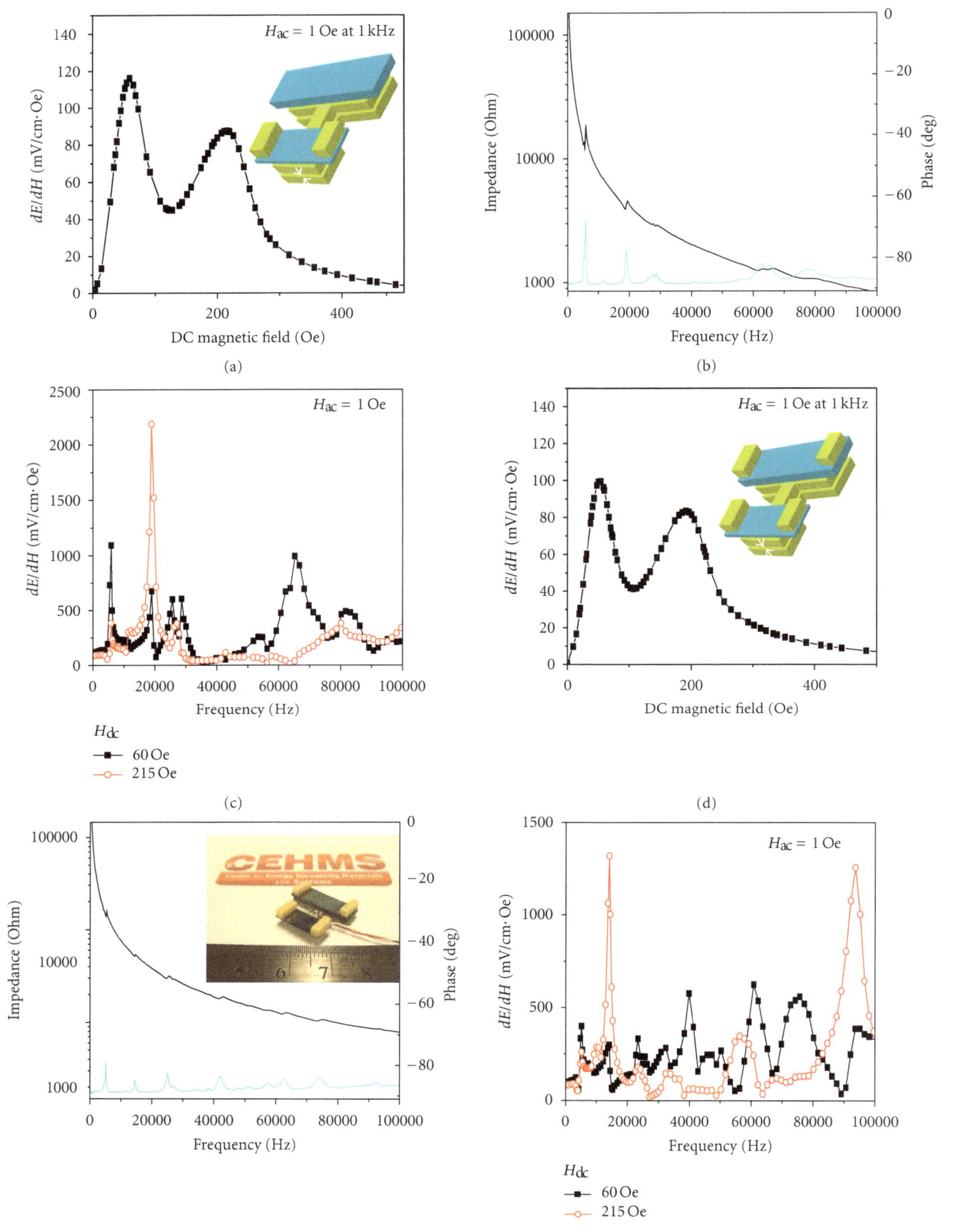

FIGURE 13: (a) ME coefficient of the laminate composite with tip mass on the smaller rectangular area as a function of magnetic DC bias at 1 kHz, (b) impedance and phase spectrums with tip mass on the smaller section as a function of frequency, (c) ME sensitivity with tip mass on the smaller section as a function of frequency, (d) ME coefficient of the laminate with tip mass on both smaller and larger rectangular sections, (e) impedance and phase spectrums with tip mass on both sections, and (f) ME sensitivity as a function of frequency with tip mass on both sections [12].

of peaks shifted towards lower frequencies, in comparison between Figures 11(b) and 12(b). The weight of Metglas on PZNT plates also brought the four peaks initially found at 7.3, 16.3, 18.3, and 20.5 kHz, which originated from the peaks at 10.09, 13.08, 23.32, 33.81, and 43.05 kHz in Figure 11(b), closer to each other. More specifically, the peak at 7.3 kHz in Figure 12(b) was combined with peaks at 10.09 and 13.08 kHz in Figure 11(b).

Consequently, a wideband response was successfully formed in the frequency range of 7–22 kHz, as shown in Figure 12(c) and Figure 12(d). This band comprises of peaks in ME coefficient at 7.5, 15.95, and 20.43 kHz at magnetic DC bias of 60 Oe. Under magnetic DC bias of 215 Oe, the peaks occur at 7.5 and 20.43 kHz. This indicates that the peak at 15.95 kHz was related to magneto-mechanical coupling and not to the electromechanical coupling. The ME sensitivity of the sensor in this band was measured to be higher than 200 mV/cm·Oe which is independent of the magnitude of applied magnetic DC bias.

3.3. Effect of Tip Mass on Wideband Sensor. In order to further reduce the magnitude of resonance frequency, the laminate composite was loaded with a tip mass. According to (2), it can be easily seen that this will result in lowering of the resonance frequency. A tip mass of 0.2 g was placed on both edges of smaller rectangular area, as schematically shown in the inset of Figure 13(a). This figure also shows the effect of tip mass on the ME response as a function of magnetic DC bias at 1 kHz. Figure 13(a) shows no significant change in the ME response as a function of DC magnetic field compared to that in Figure 12(a). However, significant changes were found to occur in ME response as a function of frequency where intensity of peaks in impedance spectrum increased as shown in Figure 13(b), and correspondingly peaks were observed in the ME spectrum shown in Figure 13(c). The magnitude of ME coefficient under 60 Oe magnetic DC bias and frequency of 5.9 kHz was found to be 1092 mV/cm·Oe, and under 215 Oe magnetic DC bias and frequency of 18.9 kHz was found to be 2184 mV/cm·Oe. Comparing with Figure 12(c), this is large increase in ME response. Further, it is interesting to note that the peaks in ME coefficient are at different position under different magnetic DC bias. In terms of wideband width, the tip mass did not result in improvement. For the next step, tip mass of 0.2 g was additionally added on the edges of larger rectangular area, as shown in the inset of Figure 13(d). The first resonance in this case was shifted to ~5 kHz and the numbers of peaks in the impedance spectrum increased as shown in Figure 13(e). As a result, a wideband ME response was obtained in the frequency range of 5–14 kHz with near-flat magnitude of ME coefficient in the range of ~150 mV/cm·Oe, as shown in Figure 13(f). The results in Figures 11–13 clearly show that by using tip mass the resonance behavior can be shifted to lower frequency ranges with wideband response.

4. Summary

Various structural and geometrical parameters were investigated to understand the variation in ME coefficient with applied DC magnetic bias. By controlling laminate dimensions, number of Metglas layers, and gradient geometry, the correlation between the overall deformation and peak position in ME coefficient was found. The broad/wideband ME behavior with near flat ME response was demonstrated by designing a dimensionally gradient bimorph structure and combining with laminate configuration. Using tip masses at the ends of the sensor, wideband frequency response was obtained in lower-frequency range. These results are quite promising for practical applications such as current probe, magnetic field sensing, and ME energy harvester.

Acknowledgment

The authors gratefully acknowledge the financial support from Office of Basic Science, Department of Energy (DE-FG02-08ER46484).

References

[1] N. A. Spaldin and M. Fiebig, "The renaissance of magnetoelectric multiferroics," *Science*, vol. 309, no. 5733, pp. 391–392, 2005.

[2] Y. Lin, N. Cai, J. Zhai, G. Liu, and C. W. Nan, "Giant magnetoelectric effect in multiferroic laminated composites," *Physical Review B*, vol. 72, no. 1, Article ID 012405, 4 pages, 2005.

[3] D. V. Chashin, Y. K. Fetisov, K. E. Kamentsev, and G. Srinivasan, "Resonance magnetoelectric interactions due to bending modes in a nickel-lead zirconate titanate bilayer," *Applied Physics Letters*, vol. 92, no. 10, Article ID 102511, 2008.

[4] J. Ryu, A. V. Carazo, K. Uchino, and H. E. Kim, "Piezoelectric and magnetoelectric properties of lead zirconate titanate/Ni-ferrite particulate composites," *Journal of Electroceramics*, vol. 7, no. 1, pp. 17–24, 2001.

[5] M. Avellaneda and G. Harshe, "Magnetoelectric effect in piezoelectric/magnetostrictive multilayer (2-2) composites," *Journal of Intelligent Material Systems and Structures*, vol. 5, no. 4, pp. 501–513, 1994.

[6] G. Srinivasan, E. T. Rasmussen, B. J. Levin, and R. Hayes, "Magnetoelectric effects in bilayers and multilayers of magnetostrictive and piezoelectric perovskite oxides," *Physical Review B*, vol. 65, no. 13, Article ID 134402, 7 pages, 2002.

[7] J. Ryu, S. Priya, A. Vázquez Carazo, K. Uchino, and H. E. Kim, "Effect of the magnetostrictive layer on magnetoelectric properties in lead zirconate titanate/Terfenol-D laminate composites," *Journal of the American Ceramic Society*, vol. 84, no. 3–12, pp. 2905–2908, 2001.

[8] S. Dong, J. F. Li, and D. Viehland, "Characterization of magnetoelectric laminate composites operated in longitudinal-transverse and transverse-transverse modes," *Journal of Applied Physics*, vol. 95, no. 5, pp. 2625–2630, 2004.

[9] S. Dong, J. Zhai, J. Li, and D. Viehland, "Near-ideal magnetoelectricity in high-permeability magnetostrictive/ piezofiber laminates with a (2-1) connectivity," *Applied Physics Letters*, vol. 89, no. 25, Article ID 252904, 2006.

[10] R. A. Islam and S. Priya, "Large magnetoelectric coefficient in Co-fired Pb $(Zr_{0.52}Ti_{0.48})O_3$-Pb $(Zn_{1/3}Nb_{2/3})O_3$-$Ni_{0.6}Cu_{0.2}Zn_{0.2}Fe_2O_4$ trilayer magnetoelectric composites," *Journal of Materials Science*, vol. 43, no. 6, pp. 2072–2076, 2008.

[11] C. S. Park, C. W. Ahn, J. Ryu et al., "Design and characterization of broadband magnetoelectric sensor," *Journal of Applied Physics*, vol. 105, no. 9, Article ID 094111, 2009.

[12] C. S. Park, C. W. Ahn, S. C. Yang, and S. Priya, "Dimensionally gradient magnetoelectric bimorph structure exhibiting wide frequency and magnetic dc bias operating range," *Journal of Applied Physics*, vol. 106, no. 11, Article ID 114101, 2009.

[13] S. Dong, J. F. Li, and D. Viehland, "Longitudinal and transverse magnetoelectric voltage coefficients of magnetostrictive/piezoelectric laminate composite: theory," *IEEE Transactions on Ultrasonics, Ferroelectrics, and Frequency Control*, vol. 50, no. 10, pp. 1253–1261, 2003.

[14] S. Dong, J. Cheng, J. F. Li, and D. Viehland, "Enhanced magnetoelectric effects in laminate composites of Terfenol-D/Pb(Zr,Ti)O_3 under resonant drive," *Applied Physics Letters*, vol. 83, no. 23, pp. 4812–4814, 2003.

[15] S. Dong, J. F. Li, and D. Viehland, "Magnetoelectric coupling, efficiency, and voltage gain effect in piezoelectric-piezomagnetic laminate composites," *Journal of Materials Science*, vol. 41, no. 1, pp. 97–106, 2006.

[16] C. M. Chang and G. P. Carman, "Modeling shear lag and demagnetization effects in magneto-electric laminate composites," *Physical Review B*, vol. 76, no. 13, Article ID 134116, 2007.

[17] M. I. Bichurin, D. A. Filippov, V. M. Petrov, V. M. Laletsin, N. Paddubnaya, and G. Srinivasan, "Resonance magnetoelectric effects in layered magnetostrictive-piezoelectric composites," *Physical Review B*, vol. 68, no. 13, Article ID 132408, 4 pages, 2003.

[18] Y. X. Liu, J. G. Wan, J. M. Liu, and C. W. Nan, "Numerical modeling of magnetoelectric effect in a composite structure," *Journal of Applied Physics*, vol. 94, no. 8, pp. 5111–5117, 2003.

[19] H. Yu, M. Zeng, Y. Wang, J. G. Wan, and J. M. Liu, "Magnetoelectric resonance-bandwidth broadening of Terfenol-D/epoxy- Pb (Zr,Ti)O_3 bilayers in parallel and series connections," *Applied Physics Letters*, vol. 86, no. 3, Article ID 032508, 3 pages, 2005.

[20] M.G. Muriuki, *An investigation into the design and control of tunable piezoelectric resonators*, Ph.D. dissertation, University of Pittsburgh, 2004.

[21] D. Charnegie, *Frequency tuning concepts for piezoelectric cantilever beams and plates for energy harvesting*, M.S. thesis, University of Pittsburgh, 2005.

[22] H. Xue, Y. Hu, and Q. M. Wang, "Broadband piezoelectric energy harvesting devices using multiple bimorphs with different operating frequencies," *IEEE Transactions on Ultrasonics, Ferroelectrics, and Frequency Control*, vol. 55, no. 9, pp. 2104–2108, 2008.

Nonlocal Electron-Phonon Interaction as a Source of Dynamic Charge Stripes in the Cuprates

Claus Falter and Thomas Bauer

Institut für Festkörpertheorie, Westfälische Wilhelms-Universität, Wilhelm-Klemm-Straße 10, 48149 Münster, Germany

Correspondence should be addressed to Claus Falter, falter@uni-muenster.de

Academic Editor: Jörg Fink

We calculate for La_2CuO_4 the phonon-induced redistribution of the electronic charge density in the insulating, the underdoped pseudogap, and the more conventional metallic state as obtained for optimal and overdoping, respectively. The investigation is performed for the anomalous high-frequency oxygen bond-stretching modes, which experimentally are known to display a strong softening of the frequencies upon doping in the cuprates. This most likely generic anomalous behaviour of these modes is shown to be due to a strong nonlocal electron-phonon interaction mediated by charge fluctuations on the ions. We demonstrate that the softening of the modes is caused by nonlocally induced dynamic charge inhomogeneities in form of charge stripes along the CuO bonds with different orbital character. The dynamic charge inhomogeneities may in turn be considered as precursors of static charge stripe order as recently observed in $La_{2-x}Ba_xCuO_4$ in a broad range of doping around $x = 1/8$. The latter may trigger a reconstruction of the Fermi surface into small pockets with reduced doping. We argue that the incompressibility of the Cu3d orbital and simultaneously the compressibility of the O2p orbital in the pseudogap state seem to be required to nucleate dynamic stripes.

In recent work [1, 2], we have developed a description of the electronic charge response and phonon dynamics for p-doped and n-doped cuprate high-temperature superconductors (HTSCs). This approach is based upon a microscopic modeling of the change of the charge response across the insulator-metal transition in terms of the compressibility of the electronic system as a primary tool to characterize the corresponding competing ground states depending on doping. This is achieved starting from the insulating state via a metallic pseudogap state to a more conventional metallic state by consecutive orbital selective incompressibility-compressibility transitions in terms of strict sum rules of the charge response given in [1] and listed below. These sum rules are a proper instrument to describe the competition between the localisation and delocalisation of the electrons, which is at the heart of the physics in the cuprates.

In the conventional metallic state, the electronic partial density of states (PDOS) at the Fermi level $Z_\kappa(\varepsilon_F)$ is related to the polarizability matrix in the longwavelength limit ($\mathbf{q} \to \vec{\mathbf{0}}$) according to

$$\sum_{\kappa'} \pi_{\kappa\kappa'}\left(\vec{\mathbf{q}} \longrightarrow \vec{\mathbf{0}}\right) = Z_\kappa(\varepsilon_F), \qquad (1)$$

and the total density of single particle states at energy ε is given by

$$Z(\varepsilon) = \sum_{\kappa} Z_\kappa(\varepsilon). \qquad (2)$$

κ, κ' are the orbital degrees of freedom in the elementary cell of the crystal. In our calculation for LaCuO, Cu3d, Cu4s, and O2p orbitals in the CuO plane are taken into account. $\mathbf{q}$ is a wavevector from the first Brillouin zone (BZ). On the other hand, for the insulating state we obtain the sum rules

$$\begin{aligned} &\sum_{\kappa'} \pi_{\kappa\kappa'}\left(\vec{\mathbf{q}} \longrightarrow \vec{\mathbf{0}}\right) = O(q), \\ &\sum_{\kappa,\kappa'} \pi_{\kappa\kappa'}\left(\vec{\mathbf{q}} \longrightarrow \vec{\mathbf{0}}\right) = O(q^2). \end{aligned} \qquad (3)$$

The sum $\sum_{\kappa,\kappa'} \pi_{\kappa\kappa'}(\mathbf{q} \rightarrow \mathbf{0})$ is equal to $\rho^2 K$ with ρ the average density and K the compressibility of the electronic system. The latter provides a measure of the charge gap in the electronic spectrum because K vanishes as a function of the chemical potential in the gap region. Moreover, $1/K$ provides a measure of the change of the chemical potential with particle number.

Approaching the delocalization-localization transition from the conventional metallic region representative for optimal and overdoping, when p-doping is decreased the Cu3d component of the wave function is admitted to become incompressible, insulator-like in the underdoped state according to our modeling. This means that the correlated 3d electrons are allowed to become more localized as compared with the conventional metallic state and disappear from the Fermi surface (FS). Moreover, this effectively models a decoupling of the 3d electrons from the holes on the O2p orbitals quite in contrast to the conventional metallic state. In terms of the sum rule, we then have

$$\sum_{\kappa'} \pi_{\kappa\kappa'}\left(\vec{\mathbf{q}} \longrightarrow \vec{\mathbf{0}}\right) = \begin{cases} O(q), & \text{Cu3}d \\ Z_\kappa(\varepsilon_F), & \text{O2}p. \end{cases} \tag{4}$$

Accordingly, the partial density of states is suppressed at the Fermi level for the correlated, localized Cu3d orbitals (orbital intrinsic incompressibility due to correlation) but not for the more delocalized O2p orbitals at the O_{xy} sublattices in the CuO plane, where the holes are predominantly injected in the p-type cuprates. So, we have an orbital selective compressible, metallic charge response only for the O2p states with a renormalized PDOS $Z_\kappa(\varepsilon_F)$ and a loss in the total density of state at ε_F (*pseudogap*) because of the incompressibility of the Cu3d states according to (4). The loss in spectral weight also is never recovered on entering the superconducting state. The incompressible regions in the pseudogap state compete with overall metallic behaviour and with superconductivity.

Our modeling of the pseudogap state in terms of compressible, metallic O2p states and Cu3d states treated as incompressible and disappearing from the FS because of strong correlation is supported by calculations performed within the self-interaction-corrected (SIC) local spin density (LSD) approximation for La_2CuO_4 [3]. Here it is shown that the self-interaction pulls the occupied localized Cu3d states below the O2p band, only little Cu3d admixture is found in the O2p band.

Furthermore, our modeling of the pseudogap state corresponds with a smaller Fermi volume as compared with the conventional metallic Fermi-liquid state because the Fermi volume is determined by the density of the compressible metallic charge carriers (holes) of dominantly O2p-type alone while the Cu3d states are incompressible as in the Mott insulator and cannot contribute to the Fermi volume quite in contrast to the conventional metallic state where both the O2p and the Cu3d states are compressible metallic leading to a large Fermi volume. The pseudogap state is separated from the conventional overall compressible Fermi-liquid state by an orbital selective compressibility-incompressibility transition of the Cu3d orbital. In the context of our modeling of the pseudogap state, it is interesting to point to another modeling of this state within the framework of the so-called fractionalized Fermi liquid. For a recent discussion see for example, [4] and references therein. In this approach, an idea is important which is based on the concept of an orbital-selective Mott transition. For the latter, for example, in a two-band model, we have one metallic band with Fermi-liquid-like quasiparticles while the other band, for example, related to localized d electrons, undergoes a Mott transition to an insulating state due to strong electronic correlations. Such a transition is analogous to the compressibility-incompressibility transition of the Cu3d orbital in our approach.

Because particle density fluctuations scale with the compressibility of the system, the sum rules introduce varying and orbital dependent fluctuations of the particle number in the different metallic states of the HTSC according to our modeling. The tuning parameter for these fluctuations of the particle number is doping, and the pseudogap state may be considered as a state with reduced particle density fluctuations as compared with the optimal and overdoped state. Moreover, the partial incompressibility of the Cu3d orbitals competes with superconductivity because the phase fluctuations $\Delta\phi$ of the order parameter grow with decreasing particle fluctuations ΔN due to the uncertainty relation $\Delta N \Delta\phi \sim 1$.

Qualitatively, the incompressibility of the Cu3d states also corresponds to a compartmentalisation of configuration space. This means that some parts of direct space cannot be approached or are hardly accessible to the charge carriers. On the other hand, this should be accompanied by a reciprocal compartmentalisation of momentum space, according to the uncertainty relationship $\Delta x \Delta k \sim 1$. As a consequence, the momentum should have corresponding fluctuations and the FS of the conventional metallic state becomes fuzzy when passing to the underdoped state and possibly reconstructs. ARPES studies and quantum-oscillation measurements seem to reflect this in form of a reconstruction of the large FS in the optimal and overdoped state into small Fermi surface pockets with Fermi-liquid-like charge carriers in the underdoped pseudogap state characterized by a loss of density of states at the Fermi level consistent with our modeling.

In our orbital-based local picture, the pseudogap state of the cuprates looks like a "two-component" electronic structure where a real space organisation of the low lying charge excitations is achieved in form of a metallic charge response by mobile holes on the oxygen network in the CuO plane that is blocked at the incompressible insulator-like Cu sites. Inspection of the sum rule in (4) shows that the incompressible Cu3d states and the compressible O2p states are entangled by the off-diagonal elements of the polarizability matrix. So our modelling of the pseudogap state is consistent with a strange not overall compressible metallic state coexisting with atom-like, incompressible Cu3d orbitals same as in the undoped, insulating state favoring short-ranged magnetic spin correlations and corresponding excitations, typically spin fluctuations. The latter should be damped by electron-hole charge excitations of the compressible metallic O2p states.

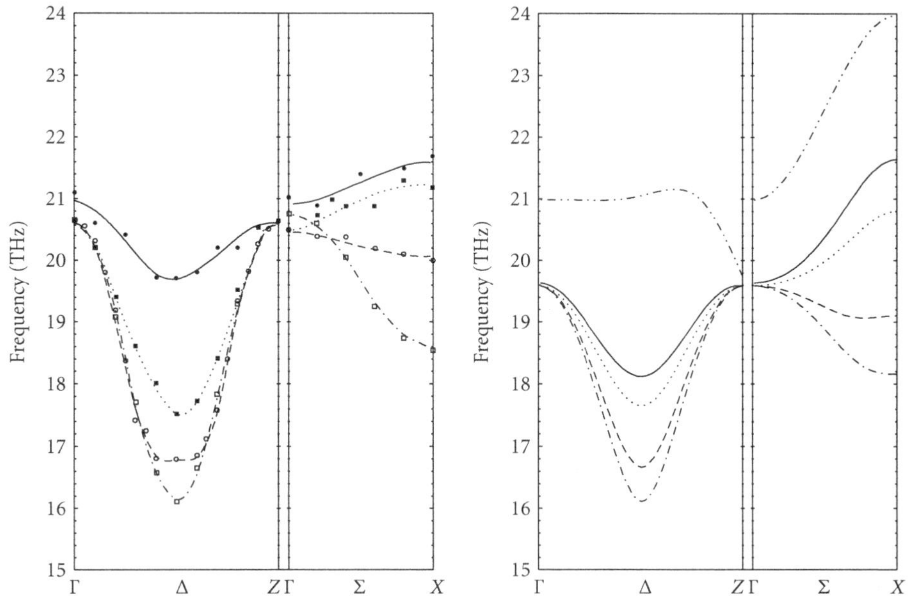

(a) Experimental Results (b) Calculated Results

FIGURE 1: (a) Experimental results for the highest Δ_1 and Σ_1 branch of $La_{2-x}Sr_xCuO_4$ for various doping levels [6]. •: $x = 0.00$, ■: $x = 0.10$, ∘: $x = 0.15$, □: $x = 0.30$. The lines are a guide to the eye. (b) Calculated results of the phonon branches shown in (a) as obtained with our model approach for the electronic state of the HTSCs [1]. —: insulating state, · · ·: underdoped state, — — —: optimally doped state, — · —: overdoped state. For comparison, the results for the RIM are shown (— · · —) in order to demonstrate the crucial influence of the *nonlocal* EPI effects in form of CFs on the dispersion.

In k-space language such a strange mixed locally compressible-incompressible state should result in a heavily renormalized electronic bandstructure and Fermi surface (FS) as compared with the more conventional overall compressible Fermi liquid state with a large FS at higher doping. The pseudogap state displaying properties characteristic of both insulators and correlated metals evolves from the incompressible insulating state with no low energy charge excitations and seems to promote the tendency to charge and spin ordering that breaks rotational and translational symmetry, for example, in form of stripes. The latter appear to be quite common in underdoped cuprates. In this context it is interesting to point out that the stripe-like modulation observed in [5] is strongest on the oxygen orbitals. So, the latter are very essential to understand the physics of charge order in the cuprates and also the charge response as shown below. Finally, the enhanced phase fluctuations due to the incompressibility of the Cu3d orbital favour pairing correlations in the normal state, possibly incoherent preformed pairs as discussed in the literature.

Our modeling of the density response is strongly supported by corresponding calculations of phonon dynamics [1, 2], which compare well with the experimental data. This is shown for p-doped LaCuO in case of the strongly doping-dependent oxygen bond-stretching mode (OBSM) anomalies, see Figure 1, where the experimental results for the highest Δ_1 and Σ_1 branches of $La_{2-x}Sr_xCuO_4$ are shown for various doping levels [6] and compared with the calculated results. The high-frequency OBSMs are generally known to display an anomalous softening upon doping in the cuprates [7–18]. Such a behaviour of the experimental phonon dispersion is clearly visible for LaCuO in Figure 1(a) and is very well accounted for by the calculation in our model for the electronic state in terms of incompressibility-compressibility transitions, see Figure 1(b). We also have included in Figure 1(b) the phonon dispersion of a rigid-ion model (RIM) for LaCuO.

It seems appropriate to explain at least the ideas of our modeling of phonon dynamics and charge response of the cuprates. A quantitative treatment of the theory and the modeling would be too extensive and can be found in [1, 2].

The rigid part of the electronic charge response and the electron-phonon interaction (EPI) is approximated by an ab initio rigid-ion model (RIM), taking into account covalent ion softening in terms of (static) effective ionic charges calculated from a tight-binding analysis. In addition, scaling of the short-range part of certain pair potentials between the ions is performed to simulate further covalence effects in the calculation in such a way that the energy-minimized structure is as close as possible to the experimental one. The RIM with the corrections just mentioned then serves as an unbiased reference system for the description of the cuprate superconductors and can be considered as a first approximation for the insulating state of these compounds because of the strong ionic nature of bonding in the cuprates. Starting with such a rigid reference system nonlocal, electronic polarisation processes are introduced in the form of more or less localized electronic charge fluctuations (CF's) at the outer shells of the ions. In the metallic state of the cuprates, especially the latter dominate the nonlocal

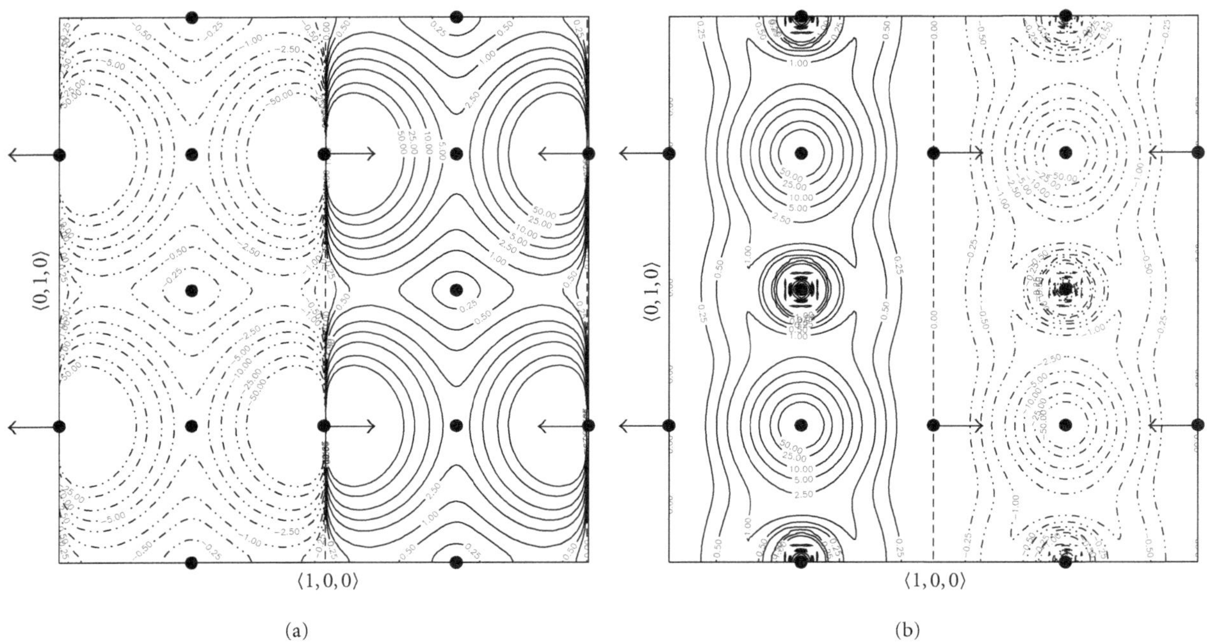

FIGURE 2: (a) Contour plot in the CuO plane of the displacement-induced charge density redistribution $\delta\rho$ for the $\Delta_1/2$ mode within the rigid ion model (RIM) for La_2CuO_4. (b) Same as (a), but for the nonlocal part of the charge response in the optimally doped state due to nonlocal EPI effects in the half-breathing mode $\Delta_1/2$. The displaced ions are indicated by a dot with an arrow. Full lines (—) mean that electrons are accumulated in the corresponding region of space and the broken dotted lines (– · · –) indicate regions where the electrons are pushed away. $\delta\rho$ is given in units of $10^{-4}e^2/a_B^3$.

contribution of the charge response and the EPI and are particularly important in the CuO planes. In addition, anisotropic dipole fluctuations are admitted in our approach. They prove to be specifically of interest for the ions in the ionic layers.

Comparing with Figure 1(b), the results of the phonon dispersion of the RIM, where only local EPI effects are considered, with the full calculation we can extract the crucial influence of the nonlocal EPI effects generated by the charge fluctuations. For example, we find a very strong softening of the $\Delta_1/2$ bond-stretching mode (half-breathing mode) increasing with doping up to about 5 THz. Comparing the calculated results with the experiments in Figure 1(a), we find a good agreement. In order to achieve such a realistic description of the phonon modes, both, nonlocal EPI effects and multiorbital effects are required. The comparison with the results from the RIM demonstrates that nonlocality is needed and from our detailed calculations we find that the CF's of the more delocalized O2p and Cu4s orbital contribute significantly to the softening of the modes, besides the more localized Cu3d orbital. The strong softening of the phonon anomalies is accompanied by corresponding dynamic charge inhomogeneities in form of charge stripes along the CuO bonds. The modes are not really soft but one may consider these dynamic stripes as precursors of static charge stripe order with ordering vectors corresponding to the wavevectors of the phonon anomalies.

Finally, it should be remarked that a constructive interplay of electron-ion interaction with short-ranged antiferromagnetic order is important for the Cu sites in the underdoped and insulating state, because electrons are transferred from the Cu ion where the Cu bond is compressed to the Cu ion where the bond is stretched, see for example, Figure 2. Both ions have opposite spin in the antiferromagnetic case. In contrast if a ferromagnetic order would be present, the Pauli blocking would suppress the transport of electrons with the same spin and consequently also the charge fluctuations and the corresponding softening of the phonon modes.

Modelling of the electronic polarizability $\pi_{\kappa\kappa'}(\mathbf{q})$ for the insulating and pseudogap state is needed because a quantitative calculation of $\pi_{\kappa\kappa'}(\mathbf{q})$ is not available due to the missing of a rigorous knowledge of this state. On the other hand, for the more conventional metallic state $\pi_{\kappa\kappa'}(\mathbf{q})$ can be calculated within the local density approximation (LDA), however, with some important corrections concerning the electronic c-axis response which is overestimated in typical LDA-based calculations [19]. Concerning this case it should be remarked that a comparison of the calculated OBSM anomalies from the model in Figure 1(b) with corresponding results obtained within the framework of the corrected LDA [19], taking into account the full $\mathbf{q}$ dependence of the polarizability matrix, leads to a good agreement. This also speaks in favour of our modeling of the charge response in terms of the compressibility sum rules, in which only the longwavelength limit of the polarizability matrix is considered.

In Figure 2 we display the calculated phonon-induced charge redistribution of the bond-stretching mode at $\Delta_1/2$ for the optimal doped state. Figure 2(a) illustrates the effect

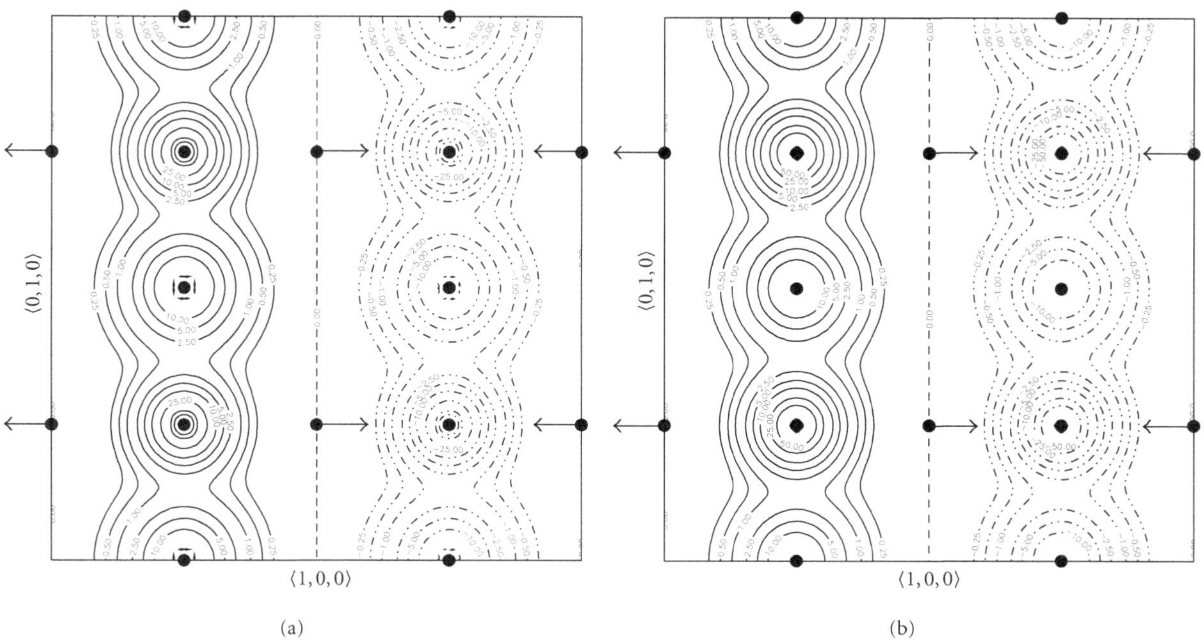

FIGURE 3: (a) Same as Figure 2 but for the nonlocal part of the charge response $\delta\rho$ for $\Delta_1/2$ in the insulating state ($x = 0$). (b) Same as Figure 2 but for the nonlocal part of the charge response $\delta\rho$ for $\Delta_1/2$ in the pseudogap state ($x = 0.1$).

of the local EPI within the RIM on the charge redistribution connected with the displacement of the O_x ions in the CuO plane, that is, a stretching or compression of the CuO bond, respectively, (half-breathing mode (HBM)). The local EPI reveals a charge redistribution of dipole character localized only at the ions as can be expected from the rigid shift of the ionic densities in the RIM during the phonon mode.

From Figure 2(a) we also obtain the message that no dynamical charge stripes are excited by local EPI. On the other hand, Figure 2(b) that displays the nonlocal part of the charge response due to nonlocal EPI effects of charge fluctuation type demonstrates how the strong nonlocal EPI generates dynamic charge stripes in the half-breathing mode. Thus, the moving O_x ions in the HBM, which are linked up with strong CF's on the silent Cu and O_y ions, induce dynamical stripes in the y direction with a recurrence periond of two lattice constants along the orthogonal x-direction. Note, that there are no changes of the transfer integral between d and p orbitals for the silent Cu and O_y ions, nevertheless there is a charge transfer.

In case of the full-breathing mode O_B^X, at the the X point of the BZ where the O_x and O_y ions vibrate in phase along the bonds against the silent Cu ion [1] nonlocal EPI generates CF's at the Cu and we obtain an electronic charge transfer from that Cu ion where the bonds are compressed to the Cu where the bonds are stretched [20]. While in case of the HBM the charge stripes point along the x- or y-direction, respectively, for O_B^X the stripes point along the diagonals in the CuO plane.

The self-consistent changes of the potential felt by an electron due to nonlocal EPI also have been calculated for the OBSMs and yield a strong coupling of the order 100 meV [19]. It should be remarked that such a strong nonlocal EPI associated with the OBSMs, which can reliably be calculated in the adiabatic approximation, is further enhanced by an order of magnitude because of the poorly screened long range Coulomb interaction and the slow charge dynamics for the polar modes from the nonadiabatic sector of phase space around the c-axis where phonon-plasmon mixing is relevant [19].

In Figures 3(a) and 3(b), we present the calculated results for the nonlocally induced charge redistribution of the HBM for the insulating and the pseudogap state, which should be compared to the corresponding results for the optimal doped state in Figure 2(b). First of all we find for all electronic states dynamical stripes along the bond direction excited by the nonlocal EPI being simultaneously responsible for the corresponding phonon softening in Figure 1. This is quite interesting because dynamic charge inhomogeneities of this type may be considered as precursors of static charge stripe order as recently observed in $La_{2-x}Ba_xCuO_4$, in a broad range of doping around $x = 1/8$ with $0.095 \leq x \leq 0.155$ [21] or in $YBa_2Cu_3O_y$, respectively, [22].

To the best of our knowledge in overdoped samples static stripe order has not been observed so far. This would mean that the pseudogap in the underdoped state may be required to nucleate dynamic stripes. In our approach this implies that for the emergence of static stripe order the incompressibility of the Cu3d orbitals seems to be essential. The latter brings about a separation of mobile holes populating the compressible O2p states and regions with local Mott- and spin correlations related to the insulator-like Cu3d states. Such stripes producing the onset of a new periodicity that breaks translational and rotational symmetry also may help to

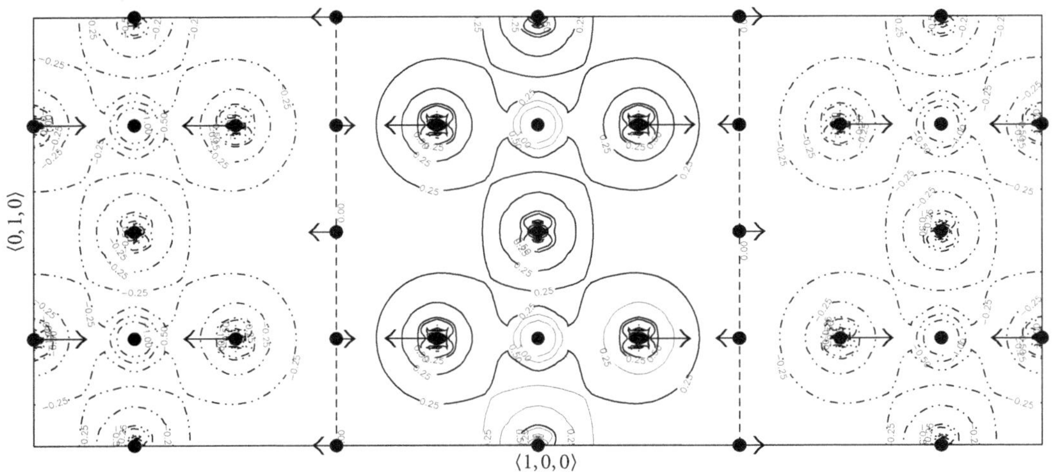

FIGURE 4: Same as Figure 2 but for the nonlocal part of the charge response $\delta\rho$ for $\Delta_1/4$ in the optimally doped state.

trigger a reconstruction of the large hole Fermi surface (FS) into small pockets as detected by quantum oscillation measurements in underdoped $YBa_2Cu_3O_{6+x}$, see for example, [23] and references therein. Also from an energetic point of view, our modeling of the pseudogap state is susceptible for the tendency to establish broken symmetry states with hole-rich compressible and hole-poor incompressible regions. In general terms, lowering of the kinetic energy occurs in the compressible metallic regions and lowering of the interaction energy is possible by antiferromagnetic spin correlations and the Coulomb repulsion between the electrons in the region of the incompressible insulator-like Cu3*d* states. Altogether, in the pseudogap state the doped holes and the superexchange-coupled spins arrange themselves in a cooperative way to promote both charge mobility and locally antiferromagnetic correlations.

Comparing the stripe patterns of the insulating and underdoped state with that of the more conventional metallic optimal and overdoped state (not shown) with a large FS as the locus of the crucial low energy excitations, we find a relatively small contribution of the O2*p* orbitals degrees of freedom to the density redistribution in the latter case. The weight of the O2*p* contribution is significantly enhanced in the insulating and pseudogap state, see Figures 3(a) and 3(b) reflecting in particular the importance of the O2*p* orbital in the pseudogap state. Holes in the pseudogap state primarily populate the O2*p* orbitals, while the holes in the more conventional metallic state gain more and more Cu3*d* character by an increased hybridization as can be seen from a large PDOS for Cu3*d* leading to a dominating contribution of the charge redistribution at the Cu ion where the charge response is now metallic, too. See also [24] where it is investigated how the nature of the hole state evolves with doping. Overall, the stripe pattern is more delocalized in the metallic state and localization increases towards the insulating state via the pseudogap state.

In the overdoped state, a growing importance of the Cu4*s* orbital and accordingly an increase of the Cu4*s* element of the polarizability matrix leads to an enhanced softening of the anomalous OBSMs, see Figure 1. The corresponding calculated dynamical stripe pattern is very similar to that of the optimally doped state in Figure 2(b) but somewhat more delocalized because of the increased influence of the more delocalized contribution due to the Cu4*s* orbital.

From our calculations it becomes obvious that a multiorbital approach is needed that includes besides Cu3*d* at least O2*p* orbitals and for the optimally doped and overdoped state in addition Cu4*s* orbitals in order to give a reliable representation of phonon dynamics and charge response in the cuprates. Single-band models for the CuO plane believed by many people to describe the cuprates are not sufficient, in particular in context with the discussion of stripes. Altogether, in the optimal and overdoped state increasing of doping enhances the hybridization between Cu3*d*, Cu4*s*, and O2*p* orbitals, respectively, with a dominant weight for Cu3*d*. This state becomes compressible metallic with a large PDOS at ε_F. So, the particle number fluctuations that characterize the different competing ground states in our model are enhanced as compared to the ground state in the pseudogap regime. Here the hybridization is weaker due to the incompressible insulator-like Cu3*d* orbital with a vanishing PDOS at ε_F.

Translational symmetry-breaking models of the electronic structure of the CuO plane generated by charge density wave order of a certain wavevector are discussed in the literature and used to characterize Fermi surface reconstructions of a large FS into small pockets, for recent work see [25, 26]. From scanning tunneling microscopy (STM), neutron diffraction experiments, and nuclear magnetic resonance measurements, the ordering in many cases occurs with a period of about 4 lattice constants [26]. This corresponds to the wavevector $\pi/2a$ $(1,0,0)$ of the $\Delta_1/4$ mode. Therefore, it is meaningful to calculate the nonlocally induced charge redistribution also for this mode. The result for the optimal doped state is shown in Figure 4.

We again obtain a stripe-like charge pattern with translational symmetry in the $(0,1,0)$ direction but a four-unit cell repeat distance along the orthogonal direction $(1,0,0)$.

Contrasting this result with the charge stripe pattern as found for the HBM in Figure 2(b), the stripes for $\Delta_1/4$ are expanded over the whole unit cell, because the CFs do not vanish at the sites of the moving oxygens, unlike $\Delta_1/2$. However, the amplitudes of the CFs are considerably smaller. According to our calculation of the eigenvectors, this is caused by the strongly reduced bond-stretching amplitude of the oxygen ions in the plane for $\Delta_1/4$, while the displacement of the apex oxygen ions is strongly enhanced in the $\Delta_1/4$ mode as compared with the HBM.

In summary, we have applied our modeling of the doping-dependent electronic states of the cuprates to the calculation of the phonon-induced charge inhomogeneities nonlocally induced by the $\Delta_1/2$ and $\Delta_1/4$ mode in LaCuO. The calculations have been performed for the insulating-, pseudogap, and the more conventional metallic state. The compressibility is used as a primary term to characterize the competing ground states controlled by doping. The strong nonlocal EPI of CF type excites in all cases investigated dynamical charge stripes along the CuO bonds. Differences in orbital character of the charge response in the various states are discussed at hand of the stripe patterns. Experimentally, charge stripe order has been observed in a broad range of doping in $La_{2-x}Ba_xCuO_4$ and other cuprates, which supports the point of view that dynamical charge stripes generated by certain strongly coupling OBSM may be considered as precursors of static stripe order at least outside the overdoped regime. Such a charge order breaks translational symmetry and may enforce a reconstruction of a larger hole FS into small pockets.

The interplay between the reduced particle density fluctuations related to the insulator-like behaviour of the localized atomic-like Cu3*d* orbital that remains incompressible for low enough *p*-doping and on the other hand the compressible, metallic O2*p* orbitals appears to play an important role for the pseudogap state and the tendency to charge ordering in the cuprates.

References

[1] C. Falter, T. Bauer, and F. Schnetgöke, "Modeling the electronic state of the high- T_c superconductor LaCuO: phonon dynamics and charge response," *Physical Review B*, vol. 73, no. 22, Article ID 224502, 2006.

[2] T. Bauer and C. Falter, "Microscopic modeling of phonon dynamics and charge response in NdCuO," *Physical Review B*, vol. 77, no. 14, Article ID 144503, 21 pages, 2008.

[3] A. Svane, "Electronic structure of La_2CuO_4 in the self-interaction-corrected density-functional formalism," *Physical Review Letters*, vol. 68, no. 12, pp. 1900–1903, 1992.

[4] M. Vojta, "Stripes and electronic quasiparticles in the pseudogap state of cuprate superconductors," *Physica C*, vol. 481, pp. 178–188, 2012.

[5] Y. Kohsaka, C. Taylor, K. Fujita et al., "An intrinsic bond-centered electronic glass with unidirectional domains in underdoped cuprates," *Science*, vol. 315, no. 5817, pp. 1380–1385, 2007.

[6] L. Pintschovius, "Electron-phonon coupling effects explored by inelastic neutron scattering," *Physica Status Solidi B*, vol. 242, no. 1, pp. 30–50, 2005.

[7] H. Uchiyama, A. Q. R. Baron, S. Tsutsui et al., "Softening of Cu-O bond stretching phonons in tetragonal $HgBa_2CuO_{4+\delta}$," *Physical Review Letters*, vol. 92, no. 19, pp. 197005–1, 2004.

[8] J. Graf, M. D'Astuto, C. Jozwiak et al., "Bond stretching phonon softening and kinks in the angle-resolved photoemission spectra of optimally doped $Bi_2Sr_{1.6}La_{0.4}Cu_2O_{6+\delta}$ superconductors," *Physical Review Letters*, vol. 100, no. 22, Article ID 227002, 2008.

[9] L. Pintschovius and W. Reichhardt, "Neutron scattering in layered copper-oxide superconductors," in *Physics and Chemistry of Materials With Low Dimensional Structures*, A. Furrer, Ed., vol. 20, Kluwer Academic Publisher, Dordrecht, The Netherlands, 1998.

[10] L. Pintschovius and M. Braden, "Anomalous dispersion of LO phonons in $La_{1.85}Sr_{0.15}CuO_4$," *Physical Review B*, vol. 60, no. 22, pp. R15039–R15042, 1999.

[11] W. Reichardt, "Cu-O bond-stretching vibrations in $YBa_2Cu_3O_7$ studied by inelastic neutron scattering," *Journal of Low Temperature Physics*, vol. 105, no. 3-4, pp. 807–812, 1996.

[12] T. Fukuda, J. Mizuki, K. Ikeuchi, K. Yamada, A. Q. R. Baron, and S. Tsutsui, "Doping dependence of softening in the bond-stretching phonon mode of $La_{2-x}Sr_xCuO_4$ $(0 \leq x \leq 0.29)$," *Physical Review B*, vol. 71, no. 6, Article ID 060501, 2005.

[13] L. Pintschovius, D. Reznik, and K. Yamada, "Oxygen phonon branches in overdoped $La_{1.7}Sr_{0.3}Cu_3O_4$," *Physical Review B*, vol. 74, no. 17, Article ID 174514, 2006.

[14] R. J. McQueeney, J. L. Sarrao, P. G. Pagliuso, P. W. Stephens, and R. Osborn, "Mixed lattice and electronic states in high-temperature superconductors," *Physical Review Letters*, vol. 87, no. 7, Article ID 077001, 4 pages, 2001.

[15] M. D'Astuto, P. K. Mang, P. Giura et al., "Anomalous dispersion of longitudinal optical phonons in $Nd_{1.86}Ce_{0.14}CuO_{4+\delta}$ determined by inelastic X-ray scattering," *Physical Review Letters*, vol. 88, no. 16, Article ID 167002, 4 pages, 2002.

[16] M. D'Astuto, P. K. Mang, P. Giura et al., "Electron-phonon interaction in N-doped cuprates: an inelastic x-ray scattering study," *International Journal of Modern Physics B*, vol. 17, no. 4-6 I, pp. 484–492, 2003.

[17] M. Braden, L. Pintschovius, T. Uefuji, and K. Yamada, "Dispersion of the high-energy phonon modes in $Nd_{1.85}Ce_{0.15}CuO_4$," *Physical Review B*, vol. 72, Article ID 184517, 10 pages, 2005.

[18] T. Bauer and C. Falter, "Dynamic charge inhomogeneity in cuprate superconductors," *Journal of Physics Condensed Matter*, vol. 22, no. 14, Article ID 142201, 2010.

[19] T. Bauer and C. Falter, "Impact of dynamical screening on the phonon dynamics of metallic La_2CuO_4," *Physical Review B*, vol. 80, no. 9, Article ID 094525, 2009.

[20] C. Falter, "Phonons, electronic charge response and electron-phonon interaction in the high-temperature superconductors," *Physica Status Solidi B*, vol. 242, no. 1, pp. 78–117, 2005.

[21] M. Hücker, M. V. Zimmermann, G. D. Gu et al., "Stripe order in superconducting $La_{2-x}Ba_xCuO_4$ $(0.095 \leq x \leq 0.155)$," *Physical Review B*, vol. 83, Article ID 104506, 16 pages, 2011.

[22] T. Wu, H. Mayaffre, S. Krämer et al., "Magnetic-field-induced charge-stripe order in the high-temperature superconductor $YBa_2Cu_3O_y$," *Nature*, vol. 477, no. 7363, pp. 191–194, 2011.

[23] D. LeBoeuf, N. Doiron-Leyraud, B. Vignolle et al., "Lifshitz critical point in the cuprate superconductor $YBa_2Cu_3O_y$ from high-field Hall effect measurements," *Physical Review B*, vol. 83, Article ID 054506, p. 14, 2011.

[24] Y. Sakurai, M. Itou, B. Barbiellini et al., "Imaging doped holes in a cuprate superconductor with high-resolution compton scattering," *Science*, vol. 332, no. 6030, pp. 698–702, 2011.

[25] N. Harrison and S. E. Sebastian, "Protected nodal electron pocket from multiple-Q ordering in underdoped high temperature superconductors," *Physical Review Letters*, vol. 106, no. 22, Article ID 226402, 2011.
[26] H. Yao, D.-H. Lee, and S. Kivelson, "Fermi-surface reconstruction in a smectic phase of a high-temperature superconductor," *Physical Review B*, vol. 84, no. 1, Article ID 012507, 2011.

Microwave Magnetoelectric Devices

A. S. Tatarenko and M. I. Bichurin

Institute of Electronic and Information Systems, Novgorod State University, B. S. Peterburgskaya Street 41, Veliky Novgorod 173003, Russia

Correspondence should be addressed to A. S. Tatarenko, alexandertatarenko@yahoo.com

Academic Editor: Vladimir Petrov

Tunable microwave magnetoelectric devices based on layered ferrite-ferroelectric structures are described. The theory and experiment for attenuator, band-pass filter and phase shifter are presented. Tunability of the ME devices characteristics can be executed by application of an electric field. This electric tuning is relatively fast and is not power-consuming. The attenuator insertion losses vary from 26 dB to 2 dB at frequency 7251 MHz. The tuning range of 25 MHz of band-pass filter at frequency 7360 MHz was obtained. A maximum phase shift of 30–40 degree at the frequency region 6–9 GHz was obtained.

1. Introduction

Multiferroics are materials or structures where different ferroic orders such as ferroelectric, ferromagnetic, and ferroelastic coexist in one material. In multiferroic magnetoelectrics, a dielectric polarization can be induced by an external magnetic field (direct effect), a magnetic moment can be induced by an external electric field (inverse effect). This phenomenon is described as the magnetoelectric effect (ME). The ME effect in single crystals is generally described using the Landau theory by writing the expansion of free energy. The ME effect is observed in two classes of materials: single-phase multiferroic materials possessing simultaneously both ferroelectric and ferromagnetic properties and composites consisting of ferroelectric and ferromagnetic phases. The ME effect in the single-phase materials [1] arises from the long-range interaction between the atomic moments and electric dipoles in ordered magnetic and ferroelectric sublattices. The realizable ME coefficient in single-phase materials is, however, very small (1–20 mV/cm·Oe) and not sufficient for practical applications. Moreover, ME effect in most of these single-phase materials is observed only at low temperatures as either ferromagnetic (or antiferromagnetic) or ferroelectric transition temperature is very low.

Composite materials, on the other hand, provide an alternative strategy which makes use of indirect coupling via mechanical strain between the materials of two different phases: ferroelectric and ferromagnetic (or antiferromagnetic or ferrimagnetic such as ferrites). Mechanical deformation of the magnetostrictive phase results in polarization in the piezoelectric phase. ME effect in composites may be more than several orders of magnitude higher than that in single-phase materials, the latter forming actually the basis for multiferroics. Such composites can be implemented in the form of "multilayers" consisting of alternating layers of the ferroelectric phase and the ferro- or ferri-magnetic phase. Bilayers and multilayers of composites are especially promising due to their low leakage current and superior poling properties. In multilayers, however, the magnetoelectric coupling effect is weakened due to the clamping effect of the substrate, unless the multilayers are fabricated as free-standing membranes.

In the discussion of linear and higher-order magnetoelectrical coupling, we have so far ignored the effects of strain. However, inclusion of strain arising from piezomagnetism and magnetostriction would introduce cross-terms that are proportional to strain and have linear and/or quadratic dependence on magnetic field H. Similarly, additional terms would arise from piezoelectricity and electrostriction, having dependencies on electric field E. Even mixed terms involving strain, H and E, would appear. The effects of strain are significant and in some cases may even dominate which actually desired in laminated two-layer composites. Most ferromagnetic materials exhibit magnetostriction, which describes a change in strain as a quadratic function of the applied magnetic field. This change in the strain may induce polarization in the ferroelectric phase, which is in contact with the ferromagnetic/antiferromagentic/ferromagnetic phase (ideally in atomic registry), through the piezoelectric effect.

For the two-phase systems, the physical properties are determined by the interaction between the constituents as well as by their individual properties. Some effects, which are already present in the constituents may be averaged or enhanced for the overall system. However, the ME effect is among the novel effects that arises from the product properties originating through the interaction between the two phases. The ME effect can be achieved by coupling a piezoelectric (or electrostrictive) material and a piezomagnetic (or magnetostrictive) material via a good mechanical contact. Such a two-phase system may be achieved in the form of composites, laminates, or epitaxial layers. In the direct ME effect, an applied magnetic field induces strain in the magnetostrictive constituent, which can be transferred to the piezoelectric constituent to the extent allowed by the mechanical coupling between the two, and the transferred strain induces an electric polarization in the piezoelectric constituent. The inverse effect is also possible, where an electric field may be applied to the piezoelectric material to induce strain, which is then transferred to the magnetostrictive material, where it affects the magnetization.

Even though the ferromagnetic resonance line broadening problem can be overcome by using layered structures such as laminated composites, the narrowest resonance linewidth is naturally possible in single-crystal material. Therefore, using single-crystal ferrites enhances the resolution in electric field tunability. Using ferroelectrics of single crystal form is also beneficial as they would produce higher strain, and therefore, a larger ME effect will result in ferroelectric/ferromagnetic bilayer structures. As single-crystal YIG has been shown to exhibit the narrowest ferromagnetic resonance linewidth, most of the efforts on ME coupling-based devices have focused on incorporating single-crystal YIG with different types of ferroelectric layers, either in single crystal or ceramic form. The two constituents are bonded using epoxy to form the bilayer structures.

An electric field-tunable yttrium iron garnet (YIG: $Y_3Fe_5O_{12}$) on gadolinium gallium garnet (GGG: $Gd_3Ga_5O_{12}$) substrate—lead zirconate titanate (PZT: $Pb(Zr, Ti)O_3$) microwave resonator based on ferromagnetic resonance for YIG has been demonstrated [2, 3] The finite lateral dimensions of the ferrite film, metallization on the ferroelectric surfaces, and the anisotropy fields in the ferrite may result in slight deviations from the theoretical estimates; however, FMR frequencies were shown to be in good agreement with measurements. The corresponding ME coefficients YIG/PZT structure extracted from magnetostriction-induced magnetic field shifts are $\alpha 1 = \delta H1/\delta E \approx 0.58$ Oe·cm/kV and $\alpha 2 = \delta H2/\delta E \approx 0.88$ Oe·cm/kV (δH is the ferromagnetic resonance shift due to an applied electric field δE), for magnetic-bias field applied parallel and perpendicular to the sample plane, respectively, which are comparable to 1–4 Oe·cm/kV measured in similar YIG/PZT bilayers [4]. The range of the frequency tuned by electric field could potentially be increased by either decreasing the thickness of the GGG substrate or by using a ferroelectric with higher piezoelectric coefficients, such as single-crystal PMN-PT. The frequency shift, which arises due to magnetoelectric interactions, is relatively large for out-of-plane H in accordance with the theory [5, 6]. From the analysis of data, it was concluded that the primary cause of the frequency shift was variation of the effective internal magnetic field in YIG with applied electric field.

To study the resonance effects, for simplicity the ME response is first measured as a function of the static magnetic field, for simplicity, to find the electric field value for peak response. This is the inverse of applications sought in microwave passive components. The static field is then set to this value, and the frequency of the AC magnetic field is swept to observe the resonance effect.

The theoretical analysis of the resonance ME effects may be performed for multilayers in the form of thin disks or thin plates, where the thickness is small compared to the wavelengths of the acoustic modes. The thickness is also assumed to be much smaller than the lateral dimensions of the multilayer (radius, R, in the case of disks) so that only the radial modes are considered. The outer surfaces of the multilayer system are coated with metallic contacts and are equipotential; therefore, only the normal component of the electric field is nonzero. Moreover, for thin layers the normal component of the stress tensor may be taken as zero in the entire volume.

Ferrites are an important class of materials for use in microwave passive devices such as phase shifters, circulators, filters, isolators, and resonators [5, 7]. Depending on the frequency of operation, different types of ferrite materials are needed and used. Because ferrites are magnetic dielectric materials (unlike magnetic metals which are conductive) they permit electromagnetic penetration and thus interaction between the electromagnetic wave and magnetization within the ferrite. Ferrite devices permit the control of microwave propagation by a static or switchable DC magnetic field. The devices can be reciprocal or nonreciprocal and linear or nonlinear. What is certain is that their development requires a good deal of knowledge of magnetic materials, electromagnetic theory, and microwave circuit theory.

Miniaturization of microwave devices is one of the fundamental requirements in communication systems. Therefore, small-size and high-performance devices are always necessary to reduce cost and enhance system performance. The advantages of such devices include rapid tuning with an electric field, zero power consumption, noise elimination due to the absence of magnetic tuning, compatibility with integrated circuit technology, and resistance to nuclear radiation. Today there is the diversified production engineering of forming of electric circuits with the small losses, manifesting frequency-dependent properties. It is possible to follow constructive alternatives.

Ferrites are used in tunable microwave and millimeter-wave devices, and the tunability is traditionally realized through the variation of a bias-magnetic field [5, 7, 8]. This magnetic tuning could be achieved over a very wide frequency range, but is relatively slow, noisy and requires high power for operation. Similar devices but with some unique advantages could be realized by replacing the ferrite with a ferrite-ferroelectric composite [9–11]. Such heterostructures are magnetoelectric due to their response to elastic and electromagnetic force fields.

Layered structure based on the magnetostrictive and piezoelectric materials are perspective for designing new microwave devices. The given materials contain magnetostrictive and piezoelectric components and possess magnetoelectric interaction that will allow creating essentially new devices on their basis.

2. Theory

2.1. Coupling between ME Resonators. For the estimation of device characteristics, we consider the transmission lines and the resonator as a single system that is described by a coupling coefficient. We follow a two-stage procedure for the device design: first, we obtain the coefficients of reflection, transmission, and absorption for the case when an ME resonator is considered as an irregularity in the transmission line. Then, we take into account the number of resonators to estimate the coupling coefficient.

Application of microstrip lines is usually limited to use of the basic wave of type T. The ratio for coupling coefficient in this case can be presented in the following:

$$k = \frac{2V\chi''_{+} z_0 \varepsilon}{\pi h^2 \lambda_y Z}\left(\operatorname{arc} tg \frac{Z}{z_0\sqrt{\varepsilon}} + \frac{1}{3}\operatorname{arc} tg \frac{3Z}{z_0\sqrt{\varepsilon}}\right)$$

$$\text{with } \chi''_{+} = \frac{8\pi M_0}{\Delta H}, Z = 120\pi \,\text{Ohm}, z_0 = 50\,\text{Ohm}, \tag{1}$$

where V is the ME resonator volume, M_0 *is* the YIG saturation magnetization, λ_y is the wavelength in transmission line, z_0 is the microstrip impedance, ε is the substrate permittivity, and h is thickness of substrate.

2.2. Insertion Loss and Phase Characteristic. The insertion losses for single-resonator devices are determined by the following expression [12]:

$$L = -20\log T1, \tag{2}$$

where

$$T1 = \frac{k}{\sqrt{(1+k)^2 + \xi^2}},$$
$$\xi = \frac{H_r - H + \delta H}{\Delta H}. \tag{3}$$

The phase characteristic is determined from the expression for the complex transmission coefficient

$$\varphi = \operatorname{arc} tg\left(\frac{2k\xi}{1 - k^2 + \xi^2}\right). \tag{4}$$

Here, $T1$ is transmission gain of the devices, H_r is the resonance field, δH is the FMR line shift due to the electrical field, H is the dc magnetic field, ΔH is the half-width of FMR line, ξ is the combined detuning, and k is the single-resonator filter coupling coefficient.

The insertion loss is the sum of losses in the ferrite α_m and piezoelectric layer α_p, metal conductors α_{Me}, dielectric substrate α_d, and due to less-than-ideal coupling between the resonator and transmission lines α_c:

$$L = \alpha_m + \alpha_p + \alpha_{\text{Me}} + \alpha_d + \alpha_c. \tag{5}$$

Under ideal conditions, with $\alpha_m = 0,5$ dB, $\alpha_p = 1,0$ dB, $\alpha_{\text{Me}} = 0,5$ dB, $\alpha_d = 0,2$ dB, and $\alpha_c = 0,3$ dB, a total insertion loss as low as 2.5 dB could be realized in the one-cavity system.

2.3. Two-Resonator Devices. As model of the multiresonator devices the single-resonator devices joined by a piece of transmission line of nonresonance length, carrying out a role of connection device between parts is accepted.

Thus, the transfer ratio of the cascade-type devices was obtained from a transfer ratio for the single-resonator devices:

$$T2 = \frac{2KK^*}{\Delta}; \quad K = \frac{k}{1+i\xi}; \; K^* = -i\frac{k}{1+i\xi};$$
$$\Delta = (1+K)^2 - K^{*2}. \tag{6}$$

The insertion losses for two-resonator devices are determined by the following expression:

$$L_2 = -20\log T2. \tag{7}$$

Here, $T2$ is transmission coefficient of two-resonator devices, K and K^* are complex coupling coefficients of two-resonator devices, k is coupling coefficient of one-resonator devices.

A single-cavity device suffers from many disadvantages including insufficient pass-band and small out-of-band characteristics. It is, however, possible to improve the characteristics with increasing the numbers of resonators.

3. Magnetoelectric Attenuator

The electrically tunable attenuator designed on the base of microstrip transmission line with layer resonator based on YIG film and lead magnesium niobate-lead titanate (PMN-PT) is described. Figure 1 shows typical samples of PMN-PT. The chemical structural formula of PMN-PT is (1-x)[$Pb(Mg_{1/3}Nb_{2/3})O_3$]-x[$PbTiO_3$]. The PMN-PT is formulated to exhibit high piezoelectric coefficient, large electromechanical coupling coefficient, high dielectric constants and low dielectric losses. As a rule, the piezoelectric coefficient is higher than PZT ceramics, which results in improving band width, sensitivity, and source level in applications.

The unique combination of magnetic, electrical, and ME interactions, therefore, opens up the possibility of reciprocal and nonreciprocal ME signal processing devices. Here, we represent the ME microwave resonance attenuator. With the

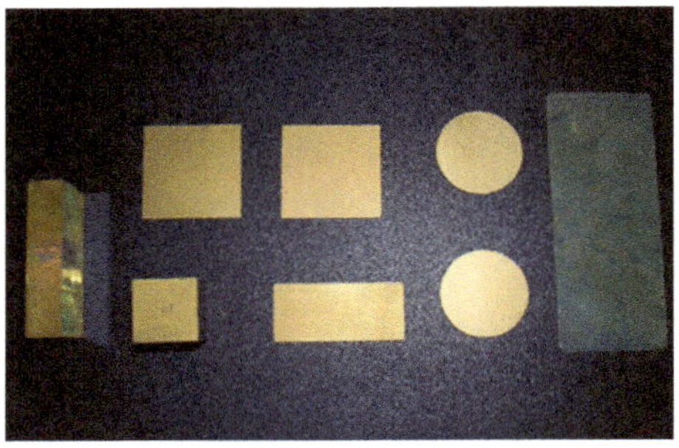

Figure 1: Typical samples of PMN-PT.

purpose of tuning the attenuator parameters in the given paper, the microwave ME effect is used [13, 14]. The effect consists in shift of FMR line on influence of an external electrical field. The used sample represents layered structure, which the magnetic part consists of the YIG thin film placed on the GGG film. A piezoelectric is the thin PMN-PT plate. For obtaining three-layer structure magnetic and piezoelectric parts are jointed together by a thin film of epoxy glue. For calculation of tunable filter parameters, it is necessary to take into account the microwave ME effect.

The base of the attenuator is a microstrip transmission line on dielectric substrate and ME resonator (Figure 2).

Work of the attenuator is based on resonance ME effect phenomena. Applying the control voltage to electrodes of the ME resonator results in a shift of FMR line due to the resonance ME effect, and so electrical tuning is realized.

Theory and Experiment. At the first approach, this curve can be written in the following expression:

$$P_{ab}(H) = P_0 \exp\left(-\frac{(H-H_0)^2}{2\sigma^2}\right), \tag{8}$$

where $P_{ab}(H)$ is the absorbed power versus bias magnetic field, P_0 is the accepted power at magnetic resonance, H_0 is the value of bias magnetic field at the magnetic resonance, and σ-parameter is determinated from the width of resonance line.

The electric field leads to the shift resonance line

$$\delta H = AE. \tag{9}$$

Substituting (9) into (8) we get

$$P_{ab} = P_0 \exp\left(-\frac{A^2E^2}{2\sigma^2}\right). \tag{10}$$

The insertion loss determined from ME resonator:

$$\begin{aligned} L_R &= 10\lg\left(\frac{P_{out}}{P_{in}}\right) \\ &= 10\lg\left(\frac{P_{in}-P_{ab}}{P_{in}}\right) \\ &= 10\lg\left(1-\frac{P_0}{P_{in}}\exp\left(-\frac{A^2E^2}{2\sigma^2}\right)\right). \end{aligned} \tag{11}$$

As the insertion loss, experimentally was measured parameter s_{21}.

The output power $P_{out} = P_{in} \cdot |s_{21}|^2$.

At the magnetic resonance

$$P_{ab} = P_0 = P_{in}\left(1-|s_{21}(H_0)|^2\right). \tag{12}$$

Using (12) we get

$$\frac{P_0}{P_{in}} = \left(1-|s_{21}(H_0)|^2\right). \tag{13}$$

Substituting (13) into (11) we get an equation for the insertion loss in the magnetoelectric resonator in following form:

$$\alpha_R = 20\lg\left(1-\left(1-|s_{21}(H_0)|^2\right)\exp\left(-\frac{A^2E^2}{2\sigma^2}\right)\right). \tag{14}$$

For the calculation was used the following experimental data $A = 4\,\text{MHz/kV}\cdot\text{cm}$, $\sigma = 14\,\text{MHz}$, and $s_{21} = 0,13$.

The insertion loss of attenuator is the sum of losses in the magnetoelectric resonator (α_R) and piezoelectric layer α_p, metal conductors (α_{Me}), dielectric substrate (α_d), and due to nonideal coupling between the resonator and transmission lines (α_c)

$$L = \alpha_R + \alpha_p + \alpha_d + \alpha_{Me} + \alpha_c. \tag{15}$$

Under ideal conditions, with $\alpha_R = 0,5\,\text{dB}$, $\alpha_p = 1,0\,\text{dB}$, $\alpha_{Me} = 0,5\,\text{dB}$, $\alpha_d = 0,2\,\text{dB}$, and $\alpha_c = 0,3\,\text{dB}$, a total insertion loss as low as 2.5 dB could be realized in the one-cavity system.

If the attenuator is worked on a resonance frequency (Figure 3), thus the attenuator insertion losses on influence of control electrical field vary from 26 dB to 2 dB (Figure 4).

The obtained results proved the possibility for designing resonance ME attenuator with electrical control.

Thus, the microwave attenuator on the basis of composite ferrite-piezoelectric resonators are offered. Calculation of attenuator characteristics with single-crystal ME resonators on the trilayer GGG/YIG/PMN-PT is performed.

The electric tuning by ME device parameters opens new capabilities for their application and manufacturing of devices with high-speed operation, small size, and integral technologies.

4. Magnetoelectric Band-Pass Filter

A microstrip band-pass filter is designed and characterized [15]. The filter is based on ferromagnetic resonance (FMR) of ferrite component. The device operating at 5–10 GHz can be tuned over a wide frequency band with a bias magnetic field and over a narrow band with a voltage applied across piezoelectric component. The voltage tuning of the device is possible through ME interactions that are mediated by mechanical deformation and manifests as a shift in FMR. Data on tuning range, insertion loss, and device characteristics are presented for filters with single- and double-ME resonators.

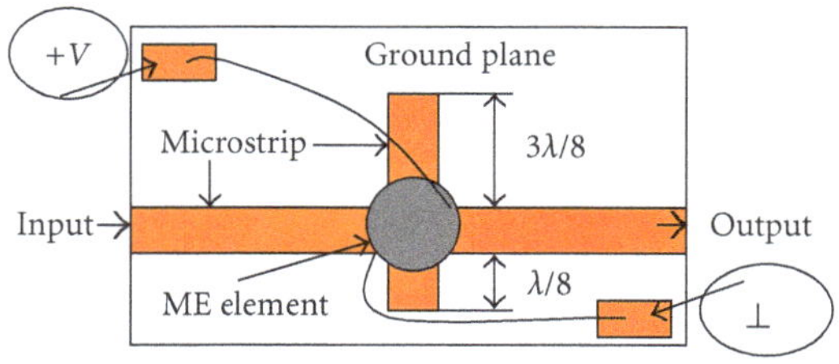

FIGURE 2: Design of microstrip ME attenuator and ME resonator.

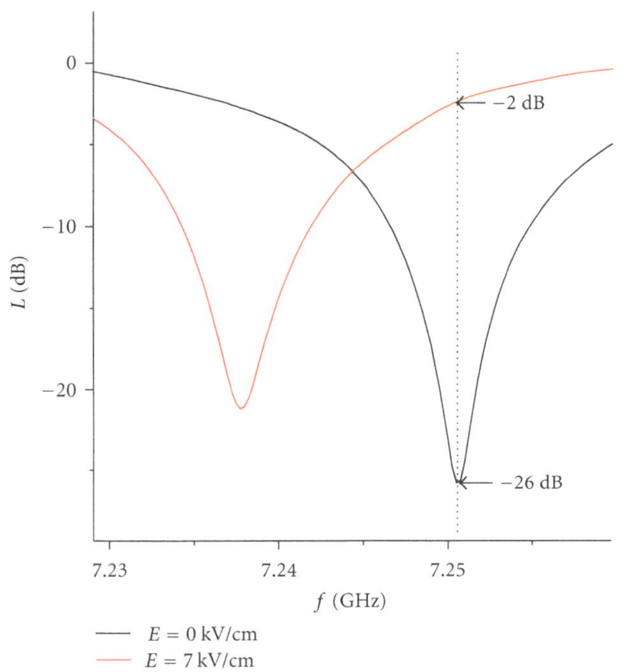

FIGURE 3: Experimental curves of insertion loss versus frequency. YIG [111] thickness is 110 μm; diameter is 2,5 mm; PMN-PT thickness is 0,5 mm; diameter is 6 mm; field $H = 1910$ Oe; field is parallel to plane of sample; central frequency is 7251 MHz.

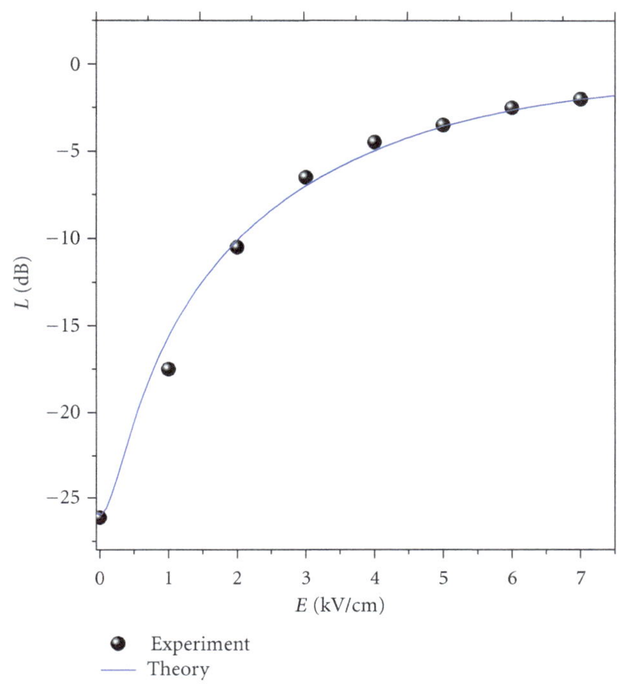

FIGURE 4: Comparison of experimental and theoretical data. YIG [111] thickness is 110 μm; diameter is 2,5 mm; PMN-PT tickness is 0,5 mm; diameter is 6 mm. Field $H = 1910$ Oe; field is parallel to plane of sample; average is 30; central frequency is 7251 MHz.

As objects used magnetostrictive component such as YIG and piezoelectric component such as PZT or PMN-PT. Designing band-pass filters utilizing ferrite-piezoelectric structure, which is ME involves calculations of the energy's coupling of the first ME resonator and of coupling of energy between resonators. Tunable resonators are widely used in microwave circuits as the frequency controlling element in reference oscillators, as the tunable resonator tank for voltage-control oscillators (VCOs), and as building elements for tunable filters. Traditional MMIC varactor-tuned resonators need to use GaAs MOSFETs or a *pin* diode as tuning elements, which will lead to high insertion loss and also small tuning range [16]. Microelectromechanical systems (MEMSs) technology has been successfully applied in developing a tunable resonator to achieve large tuning range from 3.5 to 7 GHz [12]. However, large difference in S_{11} response is found in the whole tuning range. Tunable ME resonators are used in microwave range. The resonant frequency of the ME resonator can be shifted at a tunable range from 5 to 10 GHz. There are also various other considerations of importance such as temperature properties, unwanted mode suppression, and signal-handling capability. What is not available is adequate information on coupling between ME resonators and means of designing filters for the minimum excitation and coupling of unwanted modes. One of the goals of this paper was the determination of those physical parameters which affect the coupling of the fundamental mode.

We recently studied ME interactions at FMR in layered ferrite-ferroelectric composites [17, 18]. The studies were done on bilayers of single crystal YIG films and PMN-PT. Microwave ME measurements at 9.3 GHz were performed using a traditional FMR spectrometer. An electric field E applied to the sample produced a mechanical deformation in the piezoelectric phase that in turn was coupled to the ferrite and manifested as a shift in FMR profiles. Profiles of

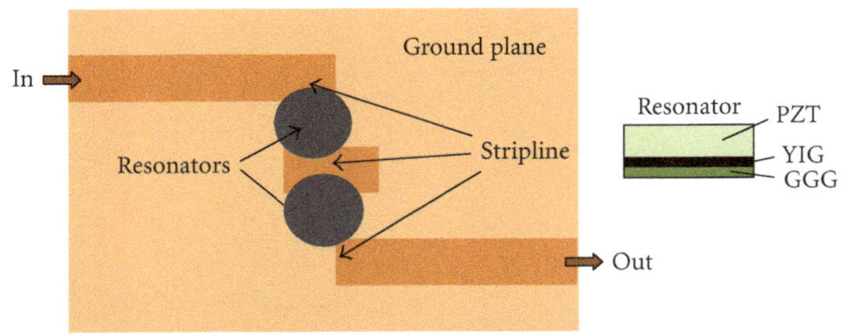

Figure 5: Schematic diagram showing a dual-cavity magnetoelectric filter and the ME resonators.

FMR absorption versus bias magnetic field H were obtained for a series of electric voltage across PMN-PT. With the application of E, we measured a shift in the profile by δH and the ME constant $A = \delta H/E$ varing over the range 1/5 Oe·cm/kV, depending on the YIG film thickness. Since 1 Oe of field shift corresponds to a frequency shift of 2.8 MHz, frequencies of FMR-based microwave devices can be tuned with E. For a nominal field of 30 kV/cm, an YIG/PMN-PT resonator, for example, can be tuned by 80/400 MHz. Thus, the ME effect would facilitate rapid voltage tuning for any FMR-based microwave device. Other advantages of microwave ME devices include miniaturization, near-zero power consumption, noise reduction and compatibility with integrated circuit technology.

This paper is on the fabrication and characterization of an electric field tunable microwave band-pass filter with single- and dual-ME resonators. Since the frequency of operation is 5/10 GHz, we choose low-loss single-crystal YIG films for the magnetic phase of the resonator and polycrystalline lead zirconate titanate (PZT) for the ferroelectric phase. Our studies indicate very good tunability due to a strong ME interaction and an acceptable insertion loss. Theoretical models account very well for the observations.

Filter Design. Both single- and dual-ME cavity filters were studied. Microstrip transducers fabricated by on substrates of foil-clad microwave material FLAN-10 (permittivity at 10 GHz, not more than 10,0 ± 0,8; tangent to dielectric loss angle at 10 GHz, not more than 0,0045; sheet thickness (including copper foil) from 1,0 ± 0,1 mm). It is a sheet microwave material manufactured from the composition on the basis of filled polyphenylene oxide with electrolytic galvano-resistant 35-micron copper foil bonded to both sides. It possesses a high stability value of dielectric permittivity and low dielectric losses in the microwave range. The base of the band-pass filter is a microstrip transmission line on dielectric substrate and ME resonators The single-cavity ME filter consisted of a 1 mm thick ground plane, input and output microstrips of nonresonance lengths, and an ME element. The microstrip transducers, 1 mm in width and 18 mm in length, were separated by 1.5 mm. The input-output decoupling is determined by this gap between the microstrips. Power is coupled from input to output under FMR conditions in the ME element. A dual-cavity filter shown in Figure 5 consisted of a 1 mm thick ground plane, input and output microstrips of nonresonance lengths equal 15 mm, section of a nonresonance length microstrip (3 mm), carrying out a role of an coupling element between two resonators, and a two-ME element. Width of all lines is 1 mm, gaps between microstrips are 1 mm also. The ME element consisted of epitaxial YIG film bonded to PZT. A YIG film grown by liquid-phase epitaxy on a (111) GGG was used. The ME element was made from YIG film 110 μm thick on GGG substrate 210 μm thick with diameter 1,5 mm and PZT plate (dimensions $4 \times 0{,}9 \times 0{,}5\ \text{mm}^3$) with electrodes.

The film saturation induction $4\pi M$ of 1750 G, and FMR line width of 1 Oe. A PZT plate was initially poled at oil by heating up to 110°C within 5 hours and cooling back to room temperature in an electric field of 4 kV/mm perpendicular to the sample plane. The layered structure was made by bonding the YIG film surface to PZT with thin layer of ethyl cyanoacrylate, a fast-dry epoxy. The layered structure was placed between the transducers and was subjected to a field H parallel and perpendicular to the sample plane.

Results. The device characterization was carried out with a vector network analyzer (PNA E-8361). An input continuous wave signal $P_{\text{in}}(f) = 1$ mW was applied to the filter. The frequency f dependence of the insertion loss L, that is, the transmitted power through the ME element, was measured at 4/10 GHz as a function of H and E applied across the PZT layer.

First, we are interested in the frequency dependencies of the insertion losses were measured as a function of magnetic field H (applied in the plane ferrite layer) and the electrical field E (applied across the piezoelectric layer). Next, we measured other characteristics of filter like bandwidth of pass-band (level of −3 dB), stop-band, and shape factor—$K_p = \text{BW}_{\text{stop-band}}/\text{BW}_{\text{pass-band}}$, Q-factor.

The insertion loss dependencies of the frequency (Figure 6) were measured for one-resonator filter as compared to two-resonator filter. The data are for a central frequency $f = 7{,}36$ GHz (corresponding to $H_0 = 1915$ Oe, H is parallel to plane of resonator).

Characteristics of one-resonator and two-resonator band-pass filters are presented in Table 1.

Figure 7 shows the dependence of the frequency shifts δf on YIG/PZT structure versus intensity of E for case that magnetic field is parallel and perpendicular to the plane of resonator.

TABLE 1: Characteristics of one-resonator and two-resonator band-pass filter.

Type of filter	$f_{central}$ (MHz)	$BW_{-3\,dB}$ (MHz)	IL_{pass} (dB)	IL_{stop} (dB)	K_p	Q
One-resonator YIG filter	7356	8	3.0	12	3.0	660
Two-resonator YIG filter	7355	16	8	24	3.0	450
Two-resonator YIG/PZT filter	7370	30	11.4	25	4.0	240

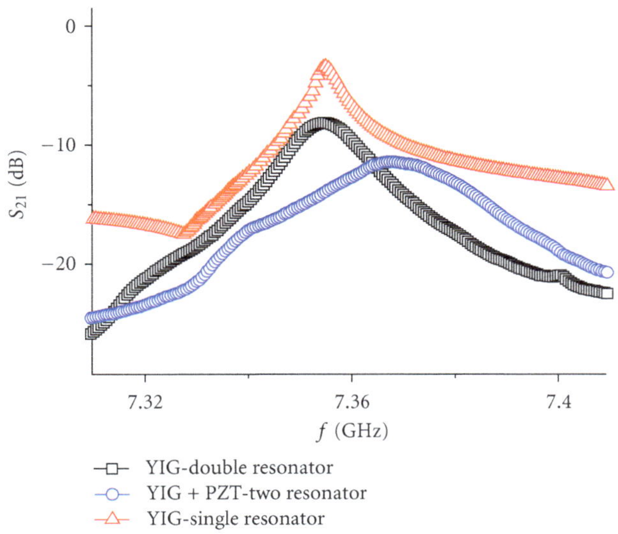

FIGURE 6: One-resonator filter as compared to two-resonator band-pass filter.

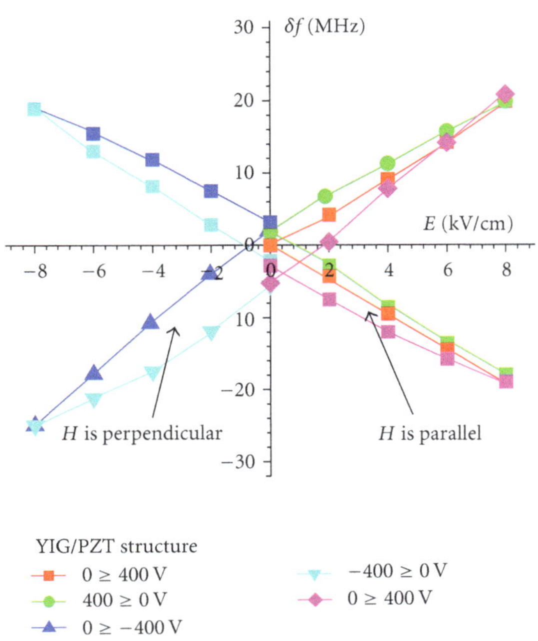

FIGURE 7: The dependence of the frequency shifts δf versus intensity of E in case that magnetic field is parallel and perpendicular to the plane of resonator.

One observes a near-linear variation in δf with small hysteresis. The observable linear dependence of the shift may be explained by internal distortions of YIG films and indicates the necessity of application in practice of thin YIG films.

Thus, the electrically tunable band-pass filter designed in microstrip transmission line and with layer resonator based on YIG-film and PZT is described. First, we measured the insertion loss as a function of magnetic field H (applied in the plane and across the ferrite layer) and the electrical field E (applied across the piezoelectric layer). A minimum in the insertion loss is 3 dB for one-resonator YIG filter and 8/11 dB for two-resonator band-pass filter in the frequency region 5/10 GHz. The insertion loss is up to 25 dB for out-of-band frequencies.

Next, we measured the frequency shift δf on YIG-PZT layered structure for different electric field E (magnetic field H is parallel and perpendicular to plane of resonator). In magnetoelectric composite frequency, tuning is realized through the variation of the applied electric field that creates effect of magnetostriction. Tunability of the devices characteristics in the chosen working range can be executed by application of an electric field. This electric tuning is possible in a narrow frequency range, but it is relatively fast and is not power-consuming. It will allow carrying out tuning in a range of 25 MHz. The electric tuning would facilitate high-speed operation, small size, and compatibility with integrated circuit technology.

5. Magnetoelectric Phase Shifter

Traditional ferrite phase shifter use magnetic tuning systems that are slow, demand high power, and are not miniature in size. Here, we discuss the design and characterization of a new electric field tunable phase shifter based on ferrite-piezoelectric layer composite. The electrical control of the phase shifter is realized through microwave magnetoelectric effect [19]. The phase shifter is capable of rapid tuning and miniature in size. This microwave phase shifter is of interest for phased array antenna system.

Experiment. Here, we presented the design and characterization of a new electric field tunable phase shifter based on ferrite-piezoelectric layer composite. The phase shifter is capable of rapid tuning and miniature in size. This microwave phase shifter is of interest for phased array antenna system. The base of the phase shifter is a microstrip transmission line on alumina substrate ($\varepsilon = 9.8$, thickness = 1 mm) and ME resonator. At researches, there were used ME

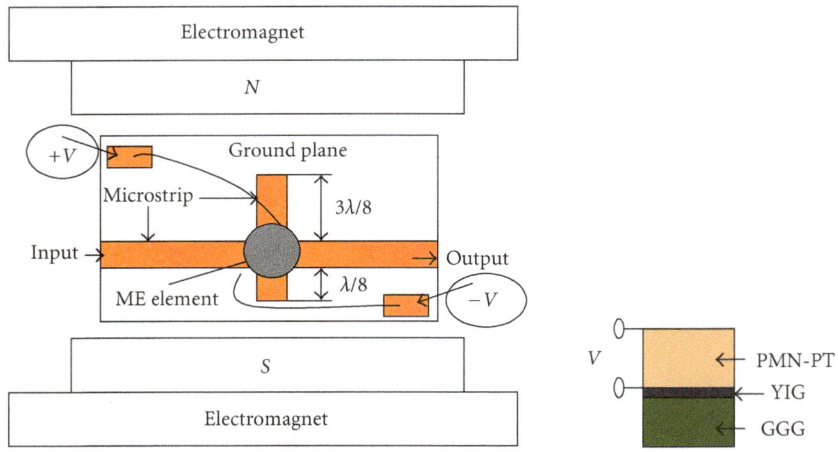

FIGURE 8: Schematics of a ME microwave phase shifter and ME resonator of YIG film on GGG bonded to PMN-PT.

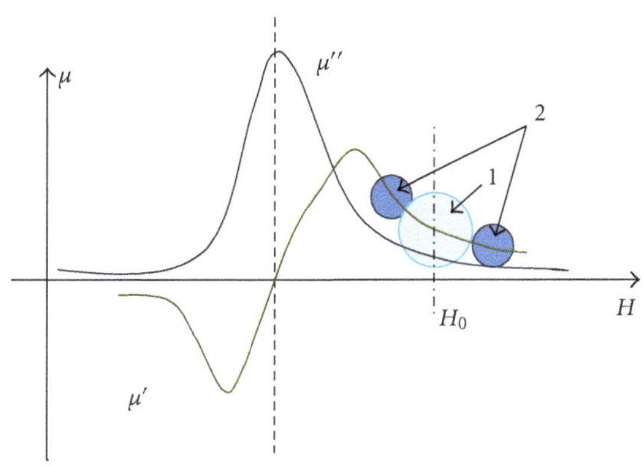

FIGURE 9: Typical ferromagnetic resonance regime for YIG showing the real and imaginary parts of the permeability as a function of bias field H_0. The change in the permeability and phase with H_0 or E is linear in region 1 and is nonlinear in region 2.

composite consisting of epitaxial single-crystal (111) YIG films on GGG substrates and 0.44 mm thick (001) PMN-PT. Metal electrodes (200 nm in thickness of gold and 30 nm in thickness of chromium) were deposited on PMN-PT for electrical contacts, and the crystal was initially poled by heating to 373 K and cooling it back to room temperature in $E = 2\,\text{kV/cm}$. A thin layer (<0.08 mm) of an epoxy, ethylcyanoacrylate, was used to bond YIG to PMN-PT. Circular polarization of the magnetic field in volume of the magnetoelectric resonator results from using the microstrip stubs of one eighth and three eighth wavelength. A bias field corresponding to FMR is applied to the ME resonator. Electrical tuning is realized with the application of a control voltage to electrodes leading to a shift of FMR line.

The ferrite film used is a 26.8 μm thick disk with diameter 4 mm. A PMN-PT plate with diameter 6 mm was used as piezoelectric element. Device characterization was carried out with a vector network analyzer PNA series E8361.

Schematics of a ME microwave phase shifter and ME resonator are represented in Figure 8.

Basis of a design of the phase shifter is the microstrip line on a substrate from an alumina and a magnetoelectric composite, serving by the resonator. In volume of the resonator by means of microstrip loops of $\lambda/8$ and $3\lambda/8$ wave lengths, a circular polarization of a magnetic field is created. In Figure 7, the design of the microstrip microwave phase shifter is presented.

The resonator is installed in area of circular polarization of a magnetic field, constant field H in parallel to a sample plane. For reduction of losses the working point gets out on a dispersive curve outside of resonance area.

Work of the phase shifter is based on the following: for a choice of the working point laying on a dispersive curve, to the resonator on frequency close the external magnetic field is put to a resonance. Under influence of the operating voltage put to electrodes, located at end faces of the resonator, owing to microwave effect there is a shift of line FMR and electric management in parameters of the phase shifter is realized.

Calculation is lead for the phase shifter with resonant frequency of 6/9 GHz. The ME composite is included as heterogeneity in area of circular polarization of a magnetic making microwave of a field. Change of a phase corner is achieved by a choice of strong not mutual connection of the resonator with a line of transfer. Thus, regulation of a phase corner is achieved by change of an operating voltage on the electrodes put on the resonator.

The working condition of the phase shifter will be chosen on dispersion curve of permeability at external magnetic field. Figure 9 shows typical permeability μ versus H profiles for YIG; FMR is expected for $H = H_r$. We choose $H_o > H_r$, to obtain low insertion loss and a linear or nonlinear variation of μ' with H.

Results. Representative data on *phase* versus *frequency* for different frequency and magnetic field H are shown in Figures 10 and 11. We measured the phase characteristics as a function of the electrical field E (applied across the piezoelectric layer) at frequency region 6/9 GHz. A magnetic field

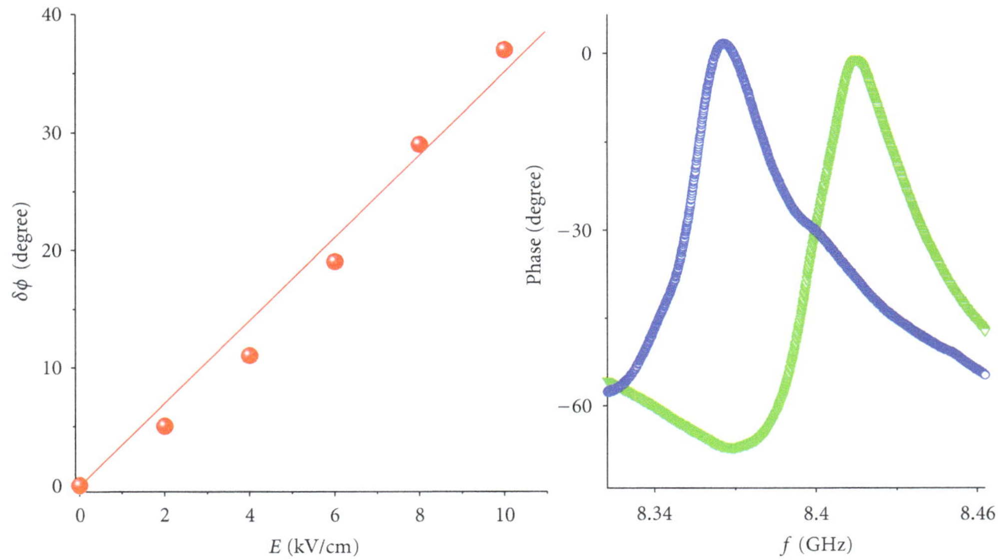

FIGURE 10: Differential phase shift versus electric field E (left) and typical view of phase characteristics versus frequency (right). H is parallel to sample $H_0 = 2277$ Oe; $L = -4$ dB; $f_0 = 8420$ MHz YIG 26.8 μm; $\phi = 4$ mm; $t_{GGG} = 440\,\mu$m; PMN-PT 0.5 mm; $\phi = 6$ mm.

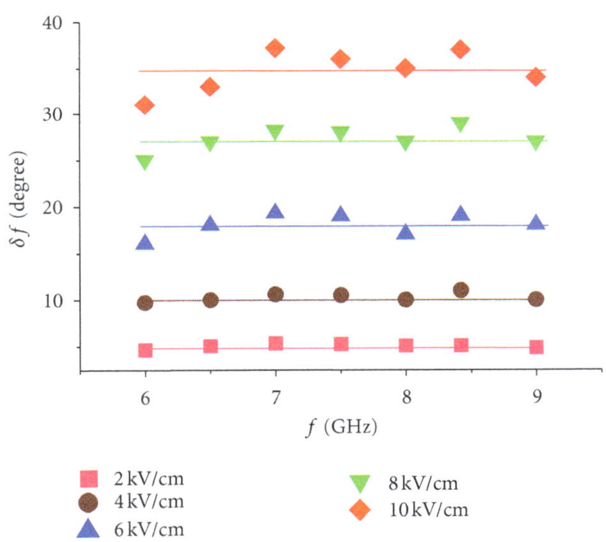

FIGURE 11: Differential phase shift versus frequency at different electric field E. H is parallel to sample $L = -3/4$ dB; YIG 26.8 μm; $\phi = 4$ mm; $t_{GGG} = 440\,\mu$m; PMN-PT 0.5 mm; $\phi = 6$ mm.

H was used for choice of working point. We used parallel magnetic field. The loss was 3/4 dB at the frequency region.

The data in Figures 10 and 11 constitute demonstration of the electrical tunability of the phase shift in a microwave ME microstrip device. The tunability is due to the influence of mechanical deformation on the piezoelectric phase. A near-linear variation in phase with E is evident from the data. A maximum phase shift is 30/40 degree at the frequency region 6/9 GHz. Obtained data allow to speak about an opportunity of practical application of our device. Optimization of the material, both ferrite and piezoelectric phase, and device parameters will allow a further reduction in the insertion loss and provide a high electric tunability.

Thus, we designed, fabricated, and tested microwave ME phase shifter in a standard thin film passive technology. Experimental results showed that 30/40 degree phase shift was achieved at ranging the 6/9 GHz over a range from 0 to 10 kV/cm. Our technology can significantly reduce the cost of phased array systems, where the phase shifter circuit is a large fraction of the overall antenna cost.

6. Conclusion

Ferrite-piezoelectric heterostructures are ideal for studies on wideband magnetoelectric interactions between the magnetic and electric subsystems that are mediated by mechanical forces. Such structures show a variety of magnetoelectric phenomena including microwave ME effects. The phenomenon can be used for creating electrical tuning microwave ME resonators and devices on their basis. The specific focus is on ME effects when electromagnetic modes close to FMR are excited in the ferrite. In the presence of an electric field, the piezoelectric deformation in the bilayer structures will manifest as an internal magnetic field in the ferrite. Tunability of the ME devices characteristics in the chosen working range can be executed by application of an electric field. This electric tuning is relatively fast and is not power-consuming.

The central problem of microwave technology is the design of receiver, transmitter, and receiver-transmitter modules, phased array system (PAS) with specified characteristics. They must address diverse requirements of PAS in sensitivity, power, frequency range, and so forth. The use of electromechanical, ferromagnetic, and magnetoacoustic resonances, magnetic dipole and electric dipole transitions in these structures allows the design of new microwave

ME devices [6, 20] and manufacture on them through various modules [21]. The results of our research show promise reviewed microwave ME devices for phased array modules [22]. Development and application of new layered ferrite-piezoelectric structures with advanced properties will significantly improve the performance of modules and PAS as a whole.

References

[1] M. I. Bichurin and V. M. Petrov, "Magnetic resonance in layered ferrite-ferroelectric structures," *Soviet Physics—Technical Physics*, vol. 33, p. 1389, 1989.

[2] M. I. Bichurin, V. M. Petrov, I. A. Kornev et al., "Resonance magnetoelectric effect in multilayer composites," *Ferroelectrics*, vol. 280, pp. 187–197, 2002.

[3] M. I. Bichurin, V. M. Petrov, R. V. Petrov et al., "Magnetoelectric microwave devices," *Ferroelectrics*, vol. 280, pp. 211–218, 2002.

[4] M. I. Bichurin, V. M. Petrov, Y. V. Kiliba, and G. Srinivasan, "Magnetic and magnetoelectric susceptibilities of a ferroelectric/ferromagnetic composite at microwave frequencies," *Physical Review B*, vol. 66, no. 13, Article ID 134404, pp. 1–10, 2002.

[5] C. S. Tsai and J. Su, "A wideband electronically tunable microwave notch filter in yttrium iron garnet-gallium arsenide material structure," *Applied Physics Letters*, vol. 74, no. 14, pp. 2079–2080, 1999.

[6] M. I. Bichurin, I. A. Kornev, V. M. Petrov, A. S. Tatarenko, Y. V. Kiliba, and G. Srinivasan, "Theory of magnetoelectric effects at microwave frequencies in a piezoelectric/magnetostrictive multilayer composite," *Physical Review B*, vol. 64, no. 9, Article ID 094409, pp. 1–6, 2001.

[7] J. D. Adam, L. E. Davis, G. F. Dionne, E. F. Schloemann, and S. N. Stitzer, "Ferrite devices and materials," *IEEE Transactions on Microwave Theory and Techniques*, vol. 50, no. 3, pp. 721–737, 2002.

[8] F. A. Miranda and G. Subramanyam, "Design and development of ferroelectric tunable microwave components for ku- and k-band satellite communication systems," *IEEE Transactions on Microwave Theory and Techniques*, vol. 48, no. 7, pp. 1181–1189, 2000.

[9] M. I. Bichurin, R. V. Petrov, and Y. V. Kiliba, "Magnetoelectric microwave phase shifters," *Ferroelectrics*, vol. 204, no. 1–4, pp. 311–319, 1997.

[10] Y. K. Fetisov and G. Srinivasan, "Electric field tuning characteristics of a ferrite-piezoelectric microwave resonator," *Applied Physics Letters*, vol. 88, no. 14, Article ID 143503, 2006.

[11] G. Srinivasan, A. S. Tatarenko, V. Mathe, and M. I. Bichurin, "Microwave and MM-wave magnetoelectric interactions in ferrite-ferroelectric bilayers," *European Physical Journal B*, vol. 71, pp. 1–5, 2009.

[12] T. Ketterl, T. Weller, and D. Fries, "A micromachined tunable CPW resonator," in *Proceedings of the International Microwave Symposium Digest (IEEE-MTT-S '01)*, pp. 345–348, May 2001.

[13] M. I. Bichurin and V. M. Petrov, "Composite magnetoelectrics: their microwave properties," *Ferroelectrics*, vol. 162, p. 33, 1994.

[14] M. I. Bichurin, I. A. Kornev, V. M. Petrov, and I. V. Lisnevskaya, "Investigation of magnetoelectric interaction in composite," *Ferroelectrics*, vol. 204, no. 1–4, pp. 289–297, 1997.

[15] A. S. Tatarenko, V. Gheevarughese, and G. Srinivasan, "Magnetoelectric microwave bandpass filter," *Electronics Letters*, vol. 42, no. 9, pp. 540–541, 2006.

[16] K. K. M. Cheng and J. K. A. Everard, "X-band monolithic tunable resonator/filter," *Electronics Letters*, vol. 25, no. 23, pp. 1587–1589, 1989.

[17] G. Srinivasan, A. S. Tatarenko, and M. I. Bichurin, "Electrically tunable microwave filters based on ferromagnetic resonance in ferrite-ferroelectric bilayers," *Electronics Letters*, vol. 41, no. 10, pp. 596–598, 2005.

[18] S. Shastry, G. Srinivasan, M. I. Bichurin, V. M. Petrov, and A. S. Tatarenko, "Microwave magnetoelectric effects in single crystal bilayers of yttrium iron garnet and lead magnesium niobate-lead titanate," *Physical Review B*, vol. 70, no. 6, Article ID 064416, 2004.

[19] A. S. Tatarenko, G. Srinivasan, and M. I. Bichurin, "Magnetoelectric microwave phase shifter," *Applied Physics Letters*, vol. 88, no. 18, Article ID 183507, 2006.

[20] M. I. Bichurin, V. M. Petrov, O. V. Ryabkov, S. V. Averkin, and G. Srinivasan, "Theory of magnetoelectric effects at magnetoacoustic resonance in single-crystal ferromagnetic-ferroelectric heterostructures," *Physical Review B*, vol. 72, no. 6, Article ID 060408, 2005.

[21] M. I. Bichurin and R. V. Petrov, "Magnetoelectric phasers for PAS," in *Proceedings of the 2nd International Conference on Satellite Communications (ICSC '96)*, pp. 172–176, September 1996.

[22] M. I. Bichurin, R. V. Petrov, and A. S. Tatarenko, "Magnetoelectric microwave devices for PAS. Modern problems in science and education, no 4," 2011, http://www.science-education.ru/98-4727.

14

Gap Structure and Gapless Structure in Fractional Quantum Hall Effect

Shosuke Sasaki

KYOKUGEN (Center for Quantum Science and Technology under Extreme Conditions), Osaka University, 1-3 Machikaneyama, Toyonaka, Osaka 560-8531, Japan

Correspondence should be addressed to Shosuke Sasaki, sasaki@ns.sist.ac.jp

Academic Editor: Roberto Zivieri

Higher-order composite fermion states are correlated with many quasiparticles. The energy calculations are very complicated. We develop the theory of Tao and Thouless to explain them. The total Hamiltonian is $(H_D + H_I)$, where H_D includes Landau energies and classical Coulomb energies. We find the most uniform electron configuration in Landau states which has the minimum energy of H_D. At $\nu = (2j-1)/(2j)$, all the nearest electron pairs are forbidden to transfer to any empty states because of momentum conservation. Therefore, perturbation energies of the nearest electron pairs are zero in all order of perturbation. At $\nu = j/(2j-1)$, $j/(2j+1)$, all the nearest electron (or hole) pairs can transfer to all hole (or electron) states. At $\nu = 4/11$, 4/13, 5/13, 5/17, 6/17, only the specific nearest hole pairs can transfer to all electron states. For example, the nearest-hole-pair energy at $\nu = 4/11$ is lower than the limiting energies from both sides (the left side $\nu = (4s+1)/(11s+3)$ and the right side $\nu = (4s-1)/(11s-3)$ for infinitely large s). Thus, the nearest-hole-pair energy at specific ν is different from the limiting values from both sides. The property yields energy gap for the specific ν. Also gapless structure appears at other filling factors (e.g., at $\nu = 1/2$).

1. Introduction

Precise experiments on ultra-high-mobility samples revealed many local minima of diagonal resistivity ρ_{xx} [1, 2]. Therein small local minima are detected at the filling factors of $\nu =$ 3/8, 3/10, 7/11, 4/11, 4/13, 5/13, 5/17, and 6/17,.... These states cannot be understood by use of the standard composite fermion (CF) model [3, 4]. Jain has originally considered multiflavor composite fermion picture with coexistence of composite fermions carrying different numbers of fluxes. Wójs et al. [5, 6], Smet [7], and Peterson and Jain [8] and Pashitskii [9] investigated these states and described the states in their extended systematics. Smet has explained these states in terms of the multiflavor composite fermion picture. Pashitskii presented expanded systematics based on Halperin's conjecture of coexistence of free electrons and bound electron pairs, with predicted new exotic fractions of ν = 5/14, 5/16, and 3/20. These investigations are complicated to explain the stability of expanded states with ν = 3/8, 3/10, 7/11, 4/11, 4/13, 5/13, 5/17, and 6/17,..., because many quasiparticles are correlated with each others. The results of various theories depend upon what kinds of quasi-particles are combined with each others. Therefore, it is preferable that the same logic is applied to any kind of filling factors (including both standard and nonstandard filling factors).

We study the other description in order to remove these ambiguities. Tao and Thouless [10, 11] examined the case that the lowest Landau levels are partially filled with electrons. They cannot lead which states are stable in comparison with the other states. However, their method is very important to investigate the FQH states. We have carefully examined the Coulomb interactions and develop their theory [12, 13]. Then the Coulomb transitions conserve the x component of the total momentum where the x direction indicates the current direction. This property produces energy gaps for the specific filling factors. Also the momentum conservation produces no binding energy for some filling factors. We study these gap structure or gapless structure for various filling factors, respectively. Quantum hall devices have an ultra thin layer of electron conducting channel as in Figure 1.

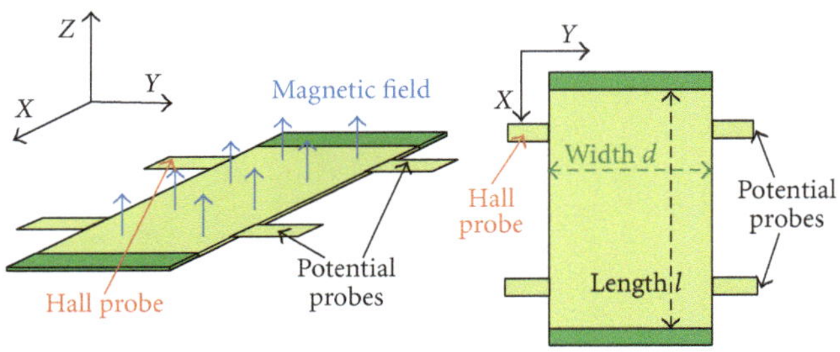

Figure 1: Quantum hall device.

That is to say the potential well in the z direction is deep and extremely narrow. Then the electron state of the z-direction is the ground state $\phi(z)$ at a low temperature because the probability of excited states is negligibly small caused by large excitation energy. The remaining freedoms are the x and y directions. If we neglect the Coulomb interactions between electrons, the single-electron eigen-states are the Landau states. Although these states are well known, we shortly write them for applying them below. Landau wave function is given as follows:

$$\psi_{L,J}(x,y,z) = \sqrt{\frac{1}{\ell}}\exp(ikx)u_L H_L\left(\sqrt{\frac{m\omega}{\hbar}}(y-\alpha_J)\right) \times \exp\left(-\frac{m\omega}{2\hbar}(y-\alpha_J)^2\right)\phi(z). \quad (1)$$

Therein, L is the Landau level number, k is the angular wave-number, and α_J is the central position in the y-direction which is related as

$$\alpha_J = \frac{2\pi\hbar J}{(eB\ell)} \qquad k = \frac{2\pi J}{\ell}, \quad (2)$$

where J indicates the integer assigned to each momentum of the x-direction. The eigen energy is

$$E_L = \lambda + \hbar\omega\left(L + \frac{1}{2}\right) \qquad (L = 0, 1, 2, 3, \ldots), \quad (3)$$

where λ is the ground state energy of the z-direction. We count the number of states with a fixed value of L by (2)

$$0 \le \alpha_J \le d \longrightarrow 0 \le \frac{2\pi\hbar J}{(eB\ell)} \le d \longrightarrow 0 \le J \le \frac{eB\ell d}{(2\pi\hbar)}, \quad (4)$$

where d is the width of the device in Figure 1. Then the total number of Landau states with the same value of L is equal to $eB\ell d/(2\pi\hbar)$.

Next we consider the many electron states $\Psi(L_1,\ldots,L_N;p_1,\ldots,p_N)$ described by the Slater determinant where L_j and p_j indicate the Landau level number and the momentum of jth electron, respectively. All the states composed of $\Psi(L_1,\ldots,L_N;p_1,\ldots,p_N)$ make the complete set. The total Hamiltonian is rewritten by use of this complete set. The diagonal part is described by H_D which is given as

$$H_D = \sum_{L_1,\ldots,L_N}\sum_{p_1,\ldots,p_N} |\Psi(L_1,\ldots,L_N;p_1,\ldots,p_N)\rangle \times W(L_1,\ldots,L_N;p_1,\ldots,p_N) \times \langle\Psi(L_1,\ldots,L_N;p_1,\ldots,p_N)|, \quad (5)$$

where $W(L_1,\ldots,L_N;p_1,\ldots,p_N)$ is the diagonal matrix element defined by

$$W(L_1,\ldots,L_N;p_1,\ldots,p_N) = \sum_{i=1}^{N} E_{L_i} + C(L_1,\ldots,L_N;p_1,\ldots,p_N). \quad (6)$$

Therein, E_{L_i} is the eigen-energy (3) of single electron with Landau level number L_i. Also C is the expectation value of Coulomb interactions which is defined by

$$C(L_1,\ldots,L_N;p_1,\ldots,p_N) = \int\cdots\int \Psi(L_1,\ldots,L_N;p_1,\ldots,p_N)^* \times \sum_{i=1}^{N-1}\sum_{j>i}^{N} \frac{e^2}{4\pi\varepsilon\sqrt{(x_i-x_j)^2+(y_i-y_j)^2+(z_i-z_j)^2}} \times \Psi(L_1,\ldots,L_N;p_1,\ldots,p_N) \times dx_1dy_1dz_1\cdots dx_Ndy_Ndz_N. \quad (7)$$

Hereafter, we call $C(L_1,\ldots,L_N;p_1,\ldots,p_N)$ "classical Coulomb energy." The total Hamiltonian is divided into two parts H_D and H_I. The residual part H_I is obtained by

$$H_I = H_T - H_D, \quad (8)$$

where H_I is constructed only by the off-diagonal elements. The residual Hamiltonian H_I is the interaction between two electrons depending upon only the relative coordinate. Therefore, the total momentum of the x-direction conserves in this system. That is to say the sum of the initial momenta p_i and p_j is equal to the sum of the final momenta p_i' and p_j' via Coulomb transition as follows:

$$p_i' + p_j' = p_i + p_j. \quad (9)$$

At a filling factor smaller than 1, the ground state of H_D is the following many-electron state as:

(1) N electrons exist in the lowest Landau levels of $L_1 = L_2 = \cdots = L_N = 0$.

(2) The electrons should most uniformly occupy the lowest Landau levels so that the classical Coulomb energy has the lowest value. The electron momenta $p_1, p_2, \ldots p_N$ are related to each centre positions as in (2). Therefore, the most uniform electron configuration determines the electron momenta $p_1, p_2, \ldots, p_N$ for each filling factor ν.

In the next section we will schematically draw the most uniform electron configuration at several filling factors. Then it is clarified that the most uniform electron configuration is unique. The most uniformity yields the minimum eigen energy of H_D, and the uniqueness produces a nondegenerate ground state although the ground states of the single electron Hamiltonian H_0 are degenerate. The electron configurations are examined in Section 2.

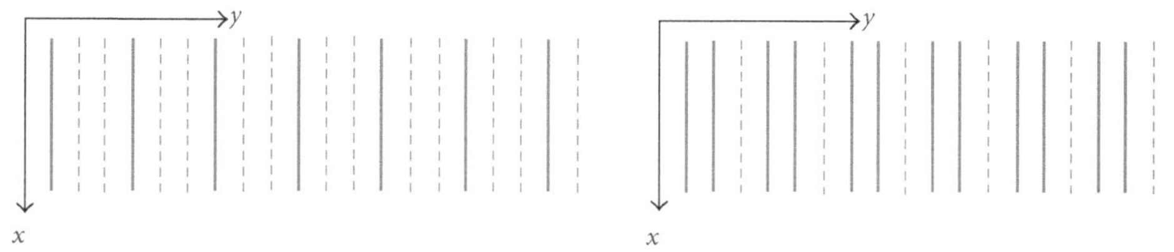

FIGURE 2: (a) Electron configuration at $\nu = 1/3$. (b) Configuration at $\nu = 2/3$.

We can estimate the perturbation energy via the Coulomb transitions by using the usual perturbation method of nondegenerate case. Electron (or hole) pairs in the nearest Landau orbitals are most affected by the Coulomb interaction. The perturbation energies of the nearest electron (or hole) pairs are sensitively dependent upon each electron configurations. The sensitivity is caused by Fermi-Dirac statistics and the momentum conservation of the x-direction. The most uniform electron configuration produces the following properties: all the *nearest electron* (or *hole*) *pairs* are allowed to transfer to all vacant (or filled) states at the filling factors of $\nu = j/(2j \pm 1)$. The property produces energy gap structure. For example, the perturbation energy of the *nearest electron pair* at $\nu = j/(2j-1)$ is lower than the limiting value from both sides as $s \rightarrow \nu \pm \varepsilon$ where ε is an infinitesimally small value. The energy gap is defined as

$$\text{Energy gap} = \Delta E_{\text{nearest pair}} = \left(E(\nu) - \lim_{s \to \nu+\varepsilon} E(s)\right). \quad (10)$$

The ratio of the energy gap and the original nearest electron pair energy is equal to

$$\begin{aligned} \frac{\Delta E_{\text{nearest pair}}}{E(\nu)} &= \frac{(E(\nu) - \lim_{s \longrightarrow \nu+\varepsilon} E(s))}{E(\nu)} \\ &= \frac{1}{2(j-1)} \quad \text{for } \nu = \frac{j}{2j-1}. \end{aligned} \quad (11)$$

It is noteworthy that both the energy $E(\nu)$ and the energy gap $E(\nu) - \lim_{s \to \nu+\varepsilon} E(s)$ are negative values. This mechanism is clarified in Section 3.

All the *nearest electron* (or *hole*) *pairs* are forbidden to transfer to any vacant (or filled) state at the filling factors of $\nu = (2j-1)/(2j)$ (or $\nu = 1/(2j)$). This mechanism is examined in Section 4.

At the filling factors of $\nu = 7/11, 4/11, 4/13, 5/13, 5/17$, and $6/17$, some of the *nearest electron* (or *hole*) *pairs* are allowed to transfer to all vacant (or filled) states. This case yields small energy gaps, for example,

$$\begin{aligned} \frac{\Delta E_{\text{nearest pair}}}{E(\nu)} &= \frac{(E(\nu) - \lim_{s \longrightarrow \nu+\varepsilon} E(s))}{E(\nu)} \\ &\approx \frac{0.006493}{0.103896} \quad \text{for } \nu = \frac{7}{11} \end{aligned} \quad (12)$$

which is investigated in Section 5. Thus, the present theory produces a gap structure or a gapless structure for each fractional filling factor. The theoretical results are in a good agreement with the experimental data.

We examine another type of gap which indicates the excitation energy gap from the ground state to excited states. This new gap is highly correlated with the gap in the spectrum of energy versus filling factor. The mechanism is studied in Section 6.

2. Most Uniform Configuration of Electrons

We can find out the many-electron states with the minimum energy of H_D for any filling factor. As most easy examples, we show the case of $\nu = 1/3$ and $\nu = 2/3$. Figures 2(a) and 2(b) indicate the configurations of electrons with the minimum classical Coulomb energy at $\nu = 1/3$ and $2/3$, respectively.

Therein the bold line indicates a Landau state filled with electron, and the dashed line means an empty state. It is noteworthy that the current direction is described by the x-direction, and the Hall voltage direction is described by the y-direction (same as in Figure 1). Figure 2(a) shows the electron configuration with repeating of the arrangement (empty, filled, empty). This filling way is the most uniform configuration at $\nu = 1/3$ and then has the minimum classical Coulomb energy. The filling way in Figure 2(b) also has the minimum classical Coulomb energy at $\nu = 2/3$. It is easily seen that Figures 2(a) and 2(b) indicate the most uniform filling ways at $\nu = 1/3$ and $\nu = 2/3$, respectively.

We explain the searching method to find the electron configuration with the minimum Coulomb energy, because it is nontrivial to find the filling way for any fractional filling factor $\nu < 1$. In order to clarify the explanation, we consider one example of $\nu = 3/5$. We compare the classical Coulomb energies of the following two cases.

Case 1. In the whole region, three electrons exist inside every 5 sequential landau states. Then the filling factor becomes 3/5.

Case 2. Two electrons exist in 5 sequential Landau states for some parts, and four electrons exist in 5 sequential Landau states for some other parts. And the average filling factor is equal to 3/5.

The Coulomb energy of Case 1 is smaller than one of Case 2 because the filling way of Case 1 is more uniform than one of Case 2. Therefore, it is sufficient to consider all the filling ways inside 5 sequential states. They are 10 filling ways as shown in Figure 3.

The five filling ways (a-1, a-2, a-3, a-4, and a-5) give the same electron configuration A by numerous repeating of themselves except both end parts. The electron configuration A is shown in Figure 4(a). The both end parts can be neglected for macroscopic number of electrons in a quantum Hall device.

Similarly the five filling ways (b-1, b-2, b-3, b-4, and b-5) in Figure 3 give the same electron configuration B as Figure 4(b). It is clear that the electron configuration A of Figure 4(a) has the classical Coulomb energy smaller than one of configuration B at $\nu = 3/5$.

It is noteworthy to examine the connections between different arrangements in Figure 3. We draw the connections of (a-1 and a-2), (a-1 and a-3), (a-1 and a-4), and (a-1 and a-5) in Figure 5 where the first filling way is red coloured, and the second filling way is blue coloured. All the connections include green areas where 2 electrons or 4 electrons exist inside five sequential Landau sates as in Figure 5.

Therefore, these connections belong to Case 2 and then have a classical Coulomb energy larger than one of configuration A in Figure 4(a).

For any filling factor $\nu = r/q$, r electrons should exist in q sequential Landau states everywhere. All filling ways have the number of $q!/(r!(q-r)!)$. We can draw all the filling ways and then find out the most uniform configuration of electrons. This procedure is applied to the cases with denominator $q = 2, 3, 4, 5, 6, 7, 8$. Then, we get the filling ways with the minimum classical Coulomb energy as drawn in Figure 6 (We abbreviate equivalent filling ways. For example only the filling way a-1 is drawn, and other filling ways a-2, a-3, a-4, a-5 are abbreviated for $\nu = 3/5$). These filling ways are called unit arrangements for each filling factors.

When we repeat the unit arrangement in Figure 6, we obtain the electron configuration with the minimum classical Coulomb energy at each filling factor.

We draw some examples with higher Coulomb energy for the denominator $q = 7, 8$ in Figure 7. Comparison of Figure 6 with Figure 7 reveals the fact that the unit arrangements in Figure 6 have more uniformity than ones in Figure 7.

Thus only one configuration has the minimum classical Coulomb energy among the enormous many configurations. The whole-electron configuration is created by repeating of only one unit arrangement of electron at any fractional number of ν.

3. Gap Structure in the Neighbourhood of $\nu = j/(2j \pm 1)$

3.1. Calculation of Binding Energy at $\nu = j/(2j-1)$. The shape of Landau wave function with $L = 0$ is schematically drawn by a straight line. We draw the most uniform electron configuration at $\nu = j/(2j-1)$ in Figure 8 where solid lines indicate the Landau orbitals filled with electron, and dashed lines indicate the vacant Landau orbitals. Therein, *nearest electron pairs* are red-coloured, and single electrons are blue-coloured. The Coulomb transitions from the nearest electron pair AB are illustrated by black arrows in Figure 8. The x and y directions are drawn in the upper left corner.

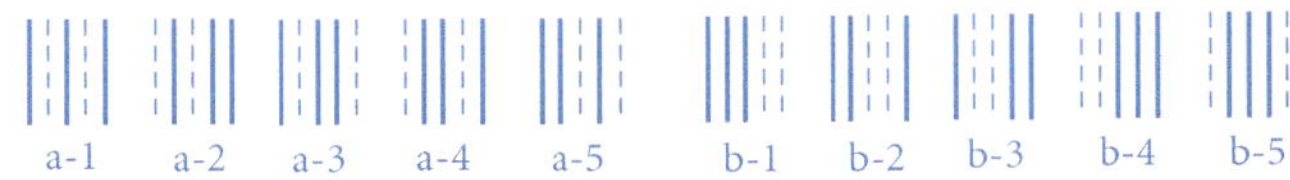

FIGURE 3: All unit arrangements of electron configurations for $\nu = 3/5$.

The electron configurations in Figure 8 have the minimum classical Coulomb energy for each filling factors. In order to explain the calculation of the second-order perturbation energy, we draw again the most uniform electron configuration of $\nu = 2/3$ in Figure 9. Therein the first transition is expressed by black arrows, and the second transition is expressed by green arrows and so on.

The momenta of the *nearest electron pair* AB are described by p_A, p_B, respectively. When electron A transfers to the first orbital to the left, the initial momentum p_A decreases by $2\pi\hbar/\ell$ according to (2). After the transition, the electron A has a new momentum p'_A as

$$p'_A = p_A - \frac{2\pi\hbar}{\ell}. \quad (13)$$

The other electron B transfers to the momentum p'_B. Then the total momentum conservation in the x-direction gives the following relation:

$$p'_A + p'_B = p_A + p_B. \quad (14)$$

Substitution of (13) into (14) yields the momentum p'_B as

$$p'_B = p_B + \frac{2\pi\hbar}{\ell}. \quad (15)$$

This momentum increment means that the electron B should transfer from its original orbital to the first orbital to the right because of (2) and (15). This transition is allowed because the right orbital is empty as in Figure 9.

Similarly the *nearest electron pair* AB can transfer to the other empty states, and the momenta after the transition are given by

$$p'_A = p_A - \Delta p,$$
$$p'_B = p_B + \Delta p, \quad (16)$$

where the momentum transfer Δp at $\nu = 2/3$ has the values as

$$\Delta p = \frac{(3j+1)2\pi\hbar}{\ell} \quad \text{for } j = 0, \pm 1, \pm 2, \pm 3, \ldots \quad (17)$$

because the electrons are possible to transfer to empty orbitals only. (In this paper we investigate the case that all the electron spins have an opposite direction of magnetic field. Another case is discussed in the other articles [14–16]). The second-order perturbation energy of the *nearest electron pair* AB is given by

$$\varsigma_{\nu=2/3} = \sum_{\Delta p=(3j+1)2\pi\hbar/\ell \text{ for } j=0,\pm1,\pm2,\ldots} R$$
$$R = \frac{\langle p_A, p_B | H_I | p'_A, p'_B \rangle \langle p'_A, p'_B | H_I | p_A, p_B \rangle}{W_G - W_{\text{excite}}(p_A \to p'_A, p_B \to p'_B)}. \quad (18)$$

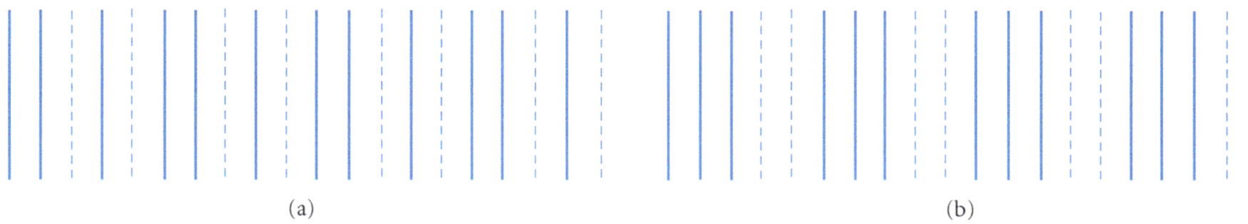

Figure 4: (a) Configuration A at $\nu = 3/5$. (b) Configuration B at $\nu = 3/5$.

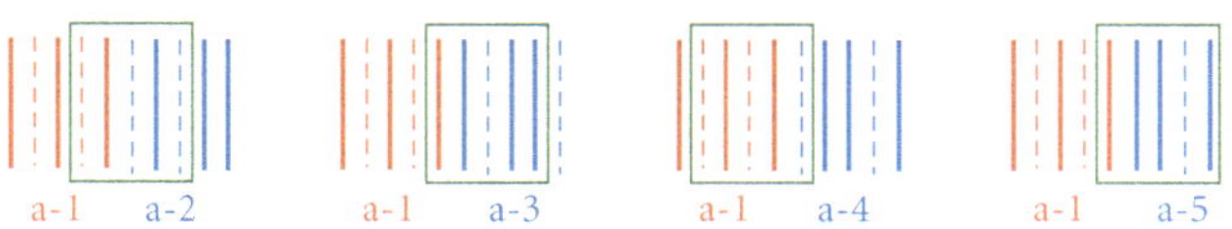

Figure 5: Connections between different arrangements.

We introduce the following summation Z as

$$Z = -\sum_{\Delta p \neq 0, -2\pi\hbar/\ell}^{\text{all}} \frac{\langle p_A, p_B | H_I | p'_A, p'_B \rangle \langle p'_A, p'_B | H_I | p_A, p_B \rangle}{W_G - W_{\text{excite}}(p_A \to p'_A, p_B \to p'_B)}, \tag{19}$$

where the momentum transfer Δp takes all values $(2\pi\hbar/\ell) \times$ integer except $\Delta p = 0$ and $\Delta p = -2\pi\hbar/\ell$. The transferred states with $\Delta p = 0$ and $\Delta p = -2\pi\hbar/\ell$ are the same state as the initial state. This state is eliminated in the summation of (19) because the diagonal matrix element of H_I is zero. Therefore, the denominator is not zero and has a negative value. The definition of Z includes (-1) in the right-hand side, and the denominator is a negative value. Accordingly the value of Z is positive.

We can systematically describe the perturbation energies at any filling factors by using the value of Z. We compare Z with the perturbation energy $\varsigma_{\nu=2/3}$ of (18). The interval of the momentum transfer is $3 \times 2\pi\hbar/\ell$ in the summation $\varsigma_{\nu=2/3}$. On the other hand, the summation Z is performed by the momentum interval $2\pi\hbar/\ell$. The interval value is very small for a quantum Hall device with a macroscopic size and then we get

$$\varsigma_{\nu=2/3} = -\frac{1}{3} Z. \tag{20}$$

Thus, we can express the perturbation energy $\varsigma_{\nu=2/3}$ of the *nearest electron pair* by the summation Z. The perturbation energies depend upon ℓ and B, namely, device size and magnetic field strength. These dependences are included in the summation Z.

The number of *nearest electron pairs* is $(1/2)N$ at $\nu = 2/3$, where N is the total number of electrons. Accordingly the total perturbation energy $E_{\text{nearest pair}}$ of all *nearest electron pairs* is obtained as

$$E_{\text{nearest pair}} = \frac{1}{2} N \times \varsigma_{\nu=2/3} = -\frac{1}{6} ZN, \quad \text{for } \nu = \frac{2}{3}. \tag{21}$$

Similar calculation leads the second order perturbation energy of *nearest electron pairs* at any filling factor of $\nu = j/(2j-1)$ as follows [16].

Therein, j electrons partially occupy each sequential $(2j - 1)$ Landau states, and then $(j - 1)$ Landau states are empty for each sequential $(2j - 1)$ Landau states. Any *nearest electron pair* can transfer to all the empty states as in Figure 8. The number of transitions is equal to $(j - 1)$ for each $(2j - 1)$ Landau states. Therefore, the second-order perturbation energy per pair is given by

$$\varsigma_{\nu=j/(2j-1)} = -\frac{j-1}{2j-1} Z, \quad \text{for } \nu = \frac{j}{(2j-1)}. \tag{22}$$

The number of *nearest electron pairs* is $(1/j)N$ at $\nu = j/(2j - 1)$, where N is the total number of electrons. Accordingly the total perturbation energy $E_{\text{nearest pair}}$ from all *nearest electron pairs* is obtained as

$$\begin{aligned} E_{\text{nearest pair}} &= \frac{N}{j} \varsigma_{\nu=j/(2j-1)} \\ &= -\frac{j-1}{j(2j-1)} ZN, \quad \text{for } \nu = \frac{j}{(2j-1)}. \end{aligned} \tag{23}$$

The *nearest electron pair* energy per electron is given by

$$\chi\left(\nu = \frac{j}{2j-1}\right) = -\frac{(j-1)Z}{(j(2j-1))}, \quad \text{for } \nu = \frac{j}{(2j-1)}. \tag{24}$$

3.2. Gap Structure at $\nu = j/(2j - 1)$. It has been clarified in the previous subsection that all the nearest electron pairs can transfer to all empty Landau orbitals at the filling factor of $\nu = j/(2j - 1)$. We next examine the perturbation energies in the neighbourhood of $\nu = j/(2j - 1)$. The fraction $\nu = (j(2s) - 1)/((2j - 1)(2s) - 2)$ approaches $\nu = j/(2j - 1)$ in the limit of infinitely large s. In the case of $j = 4, s = 2$, the fraction is equal to

$$\nu = \frac{(j(2s) - 1)}{((2j-1)(2s) - 2)} = \frac{15}{26}. \tag{25}$$

The most uniform electron configuration is illustrated in Figure 10.

It is noteworthy that some nearest electron pairs cannot transfer to some empty states because of momentum conservation and Pauli's exclusion principle. We take this prohibition of the transitions into consideration and obtain the perturbation energy of nearest electron pairs as follows:

$$\begin{aligned} E_{\text{nearest pair}} &= \left[-2 \times \left(\frac{8}{26}\right) Z - 2 \times \left(\frac{10}{26}\right) Z\right] \\ &\quad \times \left(\frac{1}{15}\right) N = -\frac{36}{26 \times 15} ZN, \quad \text{for } \nu = \frac{15}{26}. \end{aligned} \tag{26}$$

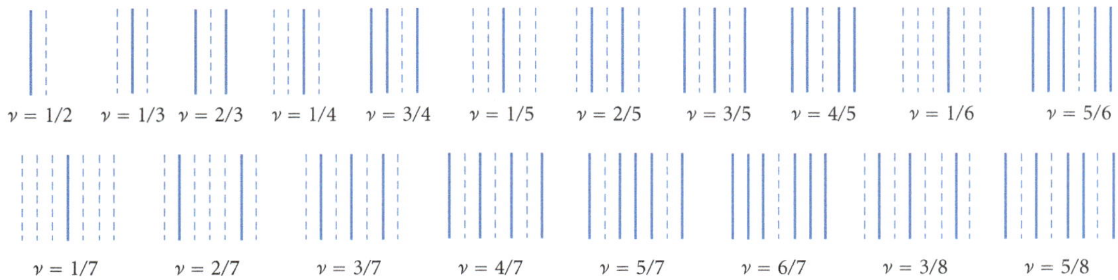

FIGURE 6: Most uniform unit arrangements of electron configuration.

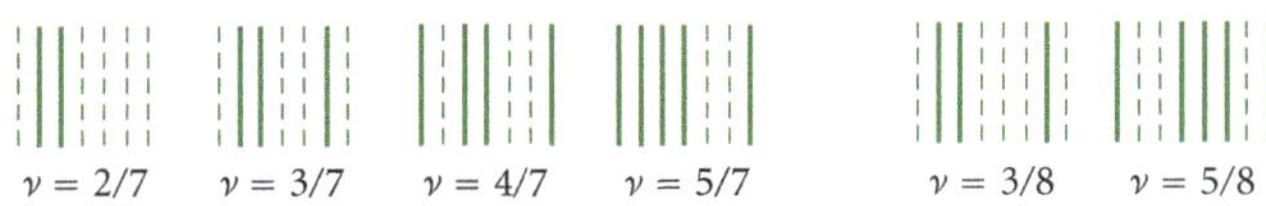

FIGURE 7: Filling ways with higher classical Coulomb energy.

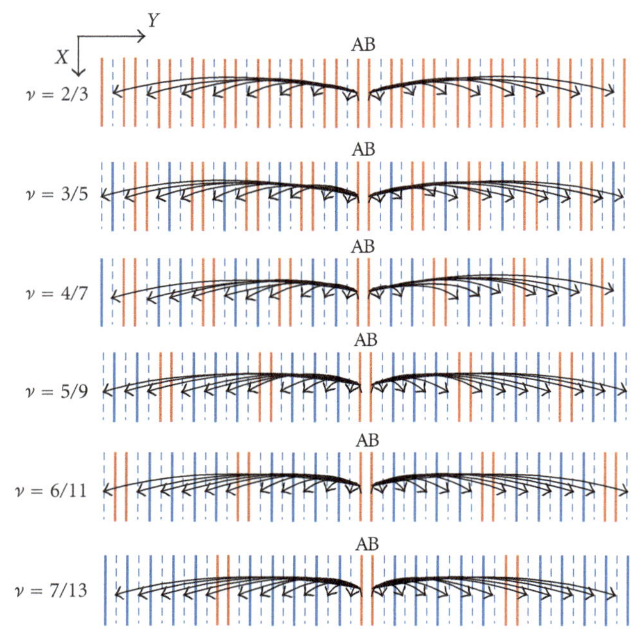

FIGURE 8: Allowed transitions of *nearest electron pair* at $\nu = j/(2j-1)$.

This calculation process can be extended to any integer s as follows:

$$E_{\text{nearest pair}} = \left[\begin{array}{l} -2\times\left(\dfrac{(4s)}{(14s-2)}\right)Z-2\times\left(\dfrac{(4s+2)}{(14s-2)}\right)Z\cdots \\ \cdots-2\times\left(\dfrac{(6s-2)}{(14s-2)}\right)Z \end{array} \right] \times\frac{N}{(8s-1)}. \quad (27)$$

The result indicates the perturbation energy of nearest electron pairs in the neighbourhood of $\nu = 4/7$.

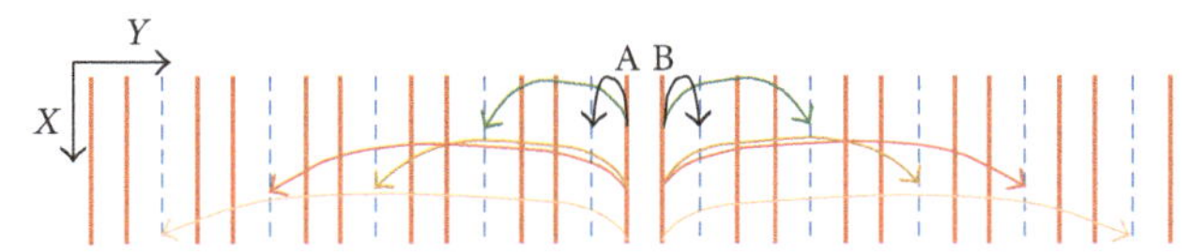

FIGURE 9: Coulomb transitions from nearest electron pair AB at $\nu = 2/3$. (The electric current flows along the x-direction. The momentum value is related to its central position of the y direction as in (2)).

Next we examine the neighbourhood of $\nu = j/(2j-1)$ for any integer of j. That is to say we consider the filling factor of $\nu = (j(2s)-1)/((2j-1)(2s)-2)$. We introduce new three parameters as follows:

$$\alpha = ((2j-1)(2s)-2), \qquad \beta = (j(2s)-1), \qquad \gamma = ((j-1)(2s)-1). \quad (28)$$

The parameter α is the number of orbitals in unit arrangement, β is the number of electrons in unit arrangement, and γ is the number of empty orbitals in unit arrangement. Accordingly we replace $(8s-1)$ into β and also replace $(14s-2)$ into α in (27). Then we get

$$E_{\text{nearest pair}} = \left[-2\times\left(\frac{(4s)}{\alpha}\right)Z-2 \times\left(\frac{(4s+2)}{\alpha}\right)Z\cdots-2\times\left(\frac{(6s-2)}{\alpha}\right)Z\right]\times\frac{N}{\beta}, \quad (29)$$

where $(6s-2)$ indicate $\gamma-1$. Accordingly $(4s)$ is replaced to $\gamma+1-2s$, and then we obtain that

$$E_{\text{nearest pair}} = \left[-2\times\left(\frac{(\gamma+1-2s)}{\alpha}\right)Z-2 \times\left(\frac{(4s+2)}{\alpha}\right)\times Z\cdots-2\times\left(\frac{(\gamma-1)}{\alpha}\right)Z\right]\times\frac{N}{\beta},$$

$$E_{\text{nearest pair}} = \left[-2\times(\gamma+1-2s)-2 \times(\gamma+3-2s)\cdots-2\times(\gamma-1)\right]\frac{Z}{\alpha}\times\frac{N}{\beta},$$

$$E_{\text{nearest pair}} = \left[-(2\gamma-2s)s\right]\frac{Z}{\alpha}\times\frac{N}{\beta} \quad \text{at } \nu=\frac{\beta}{\alpha}. \quad (30)$$

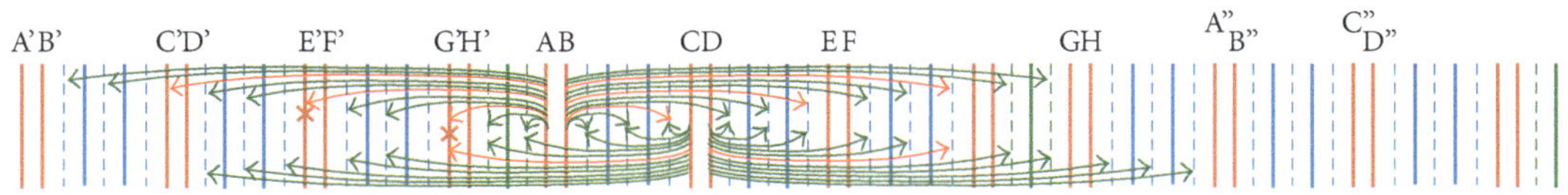

FIGURE 10: Transitions from nearest electron pairs for $\nu = 15/26$.

Substitution of (28) into (30) yields

$$E_{\text{nearest pair}} = -\frac{((2j-3)2s-2)s}{((2j-1)(2s)-2)}\frac{ZN}{(j(2s)-1)} \quad \text{at } \nu = \frac{(j(2s)-1)}{((2j-1)(2s)-2)}. \tag{31}$$

Thus, we have obtained the perturbation energy of the nearest electron pairs for any integers of j and s. Equation (31) gives the limiting value from the right for $s \to \infty$ because $\nu = \beta/\alpha$ is larger than $j/(2j-1)$. We have

$$\nu = \frac{\beta}{\alpha} = \frac{(j(2s)-1)}{((2j-1)(2s)-2)} \xrightarrow[s\to\infty]{} \frac{j}{(2j-1)}, \tag{32a}$$

$$E_{\text{nearest pair}} = -\frac{((2j-3)2s-2)s}{((2j-1)(2s)-2)}\frac{ZN}{(j(2s)-1)} \xrightarrow[s\to\infty]{} -\frac{(2j-3)ZN}{2j(2j-1)} \quad \text{at } \nu = \frac{j}{2j-1}+\varepsilon, \tag{32b}$$

where ε indicates an infinitesimally small positive value. Equation (23) indicates the perturbation energy of nearest electron pairs at $\nu = j/(2j-1)$ as

$$E_{\text{nearest pair}} = -\frac{(j-1)ZN}{j(2j-1)} \quad \text{at } \nu = \frac{j}{2j-1}, \tag{33}$$

$$\chi\left(\frac{j}{2j-1}\right) = \frac{E_{\text{nearest pair}}}{N} = -\frac{(j-1)Z}{(j(2j-1))}. \tag{34}$$

Equation (32b) means the limiting value of *nearest pair* perturbation energy from the right at $\nu = j/(2j-1)$. The perturbation energy (33) at $\nu = j/(2j-1)$ is lower than the limiting value (32b), and, therefore, the energy gap is equal to

$$\Delta E_{\text{nearest pair}} = E(\nu) - \lim_{s\to\nu+\varepsilon} E(s) = -\frac{ZN}{2j(2j-1)} \quad \text{for } \nu = \frac{j}{2j-1}. \tag{35}$$

Next we calculate the limiting value from the left. We consider the fraction $\nu = (j(2s)+1)/((2j-1)(2s)+2)$ which is smaller than $\nu = j/(2j-1)$. New three parameters α', β', and γ' are defined as

$$\alpha' = ((2j-1)(2s)+2), \qquad \beta' = (j(2s)+1), \qquad \gamma' = ((j-1)(2s)+1). \tag{36}$$

The parameter α' is the number of orbitals in unit arrangement, β' is the number of electrons in unit arrangement, and γ' is the number of empty orbitals in unit arrangement. The filling factor is given by $\nu = \beta'/\alpha'$ as

$$\nu = \frac{\beta'}{\alpha'} = \frac{(j(2s)+1)}{((2j-1)(2s)+2)}. \tag{37}$$

The total transition energy from all the *nearest electron pairs* is equal to

$$E_{\text{nearest pair}} = \left[-(2\gamma'-2s)s\right]\frac{Z}{\alpha'} \times \frac{N}{\beta'} \quad \text{at } \nu = \frac{\beta'}{\alpha'}. \tag{38}$$

Limiting values of (37) and (38) are

$$\nu = \frac{(j(2s)+1)}{((2j-1)(2s)+2)} \xrightarrow[s\to\infty]{} \frac{j}{(2j-1)},$$
$$E_{\text{nearest pair}} = -\frac{((2j-3)2s+2)s}{((2j-1)(2s)+2)}\frac{ZN}{(j(2s)+1)} \xrightarrow[s\to\infty]{} -\frac{(2j-3)ZN}{2j(2j-1)} \quad \text{at } \nu = \frac{j}{2j-1}-\varepsilon. \tag{39}$$

This limiting value from the left is compared with the original value at $\nu = j/(2j-1)$, and then the original value is lower than the limiting value from the left as

$$\Delta E_{\text{nearest pair}} = E(\nu) - \lim_{s\to\nu-\varepsilon} E(s) = -\frac{ZN}{2j(2j-1)} \quad \text{for } \nu = \frac{j}{2j-1}. \tag{40}$$

Equations (35) and (40) indicate that the limiting values from both sides of $\nu = j/(2j-1)$ are higher than the original value at $\nu = j/(2j-1)$.

The gap structure is produced from the property that all the *nearest electron pairs* are possible to transfer to all empty states at $\nu = j/(2j-1)$. When the filling factor ν changes from $j/(2j-1)$ by an infinitesimally small value, some of the transitions are forbidden. That is to say the number of allowed transitions decreases drastically. This property is caused by the most uniform electron configuration, Fermi-Dirac statistics, and momentum conservation. The drastic change also appears in higher order of the perturbation energy. The higher-order perturbations are studied in the other article.

Table 1 shows the discontinuous structure of the perturbation energies at the filling factors of $\nu = j/(2j-1)$, where ε expresses an infinitesimally small positive value. Therefore, $+\varepsilon$ indicates the limiting process from the right, and $-\varepsilon$ indicates the limiting process from the left.

TABLE 1: Energy gaps of *nearest electron pairs* per electron at $\nu = j/(2j-1)$.

ν	$E_{\text{nearest pair}}/N$	ν	$\lim(E_{\text{nearest pair}}/N)$	$\Delta\varepsilon_+(\nu) = \Delta\varepsilon_-(\nu)$
2/3	$-(1/6)Z$	$(2/3) \pm \varepsilon$	$-(1/12)Z$	$-(1/12)Z$
3/5	$-(2/15)Z$	$(3/5) \pm \varepsilon$	$-(3/30)Z$	$-(1/30)Z$
4/7	$-(3/28)Z$	$(4/7) \pm \varepsilon$	$-(5/56)Z$	$-(1/56)Z$
5/9	$-(4/45)Z$	$(5/9) \pm \varepsilon$	$-(7/90)Z$	$-(1/90)Z$
6/11	$-(5/66)Z$	$(6/11) \pm \varepsilon$	$-(9/132)Z$	$-(1/132)Z$
7/13	$-(6/91)Z$	$(7/13) \pm \varepsilon$	$-(11/182)Z$	$-(1/182)Z$
8/15	$-(7/120)Z$	$(8/15) \pm \varepsilon$	$-(13/240)Z$	$-(1/240)Z$

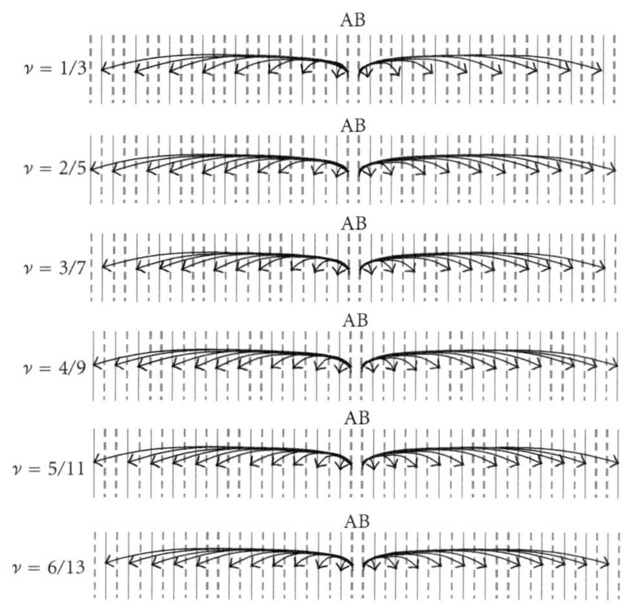

FIGURE 11: Coulomb transitions from nearest hole pair AB.

3.3. Gap Structure at $\nu = j/(2j+1)$. We draw the most uniform electron configuration at $\nu = j/(2j+1)$ in Figure 11.

Therein, solid lines indicate the Landau orbitals filled with electron, and dashed lines indicate the vacant Landau orbitals. *Nearest hole pairs* are red coloured, and single holes are blue coloured. The Coulomb transitions from the *nearest hole pair* AB are drawn by black arrows in Figure 11. Using the same method as in the previous section, we can estimate the perturbation energies of the *nearest hole pairs* $E_{\text{nearest hole pair}}$. The result is as follows:

$$E_{\text{nearest hole pair}} = -\frac{j}{(j+1)(2j+1)} Z_H N_H \quad \text{at } \nu = \frac{j}{2j+1}, \tag{41}$$

where N_H is the total number of holes. We obtain the pair energy per hole as

$$\frac{E_{\text{nearest hole pair}}}{N_H} = -\frac{j}{(j+1)(2j+1)} Z_H \quad \text{at } \nu = \frac{j}{2j+1}. \tag{42}$$

The energy per electron is

$$\frac{E_{\text{nearest hole pair}}}{N} = -\frac{1}{(2j+1)} Z_H \quad \text{at } \nu = \frac{j}{2j+1}. \tag{43}$$

We also calculate the perturbation energies in the neighbourhood of $\nu = j/(2j+1)$, and the results are

$$E_{\text{nearest hole pair}} = -\frac{((2j-1)2s-2)s}{((2j+1)(2s)-2)} \frac{Z_H N_H}{((j+1)(2s)-1)} \quad \text{at } \nu = \frac{((j)(2s)-1)}{((2j+1)(2s)-2)},$$
$$E_{\text{nearest hole pair}} = -\frac{((2j-1)2s+2)s}{((2j+1)(2s)+2)} \frac{Z_H N_H}{((j+1)(2s)+1)} \quad \text{at } \nu = \frac{((j)(2s)+1)}{((2j+1)(2s)+2)}. \tag{44}$$

The limiting values are obtained as

$$\frac{E_{\text{nearest hole pair}}}{N} \xrightarrow[s\to\infty]{} -\frac{2j-1}{2j(2j+1)} Z_H \quad \text{for } \nu = \frac{j}{2j+1} \pm \varepsilon. \tag{45}$$

We summarize the energy gaps of the *nearest hole pairs* at $\nu = j/(2j+1)$ in Table 2. The rightmost column means the energy gap per electron (not per hole).

Tables 1 and 2 show the gap structure of the fractional filling states at $\nu = j/(2j \pm 1)$ because the perturbation energy is lower than one of each neighbourhood.

3.4. Comparison between Experimental Data and Nearest Pair Energy at $\nu = j/(2j \pm 1)$. Many electrons in the electric current are scattered by impurities and thermal vibrations. These scatterings yield the diagonal resistance of the x direction.

At $\nu = j/(2j \pm 1)$, we have theoretically estimated the energy gaps in the spectrum of energy versus filling factor in Sections 3.2 and 3.3. The energy gaps in the spectrum produce the plateaus in Hall resistance curve (confinements of Hall resistance). The excitation-energy gap from the ground state is highly correlated with the energy gap in the spectrum as discussed in Section 6.

The excitation-energy gaps suppress the electron scatterings. Then the diagonal resistance becomes small. This

TABLE 2: Energy gaps of *nearest hole pairs* per electron at $\nu = j/(2j+1)$.

ν	$E_{\text{nearest pair}}/N_H$	$E_{\text{nearest pair}}/N$	ν	$\lim(E_{\text{nearest pair}}/N)$	$\Delta\varepsilon_+(\nu) = \Delta\varepsilon_-(\nu)$
1/3	$-(1/6)Z_H$	$-(1/3)Z_H$	$(1/3) \pm \varepsilon$	$-(1/6)Z_H$	$-(1/6)Z_H$
2/5	$-(2/15)Z_H$	$-(1/5)Z_H$	$(2/5) \pm \varepsilon$	$-(3/20)Z_H$	$-(1/20)Z_H$
3/7	$-(3/28)Z_H$	$-(1/7)Z_H$	$(3/7) \pm \varepsilon$	$-(5/42)Z_H$	$-(1/42)Z_H$
4/9	$-(4/45)Z_H$	$-(1/9)Z_H$	$(4/9) \pm \varepsilon$	$-(7/72)Z_H$	$-(1/72)Z_H$
5/11	$-(5/66)Z_H$	$-(1/11)Z_H$	$(5/11) \pm \varepsilon$	$-(9/110)Z_H$	$-(1/110)Z_H$
6/13	$-(6/91)Z_H$	$-(1/13)Z_H$	$(6/13) \pm \varepsilon$	$-(11/156)Z_H$	$-(1/156)Z_H$
7/15	$-(7/120)Z_H$	$-(1/15)Z_H$	$(7/15) \pm \varepsilon$	$-(13/210)Z_H$	$-(1/210)Z_H$

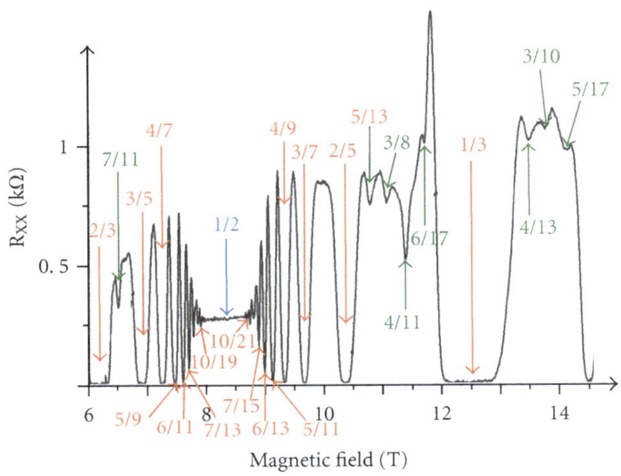

FIGURE 12: Many local minima of the diagonal resistance in the region of $2/3 \geq \nu \geq 1/3$. This experimental result has been obtained in [2].

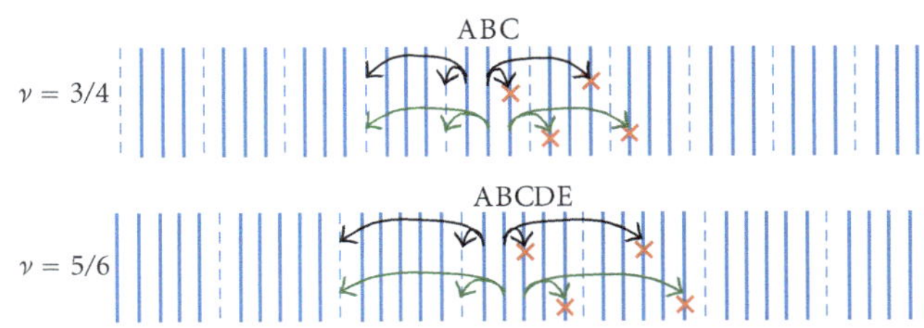

FIGURE 13: All Coulomb transitions from nearest electron pairs are forbidden.

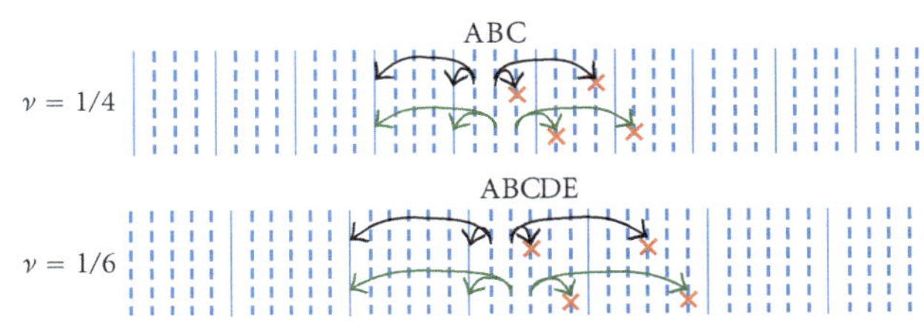

FIGURE 14: All Coulomb transitions from nearest hole pairs are forbidden.

mechanism produces local minima in the diagonal resistance curve. The theoretical results in Tables 1 and 2 are in a good accordance with the experimental data of Figure 12.

The local minima appear at the filling factors of $\nu = 2/3$, 3/5, 4/7, 5/9, 6/11, 7/13, 8/15,... and $\nu = 7/15, 6/13, 5/11$, 4/9, 3/7, 2/5, 1/3 which are indicated by red colour in Figure 12. Furthermore, there are small local minima in the curve of diagonal resistance versus magnetic field strength. The fractions are 7/11, 5/13, 3/8, 4/11, 6/17, 4/13, 3/10, and 5/17 which are coloured by green in Figure 12. We will examine the FQH states with $\nu = 7/11, 4/11, 4/13, 5/13, 5/17, 6/17$ in the later Section 5. The remaining FQH states with $\nu = 3/8, 5/8, 3/10$ have even denominators and have been already investigated in [17].

4. Gapless Structure of Nearest Pair Energy at $\nu = 1/(2j)$ and $\nu = (2j-1)/(2j)$

The most uniform electron configurations are illustrated in Figure 13 at $\nu = (2j-1)/(2j)$. The filling factors have even number of the denominator. As an example, we examine the case of $\nu = 3/4$. When electron A transfers to the fifth orbital to the left, electron B should transfer to the fifth orbital to the right. However, the fifth orbital is already filled with electron. Therefore, this transition is forbidden. Similarly all the nearest electron pairs cannot transfer to empty states at $\nu = (2j-1)/(2j)$.

Next we draw the most uniform electron configuration for $\nu = 1/(2j)$ in Figure 14. The hole pairs AB and BC cannot transfer to any filled orbitals at $\nu = 1/4$ as in Figure 14. All the nearest hole pairs cannot transfer to any filled orbitals at $\nu = 1/(2j)$ for any integer of j. Therefore, the perturbation energies of the nearest electron (or hole) pairs are zero for all order perturbations at $\nu = (2j-1)/(2j)$ (or $\nu = 1/(2j)$).

That is to say we obtain the following relations:

$$\frac{E_{\text{nearest electron pair}}}{N} = 0, \quad \text{for } \nu = \frac{(2j-1)}{(2j)},$$
$$\frac{E_{\text{nearest hole pair}}}{N_H} = 0, \quad \text{for } \nu = \frac{1}{(2j)}. \tag{46}$$

We show the perturbation energies of *nearest electron pairs* and *nearest hole pairs* in Table 3.

Consequently the states with $\nu = (2j-1)/(2j)$ (and $\nu = 1/(2j)$) are not confined in the Hall resistance curve [18–23].

TABLE 3: *Nearest pair* energies at $\nu = (2j-1)/(2j)$ and $\nu = 1/(2j)$.

ν	$E_{\text{nearest pair}}/N$	ν	$E_{\text{nearest pair}}/N_H$
1/2	0		
3/4	0	1/4	0
5/6	0	1/6	0
7/8	0	1/8	0

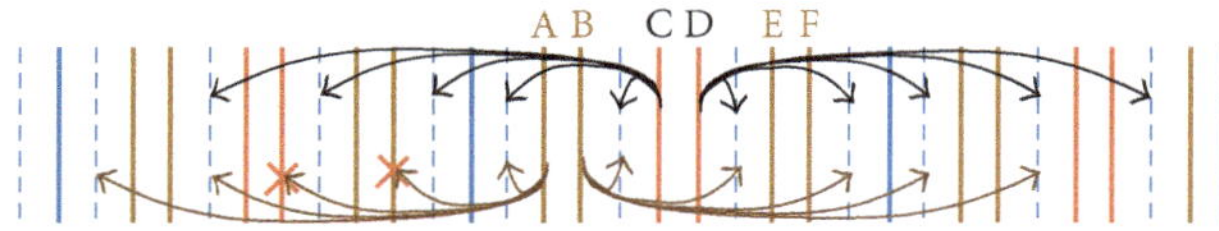

FIGURE 15: Coulomb transitions of nearest electron pairs at $\nu = 7/11$.

5. Special States with Several Fractional Filling Factors

In this section we examine the states with the filling factors $\nu = 7/11, 4/11, 4/13, 5/13, 5/17$, and $6/17$. The most uniform electron configuration with $\nu = 7/11$ is schematically drawn in Figure 15. There are three pairs of nearest electrons in every unit arrangement, namely, (filled, empty, filled, filled, empty, filled, filled, empty, filled, filled, empty). The orbitals filled with electron are illustrated by solid lines, and the empty orbitals are illustrated by dashed lines.

The *nearest electron pair* CD can transfer to all empty orbitals. Then the perturbation energy of the pair CD is equal to

$$\varsigma_{\text{CD}} = -\left(\frac{4}{11}\right)Z. \tag{47}$$

The *nearest electron pairs* AB and EF transfer to two empty states per unit arrangement and have the perturbation energies ς_{AB} and ς_{EF} as

$$\varsigma_{\text{AB}} = \varsigma_{\text{EF}} = -\left(\frac{2}{11}\right)Z, \tag{48}$$

$$\begin{aligned}\left(E_{\text{nearest pair}}\right)_{\nu=7/11} &= \left(-\frac{4}{11}Z - \frac{2}{11}Z - \frac{2}{11}Z\right)\frac{1}{7}N,\\ &= -\frac{8}{77}ZN, \quad \text{for } \nu = \frac{7}{11},\end{aligned} \tag{49}$$

$$\left(\frac{E_{\text{nearest pair}}}{N}\right)_{\nu=7/11} = -\frac{8}{77}Z \approx -0.103896\times Z, \quad \text{for } \nu = \frac{7}{11}. \tag{50}$$

We examine the neighbourhood of $\nu = 7/11$ which are $\nu = (7s-2)/(11s-3)$ and $\nu = (7s+2)/(11s+3)$. The most uniform electron configuration at $\nu = (7s \pm 2)/(11s \pm 3)$ is systematically produced by modulation of the configuration as Figure 15. Then the perturbation energy of the nearest

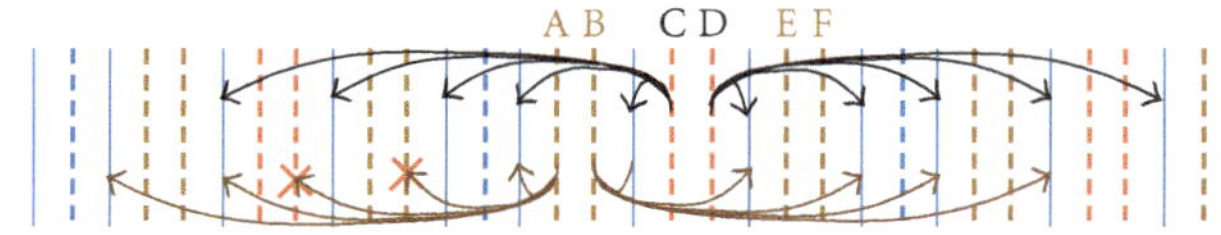

FIGURE 16: Coulomb transitions of nearest hole pairs at $\nu = 4/11$.

electron pairs is calculated by the use of the computer program. The calculation results for $s = 100$ are

$$\begin{aligned}&\left(\frac{E_{\text{nearest pair}}}{N}\right)_{\nu=698/1097}\\ &\quad = -\frac{74601}{765706}Z = -0.0974277\times Z, \quad \text{for } \nu_- = \frac{698}{1097},\\ &\left(\frac{E_{\text{nearest pair}}}{N}\right)_{\nu=702/1103}\\ &\quad = -\frac{75401}{774306}Z = -0.0973788\times Z, \quad \text{for } \nu_+ = \frac{702}{1103}.\end{aligned} \tag{51}$$

We also calculate the case of $s = 1000$ the results of which are

$$\begin{aligned}&\left(\frac{E_{\text{nearest pair}}}{N}\right)\\ &\quad = -\frac{7496001}{76957006}Z = -0.097405\times Z, \quad \text{for } \nu_- = \frac{6998}{10997},\\ &\left(\frac{E_{\text{nearest pair}}}{N}\right)\\ &\quad = -\frac{7504001}{77043006}Z = -0.0974002\times Z, \quad \text{for } \nu_+ = \frac{7002}{11003}.\end{aligned} \tag{52}$$

We compare the case of $s = 100$ with $s = 1000$ and find that the energies (51)–(52) are nearly equal to each other. Then the limiting values from both sides are approximately equal to

$$\lim_{\nu\to(7/11)\pm\varepsilon}\left(\frac{E_{\text{nearest pair}}}{N}\right) \approx -0.097403\times Z, \quad \text{for } \nu_\pm = \frac{7}{11}\pm\varepsilon. \tag{53}$$

The perturbation energy per electron is $(-0.103896 \times Z)$ at $\nu = 7/11$, and therefore the energy gap appears at $\nu = 7/11$ as in (50) and (53)

$$\left(\frac{E_{\text{nearest pair}}}{N}\right)_{\nu=7/11} - \lim_{\nu\to(7/11)\pm\varepsilon}\left(\frac{E_{\text{nearest pair}}}{N}\right) \approx -0.006493\times Z. \tag{54}$$

Next we examine the case of $\nu = 4/11$. We schematically draw the most uniform electron configuration of $\nu = 4/11$ in Figure 16.

Then the perturbation energy of the nearest hole pairs is obtained as

$$\begin{aligned}\left(E_{\text{nearest pair}}\right)_{\nu=4/11} &= -\frac{8}{77}Z_H N_H, \quad \text{for } \nu = \frac{4}{11},\\ \left(\frac{E_{\text{nearest pair}}}{N_H}\right)_{\nu=4/11} &= -\frac{8}{77}Z_H, \quad \text{for } \nu = \frac{4}{11},\\ \left(\frac{E_{\text{nearest pair}}}{N}\right)_{\nu=4/11} &= -\frac{8}{11\times 4}Z_H = -\frac{2}{11}Z_H, \quad \text{for } \nu = \frac{4}{11}.\end{aligned} \tag{55}$$

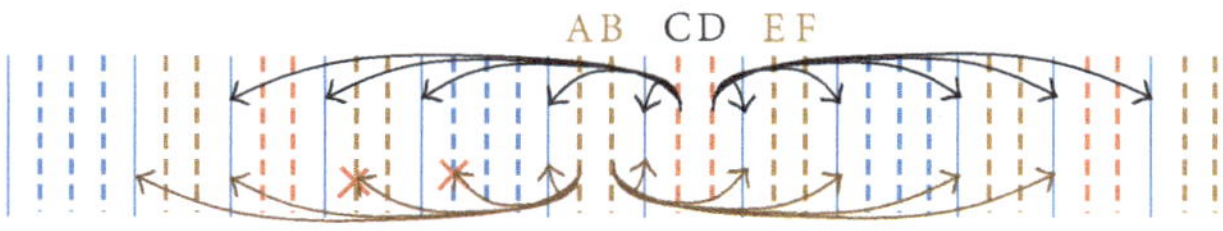

FIGURE 17: Coulomb transitions of nearest hole pair at $\nu = 4/13$.

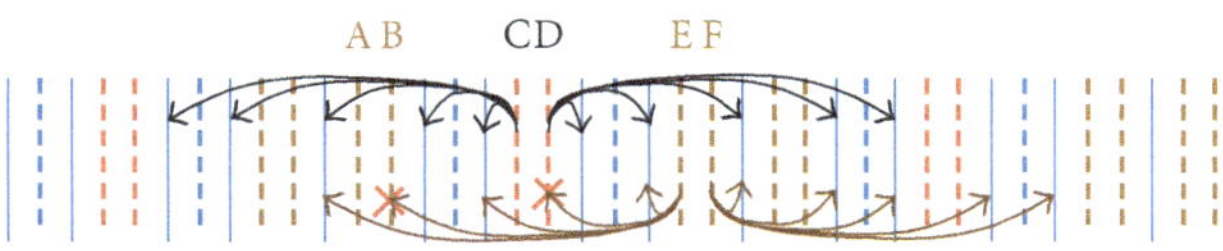

FIGURE 18: Coulomb transitions of nearest hole pair at $\nu = 5/13$.

If the configuration continues infinitely (it is a good approximation for macroscopic size of a device), then Figure 16 has the left-right symmetry for the centre of the hole pair CD. These pairs like CD are coloured red. All the red pairs can transfer to all the electron states. On the other hand, there is no left-right symmetry for the centre of the pair AB or EF. The pairs are coloured brown. They cannot transfer to some electron states because of momentum conservation. The perturbation energy of the nearest hole pairs has the energy gap at $\nu = 4/11$ as in the case of $\nu = 7/11$.

Similar configurations are schematically drawn in Figures 17, 18, 19 and 20 for $\nu = 4/13, 5/13, 5/17, 6/17$.

All the red hole pairs can transfer to all electron states as in Figures 15–20. This property is violated by infinitesimally small changing of the value ν. When the value of ν varies from the original values $\nu = 4/13, 5/13, 5/17$, and $6/17$, the number of allowed transitions becomes small. Therefore, these states with $\nu = 4/13, 5/13, 5/17, 6/17$ are stable. This property is in accordance with the experimental data as it is shown by green colour in Figure 12. (The fractions $\nu = 3/8$, 58, 3/10 have been studied in [17].)

6. Excitation Energy of FQHS

We have used the term "energy gap" for the gap in the spectrum of energy versus filling factor in the previous sections. This gap produces the plateau in the Hall resistance curve [18–23]. There is another gap which indicates the minimum value of all excitation energies from the ground state to excited states with the same filling factor ν. We call it "excitation-energy-gap." We examine the excitation-energy gap for three filling factors in this section.

We first consider one of excited states at $\nu = 2/3$ which is illustrated in Figure 21. The electron configuration is created by the excitation from the ground state as follows: one electron in the orbital B transfers to the orbital C. This excitation from the ground state is shown by green color in Figure 21. The electron configuration expresses one of eigen states for the Hamiltonian H_D. This excited state #1 has new nearest-electron pairs CE and DC which have the perturbation energies via residual Coulomb transitions. As in Figure 21, no transition is allowed from the pair CE, and only two transitions are allowed from the pair DC. Forbidden

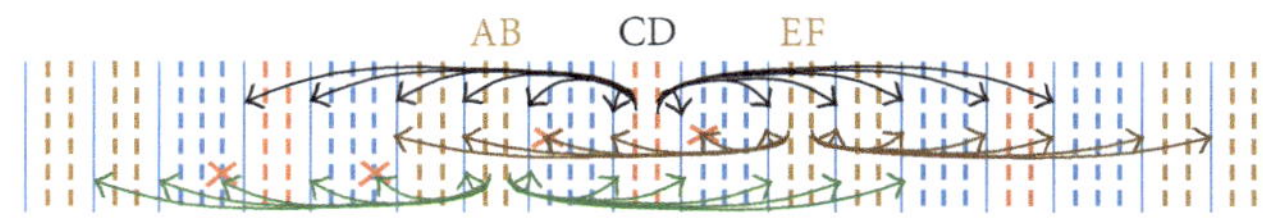

FIGURE 19: Coulomb transitions of nearest hole pair at $\nu = 5/17$.

FIGURE 20: Coulomb transitions of nearest hole pair at $\nu = 6/17$.

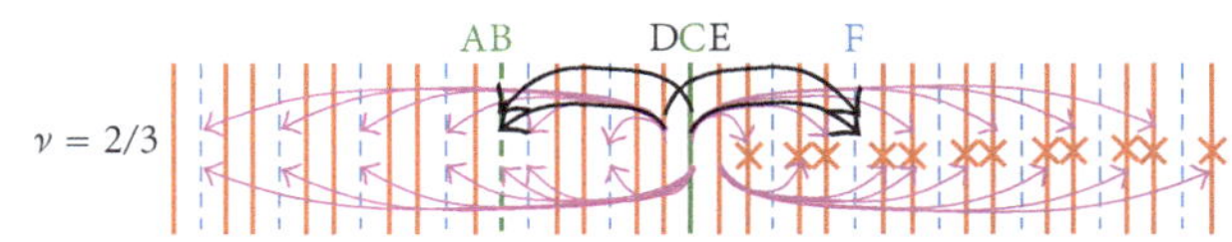

FIGURE 21: Electron configuration in excited state #1 at $\nu = 2/3$.

transitions are drawn by pink color, and allowed transitions are drawn by black color.

The perturbation energies of the pairs CE and DC are described by the symbols ς'_{CE} and ς'_{DC}, respectively. The symbol ς' (symbol prime) means the perturbation energy for the excited state #1. The second-order perturbation energy ς'_{DC} is obtained as

$$\varsigma'_{\mathrm{DC}} = \sum_{\Delta p = 6(2\pi\hbar/\ell), -7(2\pi\hbar/\ell)} \frac{\langle p_{\mathrm{D}}, p_{\mathrm{C}} | H_I | p'_{\mathrm{D}}, p'_{\mathrm{C}} \rangle \langle p'_{\mathrm{D}}, p'_{\mathrm{C}} | H_I | p_{\mathrm{D}}, p_{\mathrm{C}} \rangle}{W_{\mathrm{state\#1}} - W_{\mathrm{excite}}(p_{\mathrm{D}} \to p'_{\mathrm{D}}, p_{\mathrm{C}} \to p'_{\mathrm{C}})}. \quad (56)$$

We introduce the following summation Z' as:

$$Z' = -\sum_{\Delta p \neq 0, -2\pi\hbar/\ell}^{\mathrm{all}} \frac{\langle p_{\mathrm{D}}, p_{\mathrm{C}} | H_I | p'_{\mathrm{D}}, p'_{\mathrm{C}} \rangle \langle p'_{\mathrm{D}}, p'_{\mathrm{C}} | H_I | p_{\mathrm{D}}, p_{\mathrm{C}} \rangle}{W_{\mathrm{state\#1}} - W_{\mathrm{excite}}(p_{\mathrm{D}} \to p'_{\mathrm{D}}, p_{\mathrm{C}} \to p'_{\mathrm{C}})}. \quad (57)$$

This definition is similar to (19). We count the number of effective transitions which have the overlapping region between the initial Landau wave function and the final wave function. If the momentum transfer becomes lager than some critical value, then the transition matrix element becomes negligibly small.

The spreading width of Landau state is denoted by Δy, the value of which is about 10.5 nm for the case of $B \approx 6[\mathrm{T}]$

$$\hbar \approx 1.0546 \times 10^{-34}\,\mathrm{J\,s}, \qquad e \approx 1.6022 \times 10^{-19}\,\mathrm{C}, \qquad \Delta y \approx 10.5\ [\mathrm{nm}], \quad \text{for } B \approx 6\ [T]. \quad (58)$$

The interval width between nearest Landau orbitals is estimated as

$$\Delta\alpha = \frac{2\pi\hbar}{eB\ell} \approx 6.5 \times 10^{-4}\ [\mathrm{nm}], \quad \text{for } \ell = 1\ [\mathrm{mm}] = 10^{6}\ [\mathrm{nm}], \quad (59)$$

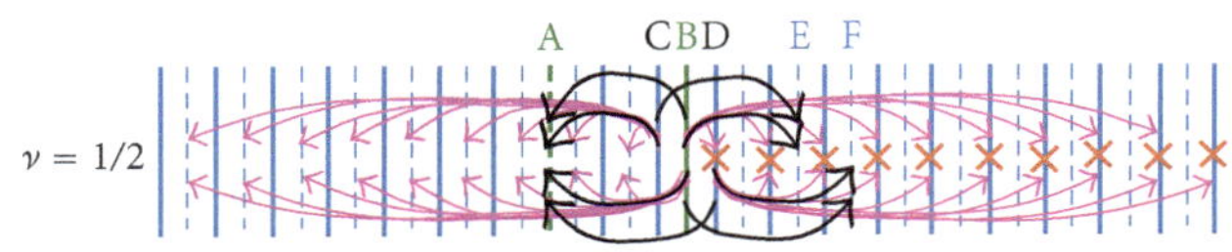

FIGURE 22: Electron configuration in excited state #2 at $\nu = 1/2$.

where we have applied $\ell = 1$ [mm] for the length of quantum Hall device. Then there are enormous many single-electron states inside the spreading width Δy. The number is

$$\frac{\Delta y}{\Delta \alpha} \approx \frac{\ell}{2\pi}\sqrt{\frac{eB}{\hbar}} \approx 2 \times 10^4. \quad (60)$$

Accordingly the effective momentum transfers satisfy the following relation:

$$\left| p'_{\mathrm{C}} - p_{\mathrm{C}} \right| \leq \frac{2\pi\hbar}{\ell} \times 2 \times 10^4. \quad (61)$$

Only two transitions are allowed from the pair DC, and therefore the ratio $-\varsigma'_{\mathrm{DC}}/Z'$ is about 10^{-4} as follows:

$$\frac{-\varsigma'_{\mathrm{DC}}}{Z'} \approx \frac{2}{(2 \times 10^4)} = 10^{-4}, \quad \text{for } \ell = 1 \text{ [mm]}. \quad (62)$$

Because this value is negligibly small, we can use the following approximation:

$$\varsigma'_{\mathrm{DC}} \approx -10^{-4} \times Z' \approx 0. \quad (63)$$

All the transitions from the pair CE are forbidden, and then the perturbation energy of the nearest pair CE is zero

$$\varsigma'_{\mathrm{CE}} = 0. \quad (64)$$

Furthermore, the nearest pair AB disappears by the excitation in Figure 21, and therefore the excitation-energy $\Delta E_{\text{excitation \#1}}$ is given by

$$\Delta E_{\text{excitation \#1}} = \varsigma'_{\mathrm{DC}} + \varsigma'_{\mathrm{CE}} - \varsigma_{\mathrm{AB}}, \quad (65)$$

where ς_{AB} indicates the summation (18) the result of which is equal to (20)

$$\varsigma_{\mathrm{AB}} = -\left(\frac{1}{3}\right)Z. \quad (66)$$

Substitution of (63), (64), and (66) into (65) yields the excitation energy as

$$\Delta E_{\text{excitation \#1}} = \varsigma'_{\mathrm{DC}} + \varsigma'_{\mathrm{CE}} - \varsigma_{\mathrm{AB}} \approx \left(\frac{1}{3}\right)Z. \quad (67)$$

It is noteworthy that the excitation energy is a positive value.

The second example is shown in Figure 22 where the electron A transfers to the orbital B at $\nu = 1/2$. The excited state #2 has two nearest electron pairs CB and BD. The excitation energy $\Delta E_{\text{excitation \#2}}$ from ground state to the state #2 is given by

$$\Delta E_{\text{excitation \#2}} = \varsigma''_{\mathrm{CB}} + \varsigma''_{\mathrm{BD}} \approx 0. \quad (68)$$

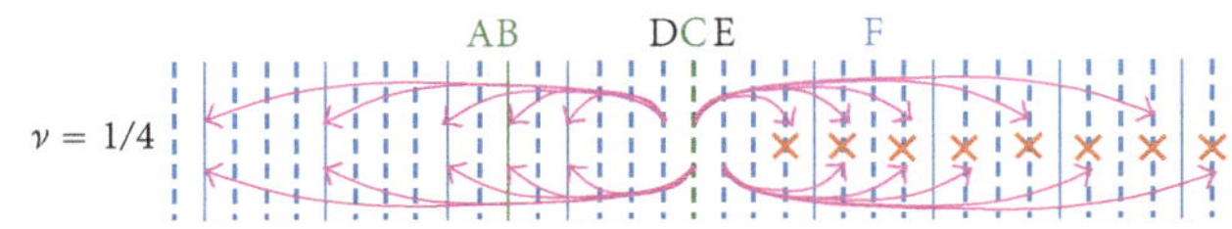

FIGURE 23: Electron configuration in excited state #3 at $\nu = 1/4$.

The third example is the case of $\nu = 1/4$. The hole B transfers to the orbital C and, then, the transition yields the excited state #3 as in Figure 23. Then the nearest hole pairs DC and CE are produced additionally. However, all the quantum transitions are forbidden from the hole pairs DC and CE. Accordingly the excitation energy $\Delta E_{\text{excitation \#3}}$ is zero as

$$\Delta E_{\text{excitation \#3}} = \varsigma'''^{\text{hole}}_{\mathrm{DC}} + \varsigma'''^{\text{hole}}_{\mathrm{CE}} - 2 \times \varsigma^{\text{hole}}_{\nu=1/4} = 0, \quad (69)$$

where $\varsigma^{\text{hole}}_{\nu=1/4}$ is the nearest hole pair energy. It is noteworthy that two nearest-hole pairs disappear by the excitation of the state #3. The value of $\varsigma^{\text{hole}}_{\nu=1/4}$ is zero derived from the estimation in Section 4.

Thus the excitation-energy gap has a correlation with the energy gap in the spectrum.

When the device size is very small, the ratio $\Delta y/\Delta \alpha$ becomes small. Therein, the energy gaps also become small for any fractional (not integer) filling factor. The size effect appears in a quantum Hall device with ultrasmall size (such as about 50 nm size for length ℓ).

7. Conclusion

We have developed the theory of Tao and Thouless. Then we have found the momentum conservation of the x direction for the Coulomb transitions. The momentum conservation law, Fermi-Dirac statistics, and the most uniform electron configuration produce gap structure at the filling factors $\nu = j/(2j \pm 1)$, $1/(2j + 1)$, $(2j)/(2j + 1)$ and gapless structure at the filling factors $\nu = 1/(2j), (2j - 1)/(2j)$. Furthermore small energy gaps are estimated for examples of $\nu = 7/11$ and $4/11$ by the use of computer program. Thus, we have applied the same procedure to FQH states with arbitrary fractional filling factors, and then we have obtained various types of energy spectra for nearest electron pairs and nearest hole pairs. It is important to use the same logic for investigation of all FQH states. This paper has explained FQH states with various filling factors by using usual quantum mechanics and usual electrons without any quasiparticle.

Acknowledgments

The author would like to acknowledge Professor H. Hori and Professor K. Oto for useful discussions. The author expresses his heartfelt appreciation for the comments of the referee and the editor. This paper has been improved and revised by the comments.

References

[1] W. Pan, H. L. Stormer, D. C. Tsui, L. N. Pfeiffer, K. W. Baldwin, and K. W. West, "Transition from an electron solid to the

sequence of fractional quantum Hall states at very low Landau level filling factor," *Physical Review Letters*, vol. 88, no. 17, Article ID 176802, pp. 1–4, 2002.
[2] W. Pan, H. L. Stormer, D. C. Tsui, L. N. Pfeiffer, K. W. Baldwin, and K. W. West, "Fractional quantum hall effect of composite fermions," *Physical Review Letters*, vol. 90, no. 1, Article ID 016801, pp. 1–4, 2003.
[3] J. K. Jain, "Composite-fermion approach for the fractional quantum Hall effect," *Physical Review Letters*, vol. 63, no. 2, pp. 199–202, 1989.
[4] J. K. Jain, "Theory of the fractional quantum Hall effect," *Physical Review B*, vol. 41, no. 11, pp. 7653–7665, 1990.
[5] A. Wójs and J. J. Quinn, "Quasiparticle interactions in fractional quantum Hall systems: justification of different hierarchy schemes," *Physical Review B*, vol. 61, no. 4, pp. 2846–2854, 2000.
[6] A. Wójs, K. S. Yi, and J. J. Quinn, "Fractional quantum Hall states of clustered composite fermions," *Physical Review B*, vol. 69, no. 20, Article ID 205322, 13 pages, 2004.
[7] J. H. Smet, "Wheels within wheels," *Nature*, vol. 422, no. 6930, pp. 391–392, 2003.
[8] M. R. Peterson and J. K. Jain, "Flavor altering excitations of composite fermions," *Physical Review Letters*, vol. 93, no. 4, Article ID 046402, 4 pages, 2004.
[9] E. A. Pashitskii, "New quantum states in the fractional quantum Hall effect regime," *Low Temperature Physics*, vol. 31, no. 2, pp. 171–178, 2005.
[10] R. Tao and D. J. Thouless, "Fractional quantization of Hall conductance," *Physical Review B*, vol. 28, no. 2, pp. 1142–1144, 1983.
[11] R. Tao, "Fractional quantization of hall conductance—II," *Physical Review B*, vol. 29, no. 2, pp. 636–644, 1984.
[12] S. Sasaki, "Energy gap in fractional quantum Hall effect," *Physica B*, vol. 281-282, pp. 838–839, 2000.
[13] S. Sasaki, "Binding energy and polarization of fractional quantum Hall state," in *Proceedings of the 25th International Conference on the Physics of Semiconductors*, pp. 925–926, Springer, 2011.
[14] S. Sasaki, "Spin polarization in fractional quantum Hall effect," in *Proceedings of the 21nd European Conference on Surface Science*, vol. 532–535, pp. 567–575, June 2003.
[15] S. Sasaki, "Spin-Peierls effect in spin-polarization of fractional quantum Hall states," in *Proceedings of the 22nd European Conference on Surface Science*, vol. 566–568, pp. 1040–1046, September 2004.
[16] S. Sasaki, *Surface Science: New Research*, chapter 4, Nova Science, 2006.
[17] S. Sasaki, "Calculation of binding energies for fractional quantum Hall states with even denominators," In press, http://arxiv.org/abs/cond-mat/0703360.
[18] S. Sasaki, "Energy gaps in fractional quantum Hall states," *Journal of Physics*, vol. 100, Article ID 042021, 4 pages, 2008.
[19] S. Sasaki, "Energy spectra for fractional quantum Hall states," In press, http://arxiv.org/abs/0708.1541.
[20] S. Sasaki, "Movement of diagonal resistivity in fractional quantum Hall effect via periodic modulation of magnetic field strength," *Journal of Physics*, vol. 100, Article ID 042022, 4 pages, 2008.
[21] S. Sasaki, "Frequency dependence of diagonal resistance in fractional quantum Hall effect via periodic modulation of magnetic field," In press, http://arxiv.org/abs/0803.0615.
[22] S. Sasaki, "Consideration of ac Josephson effect in fractional quantum Hall states," In press, http://arxiv.org/abs/0807.0288.
[23] S. Sasaki, "Tunneling effect in quantum Hall device," *Journal of Surface Science and Nanotechnology*, vol. 8, pp. 121–124, 2010.

15

Calculation of the Spontaneous Polarization and the Dielectric Constant as a Function of Temperature for KH_2PO_4

Hamit Yurtseven and Sema Şen

Department of Physics, Middle East Technical University, 06531 Ankara, Turkey

Correspondence should be addressed to Hamit Yurtseven, hamit@metu.edu.tr

Academic Editor: Biljana Stojanovic

The temperature dependence of the spontaneous polarization P is calculated in the ferroelectric phase ($T < T_C$) of KH_2PO_4 (KDP) at atmospheric pressure ($T_C = 122$ K). Also, the dielectric constant ε is calculated at various temperatures in the paraelectric phase ($T > T_C$) of KDP at atmospheric pressure. For this calculation of P and ε, by fitting the observed Raman frequencies of the soft mode, the microscopic parameters of the pure tunnelling model are obtained. In this model, the proton-lattice interaction is not considered and the collective proton mode is identified with the soft-mode response of the system. Our calculations show that the spontaneous polarization decreases continuously in the ferroelectric phase as approaching the transition temperature T_C. Also, the dielectric constant decreases with increasing temperature and it diverges in the vicinity of the transition temperature ($T_C = 122$ K) for KDP according to the Curie-Weiss law.

1. Introduction

Potassium dihydrogen phosphate (KH_2PO_4 or KDP) undergoes a phase transition at $T_C = 122$ K at atmospheric pressure [1]. In the paraelectric phase ($T > T_C$) it has the space group $I\bar{4}2d$ (D_{2d}^{12}), whereas in the ferroelectric phase ($T < T_C$) its space group is $Fdd2$ (C_{2v}^{19}). The paraelectric phase has the tetragonal structure with the four K-PO_4 groups in the unit cell and the tetrahedral PO_4 groups are connected by O–H$\cdots$O bonds. In this phase the protons are randomly distributed between the two equivalent minima of the double-well potential. Below T_C in the ferroelectric phase, the protons occupy in one minima of this double-well potential so that the transition occurs by this ordering of the protons. Thus, the spontaneous polarization, which grows as the temperature decreases below T_C in the ferroelectric phase, is parallel to the c-axis and it is mainly due to the motion of the protons in KDP. This occurs when the protons tunnel from one minima of the double-well potential to the other, which is considered in the pure tunnelling model [1–5]. However, it has been considered that the proton motion couples with an optic mode of the lattice in KDP, as studied by Kobayashi [6] in the coupled proton-optic-mode model. In the pure tunnelling model, it has been shown that the collective proton-proton tunnelling mode represents a soft mode [4], whereas in the coupled-mode model the soft mode is the lower frequency branch of the coupled system [6]. The soft mode has been observed experimentally in the Raman spectra of KDP [7–9]. At high pressures, the soft mode has also been observed experimentally in the ferroelectric phase of KDP [10]. The wavevector $\vec{q}$ dependence of the soft mode has been detected at the pressures above 40 kbar [11]. Recently, we have calculated the Raman frequency and the damping constant in KH_2PO_4 [12].

In this work, we calculate the spontaneous polarization from the observed Raman frequencies of the collective proton mode which behaves as a soft-mode response of the pure tunnelling model. By extrapolating the Raman frequencies of this mode to zero pressure, which were measured at various temperatures as a function of pressure [10] in the ferroelectric phase of KDP, the temperature dependence of the spontaneous polarization is calculated. This calculation is performed in the absence of the proton-lattice interaction for

KDP. We also calculate here the temperature dependence of the dielectric constant in the paraelectric phase of KDP using the pure tunnelling model. For this calculation, the proton-lattice interaction is also not considered and the critical behaviour of the dielectric constant is predicted close to T_C in the paraelectric phase of KDP.

Below, in Section 2, we give a theoretical outline for our calculations of the spontaneous polarization and the dielectric constant. In Section 3 we give our results. Sections 4 and 5 give our discussion and conclusions, respectively.

2. Theory

Using the microscopic theory, the mechanism of the phase transition in KDP has been explained mainly by the two models, namely, the tunnelling model of the proton and the coupled proton-optic-mode model. The ordering of the protons, which tunnels between the two minima of a double-well potential on the O–H· · · O bonds, as studied by Slater [1], causes collective proton excitations. This collective proton-proton tunnelling mode has been considered as a soft mode when the protons do not interact with the lattice [4]. Due to the fact that the spontaneous polarization (order parameter) does not have large values in this pure tunnelling model, the collective proton tunnelling mode interacts with the transverse optic mode (proton-lattice interaction) in the coupled proton-optic-mode model, as treated by Kobayashi [6]. In a previous study, the soft mode and coupled modes in the paraelectric and ferroelectric phases of KDP have been studied experimentally and the experimental results have been analyzed using both tunnelling and coupled proton-optic mode models [10]. In this study, we calculate the temperature dependence of the spontaneous polarization (order parameter) and the dielectric constant within the pure tunnelling model using the observed Raman frequencies [10] of the soft mode in KDP.

2.1. Tunnelling Model. In the tunnelling model, the proton can occupy two equivalent potential minima assigned by the pseudospin values of $\pm 1/2$. Protons can tunnel between the two minima and the energy levels of the double-well potential are separated by $\Delta E = 2\hbar\Omega$ where Ω is the tunnelling frequency.

The proton Hamiltonian H_p can be assumed to consist of a single-particle contribution and a contribution due to proton-proton interaction [3], expressed as

$$H_p = -2\Omega \sum_i S_i^x - \sum_{i,j} J_{ij} S_i^z S_j^z, \tag{1}$$

where S_i^α ($\alpha = x, y, z$) represents the α component of the pseudospin and J_{ij} is the interaction parameter between the nearest-neighbour protons. The proton Hamiltonian H_p can be solved in the mean-field approximation above the transition temperature T_C in terms of the collective proton excitation frequency Ω_0, as given in a previous study [10],

$$\Omega_0^2 = 2\Omega\left(2\Omega - J\langle S^0\rangle\right). \tag{2}$$

In (2) J is the Fourier transform of J_{ij} as the interaction parameter between the nearest-neighbour protons and $\langle S^0\rangle$ is defined as

$$\langle S^0\rangle = \frac{1}{2}\tanh\left(\frac{\Omega}{kT}\right). \tag{3}$$

As approaching the transition temperature, when there is no proton-lattice coupling, the collective excitation frequency can be assumed to disappear ($\Omega_0 \rightarrow 0$), which results in [10]

$$\frac{4\Omega}{J} = \tanh\left(\frac{\Omega}{kT_0}\right), \tag{4}$$

using (2) and (3) with the transition temperature $T \rightarrow T_0$.

According to the tunnelling model, the dielectric constant can be obtained as a function of temperature [13]

$$\varepsilon(T) = \varepsilon_L + \frac{4N\mu^2 \tanh(\Omega/kT)}{4\Omega - J\tanh(\Omega/kT)}, \tag{5}$$

where ε_L is the lattice dielectric constant independent of the temperature. The temperature dependence of the dielectric constant is governed by the second term in (5), which is the soft-mode contribution to $\varepsilon(T)$. In (5), N is the number of dipoles per unit volume and μ is the electric dipole moment.

The temperature dependence of the dielectric constant $\varepsilon(T)$ near the transition temperature T_0 in the paraelectric phase can be obtained by expanding (5) about T_0, which gives the Curie-Weiss law expressed as

$$\varepsilon(T) \cong \frac{C}{(T - T_0)}, \tag{6}$$

where the Curie constant is defined as

$$C = \frac{16N\mu^2 kT_0^2}{(J^2 - 16\Omega^2)}. \tag{7}$$

In relation to the dielectric constant $\varepsilon(T)$, the temperature dependence of the order parameter (spontaneous polarization) can also be obtained in the tunnelling model, as given in the previous study [10],

$$P = 2N\mu\langle S^z\rangle, \tag{8}$$

where S^z represents the pseudospins directed along the z-axis. Since the spontaneous polarization (order parameter) is defined in the ferroelectric phase ($T < T_C$) only, in the paraelectric phase ($T > T_C$) $P = 0$. At very low temperatures ($T << T_C$), the saturated spontaneous polarization P_S is obtained as

$$P_S = N\mu\left(1 - \frac{16\Omega^2}{J^2}\right)^{1/2}, \tag{9}$$

for the pure tunnelling model [10]. In order to obtain the temperature dependence of the spontaneous polarization (order parameter), we substitute (3) into (8) since near the transition temperature $\langle S^0\rangle$ can be replaced by $\langle S^z\rangle$ for the pseudospins orientated along the z-axis. This then gives the temperature dependence of the spontaneous polarization as

$$\frac{P}{N\mu} = \tanh\left(\frac{\Omega}{kT}\right). \quad (10)$$

The temperature dependence of the collective proton excitation frequency Ω_0 can also be obtained by identifying it with the soft-mode response frequency w_- [4, 10]. Experimentally, the pressure dependence of w_- at constant temperatures and its temperature dependence at 6.54 kbar have been observed for KDP [10]. On the assumption that there is no proton-lattice interaction, by substituting (3) and (4) into (2) the temperature dependence of Ω_0 or the soft-mode response frequency w_- can be expressed as

$$\Omega_0^2 = 4\Omega^2\left(1 - \frac{\tanh(\Omega/kT)}{\tanh(\Omega/kT_0)}\right), \quad (11)$$

where T_0 is the transition temperature, as before.

3. Calculations and Results

The temperature dependence of the spontaneous polarization and the dielectric constant for KDP were calculated here. For this calculation, the interaction parameter J and the tunnelling frequency Ω were determined by fitting (11) to the observed data for the soft-mode response frequency w_- at various temperatures at zero pressure [10]. The extrapolated values of the observed w_- for various temperatures at zero pressure as given in Table 1 for the ferroelectric phase of KDP [10], were used to determine the fitting parameters Ω and T_0 in (11). The fitting parameters Ω and T_0 determined are given in Table 2. In order to determine the interaction parameter J between the nearest-neighbour protons at the transition temperature, $\Omega = 54.5\,\text{cm}^{-1}$ and $T_C = 122\,\text{K}$ at zero pressure were used in (4) for KDP. The Curie constant was also determined using the J value at T_C and the Ω value according to (7) which is given as a ratio $C/N\mu^2$ in Table 2. Figure 1 gives the ratio P/P_S of the spontaneous polarization with respect to the saturated polarization which was calculated according to (9) and (10), where $\Omega = 54.5\,\text{cm}^{-1}$ was used, as a function of temperature below the transition temperature ($T_C = 122\,\text{K}$) in the ferroelectric phase of KDP. Finally, the dielectric constant $\varepsilon/N\mu^2$ was calculated as a function of temperature above the transition temperature ($T_C = 122\,\text{K}$) in the paraelectric phase according to (6), as plotted in Figure 2. Here, the saturation value of the spontaneous polarization was calculated as a ratio of $P_S/N\mu$ using the Ω and J values according to (9), as given in Table 2.

4. Discussion

Using the pure tunnelling model, the temperature dependence of the spontaneous polarization P in the ferroelectric phase (Figure 1) and of the dielectric function in the paraelectric phase (Figure 2) at zero pressure was calculated here. For this calculation, the observed [10] Raman frequencies of the soft-mode, which decrease with increasing temperature at zero pressure (Table 1), were used.

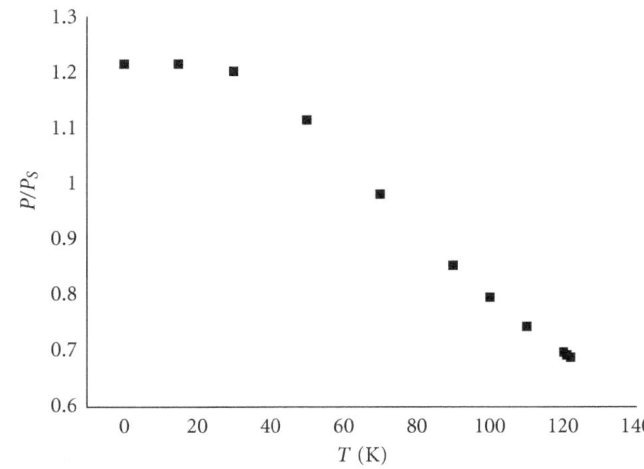

FIGURE 1: Ratio of the spontaneous polarization P/P_S with respect to the saturated polarization P_S which was calculated from (9) and (10) as a function of temperature at zero pressure in the ferroelectric phase ($T < T_C$) for KDP.

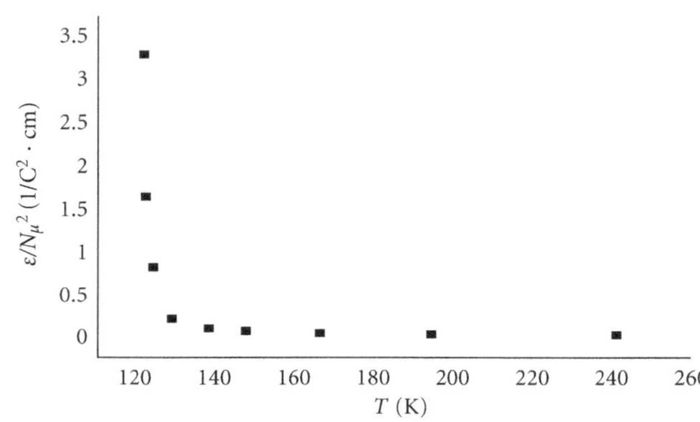

FIGURE 2: The dielectric constant ε calculated from (6) with the Curie constant C (7) as a function of temperature at zero pressure in the paraelectric phase ($T > T_C$) for KDP.

TABLE 1: Observed values of the soft-mode response frequency w_- extrapolated at zero pressure at various temperatures in the ferroelectric phase of KDP [10].

T (K)	w_- (cm^{-1})
45	150.1
60	148.9
80	145.3
90	142.7

The spontaneous polarization P decreases continuously as the temperature is increased towards the transition temperature ($T_C = 122\,\text{K}$) at zero pressure (Figure 1). This is the soft-mode behaviour where the frequency decreases abruptly as the temperature approaches T_C. In fact, it has been obtained that for the pure tunnelling model the

TABLE 2: Values of the transition temperature T_C, the microscopic parameters T_0 and the tunnelling frequency Ω, interaction parameter J (see (4)), the ratio of the Curie temperature C to $N\mu^2$ (see (7)), and the ratio of the saturated spontaneous polarization P_S to $N\mu$ (see (9)) for KDP.

T_C (K)	$-T_0$ (K)	Ω (cm^{-1})	J (cm^{-1})	$C/N\mu^2$ ($K/C^2 \cdot cm^{-1}$)	$P_S/N\mu$
122	1.35	54.5	384.6	1.65	0.824

soft mode frequency is directly related to the spontaneous polarization according to the relation

$$w = \left(\frac{J}{2\mu N}\right)P \tag{12}$$

in the ferroelectric phase of KDP [13, 14]. This indicates that the mechanism of the phase transition in KDP which is associated with the soft-mode frequency w_-, is due to the ordering of the protons in one minima of the double-well potential in this ferroelectric material.

The soft-mode response frequency w_- as it was first observed experimentally in KDP [7], is a heavily damped response centered at $w = 0$ in the Raman spectrum, as also pointed out previously [10]. The collective proton-proton tunnelling mode Ω_0 was taken here as the soft-mode response frequency w_- since Ω_0 displays a soft mode character in the absence of proton-lattice interactions [4, 10]. The temperature dependence of the soft mode frequency w_- can be described by

$$w_-^2 \alpha (T - T_-), \tag{13}$$

where T_- is different from the T_C in the presence of the optic-acoustic mode interaction [10]. It has been concluded [10] that the proton motion is strongly coupled to an optic mode of the lattice so that when the protons order, the proton-lattice interaction leads to the distortion of the K-PO_4 sublattices to produce the spontaneous polarization in the modified tunnelling model or the coupled proton-optic mode model [6]. In regard to the proton-proton tunneling model, (13) can also be used for the proton-proton tunnelling model in the case that the proton tunnelling frequency Ω is taken as the temperature dependent.

The soft-mode response also displays a temperature dependence of the Raman frequency w_- near the transition at $T_- \approx 86$ K at a constant pressure of 6.54 kbar ($T_C \approx 91$ K), as observed experimentally in KDP [10]. It has been indicated that the difference between T_- and T_C is primarily due to the optic-acoustic-mode interaction in KDP. It has also been indicated that this transition occurs at ~86 K when w_- vanishes for the coupled proton-optic-mode model, whereas for the pure tunnelling model in the absence of proton-lattice interactions the proton system undergoes a transition at ~58 K [10]. So, the temperature difference of nearly 30 K emerges due to proton-lattice interactions at 6.54 kbar in KDP. At atmospheric pressure, our fitting parameter of T_0 (Table 2) is also about 30 K lower than T_0 ~ 30 K determined previously [9]. In fact, the microscopic parameter T_0, which we obtained by fitting (11) to the observed Raman frequencies of the soft mode w_- (or collective proton excitation frequency Ω_0), was not well determined since a large ratio of $\Omega/kT_0 > 1$ was obtained in our analysis. This gives rise to $\tanh(\Omega/kT_0) \sim 1$ which means that the fit is relatively insensitive to T_0, as also pointed out previously for the analysis of KDP at 6.54 kbar [10]. So, the macroscopic properties of the ferroelectric materials are not well described by the microscopic parameters of the pure tunnelling model and also of the coupled proton-optic mode model, which both use the Raman data [10]. On this basis, in (4) we used the value of $T_C = 122$ K instead of the fitted parameter T_0 in the pure tunnelling model, which gave the ratio of $\Omega/kT_C = 0.64$ or $\tanh(\Omega/kT_C) = 0.57$ for KDP. Thus, the interaction parameter was determined as $J = 384.6$ cm^{-1} at zero pressure in this study, which can be compared with the value of 345.1 cm^{-1} at 6.54 kbar [10].

The temperature dependence of the spontaneous polarization was calculated here in the absence of the proton-lattice interactions for KDP, as stated previously. Since the spontaneous polarization does not have large values in the pure tunnelling model, the collective proton tunnelling mode can interact with the transverse optic mode (proton-lattice interaction) in the coupled proton-optic mode model introduced by Kobayashi [6], as pointed out previously [10]. The amplitude of the soft-optic mode is maximum at the phase transition temperature and there is no coupling at T_0. Below T_0, proton-lattice interaction can be considered to enhance the temperature dependence of the spontaneous polarization in KDP. However, the coupled mode features can disappear below T_0 in the ferroelectric phase in KDP, was also observed experimentally at approximately 30 K below T_0 in $BaTiO_3$ [15]. Thus, in the absence of the proton-lattice interaction, for the temperature dependence of the soft-mode frequency w_- (or collective proton excitation frequency Ω_0) can be described below T_0 in the ferroelectric phase of KDP.

In regard to the collective proton excitation frequency Ω_0 which was taken as the soft mode response frequency w_- for KDP as studied here, Ω_0 can also be considered as a soft mode for those ferroelectrics exhibiting displacive phase transition such as $BaTiO_3$. Soft mode in $BaTiO_3$ which consists of the vibration of Ti ions against oxygen [16] is the transverse optic mode of long wavelength at $k = 0$. Its frequency goes to zero ($w_T \rightarrow 0$) at the critical temperature T_0 and the lattice displacements associated with this mode cause the crystal distorting to the ferroelectric phase [17]. Very recently, we have investigated the temperature dependence of this soft-optic mode in ferroelectric barium titanate [18].

Using the pure tunnelling model, we also calculated the temperature dependence of the dielectric constant in the paraelectric phase of KDP, as shown in Figure 2. It decreases as the temperature increases below T_C and it diverges in the vicinity of the transition temperature

($T_C = 122\,\text{K}$) according to Curie-Weiss law (see (6)). This diverging behaviour of the dielectric constant is due to the soft-mode response frequency w_- or the collective proton-tunnelling frequency Ω_0 which vanishes as $T \rightarrow T_C$ in the pure tunnelling model of KDP. Thus, the temperature or pressure dependence of the soft-mode response frequency from the Raman spectra determines the temperature or pressure dependence of the dielectric properties of those ferroelectric materials such as KDP. Also, from the pressure dependence of the transition temperature T_C, the phase diagram calculated using proton- or-deuteron-tunnelling model [19] can be tested by the experimental Raman spectra for KDP and DKDP. Alternatively, a temperature-uniaxial pressure phase diagram of KH_2PO_4 which was obtained experimentally [20] can be calculated using the proton tunnelling model.

5. Conclusions

The spontaneous polarization P and the dielectric constant ε were calculated at various temperatures in the ferroelectric ($T < T_C$) and paraelectric ($T > T_C$) phases, respectively, at atmospheric pressure in KDP. The observed soft-mode frequencies from the Raman data were used to evaluate the temperature dependence of P and ε by the pure tunnelling model close to the transition temperature ($T_C = 122\,\text{K}$) in KDP.

The critical behaviour of the spontaneous polarization P and the dielectric constant ε was predicted in both the ferroelectric and paraelectric phases and they can be compared with the experimental measurements. The critical behaviour of the soft-mode response frequency w_-, spontaneous polarization P, and the dielectric constant ε can also be investigated at higher pressures in KDP.

References

[1] J. C. Slater, "Theory of the transition in KH_2PO_4," *The Journal of Chemical Physics*, vol. 9, no. 1, pp. 16–33, 1941.

[2] R. Blinc, "On the isotopic effects in the ferroelectric behaviour of crystals with short hydrogen bonds," *Journal of Physics and Chemistry of Solids*, vol. 13, no. 3-4, pp. 204–211, 1960.

[3] P. G. de Gennes, "Collective motions of hydrogen bonds," *Solid State Communications*, vol. 1, no. 6, pp. 132–137, 1963.

[4] R. Brout, K. A. Müller, and H. Thomas, "Tunnelling and collective excitations in a microscopic model of ferroelectricity," *Solid State Communications*, vol. 4, no. 10, pp. 507–510, 1966.

[5] M. Tokunaga and T. Matsubara, "Theory of ferroelectric phase transition in KH_2PO_4 type crystals I," *Progress of Theoretical Physics*, vol. 35, pp. 581–599, 1966.

[6] K. K. Kobayashi, "Dynamical theory of the phase transition in KH_2PO_4-type ferroelectric crystals," *Journal of the Physical Society of Japan*, vol. 24, no. 3, pp. 497–508, 1968.

[7] I. P. Kaminov and T. C. Damen, "Temperature dependence of the ferroelectric mode in KH_2PO_4," *Physical Review Letters*, vol. 20, no. 20, pp. 1105–1108, 1968.

[8] J. F. Scott and C. M. Wilson, "Dynamical theory of the phase transition in KH_2PO_4-type ferroelectric crystals," *Solid State Communications*, vol. 10, pp. 597–600, 1972.

[9] C. Y. She, T. W. Broberg, L. S. Wall, and D. F. Edwards, "Effect of proton-phonon coupling on the ferroelectric mode in KH_2PO_4," *Physical Review B*, vol. 6, no. 5, pp. 1847–1850, 1972.

[10] P. S. Peercy, "Measurement of the "soft" mode and coupled modes in the paraelectric and ferroelectric phases of KH_2PO_4 at high pressure," *Physical Review B*, vol. 12, no. 7, pp. 2725–2740, 1975.

[11] T. Yagi, A. Sakai, and M. Arima, "Raman scattering study of KDP under high pressure with a diamond anvill cell," *Phase Transitions*, vol. 28, pp. 63–78, 1990.

[12] H. Karacali, A. Kiraci, and H. Yurtseven, "Calculation of the raman frequency and the damping constant of a coupled mode in the ferroelectric and paraelectric phases in KH_2PO_4," *Physica Status Solidi B*, vol. 247, no. 4, pp. 927–936, 2010.

[13] R. Blinc and B. Žekš, "Dynamics of order-disorder-type ferroelectrics and anti-ferroelectrics," *Advances in Physics*, vol. 21, pp. 693–757, 1972.

[14] R. Blinc, B. Lavrencic, I. Levstek, V. Smolej, and B. Zeks, "Underdamped soft tunnelling modes in hydrogen-bonded ferroelectrics below T_C," *Physica Status Solidi B*, vol. 60, no. 1, pp. 255–259, 1973.

[15] M. Yamaguchi, M. Watanabe, K. Inoue, Y. Akishige, and T. Yagi, "Light scattering study of the coupled soft-optic and acoustic mode in hexagonal barium titanate," *Physical Review Letters*, vol. 75, no. 7, pp. 1399–1402, 1995.

[16] Y. Luspin, J. L. Servoin, and F. Gervais, "Soft mode spectroscopy in barium titanate," *Journal of Physics C*, vol. 13, no. 19, pp. 3761–3773, 1980.

[17] G. A. Samara, "Pressure and temperature dependences of the dielectric properties of the perovskites $BaTiO_3$ and $SrTiO_3$," *Physical Review*, vol. 151, no. 2, pp. 378–386, 1966.

[18] A. Kiraci and H. Yurtseven, "Temperature dependence of the Raman frequency, damping constant and the activation 3 energy of a soft-optic mode in ferro-electric barium titanate," *Ferroelectrics*, vol. 432, no. 1, pp. 14–21, 2012.

[19] H. Mashiyama, "Proton potential and a tunneling model for KDP," *Journal of the Korean Physical Society*, vol. 46, no. 1, pp. 63–68, 2005.

[20] F. E. A. Melo, S. G. C. Moreira, A. S. Chaves, I. Guedes, P. T. C. Freire, and J. Mendes-Filho, "Temperature-uniaxial pressure phase diagram of KH_2PO_4," *Physical Review B*, vol. 59, no. 5, pp. 3276–3279, 1999.

Ge/Si Quantum Dots Superlattices Grown at Different Temperatures and Characterized by Raman Spectroscopy and Capacitance Measurements

A. D. Rodrigues,[1] A. J. Chiquito,[1] G. Zanelatto,[1] A. G. Milekhin,[2] A. I. Nikiforov,[2] V. V. Ulyanov,[2] O. P. Pchelyakov,[2] D. R. T. Zahn,[3] and J. C. Galzerani[1]

[1] *Department of Physics, Federal University of São Carlos, CP 676, 13565-905 São Carlos, SP, Brazil*
[2] *Institute of Semiconductor Physics, Novosibirsk 630090, Russia*
[3] *Institut für Physik, Technische Universität Chemnitz, 09107 Chemnitz, Germany*

Correspondence should be addressed to A. D. Rodrigues, ariano@df.ufscar.br

Academic Editor: Michael C. Tringides

Ge/Si heterostructures with Ge self-assembled quantum dots (SAQDs) grown at various temperatures by molecular beam epitaxy were investigated using resonant Raman spectroscopy and capacitance measurements. The occurrence of quantum confinement effects was confirmed by both techniques. For the structures grown at low temperatures (300 – 400°C), the SAQDs optical phonon wavenumbers decrease as the Raman excitation energy is increased; this is an evidence of the scattering sensitivity to the size of the SAQDs and to the inhomogeneity in their sizes. However, the opposite behavior is observed for the SAQDs grown at higher temperatures, as a consequence of the competition between the phonon localization and internal mechanical stress effects. The E_1 electronic transition of the Ge in the SAQDs was found to be shifted towards higher energies as compared to bulk Ge, due to biaxial compressive stress and to the electronic confinement effect present in the structures. The intermixing of Si atoms in the quantum dots was found to be much more significant for the sample grown at higher temperatures. The capacitance measurements, besides confirming the existence of the dots in these structures, showed that the deepest Ge layers lose their 0D signature as the growth temperature increases.

1. Introduction

Undoubtedly, Ge/Si self-assembled quantum dots (SAQDs) present an interesting technological potential, confirmed by several studies pointing to the development of devices such as photodetectors operating in the midinfrared [1], transistors [2], memory devices [3], nonlinear optic devices [4], and lasers [5], based in these structures. It is obvious that for the production of these devices it is necessary that the growth techniques are fully controlled and that the characteristics of the obtained material are very well known. In comparison with single-layered structures, the quantum dots superlattices present superior potential for optoelectronic applications due to the larger volumetric density of islands, the modified confining potential, and the suppression of surface recombinations. Recently electroluminescence in frequencies of order of THz was observed in Si/Ge superlattices [6]. Nevertheless, in system in which the Ge islands are embedded into a Si matrix, besides the nominal thickness of the Ge layers, the growth temperature strongly influences both the size and the chemical composition of islands, due to SiGe intermixing [7–11]. The growth dynamics in a multilayer structure can be fundamentally different as compared to the monolayer case. During the process of preparation, the sample temperature is successively varied in order to establish ideal conditions for the growth of either Si or Ge layers. So, the deeper the Ge layer is, the most prominent can be the effect of intermixing, that directly induces modifications in the electronic energy spectrum. In this work we have combined Raman spectroscopy and capacitance measurements in order to study the compositional structure and the growth dynamics of Ge/Si SAQDs superlattices

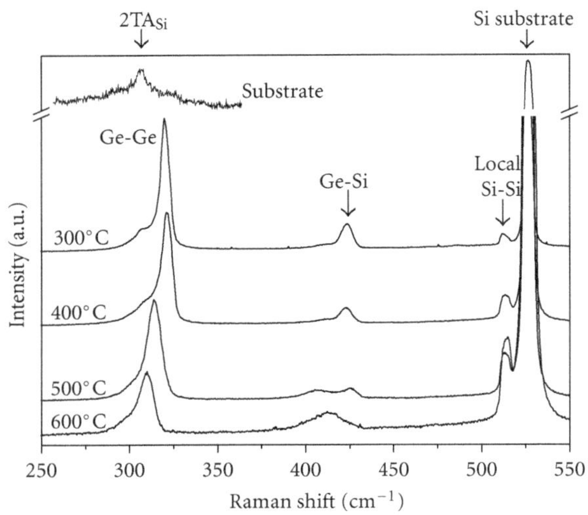

FIGURE 1: Raman spectra of the samples grown at several temperatures, measured at 8 K and excited with the 514.5 nm laser line.

grown at different temperatures. By obtaining the resonant Raman scattering profiles, we could draw some conclusions about the electronic properties of the samples as a function of growth temperatures. Furthermore, since the vibrational properties play an important role both in the charge and in the heat transport, the knowledge of the Raman spectra of these nanostructures is fundamental for their applications in devices.

2. Experimental

The samples were grown by Molecular Beam Epitaxy (MBE) following the Stranski-Krastanov regime, on (001) Si substrates. The set of four samples consists of alternate Si and Ge layers with nominal thicknesses 500 Å and 10 monolayers, respectively, repeated five times, ending with Si. The growth temperature was varied, assuming the values 300, 400, 500, and 600°C. An investigation using cross-sectional high-resolution transmission electronic microscopy (HRTEM) of samples similarly prepared [12] shows that, for the here used growth parameters, SAQDs are expected with dimensions that increase with the growth temperature: the average base sizes vary between 10 and 100 nm and the dot heights vary from 2 to 6 nm. Resonant Raman scattering was accomplished with the excitation given by two sources: a Rhodamine 6G dye laser and an Ar/Kr laser. For all measurements the power density was kept below 2×10^5 W/m^2 in order to prevent sample heating. A triple monochromator T64000 equipped with a liquid nitrogen cooled CCD detector was then used for the analysis of the scattered light. The spectra were taken at the temperature 8 K.

For the capacitance experiments, an $Au_{0.5}In_{0.5}$ alloy was used as the ohmic contact (annealed at 450°C for 120 s) and Au was evaporated in order to obtain the Schottky one (200 nm thick layer with 500 μm diameter). The measurements were realized with conventional AC techniques, the signal detection being obtained with a lock-in amplifier. These measurements were taken at 300 and 150 K, with 10 mV and 100 kHz as the excitation voltage and excitation frequency, respectively.

3. Results and Discussions

We present the unpolarized Raman spectra of the studied samples, grown at different temperatures, in Figure 1. All the spectra exhibit strong peaks in the interval 310–320 cm^{-1}. Another feature around 420 cm^{-1}, together with a shoulder in the low-energy region, could be observed for the samples grown at 300, 400, and 600°C; in the case of the sample grown at 500°C the presence of two peaks is evident in this region of energies. The lower-energy peaks present in our spectra can be attributed to the Ge-Ge vibration. In bulk materials this mode appears at 304 cm^{-1}; for our samples however this peak was found to be blue shifted. Together with them it can be observed low-energy shoulders corresponding to the second-order 2TA$_{Si}$. The spectrum of the substrate in this interval (shown for comparison in the top of the figure) confirms this assertion. The features around 420 cm^{-1} can be associated to the optical vibrations of the Ge-Si alloy present in the grown structures [13]. The frequencies of the optical vibrational modes of both Ge-Ge and Ge-Si alloy depend on two concurrent effects: the interdiffusion strength of the Si atoms through the Ge layers—provoking blue shifts—and the confinement of the optical phonon inside the SAQDs—causing a red shift due to the negative phonon dispersion relation. The resonant Raman scattering, detailed below will help us to drive some conclusions about the competition between these effects. Finally, we can notice around 525 cm^{-1} the Si bulk phonon, originated at the (001) substrate and around 512 cm^{-1}, a feature with lower intensity, that can be attributed to a local Si-Si mode [13]. Figure 2 presents the profiles of the resonant Raman scattering of the Ge-Ge vibration for the four samples. For each spectrum the Ge-Ge signal was normalized with respect to the Si signal of the substrate and plotted as a function of the excitation energy. The most intense maxima appearing in the four curves are attributed to the E_1 electronic transition of Ge. At room temperature, this transition for bulk Ge occurs at 2.23 eV [14]. However, both the electronic confinement and the compressive stress of the islands can contribute to the blue shifting in E_1 energy. The observed values of the E_1 transition agree with the data obtained by Kwok et al. [15]. In the case of the samples grown at 400, 500, and 600°C, an additional increase in the Ge-Ge signal intensity is observed, for excitation energies 2.54, 2.52, and 2.68 eV, respectively. We believe that the electronic transitions of the Ge_xSi_{1-x} alloy, present in the samples, are responsible for these features. As a matter of fact, Cerdeira et al. [16] reported resonances in this energy range for particular stoichiometries of the referred alloy. In principle, the electronic transitions of the GeSi alloy are not expected to interfere in the resonance of the Ge-Ge optical modes. However, considering that these modes are not rigorously confined in the regions of pure Ge of the islands (in other

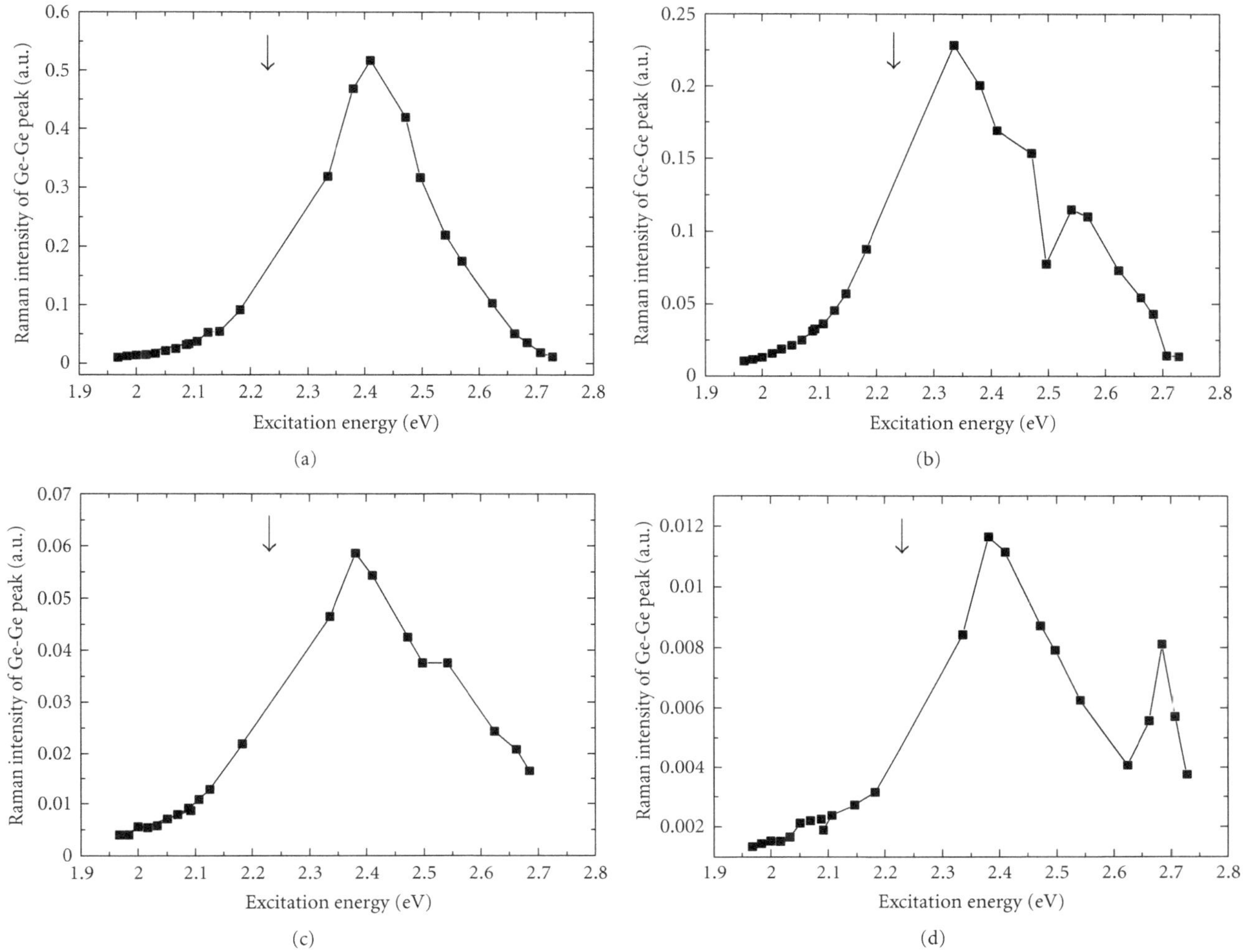

FIGURE 2: The intensity of the Ge-Ge signal as a function of the excitation energy for the samples grown at (a) 300°C, (b) 400°C, (c) 500°C, and (d) 600°C. The arrows show the bulk Ge resonance.

words, the vibration relative to these phonons can extend as far as the first atomic layers of the GeSi alloy), it is quite reasonable to suppose that the specific Ge vibrations have enough amplitudes in the alloy in order to couple with its electrons and consequently contribute to the resonance curve of the Ge optical mode, a "signature" of the electronic states of the alloy. Influences of the GeSi alloy in the intensity of the Ge-Ge mode were also observed in Ge/Si superlattices [17], where the electronic transition was attributed to the E_1 gap of a $Ge_{0.5}Si_{0.5}$ alloy. The sample grown at 300°C does not show this transition, as observed from Figure 2(a). This fact indicates a negligible Si content in the islands, due to the low diffusion rate of Si towards the Ge layers. Further referring to Figure 2, it is also quite remarkable the decrease in the intensities of the resonance curves as the growth temperature increases. This can be associated to the density of the quantum dots in the samples. Since the monitored Ge-Ge signal originates mainly from the SAQDs, the larger the number of dots per area is, the stronger the phonon signal should be and consequently the resonance curve intensity. So we can definitely affirm that the SAQDs density decreases as the growth temperature increases; at 600°C we are probably close to a two-dimensional GeSi layer, as confirmed below by the C-V measurements. Figure 3 shows the optical phonon wavenumber of the Ge SAQDs as a function of the excitation energy. Here again we should be careful in the analysis of the results, since we are dealing with the competition between two phenomena: for the SAQDs of smaller sizes the optical phonons confinement effect would contribute to the frequency of the Ge-Ge vibrational mode, while for the SAQDs of larger sizes the confinement effect could be neglected and the intrinsic stress will prevail in the frequency shift determination. Raman scattering of a size distribution of SAQDs takes place in a highly selective way. Since the quantum confinement is stronger (and, consequently, the electronic transitions are more energetic) in the smallest dots, the higher the excitation energy is, the stronger the contribution of these structures to the Raman spectrum is (which is than dominated by them). For the same reason, for the lowest excitation energies, almost the totality of

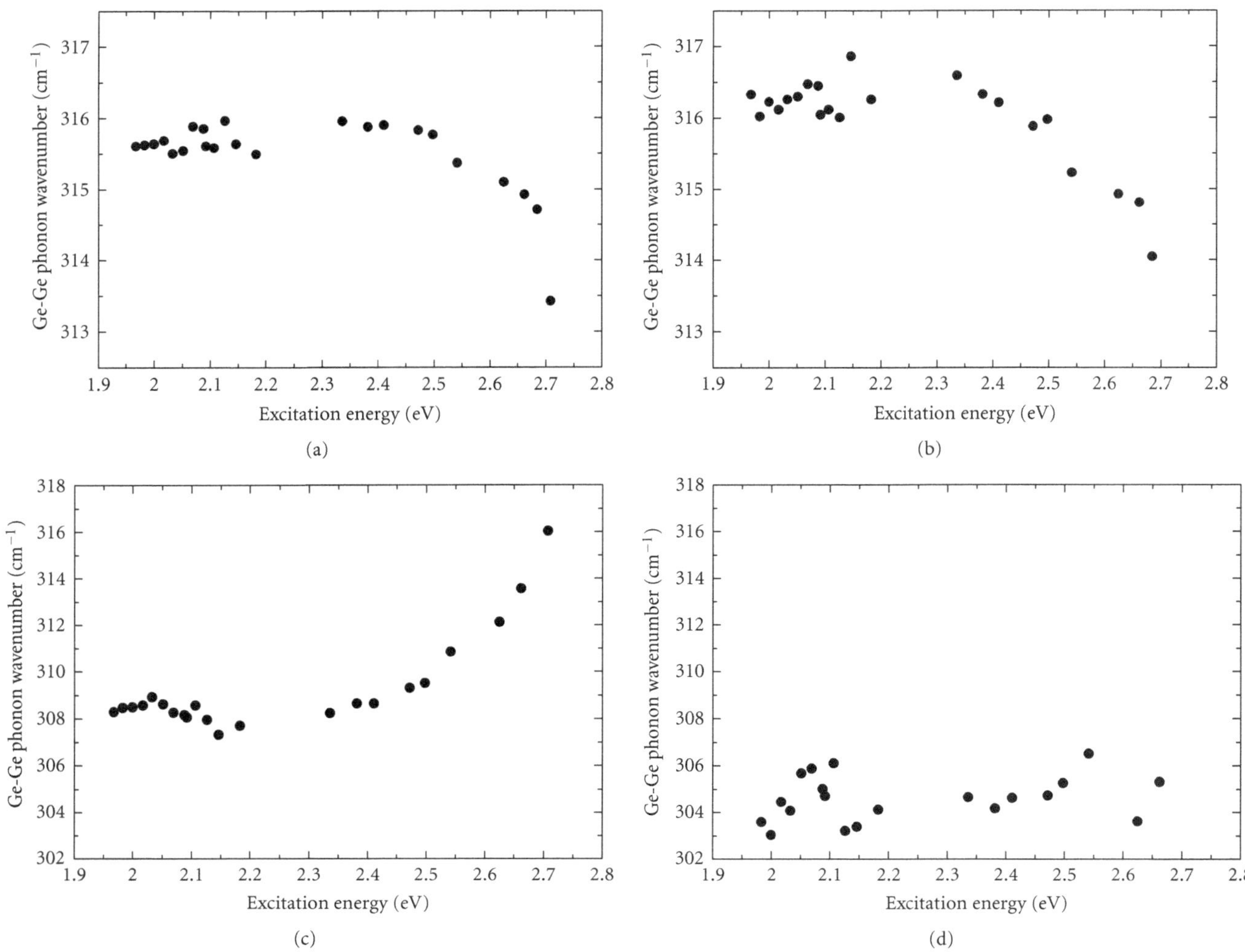

FIGURE 3: The quantum dot optical phonon wavenumbers as a function of the excitation energy for the four samples grown at (a) 300°C, (b) 400°C, (c) 500°C, and (d) 600°C.

the spectrum intensity will be originated from the SAQDs of larger dimensions, because these energies will be insufficient to excite electronic states with higher energies, like those of the smallest dots. Besides, the larger the dimension of the quantum dot is, the less important the confinement effect of its optical phonon will be. As a consequence, in a sample of this kind, the tuning of the excitation energy would select the dimensions of the quantum dots that more effectively contribute to the Raman scattering. If we refer to the sample grown at 300°C (Figure 3(a)), we observe that a change in the energy of the optical phonon occurs for excitation energies higher than 2.5 eV. The shift to lower wavenumbers—due to the confinement effect—is typical of materials presenting a negative dispersion of phonons, such as Ge; as expected, the confinement progressively increases with the photon energy, clearly indicating the existence of a size distribution of the Ge SAQDs for this growth temperature. Considering now the sample grown at 400°C (Figure 3(b)), for the energy range (1.95–2.35 eV) in which the confinement effect is not relevant, the phonon wavenumber is slightly higher as compared to the former sample indicating that here the quantum dots dimensions have increased. The curve decays in a more smooth way, showing that the size distribution is now more heterogeneous. The sample grown at 500°C presents an interesting behavior, opposite to the first two ones. We believe that the SAQDs are here distributed in two distinct regimes, as observed before by Milekhin et al. [18] for samples where a thin SiO_2 layer was grown over the Si substrates. In the first regime, the three-dimensional islands have much larger dimensions than those found in the first two samples, such that here the compressive stress is lower, as observed before for this kind of sample [12]; being so, for the lower excitation energies utilized, where these SAQDs are responsible for the scattering, the optical phonon wavenumber is quite near those expected for the bulk Ge. We can say that these islands have lost their zero-dimension signature, although the three-dimensional growth may still occur. In the second regime, the SAQDs' dimensions are comparable to those observed in the first two samples. Consequently, they conserve part of the wetting layer compressive stress, while being subjected to the phonon confinement effect. It is observed that, as the excitation energy increases, approximating the characteristic electronic transitions of these smaller SAQDs, such structures are the major responsible for the scattering, such that the optical phonon wavenumbers are quite similar to those measured in the former samples when excited with lower energies. This is an evidence that the smallest SAQDs in the sample prepared at 500°C would present dimensions close to those found in the larger islands grown at lower temperatures (300 and 400°C). The existence of two distinct regimes in the size distribution of the SAQDs can still be observed in another interval of the Raman spectrum of the sample grown at 500°C. In the region around 420 cm^{-1}, it is evident the presence of two separate peaks related to optical vibrations in the GeSi alloy. It is known that the phonon energies in binary alloys are influenced by the amount of each chemical element forming the compound. Ratto et al. [11] demonstrated that the alloying process is strongly dependent on the areas of the islands' bases, that is, the larger the SAQDs dimensions are, the more prominent the Si intermixing through the Ge quantum dots will be. Being so, we expect that the alloying process is more effective in the SAQDs of the first regime (thus producing a Si richer alloy), in comparison to those of the smaller islands. This originates Raman peaks with different energies: a more energetic one, related to the Si richer alloy present in the largest SAQDs, and a lower energetic one, due to the scattering in the smallest islands (in which the alloying process is less efficient). Considering now Figure 3(d), we observe that the Ge optical mode shows no relevant variation with the excitation energy indicating that the highest growth temperature possibly gave origin to very large three-dimensional GeSi islands. In this case the Ge content in the GeSi SAQDs is about 53% [12]. As a matter of fact, we should recall that for this sample we observed the alloy E_1 resonance at 2.68 eV (Figure 2(d)); it should also be observed that, in [17], in the case of the $Ge_{0.5}Si_{0.5}$ solid solution, it was claimed that the E_1 resonance appears at 2.7 eV. As a consequence of the light attenuation through the sample and since we are dealing with a superlattice with five periods, it is expected a major contribution of shallower layers to the information taken from the analyses of Raman spectra. In order to characterize the morphology of deeper layers, we have performed C-V measurements. Under zero field, the depletion layer is around 2000 Å thick. As a result, through this technique, we could access specifically the deepest two Ge layers. Figure 4 depicts the C-V measurements for the samples grown at 300 and 400°C, respectively. In these heterostructures, the capacitance is caused by two distinct contributions: a three-dimensional capacitance due to the electronic distribution in the volume around the quantum region and the capacitance due to the quantum confined electrons [19]. The contribution of the quantum dots to the total capacitance is a direct consequence of their characteristic density of states and their occupation at a given temperature. Taking into account that the samples were undoped the expected electron density at the SAQDs were very low: using the well-known Schottky capacitance theory [20, 21], the electron density was found to be 6.0×10^{12} cm^{-3} (sample grown at 300°C) and 4.0×10^{12} cm^{-3} (sample grown at 400°C). As previously published [22–24], the density of SAQDs can be estimated from the C-V curves. Using the model described in [18], the density of SAQDs was found to be almost the same ($\sim 10^{12}$ cm^{-2}) for both samples. This result is in accordance with the Raman analysis presented in Figures 2(a) and 2(b), from where it can be noticed that relative intensities of Ge signal have the same order for both samples. Decreasing the temperature, the localization effects and consequently the presence of SAQDs can be readily observed: additional peaks (indicated by arrows) are observed in the capacitance signal because the dots are being filled and emptied following the applied voltage. Due to the low temperature, the electrons remain in the dots and their effects are now observed. In fact, from Figure 4(a), two peaks are observed at 150 K revealing the peaked character of the zero-dimensional density of states, while at room temperature only one-layer occupation is observed. Both samples (grown at 400°C and 300°C) exhibit the same behavior. Our results are a clear indication of the presence of the quantum dots in the deepest layers of these samples. For the samples grown at the two highest temperatures, no clear signs of quantization could be observed with this technique.

4. Conclusions

The Raman spectra obtained with excitation energies near the resonance showed that the Ge optical phonons related to the quantum dots attain different wavenumbers as compared to the Ge bulk case. These wavenumber shifts are generated by the competition between two physical phenomena present in the structures: the strongest one is undoubtedly the compressive stress inside the SAQDs, but the phonon confinement effect is also relevant. The increase of the growth temperature implies in the nonhomogeneity of the vibrational modes, both due to the presence of large SAQDs—where the stress relaxation is more considerable—and to the increase in the rate and depth of the Si diffusion along the growth direction.

The measurements with the variation of the excitation energy showed that the samples grown at 300, 400, and 500°C present a size distribution of the SAQDs. However, it is clearly observed that this distribution is more heterogeneous in the second than in the first sample. Also, for the growth made at 500°C, a regime of two modes can be inferred: the first one presenting dots with sizes comparable to those found for the lower temperatures and the second with much larger three-dimensional islands. Besides confirming the presence of the quantum dots grown at the two lowest temperatures, the C-V measurements were able to estimate the electron densities and the quantum dot densities in the deepest layers of these samples.

The resonant Raman data led us to identify a principal electronic transition attributed to the Ge E_1 gap. A secondary maximum observed for the growth temperatures 400, 500, and 600°C was associated to the Ge_xSi_{1-x} alloy formed in the structures and allowed us to conclude that the Si amount is negligible for the lowest growth temperature (where

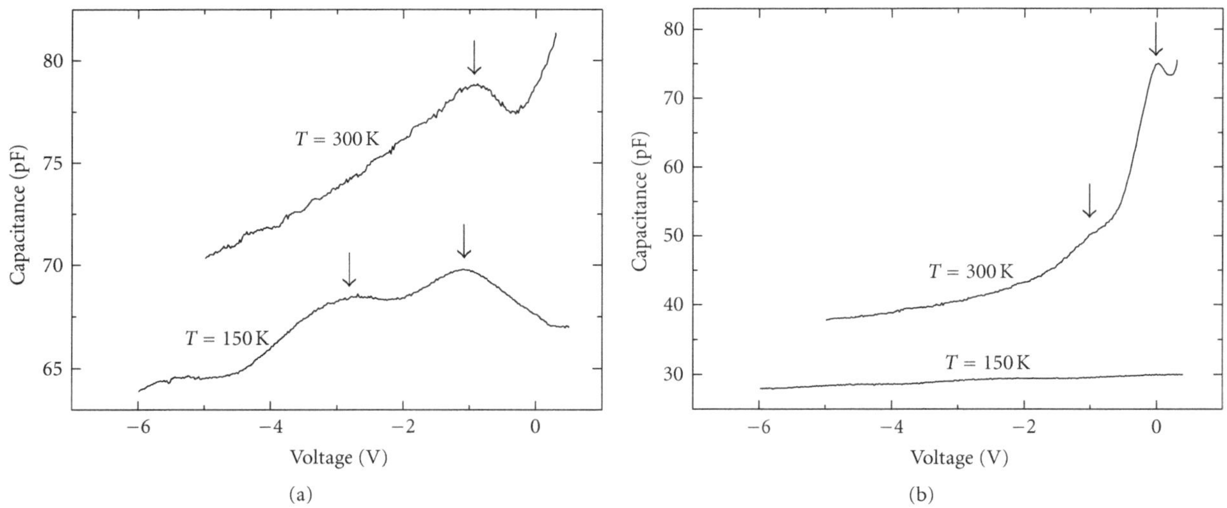

FIGURE 4: Capacitance versus Voltage measurements for the samples grown at (a) 300°C and (b) 400°C.

the feature is not present), while for the highest temperature it reaches 50% of the alloy composition.

Acknowlegments

The authors are grateful to financial support from Brazilian agencies FAPESP and CNPq.

References

[1] K. L. Wang, J. L. Liu, and G. Jin, "Self-assembled Ge quantum dots on Si and their applications," *Journal of Crystal Growth*, vol. 237–239, no. 1–4, pp. 1892–1897, 2002.

[2] S. Banerjee, S. Nozaki, and H. Morisaki, "Coulomb-blockade effect observed at room temperature in Ge nanocrystalline films deposited by the cluster-beam evaporation technique," *Applied Physics Letters*, vol. 76, no. 4, pp. 445–447, 2000.

[3] W. K. Choi, W. K. Chim, C. L. Heng et al., "Observation of memory effect in germanium nanocrystals embedded in an amorphous silicon oxide matrix of a metal-insulator-semiconductor structure," *Applied Physics Letters*, vol. 80, no. 11, pp. 2014–2016, 2002.

[4] Y. X. Jie, Y. N. Xiong, A. T. S. Wee, C. H. A. Huan, and W. Ji, "Dynamics of optical nonlinearity of Ge nanocrystals in a silica matrix," *Applied Physics Letters*, vol. 77, no. 24, pp. 3926–3928, 2000.

[5] J. L. Liu, G. Jin, Y. S. Tang, Y. H. Luo, K. L. Wang, and D. P. Yu, "Optical and acoustic phonon modes in self-organized Ge quantum dot superlattices," *Applied Physics Letters*, vol. 76, no. 5, pp. 586–588, 2000.

[6] G. Matmon, D. J. Paul, L. Lever et al., "Si/SiGe quantum cascade superlattice designs for terahertz emission," *Journal of Applied Physics*, vol. 107, no. 5, Article ID 053109, 2010.

[7] V. A. Markov, H. H. Cheng, C. T. Chia et al., "RHEED studies of nucleation of Ge islands on Si(001) and optical properties of ultra-small Ge quantum dots," *Thin Solid Films*, vol. 369, no. 1, pp. 79–83, 2000.

[8] G. Capellini, M. De Seta, and F. Evangelisti, "SiGe intermixing in Ge/Si(100) islands," *Applied Physics Letters*, vol. 78, no. 3, pp. 303–305, 2001.

[9] M. W. Dashiell, U. Denker, C. Müller et al., "Photoluminescence of ultrasmall Ge quantum dots grown by molecular-beam epitaxy at low temperatures," *Applied Physics Letters*, vol. 80, no. 7, pp. 1279–1281, 2002.

[10] M. Brehm, M. Grydlik, H. Groiss et al., "The influence of a Si cap on self-organized SiGe islands and the underlying wetting layer," *Journal of Applied Physics*, vol. 109, no. 12, Article ID 123505, 2011.

[11] F. Ratto, F. Rosei, A. Locatelli et al., "Composition of Ge(Si) islands in the growth of Ge on Si(111)," *Applied Physics Letters*, vol. 84, no. 22, pp. 4526–4528, 2004.

[12] A. G. Milekhin, A. I. Nikiforov, M. Y. Ladanov et al., "Phonons in Ge/Si quantum dot structures: influence of growth temperature," *Physica E*, vol. 21, no. 2–4, pp. 464–468, 2004.

[13] M. I. Alonso and K. Winer, "Raman spectra of c-$Si_{1-x}Ge_x$ alloys," *Physical Review B*, vol. 39, no. 14, pp. 10056–10062, 1989.

[14] M. A. Renucci, J. B. Renucci, R. Zeyher, and M. Cardona, "Second-order Raman scattering in germanium in the vicinity of the E_1, $E_1 + \Delta_1$ edges," *Physical Review B*, vol. 10, no. 10, pp. 4309–4323, 1974.

[15] S. H. Kwok, P. Y. Yu, C. H. Tung et al., "Confinement and electron-phonon interactions of the E_1 exciton in self-organized Ge quantum dots," *Physical Review B*, vol. 59, no. 7, pp. 4980–4984, 1999.

[16] F. Cerdeira, A. Pinczuk, and J. C. Bean, "Observation of confined electronic states in $Ge_xSi_{1-x}Si$ strained-layer superlattices," *Physical Review B*, vol. 31, no. 2, pp. 1202–1204, 1985.

[17] R. Schorer, G. Abstreiter, H. Kibbel, and H. Presting, "Resonant-Raman-scattering study on short-period Si/Ge superlattices," *Physical Review B*, vol. 50, no. 24, pp. 18211–18218, 1994.

[18] A. G. Milekhin, D. Tenne, and D. R. T. Zahn(, "Quantum dot structures: raman and infrared spectroscopy," in *Quantum Dots and Nanowires*, S. Bandyopadhyay and H. S. Nalwa, Eds., pp. 375–419, American Scientific, California, Calif, USA, 2003.

[19] A. J. Chiquito, Y. A. Pusep, S. Mergulhão, and J. C. Galzerani, "Carrier confinement in an ultrathin barrier GaAs/AlAs superlattice probed by capacitance-voltage measurements," *Physica E*, vol. 13, no. 1, pp. 36–42, 2002.

[20] S. M. Sze, *Physics of Semiconductors Devices*, John Wiley and Sons, New York, NY, USA, 1981.
[21] A. J. Chiquito, Y. A. Pusep, S. Mergulhão, J. C. Galzerani, and N. T. Moshegov, "Effect of photogenerated holes on capacitance-voltage measurements in InAs/GaAs self-assembled quantum dots," *Physical Review B*, vol. 61, no. 7, pp. 4481–4484, 2000.
[22] A. I. Yakimov, A. V. Dvurechenskii, A. I. Nikiforov, and O. P. Pchelyakov, "Formation of zero-dimensional hole states in Ge/Si heterostructures probed with capacitance spectroscopy," *Thin Solid Films*, vol. 336, no. 1-2, pp. 332–335, 1998.
[23] A. J. Chiquito and M. G. de Souza, "Temperature dependence of the electron distribution in a GaAs matrix with embedded InAs quantum dots," *Physica E*, vol. 25, no. 4, pp. 613–618, 2005.
[24] A. J. Chiquito, Y. A. Pusep, S. Mergulhão, J. C. Galzerani, and N. T. Moshegov, "Capacitance-voltage profile in a structure with negative differential capacitance caused by the presence of InAs/GaAs self-assembled quantum dots," *Physical Review B*, vol. 61, no. 8, pp. 5499–5504, 2000.

17

Spin-Wave Band Structure in 2D Magnonic Crystals with Elliptically Shaped Scattering Centres

Sławomir Mamica, Maciej Krawczyk, and Jarosław Wojciech Kłos

Nanomaterials Physics Division, Faculty of Physics, Adam Mickiewicz University, 61-614 Poznań, Poland

Correspondence should be addressed to Sławomir Mamica, mamica@amu.edu.pl

Academic Editor: Roberto Zivieri

Spin waves in 2D periodic magnetic nanocomposites are studied by means of the plane wave method. The effect of the ellipticity and in-plane rotation of the scattering centers on the band structure is investigated, to indicate new possibilities of fine tuning of spin-wave filter passbands.

1. Introduction

Magnetic composites with a structure modulated periodically on the nanoscale are the subject of a very intensive research activity, and the interest in their properties is increasing [1–7]. By analogy to photonic crystals (PCs), they are referred to as magnonic crystals (MCs) [8], since the role of information carriers in these materials is played by magnons, or spin-wave quanta. Magnonic crystals have properties that cannot be reduced to those of their constituent materials, as manifested, among others, by the band structure of their spin-wave spectrum. The periodic structure given to a magnetic composite strictly determines the possible occurrence of magnonic gaps, or energy ranges forbidden to propagating spin waves [9–11]. The anticipated full control over the spin waves propagating in MCs, similar to that of the electromagnetic waves in PCs, combined with a long-term stability of the programmable magnetic state makes MCs excellent for both research and application purposes [12, 13]. Moreover, the wavelength of spin waves is much shorter than that of electromagnetic waves of the same frequency. This provides additional possibilities in the miniaturization of MC-based devices [14–16]. For a broad survey of the current state of both the experimental and theoretical research in MCs and their potential applications, please refer to review papers [17, 18].

The most frequently mentioned of the numerous potential applications of MCs include microwave resonators, magnonic waveguides, spin-wave emitters, and filters [19]. These potentia applications, along with the possibility of modeling the energy spectrum of spin-wave excitations propagating in MCs, are the very reason of the intensification of the research on new magnonic materials with so far unknown properties and functionalities [14, 16]. In particular, two-dimensional (2D) MCs with band gaps in the spin-wave spectrum have potential applications in diverse magnonic devices, such as spin-wave filters or switches, or current-controlled delay lines [20]. For example, the latest results of micromagnetic simulations [21] indicate the occurrence of wide magnonic gaps, implying a possible application in spin-wave filters, in 2D Fe/YIG MCs, two-component magnetic composites with iron scattering centers embedded in a matrix of yttrium iron garnet. A particular role in the modeling of magnonic gaps is played by the deformation of the scattering centers in the plane of spin-wave propagation [22].

In this study we examine the possibilities of modeling the spin-wave spectrum of 2D MCs that could be used for fine tuning of spin-wave filter passbands [23]. We present the results of calculations of the magnonic band structure of a Co/Fe composite with scattering centers in the shape of elliptic cylinders. We find that for different filling fraction values there are specific in-plane rotation angles for which modifying the rod ellipticity can alter the position of the allowed band without changing its width or cause a substantial shrinking of two adjacent bands without changing the width of the gap between them.

2. The Model

Figure 1(a) depicts schematically a section of the system under consideration in the plane perpendicular to the rod axis (the x-y plane), which is the plane of spin-wave propagation. The system includes cobalt rods (scattering centers), which are assumed to be parallel to each other and have an infinite length. The rods are arranged in sites of a 2D square lattice and embedded in an iron matrix. The system is infinite in the plane of periodicity. The filling fraction *ff*, describing the proportion of the rod material in the whole volume of the system, in the case of elliptic cylinders is given as $ff = \pi R_a R_b / a^2$, where R_a and R_b are the semiaxes of the rod cross section, and a is the lattice constant. Let us define the ellipticity RR of the rods as the semiaxis ratio: $RR = R_a/R_b$. The angle α between the major semiaxis R_a and the x-axis of the crystallographic system is the angle of rotation of the rods (in the plane of periodicity). Applied to the system, an external magnetic field perpendicular to the plane of periodicity is assumed to be strong enough to enforce a uniform magnetization throughout the system.

We shall consider in-plane propagation of spin waves, that is, their propagation in the plane of periodicity, in 2D MCs as described above. Thus, the wave vectors to be considered are limited to the Brillouin zone of the 2D lattice. Figure 1(b) shows the high-symmetry line in such a 2D Brillouin zone. In the case of square lattice the line starts at the zone center (point Γ) to pass through X and M and return to Γ (segments C, D, and E in Figure 1(b)). The introduction of rods with elliptical cross section breaks the symmetry of the square lattice. If the semiaxes of the ellipse follow the axes of the 2D crystallographic system, the structure has the symmetry of a rectangular lattice. Points X and X' are not equivalent anymore, and the high-symmetry line in this case leads from M to X', Γ, X, and back to M (segments A, B, C, and D). For an ellipse rotated by 45° points, X and X' are equivalent symmetry points, whereas M and M' are not. Thus, the high-symmetry line leads from Γ to X, M, Γ, M', and X (segments C, D, E, F, G). In the general case, that is, for any angle of rotation of the ellipse, the full line shown in Figure 1(b) must be considered.

Our theoretical approach is based on a set of equations including the linearized Landau-Lifshitz equation and Maxwell's magnetostatic equations [24]. When the applied magnetic field is parallel to the rods, the internal static magnetic field is uniform, which allows us to only take into account the exchange and dynamic dipolar interactions. The ferromagnetic materials of the rods and the matrix are characterized by two quantities: the spontaneous magnetization M_s and the exchange stiffness constant A. For the materials considered in this study the specific values of these two magnetic parameters are for iron $M_s = 1.752e6$ A/m and $A = 2.1e - 11$ J/m, and for cobalt $M_s = 1.390e6$ A/m and $A = 2.8e - 11$ J/m [25]. Crucial for the magnonic nature of the structure under consideration is the assumption that these two material parameters are periodic functions of position, with the same periodicity as the 2D lattice on which the magnonic crystal is built. With this assumption our equations can be solved by the plane wave method. The main point of this method is the Fourier expansion of the material parameters. Bloch theorem is applied to the dynamic functions, such as the demagnetizing field potential and the dynamic component of magnetization. Thus, the equations are transformed to the reciprocal space, where their solution is equivalent to the diagonalization of a $2N \times 2N$ matrix, N being the number of plane waves used in the Fourier expansion (for more details, see [24] and references therein).

3. The Role of Magnetostatic Interactions

Figure 2 shows examples of the so-called magnonic spectra, represented by the spin-wave spectra of 2D Co/Fe magnetic composites, calculated along the high-symmetry line in the 2D Brillouin zone. The spectra shown in Figures 2(a) and 2(b) have been obtained for lattice constants $a = 50$ nm and $a = 100$ nm, respectively. A circular cross section of the rods ($RR = 1$) and a filling fraction $ff = 0.5$ are assumed in both cases. For small lattice constants the exchange interactions play a dominant role; however, as the lattice constant grows, their importance diminishes to the advantage of the magnetostatic interactions [26, 27]. As a result, the spin-wave frequency range lowers; for the lattice constant of 50 nm the frequency of the ten lowest modes is below 80 GHz (Figure 2(a)), but only ranges from 10 GHz to 35 GHz for $a = 100$ nm (Figure 2(b)). The flattening of successive bands results in the opening of a magnonic gap: for $a = 50$ nm all the bands overlap, while for $a = 100$ nm a gap occurs between the lowest band and the rest of the spectrum.

4. The Ellipticity and Rotation of the Rods

Figure 3 presents the effect of the cross-sectional ellipticity of the scattering centers and their rotation in the plane of spin-wave propagation on the magnonic spectrum in a Co/Fe composite with a lattice constant of 300 nm and a filling fraction of 0.3. For unrotated rods (Figure 3(a)) the maximum ellipticity (corresponding to touching rods) is $RR = 2.6$, which means the major and minor semiaxes R_a and R_b can range from 92.7 nm up to 149.8 nm and down to 57.4 nm, respectively. In the whole range of rod ellipticity the bottom of the lowest band is nearly constant, remaining between 10.84 GHz and 10.89 GHz. Also the top of the second band varies very slightly, only ranging from 12.04 GHz to 12.15 GHz. However, the width of both bands grows rapidly, as the top of the first band and the bottom of the second one converge. Consequently, the gap between the bands shrinks to vanish completely for $RR = 2.1$ ($R_a = 134.3$ nm, $R_b = 64.0$ nm). The two bands merge to form a single wide band, separated from the rest of the spectrum by a third gap, the width of which varies from 1.26 GHz for right circular cylinders ($RR = 1.0$) to 0.79 GHz for $RR = 2.1$ (the closing of the second gap) to 0.37 GHz for $RR = 2.6$ (the

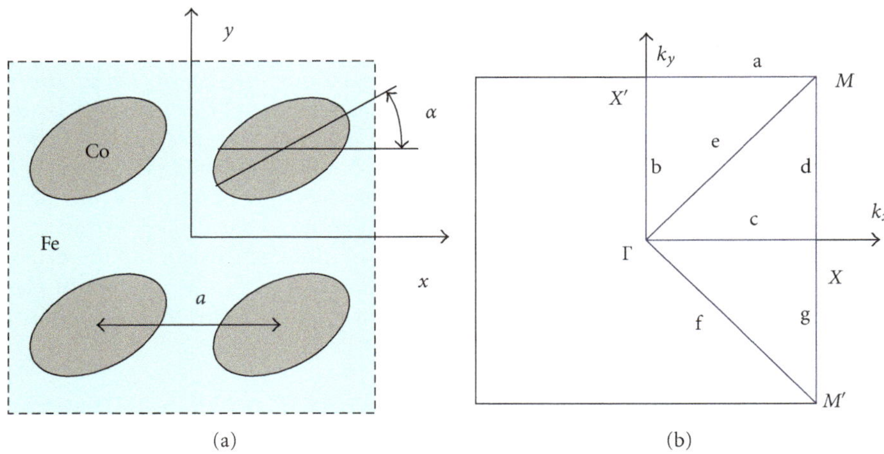

FIGURE 1: (a) Schematic view of the 2D MC under consideration, section in the plane of periodicity. (b) High-symmetry line over the 2D Brillouin zone for ellipses arranged in a square lattice.

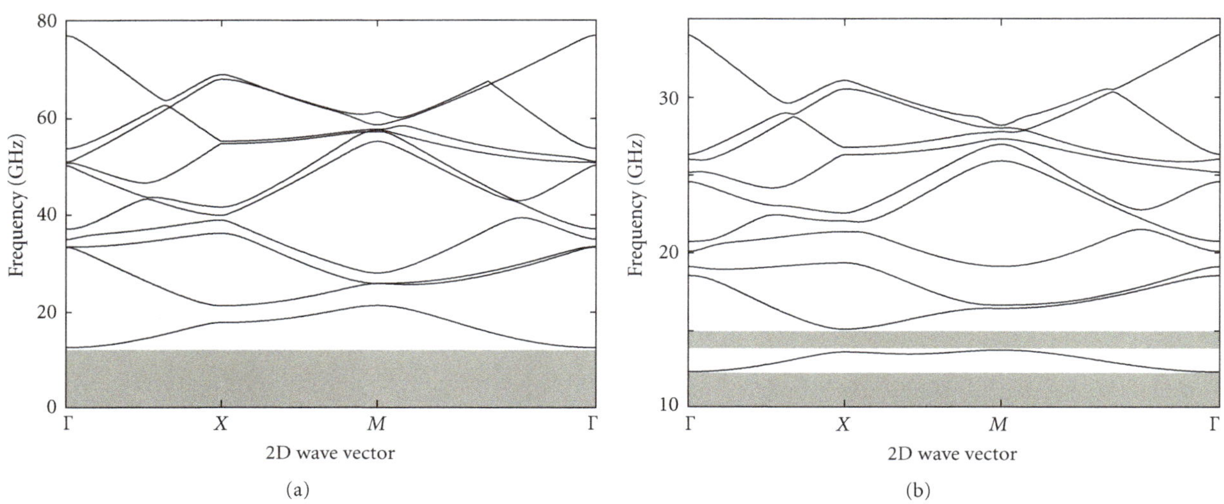

FIGURE 2: Ten lowest bands in the spin-wave spectrum of a 2D Co/Fe composite with lattice constant (a) 50 nm and (b) 100 nm, calculated along the high-symmetry line in the 2D Brillouin zone (cf. Figure 1(b)). Circular cross section of the rods ($RR = 1$) and filling fraction $ff = 0.5$ are assumed in both cases. Shaded areas represent magnonic gaps. Note the plots differ in frequency range: 0 GHz to 80 GHz in (a), and 10 GHz to 35 GHz in (b).

maximum ellipticity). Figure 3(b) shows the band and gap widths plotted versus the rod ellipticity.

Also for rods rotated by 45° (Figure 3(c)) the second gap is seen to shrink, though not as rapidly as in the case of unrotated rods. The gap closes for $RR = 3.45$, which corresponds to $R_a = 212.0$ nm and $R_b = 40.5$ nm. A significant difference with respect to the composite with unrotated rods is seen in the behavior of the second band, which moves down the frequency scale with nearly constant width as RR grows from 1.0 to 2.8 (cf. Figure 3(d)).

In a Co/Fe composite with a filling fraction of 0.5 and rods unrotated in the plane of periodicity (the major semiaxis following the x direction), the rod ellipticity can range from 1.0 (circular cross section) to 1.57, which for the assumed lattice constant $a = 300$ nm corresponds to the major semiaxis ranging from 119.7 nm up to 150.0 nm, and the minor semiaxis from 119.7 nm down to 95.5 nm. As the ellipticity of the cylinders grows, the gaps are seen to shrink and the bands to widen (Figure 4(a)). Although the bottom of the lowest band at first moves towards higher frequencies, the change is to slight to compensate the concurrent rising of the top. In the second band the top is seen to descend slightly, while the bottom moves much faster in the same direction. As a consequence, the second gap, between the first and second bands, shrinks rapidly with growing ellipticity to vanish completely for $RR = 1.2$ ($R_a = 131.1$ nm, $R_b = 109.3$ nm). The first and second bands merge to form one relatively wide band (of width ranging from 1.17 GHz for $RR = 1.2$ to 1.32 GHz for $RR = 1.57$), separated from the rest of the spectrum by a third gap, which has a maximum width of 0.65 GHz for $RR = 1.2$, and a minimum width of 0.25 GHz for $RR = 1.57$ (cf. Figure 4(b)).

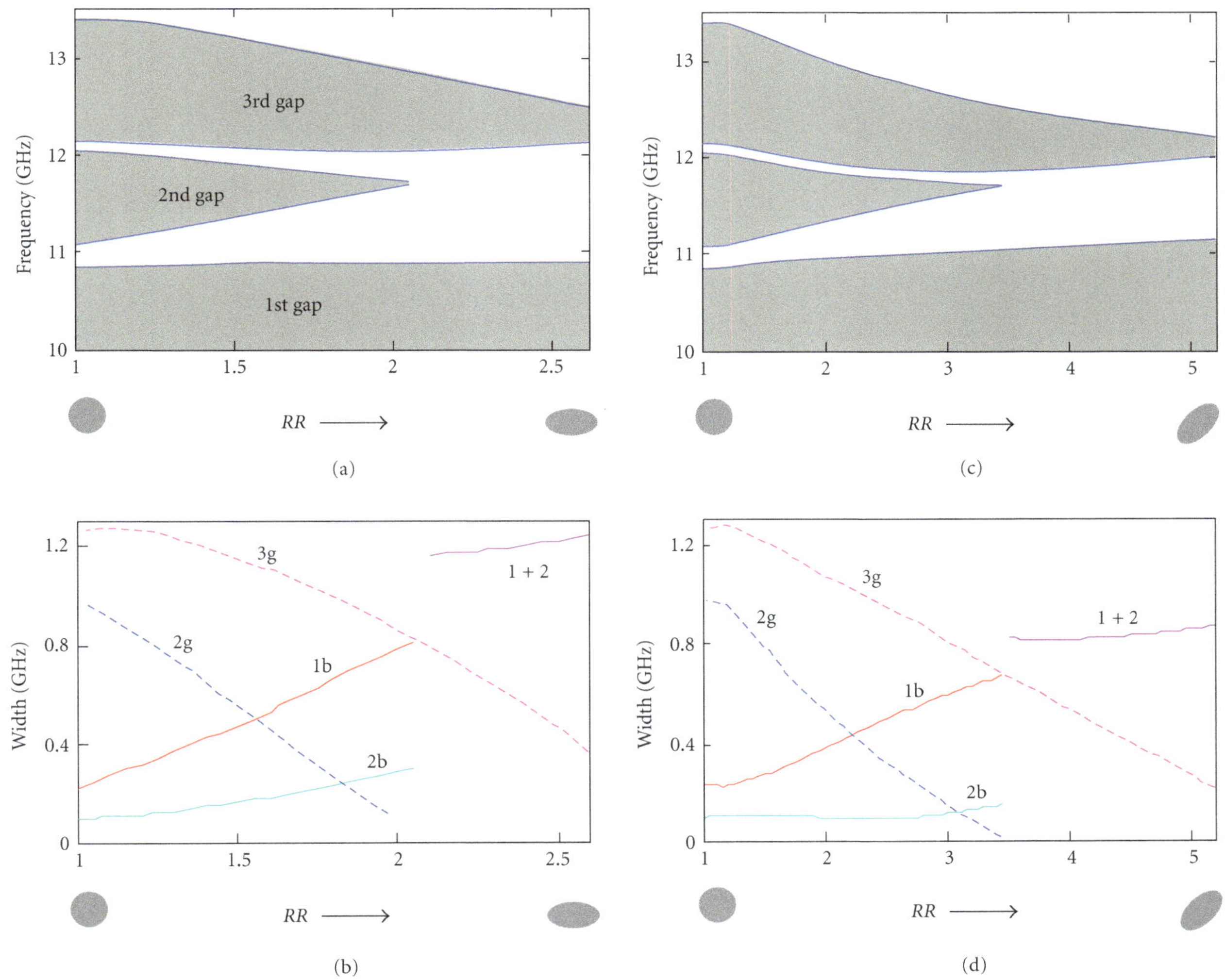

FIGURE 3: (a), (c) Three lowest magnonic gaps (shaded) versus rod ellipticity RR for 2D Co/Fe MCs with lattice constant 300 nm, filling fraction 0.3, and two angles of in-plane rotation of the rods: (a) $\alpha = 0°$ and (c) $\alpha = 45°$. (b), (d) The width of the lowest gaps and bands versus RR in (a) and (c), respectively. The width of the second (2g) and third (3g) gaps is plotted with dashed line. Solid line represents the width of the lowest bands: the first (1b), the second (2b), and, in the range of RR values in which the gap between them does not occur, the band resulting from their merging (1 + 2).

A completely different behavior of the spin-wave spectrum is seen for the same filling fraction ($ff = 0.5$), but with rods rotated by 45° in the plane of periodicity (Figure 4(c)). Two ranges of RR can be distinguished in this case. In the first range, from 1.0 to 2.0 (the major semiaxis growing from 119.7 nm to 169.3 nm, and the minor semiaxis shrinking from 119.7 nm to 84.6 nm), the width of both lowest bands decreases substantially, the first one shrinking from 0.52 GHz to 0.28 GHz, and the second from 0.34 GHz to 0.22 GHz (Figure 4(d)). At the same time, the midlevels of both bands move, from 11.0 GHz to 11.1 GHz in the case of the first band, and from 11.7 GHz to 11.6 GHz for the second. Interestingly, in this range of RR the second gap remains nearly unchanged, its width only ranging from 0.29 GHz to 0.30 GHz. In the other part of the RR dependence the spectrum behaves as in the cases considered previously: the bands widen and the gaps shrink rapidly. However, in this case even for the maximum ellipticity of the rods ($RR = 3.14$) the second gap will not close completely, though its width falls as low as 0.02 GHz.

5. Conclusions

In the 2D magnetic composites considered in this paper, with Co rods embedded in an Fe matrix, the increase in importance of the magnetostatic interactions results in the formation of band gaps in the spin-wave spectrum. These magnonic gaps are destroyed as the exchange interactions begin to play a dominant role. On the other hand, the increase in importance of the exchange interactions not only results in a widening of the bands, but also, consequently, causes the possible gaps to move towards higher frequencies. In contrast, when the magnetostatic interactions gain in importance at the cost of the exchange interactions, the

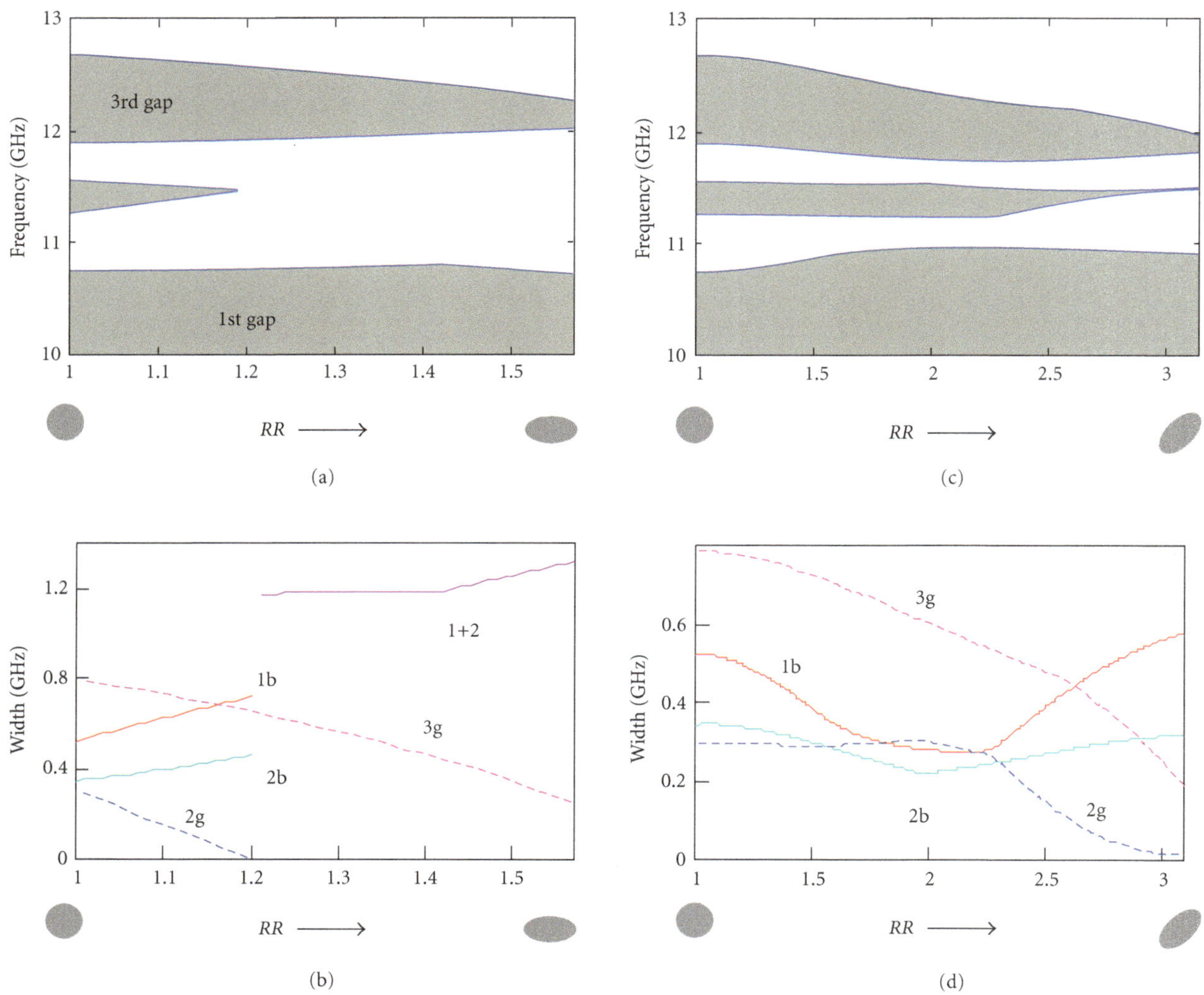

Figure 4: The same as in Figure 3, but for filling fraction 0.5.

frequency range in which magnonic gaps will occur lowers substantially. This finding is consistent with the results obtained for 2D composites with an EuO matrix [28]. However, an entirely different behavior is observed in 3D magnonic crystals, in which absolute magnonic gaps are destroyed by the magnetostatic interactions [29].

The use of rods in the shape of elliptic cylinders as scattering centers in 2D magnetic composites implies the introduction of two addition structural parameters: the cross-sectional ellipticity of the rods and the angle of their rotation in the plane perpendicular to the rod axis (the plane of spin-wave propagation). In contrast to the lattice constant, a change of which will strongly modify the spin-wave spectrum, these new parameters allow fine tuning of the width and position of the bands and gaps. For specific in-plane rotation angles, changing the rod ellipticity will modify the position of a band without changing its width or cause two adjacent bands to shrink substantially without affecting the width of the gap between them. Thus, an appropriate use of rods of elliptical cross section offers additional possibilities in the design of spin-wave filters with precisely adjusted passband.

Acknowledgments

Those results were presented at 18th International Conference on Composite Materials, Jeju Island, Korea. The research leading to these results has received funding from the European Community's Seventh Framework Programme (FP7/2007-2013) under Grant Agreement no. 228673 for MAGNONICS project.

References

[1] A. V. Chumak, A. A. Serga, B. Hillebrands, and M. P. Kostylev, "Scattering of backward spin waves in a one-dimensional magnonic crystal," *Applied Physics Letters*, vol. 93, no. 2, Article ID 022508, 2008.

[2] Z. K. Wang, V. L. Zhang, H. S. Lim et al., "Observation of frequency band gaps in a one-dimensional nanostructured

magnonic crystal," *Applied Physics Letters*, vol. 94, no. 8, Article ID 083112, 2009.

[3] S. L. Vysotskii, S. A. Nikitov, E. S. Pavlov, and Y. A. Filimonov, "The spectrum of the spin-wave excitations of the tangentially magnetized 2D hexagonal ferrite magnonic crystal," *Journal of Communications Technology and Electronics*, vol. 55, no. 7, pp. 800–809, 2010.

[4] S. Neusser, G. Duerr, H. G. Bauer et al., "Anisotropic propagation and damping of spin waves in a nanopatterned antidot lattice," *Physical Review Letters*, vol. 105, no. 6, Article ID 067208, 2010.

[5] Y. Cao, G. Yun, X. Liang, and N. Bai, "Band structures of two-dimensional magnonic crystals with different shapes and arrangements of scatterers," *Journal of Physics D*, vol. 43, no. 30, Article ID 305005, 2010.

[6] J. W. Kłos, M. Krawczyk, and M. Sokolovskyy, "Bulk and edge modes in two-dimensional magnonic crystal slab," *Journal of Applied Physics*, vol. 109, no. 7, Article ID 07D311, 2011.

[7] R. Zivieri, S. Tacchi, F. Montoncello et al., "Bragg diffraction of spin waves from a two-dimensional antidot lattice," *Physical Review B*, vol. 85, no. 1, Article ID 012403, pp. 1–6, 2012.

[8] S. A. Nikitov, P. Tailhades, and C. S. Tsai, "Spin waves in periodic magnetic structures—magnonic crystals," *Journal of Magnetism and Magnetic Materials*, vol. 236, no. 3, pp. 320–330, 2001.

[9] M. Krawczyk and H. Puszkarski, "Magnonic excitations versus three-dimensional structural periodicity in magnetic composites," *Crystal Research and Technology*, vol. 41, no. 6, pp. 547–552, 2006.

[10] S. Neusser, H. G. Bauer, G. Duerr et al., "Tunable metamaterial response of a Ni80Fe20 antidot lattice for spin waves," *Physical Review B*, vol. 84, no. 18, Article ID 184411, 2011.

[11] S. Tacchi, F. Montoncello, M. Madami et al., "Band diagram of spinwaves in a two-dimensional magnonic crystal," *Physical Review Letters*, vol. 107, no. 12, Article ID 127204, pp. 1–5, 2011.

[12] J. Topp, D. Heitmann, M. P. Kostylev, and D. Grundler, "Making a reconfigurable artificial crystal by ordering bistable magnetic nanowires," *Physical Review Letters*, vol. 104, no. 20, Article ID 207205, 2010.

[13] S. Tacchi, M. Madami, G. Gubbiotti et al., "Analysis of collective spin-wave modes at different points within the hysteresis loop of a one-dimensional magnonic crystal comprising alternative-width nanostripes," *Physical Review B*, vol. 82, no. 18, Article ID 184408, 2010.

[14] S. Tacchi, M. Madami, G. Gubbiotti et al., "Anisotropic dynamical coupling for propagating collective modes in a two-dimensional magnonic crystal consisting of interacting squared nanodots," *Physical Review B*, vol. 82, no. 2, Article ID 024401, 2010.

[15] G. Duerr, M. Madami, S. Neusser et al., "Spatial control of spin-wave modes in Ni80Fe20 antidot lattices by embedded Co nanodisks," *Applied Physics Letters*, vol. 99, no. 20, Article ID 202502, 2011.

[16] S. Neusser, G. Duerr, S. Tacchi et al., "Magnonic minibands in antidot lattices with large spin-wave propagation velocities," *Physical Review B*, vol. 84, no. 9, Article ID 094454, pp. 1–9, 2011.

[17] V. V. Kruglyak, S. O. Demokritov, and D. Grundler, "Magnonics," *Journal of Physics D*, vol. 43, no. 26, Article ID 264001, 2010.

[18] S. Neusser and D. Grundler, "Magnonics: spin waves on the nanoscale," *Advanced Materials*, vol. 21, no. 28, pp. 2927–2932, 2009.

[19] S. Choi, K. S. Lee, K. Y. Guslienko, and S. K. Kim, "Strong radiation of spin waves by core reversal of a magnetic vortex and their wave behaviors in magnetic nanowire waveguides," *Physical Review Letters*, vol. 98, no. 8, Article ID 087205, 2007.

[20] A. B. Ustinov, A. V. Drozdovskii, and B. A. Kalinikos, "Multifunctional nonlinear magnonic devices for microwave signal processing," *Applied Physics Letters*, vol. 96, no. 14, Article ID 142513, 2010.

[21] F. S. Ma, H. S. Lim, Z. K. Wang, S. N. Piramanayagam, S. C. Ng, and M. H. Kuok, "Micromagnetic study of spin wave propagation in bicomponent magnonic crystal waveguides," *Applied Physics Letters*, vol. 98, no. 15, Article ID 153107, 2011.

[22] H. T. Nembach, J. M. Shaw, T. J. Silva et al., "Effects of shape distortions and imperfections on mode frequencies and collective linewidths in nanomagnets," *Physical Review B*, vol. 83, no. 9, Article ID 094427, 2011.

[23] S. Mamica and M. Krawczyk, "Tuning of the spin-wave band structure in 2d magnetic composites," in *Proceedings of the 18th International Conference on Composite Materials*, P4-34, Jeju Island, Rpublic of Korea, 2011.

[24] M. Krawczyk, J. Klos, M. L. Sokolovskyy, and S. Mamica, "Materials optimization of the magnonic gap in three-dimensional magnonic crystals with spheres in hexagonal structure," *Journal of Applied Physics*, vol. 108, no. 9, Article ID 093909, 2010.

[25] C. A. F. Vaz, J. A. C. Bland, and G. Lauhoff, "Magnetism in ultrathin film structures," *Reports on Progress in Physics*, vol. 71, no. 5, Article ID 056501, 2008.

[26] V. S. Tkachenko, V. V. Kruglyak, and A. N. Kuchko, "Spectrum and reflection of spin waves in magnonic crystals with different interface profiles," *Physical Review B*, vol. 81, no. 2, Article ID 024425, 2010.

[27] M. Krawczyk, "Magnetostatic waves in one-dimensional magnonic crystals with magnetic and nonmagnetic components," *IEEE Transactions on Magnetics*, vol. 44, no. 11, pp. 2854–2857, 2008.

[28] J. O. Vasseur, L. Dobrzynski, B. Djafari-Rouhani, and H. Puszkarski, "Magnon band structure of periodic composites," *Physical Review B*, vol. 54, no. 2, pp. 1043–1049, 1996.

[29] M. Krawczyk and H. Puszkarski, "Plane-wave theory of three-dimensional magnonic crystals," *Physical Review B*, vol. 77, no. 5, Article ID 054437, 2008.

18

On the Travelling Wave Solution for the Current-Driven Steady Domain Wall Motion in Magnetic Nanostrips under the Influence of Rashba Field

Vito Puliafito[1] and Giancarlo Consolo[2]

[1] *Department of Matter Physics and Electronic Engineering, University of Messina, V.le F. Stagno D'Alcontres 31, 98166 Messina, Italy*
[2] *Department of Science for Engineering and Architecture, University of Messina, Contrada Di Dio, 98166 Messina, Italy*

Correspondence should be addressed to Vito Puliafito, vpuliafito@unime.it

Academic Editor: Roberto Zivieri

Spin-orbit Rashba effect applies a torque on the magnetization of a ferromagnetic nanostrip in the case of structural inversion asymmetry, also affecting the steady domain wall motion induced by a spin-polarized current. This influence is here analytically studied in the framework of the extended Landau-Lifshitz-Gilbert equation, including the Rashba effect as an additive term of the effective field. Results of previous micromagnetic simulations and experiments have shown that this field yields an increased value of the Walker breakdown current together with an enlargement of the domain wall width. In order to analytically describe these results, the standard travelling wave ansatz for the steady domain wall motion is here adopted. Results of our investigations reveal the impossibility to reproduce, at the same time, the previous features and suggest the need of a more sophisticated model whose development requires, in turn, additional information to be extracted from ad hoc micromagnetic simulations.

1. Introduction

Magnetization dynamics in nanodevices has been intensely investigated in the last decades as it provides a wide variety of technological applications in the area of storage and logic devices. Formerly, the manipulation of the magnetic configuration was typically achieved by means of external magnetic fields but the contemporary demand of miniaturizing storage devices and of increasing their capacity would have required higher and higher fields. An alternative method, realized by using spin-polarized currents, was outlined by the discovery of spin-transfer torque effect [1, 2]. Theoretical and experimental studies, therefore, examined magnetization dynamics due to the simultaneous action of external magnetic fields and electric currents [3–5] as well as current-driven dynamics at zero field [6, 7].

Among the different geometries used for spintronic devices, a more recent attention is directed to magnetic nanowires and strips [8–22]. Such thin ferromagnetic structures turn out to be relevant for the realization of oscillators and high-density memories with low energy consumption [9]. In particular, their behavior and applications are strictly connected to the motion of magnetic domain walls (DWs), namely, the continuous transition regions that separate two uniformly and oppositely magnetized domains [9, 10]. Also, in this case, DW dynamics can be activated by means of an external field or an electric current. Independently of the nature of the source term, it has been widely demonstrated that, by varying the strength of such a driving source, the DW motion experiences two different dynamical regimes. At low fields (or currents), the equilibrium wall structure is rigidly shifted along the nanostrip (or nanowire) axis, leading to a "steady" regime of high DW mobility. Above a critical field (or current) value, named Walker breakdown, a regime of lower DW mobility takes place and the internal deformation is so strong that the wall structure is altered giving rise to a periodical alternation of Bloch and Néel DW structures. Such a dynamics is generally referred to as "precessional" regime [11].

DW motion can be also strongly modified by spin-orbit interaction [12] that takes place in the case of structural inversion asymmetry of the nanodevice. Some previous works [13–18], in particular, pointed out that this effect, induced by the flow of an electric current through the material, acts as an effective field. This contribution, which was named Rashba field, produces two main consequences: (1) a (Bloch-like) DW structure is stabilized as the steady regime turns broadened up to higher current densities so allowing higher DW velocities; (2) the DW width enlarges with the increase of the current-induced Rashba field.

From the theoretical viewpoint, DW motion in ferromagnetic thin layers is ruled by the extended Landau-Lifshitz-Gilbert (ELLG) equation including the current-driven spin-torque effects [20–24]. Travelling wave solutions for the ELLG equation have been recently obtained, providing a strong theoretical support for both experimental and numerical results [20, 22]. It has been also demonstrated that the inclusion of a different dissipation function into the ELLG equation, usually referred to as "dry-friction," gives a good description of DW dynamics in the presence of crystallographic defects and structural disorder [20, 22].

In this work, we propose to analytically study, by using a one-dimensional mathematical model, the steady DW motion in ferromagnetic nanostrips subject to the action of spin-polarized currents and Rashba fields. In particular, we investigate on the appropriateness of using a standard travelling wave ansatz describing a Bloch DW structure which rigidly shifts under the action of the external source. In order to validate the developed model, these analytical results are then compared, at qualitative level, with those arising from recent numerical and experimental investigations. The model also accounts for the nonlinear dry-friction dissipation function in order to evaluate how such a contribution affects the current-driven steady DW motion in the presence of a Rashba field.

2. The Analytical One-Dimensional Model: Results and Discussion

As depicted in Figure 1, a ferromagnetic nanostrip can be represented by a rectangular prism of length l, width w, and thickness t along $\mathbf{c}_x$, $\mathbf{c}_y$, $\mathbf{c}_z$ axes, respectively, with $l \gg w > t$. Let us assume that an electric current density $\mathbf{J} = J\mathbf{c}_x$, constant in time and uniform in space, is applied to the device along the $\mathbf{c}_x$ axis. Under the hypothesis that the Rashba field does not modify the equilibrium configuration obtained in its absence, we assume that a 180° DW of width δ is nucleated at the center of the structure with the magnetization vector that rotates between the state (0,0,1) at $x \to -\infty$ and the opposite one (0,0,−1) at $x \to \infty$. These uniformly magnetized states, far away from the wall location, are hence supposed to be directed towards the easy axis $\mathbf{c}_z = \mathbf{e}$, namely, the energetically preferred direction of spontaneous magnetization.

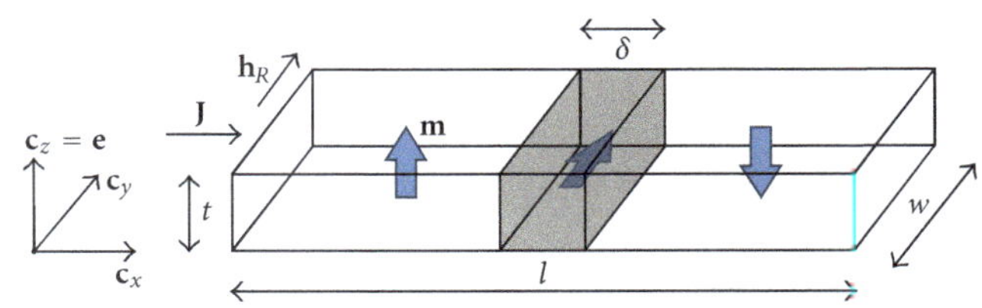

Figure 1: Schematics of a ferromagnetic nanostrip exhibiting a Bloch DW.

Current-driven DW dynamics in such a thin layer is described by the ELLG equation [20–24]:

$$\dot{\mathbf{m}} = \gamma(\mathbf{h}_{\text{eff}} \wedge \mathbf{m}) + \mathbf{t}_d + \mathbf{t}_{\text{st}}, \tag{1}$$

where the over-dot denotes time derivation, $\mathbf{m} = \mathbf{M}/M_S$ is the unit vector along the local magnetization and all the field vectors are normalized with respect to the saturation magnetization M_S. The constant $\gamma = M_S\mu_0\gamma_e$ is expressed in terms of the vacuum magnetic permeability μ_0 and the gyromagnetic ratio $\gamma_e = ge/m_e$, being g the Landé factor, e the electron charge, and m_e the electron mass. The first term in the right-hand side of (1) describes the precession of magnetization $\mathbf{m}$ around the direction of the effective magnetic field $\mathbf{h}_{\text{eff}}$, the second term $\mathbf{t}_d$ is the dissipative torque representing energy dissipation, and $\mathbf{t}_{\text{st}}$ corresponds to the current-induced spin torque.

The effective magnetic field $\mathbf{h}_{\text{eff}} = -\partial W/\partial \mathbf{m}$, calculated as the variational derivative of the free energy density W, accounts for external $\mathbf{h}_{\text{ext}}$, exchange $\mathbf{h}_{\text{exc}}$, demagnetizing $\mathbf{h}_{\text{dmg}}$, anisotropy $\mathbf{h}_{\text{an}}$, and Rashba $\mathbf{h}_R$ fields:

$$\mathbf{h}_{\text{eff}} = \mathbf{h}_{\text{ext}} + \mathbf{h}_{\text{exc}} + \mathbf{h}_{\text{dmg}} + \mathbf{h}_{\text{an}} + \mathbf{h}_R, \tag{2}$$

As our attention focuses on the influence of the current-induced Rashba field, we will limit our analysis to the zero-field configuration:

$$\mathbf{h}_{\text{ext}} = 0. \tag{3}$$

The exchange field can be written as

$$\mathbf{h}_{\text{exc}} = A\frac{\partial^2 \mathbf{m}}{\partial x^2} \tag{4}$$

being A related to the exchange constant A_{ex} of the material through

$$A = \frac{2A_{\text{ex}}}{\mu_0 M_S^2}. \tag{5}$$

The demagnetizing field can be approximated by considering the only diagonal terms of the corresponding tensor which relates the field to the magnetization [11]:

$$\mathbf{h}_{\text{dmg}} = -N_x(\mathbf{m}\cdot\mathbf{c}_x)\mathbf{c}_x - N_y\left(\mathbf{m}\cdot\mathbf{c}_y\right)\mathbf{c}_y - N_z(\mathbf{m}\cdot\mathbf{e})\mathbf{e}, \tag{6}$$

where N_x, N_y, and N_z are the demagnetizing factors satisfying the normalization condition $N_x + N_y + N_z = 1$. These coefficients depend both on the shape (assumed to be a rectangular prism) and on the dimensions ($w \times t \times \delta$) of the DW.

We also assume that the strip is made by a material exhibiting a high perpendicular magnetic anisotropy, so that we can express

$$\mathbf{h}_{\text{an}} = \beta(\mathbf{m} \cdot \mathbf{e})\mathbf{e}, \tag{7}$$

where $\beta \gg 1$ is proportional to the anisotropy constant K of the material through

$$\beta = \frac{2K}{\mu_0 M_S^2}. \tag{8}$$

Finally, the Rashba field is given by [13–18]:

$$\mathbf{h}_R = \frac{\alpha_R P}{\mu_0 \mu_B M_S^2}(\mathbf{e} \wedge \mathbf{J}) \tag{9}$$

being α_R the Rashba parameter, P the polarization factor of the current, and μ_B the Bohr magneton. The expression (9) can be also written in compact form as

$$\mathbf{h}_R = \chi u \,\mathbf{c}_y, \tag{10}$$

where χ accounts for the Rashba effect, $\chi = (2e/g\mu_0\mu_B^2 M_S)\alpha_R$, and the spin-torque velocity u accounts for the applied current, $u = (g\mu_B P/2eM_S)J$.

The dissipative torque $\mathbf{t}_d$ is here described by the phenomenological Gilbert-like expression that also includes a dry-friction damping function [19–22]:

$$\mathbf{t}_d = \mathbf{m} \wedge \left[\alpha_G + \frac{\zeta}{|\dot{\mathbf{m}}|}\right]\dot{\mathbf{m}}, \tag{11}$$

Being α_G the classical linear damping coefficient and ζ the phenomenological dry-friction parameter (for more details on the dry-friction formulation, together with its practical justification within the equation of motion, see [20–22]).

The spin transfer torque $\mathbf{t}_{\text{st}}$ is given by

$$\mathbf{t}_{\text{st}} = u(-1 + \eta\mathbf{m}\wedge)\frac{\partial \mathbf{m}}{\partial x}, \tag{12}$$

where η is the phenomenological nonadiabatic spin-torque parameter [10].

Taking into account (3)–(12), (1) becomes

$$\dot{\mathbf{m}} - \left[\alpha_G + \frac{\zeta}{|\dot{\mathbf{m}}|}\right](\mathbf{m} \wedge \dot{\mathbf{m}}) = \left(\gamma\mathbf{h}_{\text{eff}} - u\mathbf{m} \wedge \frac{\partial \mathbf{m}}{\partial x} - \eta u \frac{\partial \mathbf{m}}{\partial x}\right) \wedge \mathbf{m} \tag{13}$$

with

$$\mathbf{h}_{\text{eff}} = A\frac{\partial^2 \mathbf{m}}{\partial x^2} + (\beta - N_z)(\mathbf{m} \cdot \mathbf{e})\mathbf{e} - N_x(\mathbf{m} \cdot \mathbf{c}_x)\mathbf{c}_x - N_y(\mathbf{m} \cdot \mathbf{c}_y)\mathbf{c}_y + \chi u \mathbf{c}_y. \tag{14}$$

Let us now make a transformation from the Cartesian to the spherical frame, so that it is possible to express the local magnetization as:

$$\mathbf{m} = \cos\varphi \sin\vartheta\, \mathbf{c}_x + \sin\varphi \sin\vartheta\, \mathbf{c}_y + \cos\vartheta\, \mathbf{e}. \tag{15}$$

From (13)–(15), we therefore obtain a system of two second-order partial differential equations:

$$\begin{aligned}
&\dot{\vartheta} + \left[\alpha_G + \left(\dot{\vartheta}^2 + \sin^2\vartheta\, \dot{\varphi}^2\right)^{-1/2}\zeta\right]\sin\vartheta\dot{\varphi} \\
&\quad = A\gamma \sin\vartheta \frac{\partial^2 \varphi}{\partial x^2} + 2A\gamma\cos\vartheta\frac{\partial\varphi}{\partial x}\frac{\partial\vartheta}{\partial x} \\
&\quad + \gamma\left(N_x - N_y\right)\sin\varphi\cos\varphi\sin\vartheta \\
&\quad - \eta u \sin\vartheta\frac{\partial\varphi}{\partial x} - u\frac{\partial\vartheta}{\partial x} + \gamma\chi u\cos\varphi \\
&\sin\vartheta\,\dot{\varphi} - \left\{\left[\alpha_G + \left(\dot{\vartheta}^2 + \sin^2\vartheta\,\dot{\varphi}^2\right)^{-1/2}\zeta\right]\dot{\vartheta}\right\} \\
&\quad = -A\gamma\frac{\partial^2\vartheta}{\partial x^2} + A\gamma\sin\vartheta\cos\vartheta\left(\frac{\partial\varphi}{\partial x}\right)^2 \\
&\quad + \gamma\sin\vartheta\cos\vartheta\left(\beta - N_z + N_x\cos^2\varphi + N_y\sin^2\varphi\right) \\
&\quad - u\sin\vartheta\frac{\partial\varphi}{\partial x} + \eta u\frac{\partial\vartheta}{\partial x} - \gamma\chi u\sin\varphi\cos\vartheta.
\end{aligned} \tag{16}$$

Since it was demonstrated that the previous system admits analytical solutions in the form of travelling waves [19–22], we search for such solutions apt to describe the DW motion within the steady regime. In particular, in order to reduce the system (16) to a couple of ordinary differential equations, we study the appropriateness of adopting the commonly used travelling wave ansatz $\vartheta = \vartheta(x - vt)$, where the DW velocity v is assumed to be a positive constant, and $\varphi = \varphi_0 =$ constant. By using this strategy, we get

$$\begin{aligned}
(u - v)\vartheta' &= \gamma\cos\varphi_0\left[\left(N_x - N_y\right)\sin\varphi_0\sin\vartheta + \chi u\right] \\
(\alpha_G v - \eta u)\vartheta' + \hat{\zeta} &= -\gamma A\vartheta'' \\
&+ \gamma\sin\vartheta\cos\vartheta\left(\beta - N_z + N_x\cos^2\varphi_0 + N_y\sin^2\varphi_0\right) \\
&- \gamma\chi u\sin\varphi_0\cos\vartheta,
\end{aligned} \tag{17}$$

where $\hat{\zeta} = \zeta\,\text{sign}(v\vartheta')$, while the prime denotes the derivative with respect to the travelling wave variable $x - vt$ and the boundary conditions take the Dirichlet form $\vartheta(-\infty) = 0$, $\vartheta(+\infty) = \pi$. We can recast $(17)_1$ in the following form:

$$\vartheta' = \Gamma_0(\sin\vartheta + \rho), \tag{18}$$

where

$$\begin{aligned}
\rho &= \frac{\gamma\chi u\cos\varphi_0}{(u - v)\Gamma_0} \\
\Gamma_0 &= \frac{\gamma\left(N_x - N_y\right)\cos\varphi_0\sin\varphi_0}{u - v}.
\end{aligned} \tag{19}$$

In $(19)_1$ the parameter ρ accounts for the presence of the Rashba field. For what concerns $(19)_2$, let us remind that, in the absence of the Rashba field and nonlinear dissipations, we recover the classical definition of DW width $\Gamma_0^{-1} = \delta$ [10, 11, 20] with

$$\delta = \sqrt{\frac{A}{\beta - N_z + N_y + \left(N_x - N_y\right)\cos^2\varphi_0}}. \tag{20}$$

Let us also notice that the knowledge of the DW width requires, as shown in (20), the value of the demagnetizing factors (N_x, N_y, N_z). On the other hand, as discussed after (6), the demagnetizing factors can be computed once the DW width is known. This apparent conflict is generally solved by determining the DW width by means of alternative methods (e.g., by extracting it either from the profile of the travelling wave computed numerically and/or analytically, or from experiments).

Substituting (18) in $(17)_2$, after some algebraic steps, leads to

$$\widehat{M}\sin\vartheta + \widehat{Q}\sin\vartheta\cos\vartheta + \widehat{S}\cos\vartheta + \widehat{P} = 0, \tag{21}$$

where

$$\begin{aligned}
\widehat{M} &= \Gamma_0\left(\alpha_G v - \eta u\right)\\
\widehat{Q} &= \gamma\left[A\Gamma_0^2 - \left(\beta - N_z + N_x\cos^2\varphi_0 + N_y\sin^2\varphi_0\right)\right]\\
\widehat{S} &= \gamma\chi u\left(\sin\varphi_0 + \frac{\gamma A\Gamma_0\cos\varphi_0}{u - v}\right)\\
\widehat{P} &= \left(\alpha_G v - \eta u\right)\frac{\gamma\chi u\cos\varphi_0}{u - v} + \widehat{\zeta},
\end{aligned} \tag{22}$$

By performing the average of (21) over the DW width (i.e., for $0 \le \vartheta \le \pi$) and taking into account that the terms defined in (22) do not depend on ϑ, it is possible to derive the following expression for the DW velocity v as a function of the current-dependent spin-torque velocity u:

$$v = \frac{\left(\eta\chi\Gamma_0/\left(N_x - N_y\right)\right)u^2 + \left(2\eta\Gamma_0/\pi\right)u - \widehat{\zeta}}{\left(\alpha_G\chi\Gamma_0/\left(N_x - N_y\right)\right)u + \left(2\alpha_G\,\Gamma_0/\pi\right)}, \tag{23}$$

where it is supposed to deal with a DW of Bloch type ($\varphi_0 = \pi/2$) [19, 20].

As pointed out in some previous works [20–22], the inclusion of a dry-friction dissipation generally yields the steady DW motion to take place for values of the input stimulus which overcome a well-defined threshold. Equation (23) gives also the possibility to determine such a threshold current, defined as the minimum current value which satisfies the condition $v \ge 0$. In order to properly investigate on this aspect (and to emphasize the sole effect of the Rashba field), let us consider the two separate cases corresponding to the presence and the absence of the dry-friction dissipation function.

2.1. DW Dynamics in the Absence of Dry-Friction ($\zeta = 0$). If we exclude the additive dry friction term in the dissipation function, (23) simply reduces to

$$v = \frac{\eta}{\alpha_G}u \tag{24}$$

that, interestingly, matches exactly the current-driven steady DW velocity derived in the absence of the Rashba field [20, 22]. It is also straightforward to notice that, in the perfect adiabatic case ($\eta = 0$), no DW motion occurs. Equation (24) also implies that the threshold current is null:

$$u_{\text{th}} = 0, \tag{25}$$

so that the DW motion takes place for any nonnull value of the applied current. Results coming from (24) and (25), which clearly claim that the Rashba field does not modify the DW velocity (and, in turn, the DW mobility, defined as the ratio between the velocity and the input current) and the threshold current, satisfactorily agree with the recent experimental [14, 15] and numerical [17, 18] investigations. However, since these studies pointed out an increase of the Walker breakdown value and an enlargement of the DW width, we perform further investigations in this direction to validate the appropriateness of our initial conjectures.

In order to determine the Walker breakdown, from the definition $(19)_1$, we can write

$$\sin 2\varphi_0 = \frac{2\Gamma_0(u - v)}{\gamma\left(N_x - N_y\right)} \tag{26}$$

that implies

$$u - \frac{\gamma}{2\Gamma_0}\left|N_x - N_y\right| \le v \le u + \frac{\gamma}{2\Gamma_0}\left|N_x - N_y\right|. \tag{27}$$

Let us remember that the left and the right implications of (27) are representative of the so-called lower and upper Walker breakdown conditions, respectively [20–22]. They define the range of the input source in which the steady DW motion takes place.

By comparing the expression of the DW velocity (24) and the breakdown condition (27) with those derived in the absence of the Rashba field [20, 22], we report no differences which, at first, would lead to conclude that the inclusion of this field contribution has no influence on the DW dynamics at all. Nonetheless, as mentioned previously, the works carried out in [15, 17] have shown that, in a framework with Rashba field and no internal disorder, in spite of the DW mobility that is unchanged, the upper Walker breakdown is increased (the lower breakdown brings a nonnegligible contribution only in the presence of nonlinear dissipations [22]). Since our results do not allow to apparently satisfy this latter property, we therefore ask on the reasons of this discrepancy. We believe that the answer has to be searched in the numerical values of the parameters appearing in the expression of the Walker breakdown. In detail, we can hypothesize that the Rashba field has changed the modulus of the quantity $\gamma|N_x - N_y|/2\Gamma_0$, making it somehow larger. In particular, considering that the expression of Γ_0 (19) is not formally affected, the only way to obtain such an increase is that the demagnetizing factors, strictly related to the DW width, are varied. In order to find out if our conjecture is correct, we search for the expression of the DW profile by integrating (18). The resolving procedure strictly depends on the parameter ρ that can be recast in the form:

$$\rho = \frac{\chi u}{\left(N_x - N_y\right)}. \tag{28}$$

For completeness, let us remember first that, in the case $\rho = 0$, corresponding to the absence of Rashba field, the solution is in the classical form [19, 20]:

$$\vartheta = 2 \arctan\left[e^{\Gamma_0 (x - vt)}\right], \tag{29}$$

where Γ_0 indeed equals the inverse of the DW width.

Let us discuss, now, the solutions obtained as a function of the strength of the quantity $|\rho|$.

For $|\rho| < 1$, the solution can be expressed as

$$\vartheta = \arccos \frac{1 - f^2(x - vt)}{1 + f^2(x - vt)}, \tag{30}$$

where the expression of the function f can be found in (99)–(102) of [19] and in (35) of [20]. In this case, the DW profile (30) is distorted with respect to the classical case, as shown in Figure 4 of [19], and, in particular, the DW width increases with increasing ρ. Under this circumstance, therefore, the analytical model confirms the enlargement of the DW width which was highlighted in a previous work [18]. Nevertheless, this increase of the DW width changes, in turn, the values of the demagnetizing factors, in such a way that N_x approaches N_y leading to an overall decrease of the quantity $|N_x - N_y|$, so lowering the Walker breakdown value. This latter evidence is thus in contradiction with the other expected feature of the dynamics under investigation.

For this reason, we can hypothesize that the correct solution has to be searched for $|\rho| > 1$. The solution in this case is

$$\vartheta = 2 \arctan\left\{\frac{\sqrt{\rho^2 - 1} \tan\left[\Gamma_0 \sqrt{\rho^2 - 1}(x - vt) + \kappa\right] - 1}{\rho}\right\} \tag{31}$$

being κ the integration constant

$$\kappa = \frac{1}{\sqrt{\rho^2 - 1}} \arctan\left(\frac{\rho + 1}{\sqrt{\rho^2 - 1}}\right) \tag{32}$$

that has to be chosen in such a way the variable ϑ, evaluated at the center of the DW, is null. It should be indeed mentioned that, in this case, the travelling wave solution does not allow to satisfy the Dirichlet boundary conditions, so that the solution is only locally valid, namely, in the proximity of the center of the DW.

In order to estimate the orders of magnitude of the quantities involved in the model and to validate our assumption, we carry out a numerical evaluation of the travelling wave profile. For this reason, we take into account the parameter setup proposed in [17]. In detail, we consider a magnetic nanostrip of thickness $t = 3$ nm and $w = 120$ nm, saturation magnetization $M_S = 3 \times 10^5$ A/m, Landé factor $g = 2$, exchange constant $A = 10^{-11}$ J/m, anisotropy constant $K = 2 \times 10^5$ J/m^3, Gilbert damping constant $\alpha_G = 0.2$, polarization factor $P = 0.5$, nonadiabatic parameter $\eta = 0.4$.

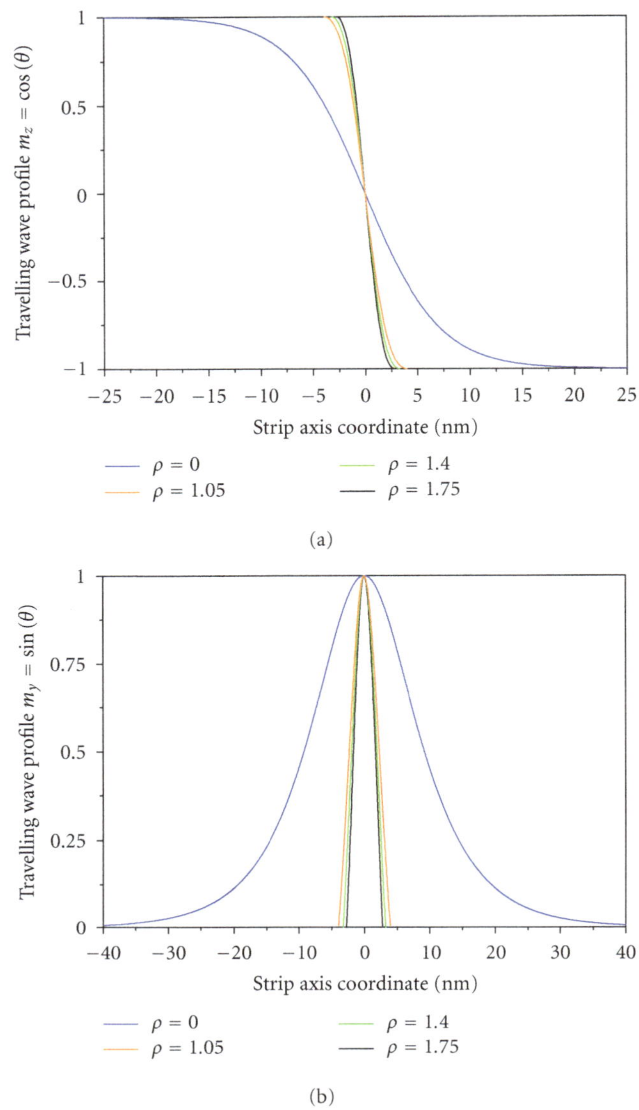

FIGURE 2: Comparison among the travelling wave profiles with and without Rashba field ($|\rho| > 1$ and $\rho = 0$, resp.) computed analytically by using (31)-(32).

Figure 2 shows the profile of the Bloch DW (characterized by $\varphi_0 = \pi/2$), as its components $m_z = \cos\vartheta$ and $m_y = \sin\vartheta$ (m_x is null everywhere), for $\rho = 0$ (namely, with no Rashba field) and $|\rho| > 1$ (with Rashba field), as deduced from (31)-(32). It is clear that, in this case, the presence of the Rashba field would strongly modify the DW profile and width, making this latter about ten times narrower. It can be also appreciated that the DW width is only slightly affected by variations of ρ, higher than unity.

Starting from these results, we evaluate the new demagnetizing factors corresponding to the modified situation.

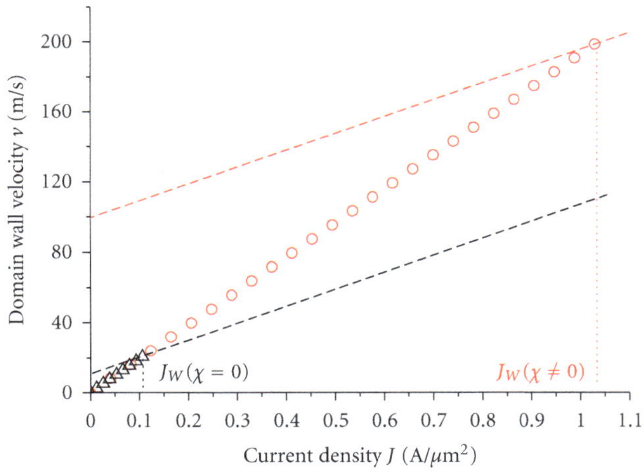

FIGURE 3: Current-driven DW velocity in the steady dynamic regime with and without Rashba field ($\chi \neq 0$ and $\chi = 0$, resp.) computed analytically by using (24) together with the upper breakdown condition (27).

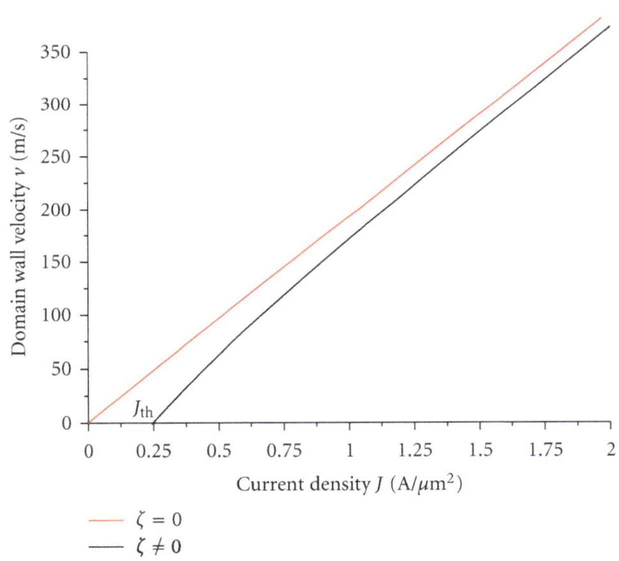

FIGURE 4: Current-driven DW velocities in the steady dynamic regime with and without the inclusion of the dry-friction dissipation function in the mathematical model ($\zeta \neq 0$ and $\zeta = 0$, resp.). The limit of Walker has not been considered in this figure in order to better emphasize the nonlinear behavior of the DW velocity influenced by internal disorder, $\zeta \neq 0$, with respect to the case $\zeta = 0$. Walker breakdown conditions, indeed, are not influenced by ζ.

Because of the DW width is reduced, the difference $|N_x - N_y|$ increases, leading to an increase of the Walker breakdown. For instance, such a difference is equal to 0.046 for $\rho = 0$ and 0.445 for $|\rho| > 1$. The resulting increase of the Walker breakdown value and the enlargement of the range of steady DW motion are depicted in Figure 3 (obtained by using (24) together with (27)). When no Rashba field is considered, namely, $\chi = 0$ (i.e., $\rho = 0$), the Walker breakdown current is $J_W = 0.11\,\text{A}/\mu\text{m}^2$. Under the influence of the Rashba effect, instead, the upper Walker limit increases, making the steady regime possible up to a higher current $J_W(\chi \neq 0) = 1.04\,\text{A}/\mu\text{m}^2$. From a direct inspection of Figure 3, it should be also noticed that, in both cases, there is no threshold current and the DW mobility is not affected by the Rashba field.

Summarizing, the case $|\rho| > 1$ leads to the expected raise of the Walker breakdown value but negates the enlargement of the DW width.

We can state, therefore, that the classical travelling wave ansatz for the ELLG equation ($\vartheta = \vartheta(x - vt)$ with $v =$ constant and $\varphi =$ constant) cannot be satisfactorily used to reproduce the overall effects of the Rashba field on the steady DW motion in ferromagnetic nanostrip.

2.2. DW Dynamics in the Presence of Dry Friction ($\zeta \neq 0$). The usage of a dry-friction dissipation function already turned to be useful to model the effects of crystallographic defects, structural disorder, including surface roughness, on the DW motion [20, 21]. In particular, it was demonstrated that the inclusion of such a friction mechanism leads to the appearance of a threshold below which no DW motion can take place, whereas the DW mobility is not affected in the above-threshold regime. In the present work, we would like to test these two properties when the dry-friction dissipation acts simultaneously to the Rashba field. To this aim, (23) is plotted in Figure 4. As it is expected, also in this case, the dry friction causes the motion to occur for current values larger than a given threshold current J_{th} (in the figure, $J_{\text{th}} = 0.25\,\text{A}/\mu\text{m}^2$ for $\zeta = 2 \times 10^{-2}\gamma$). On the other hand, the DW velocity, which followed a linear trend in the absence of Rashba field (independently of the presence of dry friction), now exhibits a nonlinear dependence on the input current which approximate, for large current values, the velocity obtained in the case $\zeta = 0$. It is interesting to notice that the same result was qualitatively obtained in experiments [16] as well as in numerical simulations [17] that accounted for thermal effects and roughness. However, the corresponding travelling wave solutions would suffer from the same incompatibility with respect to numerical and experimental observations.

3. Conclusions

In this work, we have analyzed the bias-field-free current-driven DW motion in a ferromagnetic nanostrip subject to the Rashba field and dry-friction dissipation. The study has been mathematically carried out by modifying the extended Landau-Lifshitz-Gilbert equation with the inclusion of the Rashba contribution into the effective field. The standard travelling wave ansatz generally used for the equation of motion, within the steady regime, does not succeed in confirming simultaneously both the key features revealed in recent numerical and experimental observations: increase of the Walker breakdown value and enlargement of the DW width.

This result suggests that the system (16) has to be solved by using a different approach. We believe that, due to the transversal component of the effective field induced by the Rashba field, the hypothesis $\varphi = \text{constant}$, together with the Dirichlet boundary conditions $\vartheta(-\infty) = 0$ and $\vartheta(\infty) = \pi$, does not apply in this case. For this reason, one could consider that the angle φ exhibits an analogous travelling wave dependence $\varphi = \varphi(x - vt)$ and that, due to the symmetry of the problem, Neumann boundary conditions $\vartheta'(\pm\infty) = 0$ and $\varphi'(\pm\infty) = 0$ should be satisfied instead. By imposing such constraints, we get the following nonlinear system of coupled ordinary differential equations:

$$\begin{aligned}
&(u - v)\vartheta' - \left(\vartheta'^2 + \sin^2\vartheta\varphi'^2\right)^{-1/2}\zeta v^2 \sin\vartheta\,\varphi' \\
&\quad -2A\gamma\cos\vartheta\,\vartheta'\varphi' - A\gamma\sin\vartheta\,\varphi'' \\
&\quad +(\eta u\sin\vartheta - \alpha_G v\sin\vartheta)\varphi' \\
&\quad -\gamma\left(N_x - N_y\right)\sin\varphi\cos\varphi\sin\vartheta \\
&\quad -\gamma\chi u\cos\varphi = 0 \\
&\gamma\vartheta'' + (\alpha_G v - \eta u)\vartheta' + \left(\vartheta'^2 + \sin^2\vartheta\varphi'^2\right)^{-1/2}\zeta v^2\vartheta' \\
&\quad -A\gamma\sin\vartheta\cos\vartheta\varphi'^2 + (u - v)\sin\vartheta\,\varphi' \\
&\quad -\gamma\sin\vartheta\cos\vartheta\left(\beta - N_z + N_x\cos^2\varphi + N_y\sin^2\varphi\right) \\
&\quad +\gamma\chi u\sin\varphi\cos\vartheta = 0.
\end{aligned} \tag{33}$$

However, solving this system without simplifying assumptions is not trivial at all. For example, the missing information could be argued from ad hoc micromagnetic simulations which should provide the accurate profile of the travelling wave variable and their dependence on the strength of the Rashba field. Therefore, we strongly encourage numerical investigations in this direction to overcome this issue.

Acknowledgment

G. Consolo gratefully acknowledges support from GNFM-INdAM.

References

[1] J. C. Slonczewski, "Current-driven excitation of magnetic multilayers," *Journal of Magnetism and Magnetic Materials*, vol. 159, no. 1-2, pp. L1–L7, 1996.

[2] L. Berger, "Emission of spin waves by a magnetic multilayer traversed by a current," *Physical Review B*, vol. 54, no. 13, pp. 9353–9358, 1996.

[3] A. N. Slavin and P. Kabos, "Approximate theory of microwave generation in a current-driven magnetic nanocontact magnetized in an arbitrary direction," *IEEE Transactions on Magnetics*, vol. 41, no. 4, pp. 1264–1273, 2005.

[4] G. Finocchio, O. Ozatay, L. Torres, M. Carpentieri, G. Consolo, and B. Azzerboni, "Micromagnetic modeling of magnetization reversal in nano-scale point contact devices," *IEEE Transactions on Magnetics*, vol. 43, no. 6, pp. 2938–2940, 2007.

[5] G. Consolo and V. Puliafito, "Analytical and micromagnetic study of nonlinear amplitude modulation in spintronic modulators," *IEEE Transactions on Magnetics*, vol. 46, no. 6, Article ID 5467373, pp. 2063–2066, 2010.

[6] X. Zhu and J. G. Zhu, "Bias-field-free microwave oscillator driven by perpendicularly polarized spin current," *IEEE Transactions on Magnetics*, vol. 42, pp. 2670–2672, 2006.

[7] V. Puliafito, B. Azzerboni, G. Consolo, G. Finocchio, L. Torres, and L. L. Diaz, "Micromagnetic modeling of nanocontact spin-torque oscillators with perpendicular anisotropy at zero bias field," *IEEE Transactions on Magnetics*, vol. 44, no. 11, pp. 2512–2515, 2008.

[8] N. Vernier, D. A. Allwood, D. Atkinson, M. D. Cooke, and R. P. Cowburn, "Domain wall propagation in magnetic nanowires by spin-polarized current injection," *Europhysics Letters*, vol. 65, no. 4, pp. 526–532, 2004.

[9] D. A. Allwood, G. Xiong, C. C. Faulkner, D. Atkinson, D. Petit, and R. P. Cowburn, "Magnetic domain-wall logic," *Science*, vol. 309, no. 5741, pp. 1688–1692, 2005.

[10] A. Thiaville, Y. Nakatani, J. Miltat, and Y. Suzuki, "Micromagnetic understanding of current-driven domain wall motion in patterned nanowires," *Europhysics Letters*, vol. 69, no. 6, pp. 990–996, 2005.

[11] A. Mougin, M. Cormier, J. P. Adam, P. J. Metaxas, and J. Ferré, "Domain wall mobility, stability and Walker breakdown in magnetic nanowires," *Europhysics Letters*, vol. 78, no. 5, Article ID 57007, 2007.

[12] A. Manchon and S. Zhang, "Theory of spin torque due to spin-orbit coupling," *Physical Review B - Condensed Matter and Materials Physics*, vol. 79, no. 9, Article ID 094422, 2009.

[13] K. Obata and G. Tatara, "Current-induced domain wall motion in Rashba spin-orbit system," *Physical Review B*, vol. 77, no. 21, Article ID 214429, 2008.

[14] I. M. Miron, G. Gaudin, S. Auffret et al., "Current-driven spin torque induced by the Rashba effect in a ferromagnetic metal layer," *Nature Materials*, vol. 9, no. 3, pp. 230–234, 2010.

[15] I. M. Miron, T. Moore, H. Szambolics et al., "Fast current-induced domain-wall motion controlled by the Rashba effect," *Nature Materials*, vol. 10, no. 6, pp. 419–423, 2011.

[16] J. Ryu, S. B. Choe, and H. W. Lee, "Magnetic domain-wall motion in a nanowire: depinning and creep," *Physical Review B*, vol. 84, Article ID 075469, 2011.

[17] E. Martinez, "The influence of the Rashba field on the current-induced domain wall dynamics: a full micromagnetic analysis, including surface roughness and thermal effects," *Journal of Applied Physics*, vol. 111, Article ID 07D302, 2012.

[18] E. Martinez, "Micromagnetic analysis of the Rashba field on current-induced domain wall propagation," *Journal of Applied Physics*, vol. 111, Article ID 033901, 2012.

[19] P. Podio-Guidugli and G. Tomassetti, "On the steady motions of a flat domain wall in a ferromagnet," *European Physical Journal B*, vol. 26, no. 2, pp. 191–198, 2002.

[20] G. Consolo, C. Currò, E. Martinez, and G. Valenti, "Mathematical modeling and numerical simulation of domain wall motion in magnetic nanostrips with crystallographic defects," *Applied Mathematical Modelling*, vol. 36, no. 10, pp. 4876–4886, 2012.

[21] G. Consolo and E. Martinez, "The effect of dry friction on domain wall dynamics: a micromagnetic study," *Journal of Applied Physics*, vol. 111, Article ID 07D312, 3 pages, 2012.

[22] G. Consolo and G. Valenti, "Travelling wave solutions of the one-dimensional extended Landau-Lifshitz-Gilbert equation with nonlinear dry and viscous dissipations," *Acta Applicandae Mathematicae*, 2012.

[23] L. D. Landau and E. Lifshitz, "On the theory of the dispersion of magnetic permeability in ferromagnetic bodies," *Physikalische Zeitschrift der Sowietunion*, vol. 8, pp. 153–169, 1935.

[24] W. F. Brown, *Micromagnetics*, Krieger, Huntington, NY, USA, 1963.

Electrodynamics of Metallic Superconductors

M. Dressel

1. Physikalisches Institut, Universität Stuttgart, Pfaffenwaldring 57, 70550 Stuttgart, Germany

Correspondence should be addressed to M. Dressel; dressel@pi1.physik.uni-stuttgart.de

Academic Editor: Victor V. Moshchalkov

The theoretical and experimental aspects of the microwave, terahertz, and infrared properties of superconductors are discussed. Electrodynamics can provide information about the superconducting condensate as well as about the quasiparticles. The aim is to understand the frequency dependence of the complex conductivity, the change with temperature and time, and its dependence on material parameters. We confine ourselves to conventional metallic superconductors, in particular, Nb and related nitrides and review the seminal papers but also highlight latest developments and recent experimental achievements. The possibility to produce well-defined thin films of metallic superconductors that can be tuned in their properties allows the exploration of fundamental issues, such as the superconductor-insulator transition; furthermore it provides the basis for the development of novel and advanced applications, for instance, superconducting single-photon detectors.

1. Introduction

1.1. Historical Developments. The history of superconductivity in the 1950s is one of the best examples of how the interplay of experiment and theory advances science towards a profound understanding. After having moved to Urbana in 1951 and even more intense after having received the Nobel Prize for the development of the transistor in 1956, John Bardeen was pressing to solve the longstanding problem of superconductivity. During these years he was very well aware of the experimental facts and recent developments but also triggered experimental investigations and pushed certain measurements.

The presence of an energy gap in the density of states of a superconductor (and thus in the excitation spectrum) can be inferred from the temperature dependence of the specific heat that had been measured already in Heike Kamerlingh Onnes' laboratory in Leiden [1]; however, it was Bardeen who seriously discussed this idea in more detail in 1955 [2, 3] based on considerations proposed by Welker [4] years earlier. He assumed "that in the superconducting state a finite energy gap $\Delta \approx k_B T_c$ is required to excite electrons from the surface of the Fermi sea." Cooper developed this concept further when he introduced the electron pairs in the following year [5]. In the seminal BCS paper of 1957 [6], Bardeen, Cooper and Schrieffer finally derived the fundamental ratio $2\Delta/k_B T_c = 3.5$ and the temperature dependence of the energy gap. In their calculation of the matrix elements, they explicitly refer to the ongoing NMR relaxation-rate experiments by Hebel and Slichter [7, 8], ultrasound-attenuation measurements by Morse and Bohm [9], and microwave-absorption experiments by the Tinkham group [10–12].

Concerning the high-frequency absorption, in an article submitted 1955 [13] Buckingham suggested that (for consistency reasons the notation is transformed):

> *"a gap of width $2\Delta(0)$ in the electron energy spectrum in the ground state, its width, 2Δ, at finite temperatures will decrease, eventually vanishing at the transition temperature, T_c. The normal state is regarded as one in which the energy spectrum is without a gap and is independent of temperature, the system representing the superconducting state becoming identical with the normal one for $T > T_c$. The electromagnetic properties of such a system will depend strongly on the size of the gap, but at high frequencies and near the transition*

temperature, when $\hbar\omega/2\Delta$ is large, will approach those of the normal state (without gap)."

Looking at the experimental results of Blevins et al. who measured the surface resistance of tin in the millimeter-wave range [14], he suggested some scaling of frequency and transition temperature close to T_c. This was further developed by Tinkham [11], based on his far-infrared transmission measurements on thin lead films [10] where an anomalously high transmission was observed (cf. Figure 1(b)). He suggested two contributions (superconducting and normal electrons) becoming relevant in the limiting cases of low and high frequencies:

"*We consider a model of a superconductor at absolute zero in which a gap of width $\hbar\omega_g \approx 3k_BT_c$ appears in the spectrum of one-electron energy states. [$\cdots$] The lossy conductivity σ_1 will be zero for photon energies $\hbar\omega < \hbar\omega_g$. For $\omega > \omega_g$, σ_1 will rise gradually with ω as an increasing fraction of the states originally within $\hbar\omega$ of the occupied states below the Fermi level are still available for excitation. If the gap merely excluded the states within $\hbar\omega_g$ leaving all else unchanged, this simple picture would suggest a rise as $\sigma_1 = \sigma_n(1-\omega_g/\omega)$. The experimental rise seems to be faster, cutting off at least as fast as $(1 - \omega_g^2/\omega^2)$. This might correspond to the states displaced from the gap just "piling up" on either side, so that the change averages to zero more rapidly for $\omega \gg \omega_g$.*"

These ideas are further developed and extended in later papers [12, 15], where the size of the superconducting energy gap and the shape $\sigma_1(\omega)$ at ω_g were analyzed using data of different materials [16].

The situation was more or less settled in 1958, when the Westinghouse group of Biondi et al. summarized the situation on the superconducting energy gap in a review [17]. Previously they had published several thorough microwave and millimeter-wave results on superconducting aluminum and tin [18–20], and several nice papers followed [21–23]. At the same time Tinkham gave an overview on the work of his group [24]. While Biondi and Garfunkel always measured the temperature dependence for certain frequencies, Tinkham and his collaborators focussed at the spectral properties at fixed temperatures. This allowed him to make the sum-rule argument introduced in the previous quote [11], which was finalized by Ferrell et al. in 1959 [25, 26]. In Figure 1 the frequency dependence of the electrodynamical properties of a superconductor at $T = 0$ is sketched. Upon decreasing the frequency the real part of the conductivity $\sigma_1(\omega)/\sigma_n$ gradually vanishes at the gap frequency $\omega_g = 2\Delta/\hbar$, while the imaginary part $\sigma_2(\omega)/\sigma_n$ diverges as $\omega \rightarrow 0$. The maximum in the relative transmission $\mathrm{Tr}_s/\mathrm{Tr}_n$ is observed right at the gap (with values well above the unity) before it goes to zero for $\omega = 0$. There is also a maximum of the absorption above the superconducting gap, but no absorption takes place below the gap 2Δ. It poses a real challenge to measure the reflectivity of a metallic superconductor since it approaches unity at ω_g, but in general it is well above 99% already in the normal state (Figure 1(c)).

The problem remaining was the determination of the superconducting conductivity, that is, the actual frequency and temperature dependence of $\sigma_1(\omega)$ and $\sigma_2(\omega)$ individually without the assumption of any model or Kramers-Kronig relation. Figure 2 displays temperature and frequency-dependent conductivity σ_1/σ_n as it is calculated from Mattis-Bardeen theory. σ_1/σ_n features a minimum at $\hbar\omega/2\Delta(T) = 1$ which shifts to lower frequencies and is smeared out as T approaches T_c from below. Also seen is the enormous enhancement of the conductivity $\sigma_1(\omega)/\sigma_n$ for low frequencies observed right below the superconducting gap: it became known as the coherence peak. Despite continuous efforts over the years [30–32], only in 1968 Palmer and Tinkham reached this goal of measuring the real and imaginary parts of the electrodynamic response independently [33]; simultaneous measurements of the far-infrared transmission through and reflection off thin films of lead made it possible to calculate the conductivity unambiguously. The status is summarized in several reviews [34, 35]. The precise measurement of the temperature-dependent conductivity right below the superconducting transition, however, was achieved only a quarter of a century later in the group of Grüner [36, 37]. Previous efforts by the Cambridge group of Pippard and others were less convincing [38–40]. The so-called coherence peak known from the NMR experiments of Hebel and Slichter [7, 8] is also seen in the microwave conductivity of lead, niobium, and eventually even in alumimum [29] where for the first time the complete frequency evolution of the coherence peak was measured; exactly fifty years after it was calculated. Still ultrasound absorption remains a desideratum in the exploration of the coherence effects: a peak is predicted above 2Δ, but no frequency-dependent ultrasound absorption measurements on superconductors have been carried out in the vicinity of the energy gap.

Even today, Tinkham's book [41–43] gives the best introduction to the electrodymical properties of conventional superconductors. More detailed descriptions can be found by Rickayzen [44] and Parks [45] or more recent textbooks [46].

1.2. Recent Developments. Having understood the bulk properties of metallic superconductors fairly well, there are still many open questions when going to the vortex lattice, very thin films, nanostructures, inhomogeneous materials, granular superconductors, highly disordered films and so forth, which are subject of today's research efforts. Thin films of metallic superconductors are of both technological and academic interests.

As we will see, the properties of interacting electron systems are governed by the electronic density of states [47], which changes upon reducing spatial dimensions. Thus, dimensional reduction from three dimensions to two dimensions (3D $\rightarrow$ 2D) may affect optical, electronic, and thermodynamic properties of the system. In superconductivity, studying quasi-two-dimensional thin-film systems turns out to be a particularly fruitful field and leads to remarkable findings such as insulators featuring a superconducting gap [48],

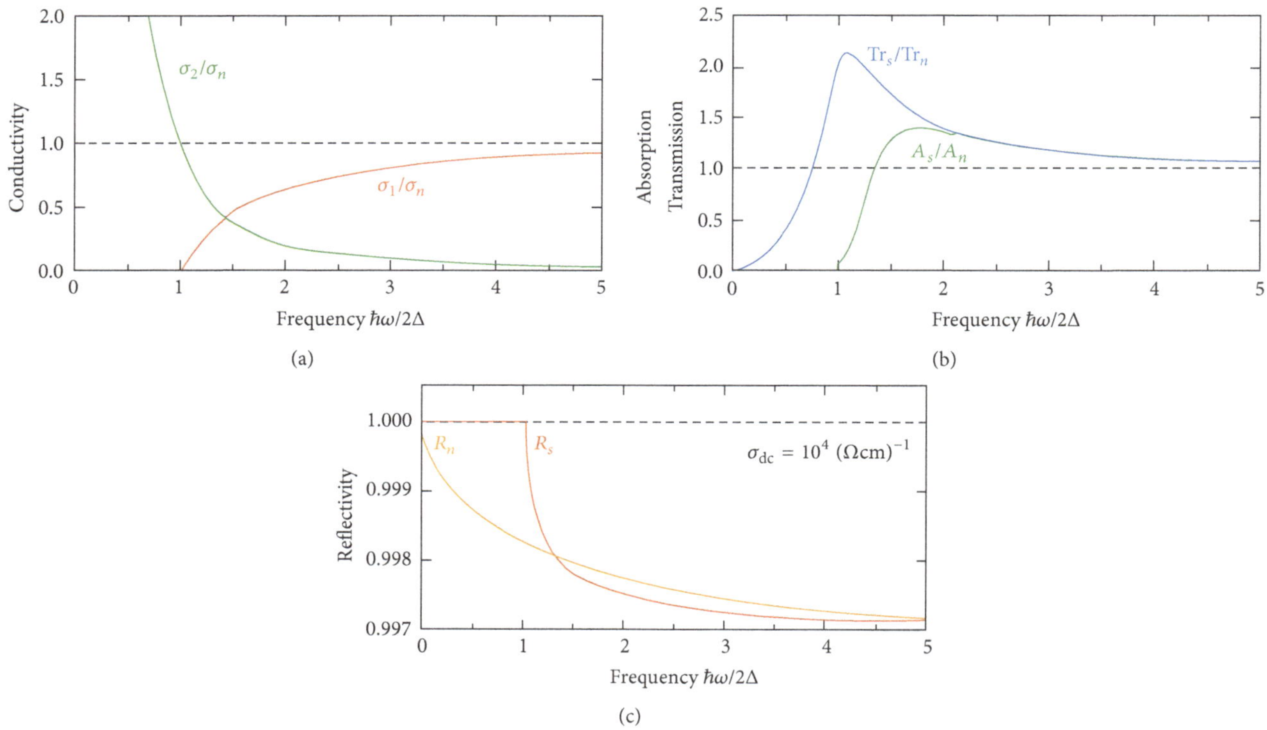

FIGURE 1: Sketch of the electrodynamic properties of a superconductor at $T = 0$: (a) in the upper panel the frequency dependence of the real and imaginary parts of the conductivity σ_1 and σ_2 is plotted, normalized to the normal state behavior σ_n, which basically can be assumed to be constant in the spectral range of interest, where $\omega \ll \Gamma$ with Γ the quasiparticle scattering rate. (b) The second panel exhibits the frequency dependence of the relative transmission $\mathrm{Tr}_s/\mathrm{Tr}_n$ and absorption A_s/A_n. (c) The optical reflectivity of a superconductor R_s is compared to the reflectivity of a normal metal R_n, assuming a dc conductivity of $10^4\ (\Omega\mathrm{cm})^{-1}$.

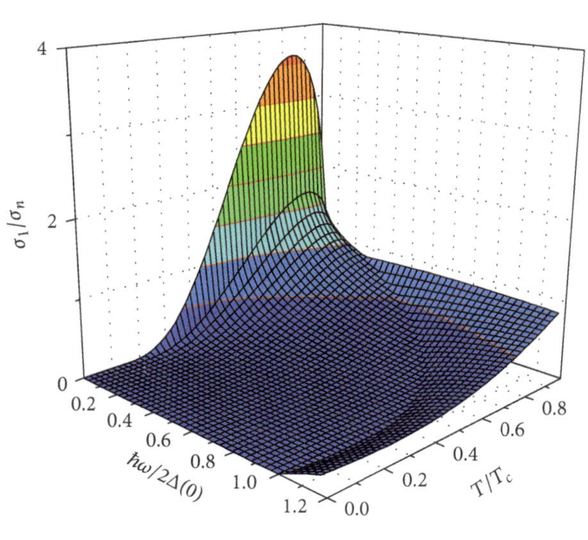

FIGURE 2: Frequency and temperature dependence of the real part of the conductivity σ_1/σ_n calculated according the BCS theory [27, 28] with the ratio of the coherence length to the mean free path $\pi\xi(0)/2\ell = 10$. The pronounced maximum for low frequencies at a temperature slightly below T_c is the coherence peak. The kink at $\hbar\omega = 2\Delta(0)$ corresponds to the superconducting gap that decreases with temperature and vanishes for $T \to T_c$ (after [29]).

pseudogap in conventional s-wave superconductors [49], or extremely strong-coupling superconductivity in Kondo lattices [50]. NbN has recently gained attention as a model system for the superconductor-insulator transition; a topic intensively investigated in thin films of Bi, InO, or TiN: tunneling studies [51] reveal a gapped density of states slightly above T_c in highly disordered films as it has been found for TiN films [49]. This peculiar state resembles the well-established pseudogap in high-T_c cuprate superconductors [52, 53]. In case of NbN, however, T_c is about one order of magnitude higher compared to TiN. Thus, measurements on NbN in the vicinity of the superconductor-insulator transition (SIT) are much less demanding [54, 55]. Consequently, results obtained for NbN might serve as a key to understand the still puzzling pseudogap state in high-T_c cuprates.

Besides pure academic interest, thin-film superconductors play a key role in many state-of-the-art applications such as superconducting quantum interference devices (SQUID), thermal and electrical switches using Josephson junctions, or microwave resonators [56]. Contrary to nonmetallic systems such as MgB_2 or cuprates, thin and ultrathin films of NbN and TaN can be produced of very high quality, for example, by reactive dc-magneton sputtering [57, 58]. They are of particular interest due to their high critical temperature that makes them suited for novel approaches to build single-photon detectors, for example [57, 59–63].

2. Theory

2.1. Mattis-Bardeen Theory. Based on the BCS theory [6], the description of the electrodynamic properties of a superconductor was given by Mattis and Bardeen in their seminal paper [27]. Miller applied these considerations on the surface impedance [64]. In parallel Abrikosov and coworkers [65, 66] also described the electromagnetic absorption of a superconductor, including the coherence factors and the nonlocal relationship between vector potential and induced currents.

The celebrated Mattis-Bardeen equations read

$$\frac{\sigma_1(\omega,T)}{\sigma_n} = \frac{2}{\hbar\omega}\int_{\Delta}^{\infty}\frac{[f(\mathcal{E})-f(\mathcal{E}+\hbar\omega)](\mathcal{E}^2+\Delta^2+\hbar\omega\mathcal{E})}{(\mathcal{E}^2-\Delta^2)^{1/2}[(\mathcal{E}+\hbar\omega)^2-\Delta^2]^{1/2}}d\mathcal{E} + \frac{1}{\hbar\omega}\int_{\Delta-\hbar\omega}^{-\Delta}\frac{[1-2f(\mathcal{E}+\hbar\omega)](\mathcal{E}^2+\Delta^2+\hbar\omega\mathcal{E})}{(\mathcal{E}^2-\Delta^2)^{1/2}[(\mathcal{E}+\hbar\omega)^2-\Delta^2]^{1/2}}d\mathcal{E} \quad (1)$$

$$\frac{\sigma_2(\omega,T)}{\sigma_n} = \frac{1}{\hbar\omega}\int_{\Delta-\hbar\omega,-\Delta}^{\Delta}\frac{[1-2f(\mathcal{E}+\hbar\omega)](\mathcal{E}^2+\Delta^2+\hbar\omega\mathcal{E})}{(\Delta^2-\mathcal{E}^2)^{1/2}[(\mathcal{E}+\hbar\omega)^2-\Delta^2]^{1/2}}d\mathcal{E}, \quad (2)$$

where for $\hbar\omega > 2\Delta$ the lower limit of the integral in (2) becomes $-\Delta$. Here $f(\mathcal{E},T) = (\exp\{\mathcal{E}/k_BT\}-1)^{-1}$ is the usual Fermi-Dirac distribution function. The expression of σ_1/σ_n describes the response of the normal carriers, that is, the dissipative part. The first term of (1) represents the effects of thermally excited quasiparticles, which vanishes for $T = 0$. The second term mainly accounts for the contribution of photon-excited quasiparticles; since it requires the breaking of a Cooper pair, it is basically zero for $\hbar\omega < 2\Delta(T)$. However, there is a small contribution possible of quasiparticles thermally excited across the gap for $T \lesssim T_c$.

The sketch in Figure 3 illustrates the electronic density of states for a normal metal (panel a), a superconductor at finite temperatures (panel b) and the superconductor at $T = 0$ (panel c). Single-particle excitations only take place from occupied states (dark colors) to empty states (light colors). In the case of a metal ($T > T_c$) this is possible for any finite photon energy $\hbar\omega$ leading to a Drude response limited only by the scattering rate $1/\tau$. When the superconducting gap opens in the density of states for $T < T_c$, the states from the gap region pile up above and below $\pm\Delta(T)$ making a large number of excitations possible. We have to distinguish excitations across the gap (solid arrows) for $\hbar\omega > 2\Delta(T)$ from those above (and correspondingly below) the gap between occupied and unoccupied states due to the finite-temperature Fermi distribution $f(\mathcal{E},T)$. The latter ones have no lower-energy limit and dominate the low-frequency response since the joint density of states is largest for $\hbar\omega \to 0$. These thermally activated contributions to the optical conductivity [first term in (1)] are shown by red dashed lines in Figure 4. They are actually observed as the coherence peak in the temperature dependent conductivity right below T_c (see Section 3.1 below). These quasiparticle excitations are limited by the occupation given by the Fermi distribution. The across-gap excitations, on the other hand, always have a clear-cut lower-frequency limit ω_g determined by the gap $2\Delta/\hbar$; this is the energy required to break a Cooper pair. In Figure 4 these excitations (second term in (1)) are plotted as blue dashed lines. Due to the case-2 coherence factor describing the quasiparticle excitations in a superconductor, the conductivity only smoothly increases for $\omega > 2\Delta/\hbar$, despite the divergency in the density of states right above and below the gap [46]. For $T = 0$ these are the only quasiparticle contributions to $\sigma_1(\omega)$. The superconducting condensate is observed as a δ-peak at $\omega = 0$ and dominates the imaginary part $\sigma_2(\omega)$. A generalized Mattis-Bardeen theory which accounts for quasiparticle scattering is derived by Zimmermann et al. [28]. The inset of Figure 4 displays how a finite scattering rate modifies $\sigma_1(\omega)$ and $\sigma_2(\omega)$.

An interesting aspect was elucidated by Chang and Scalapino [67–69], who found an enhancement of the superconducting energy gap and a change in the quasiparticle density of states upon microwave irradiation, in addition to a redistribution of quasiparticles by the microwaves. Eventually it extends to the field of nonequilibrium superconductivity that has been a widely studied field for half a century [70, 71].

Already from the London equations, a relation between the imaginary part of the conductivity and the London penetration depth is obtained [46]:

$$\sigma_2(\omega) = \frac{Ne^2}{m\omega} = \frac{c^2}{4\pi\lambda_L^2\omega}, \quad (3)$$

where the London penetration $\lambda_L = (mc^2/4\pi N_s e^2)^{1/2}$ is determined by the superconducting charge carrier mass and density, m and N_s. Through the Kramers-Kronig relation the real part of the conductivity is then given by

$$\sigma_1(\omega) = \frac{\pi}{2}\frac{N_s e^2}{m}\delta\{\omega=0\} = \frac{c^2}{8\lambda_L^2}\delta\{\omega=0\}, \quad (4)$$

where $\delta\omega = 0$ denote a delta peak centered at $\omega = 0$.

Figures 5 and 6 summarize the frequency and temperature dependence of σ_1 and σ_2 as derived from (1) and (2) assuming the finite scattering effects $\pi\xi(0)/\ell = 10$ discussed in [28, 73–77]. Although the density of states diverges at $\pm\Delta$, the conductivity $\sigma_1(\omega)$ does not show a divergency but a smooth increase which follows approximately the dependence $\sigma_1(\omega)/\sigma_n \propto 1-(\hbar\omega/k_BT_c)^{-1.65}$. The temperature dependent conductivity $\sigma_1(T)$ shows a peak just below T_c at a low frequency; the maximum occurs around $0.8T_c$ and slightly shifts up with frequency. It is called the coherence peak because it does not just reflect the enhanced density of states but is also an expression of the quantum mechanical

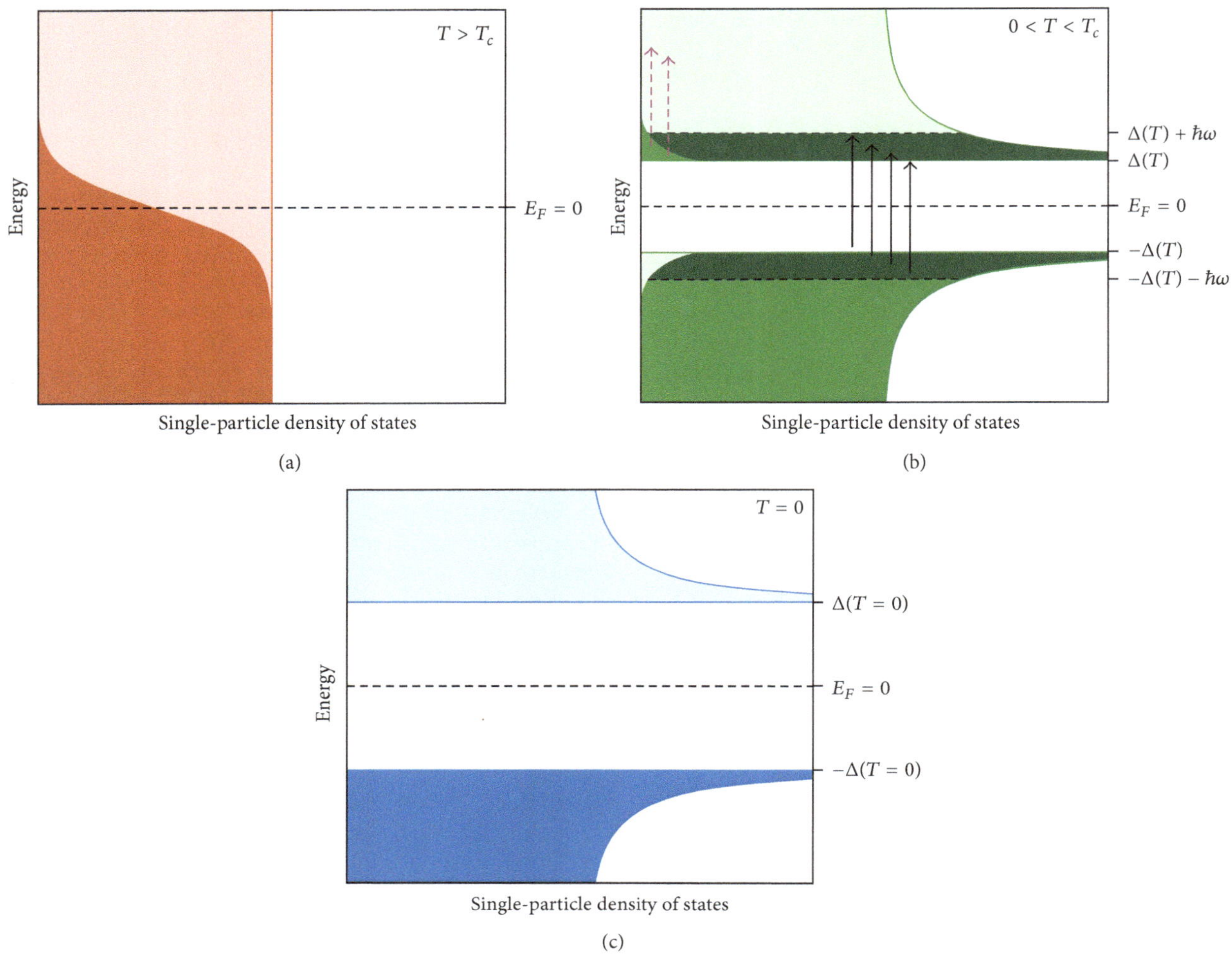

FIGURE 3: The density of electronic states in the close vicinity of the Fermi energy E_F. (a) For a normal metal, the density of states is basically constant. The dark colored area indicates the occupied states according to the Fermi-Dirac statistic at finite temperature. (b) In the case of a superconductor, an energy gap opens around E_F; it grows continuously as the temperature is reduced below T_c. The dotted arrow indicates possible excitations of the occupied states above the gap [first term in (1)], leading to a quasiparticle peak at $\omega = 0$. For the electronic excitations shown by the solid arrow, a minimum energy of 2Δ is required; their contribution is captured by the second term in (1). The dark shaded area up to $|\Delta(T) + \hbar\omega|$ indicates states that can contribute to the conductivity by absorption of photons of arbitrary energy $\hbar\omega$. (c) The full size of the superconducting energy gap is given by $2\Delta_0$ for $T = 0$. No quasiparticle peak is present, leading to absorption only above $\omega_g = 2\Delta/\hbar$. The states removed from the gap area are pilled up below and above the gap, leading to a $E/\sqrt{E^2 - \Delta_0^2}$ divergency.

factor relevant for these excitations. The so-called coherence factor $F(\mathscr{E}_{\mathbf{k}}, \mathscr{E}_{\mathbf{k}'})$ describes the scattering of a quasiparticle from a state $\mathbf{k}$ with energy $\mathscr{E}_{\mathbf{k}}$ to a state $\mathbf{k}' = \mathbf{k}+\mathbf{q}$ with energy $\mathscr{E}_{\mathbf{k}'} = \mathscr{E}_{\mathbf{k}} + \hbar\omega$ upon absorption of a photon with energy $\hbar\omega$ and momentum $\mathbf{q}$. If summed over all $\mathbf{k}$ values, it reads [41–43, 46]

$$F\left(\Delta, \mathscr{E}, \mathscr{E}'\right) = \frac{1}{2}\left(1 + \frac{\Delta^2}{\mathscr{E}\mathscr{E}'}\right). \quad (5)$$

Only for energies below the gap 2Δ, this factor is appreciable: $F \approx 1$ for $\hbar\omega \ll 2\Delta$. For $\hbar\omega \geq 2\Delta$, the coherence factors are reversed, and F vanishes in the present case. For large energies, the coherence effects become negligible since $\mathscr{E}, \mathscr{E}' \gg 2\Delta$ and $F \approx 1/2$. Hence the coherence peak is seen as a maximum in $\sigma_1(T)$ in the low-frequency limit; it becomes smaller with increasing frequency and shifts to higher temperatures. The height of the peak has the following frequency dependence:

$$\left(\frac{\sigma_1}{\sigma_n}\right)_{\max} \sim \log\left\{\frac{2\Delta(0)}{\hbar\omega}\right\}. \quad (6)$$

The peak disappears completely for $\hbar\omega \geq \Delta/2$ (well below 2Δ). At $T = 0$ and $\omega < 2\Delta/\hbar$ the complex part of the conductivity σ_2/σ_n describes the response of the Cooper pairs and is related to the gap parameter through the expression

$$\frac{\sigma_2(T)}{\sigma_n} \approx \frac{\pi\Delta(T)}{\hbar\omega}\tanh\left\{\frac{\Delta(T)}{2k_BT}\right\} \approx \lim_{T\to 0}\frac{\pi\Delta(0)}{\hbar\omega}. \quad (7)$$

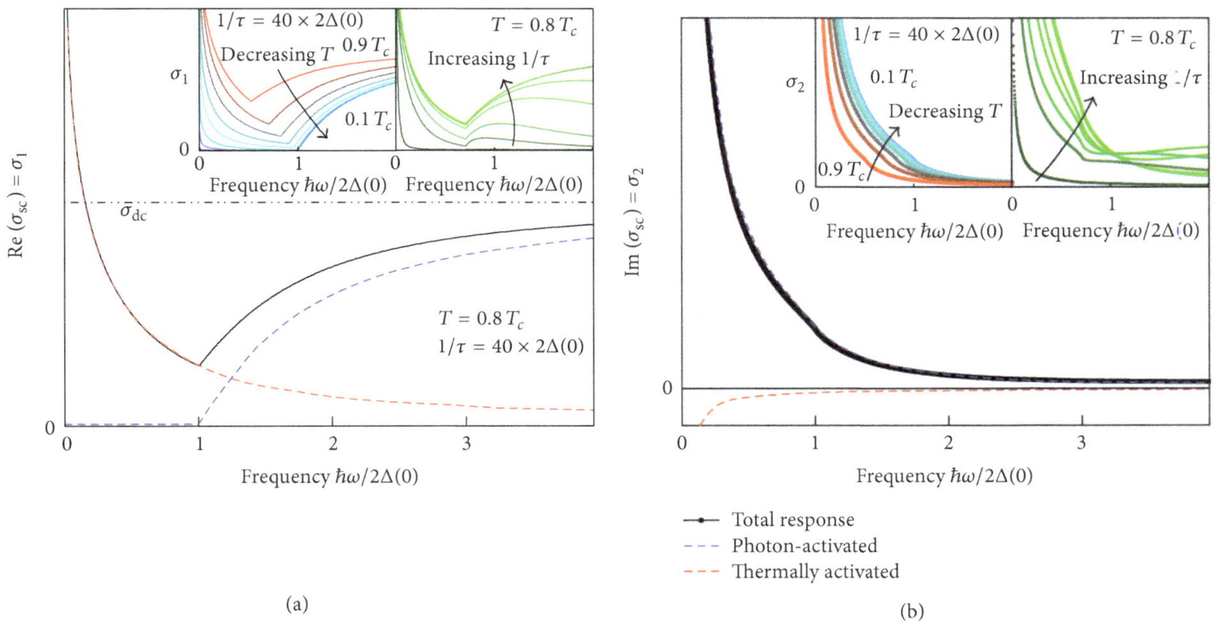

FIGURE 4: (a) Real and (b) imaginary parts of the complex optical conductivity (solid line) of a superconductor as a function of frequency based on generalized Mattis-Bardeen theory [27] for $T = 0.8T_c$ and a finite scattering rate [28]. Dashed lines disentangle the contributions of both the thermal electrons [first term in (1) and dashed arrows in Figure 3(b)] and photon-activated electrons [second term in (1) and solid arrow in Figure 3(b)] to the conductivity. Insets of (a) and (b) show σ_1 and σ_2 for different temperatures (T/T_c = 0.9, 0.8, 0.7, 0.6, 0.5, 0.4, 0.3, 0.2, and 0.1) and scattering rates ($1/\tau$ = 40, 20, 4, 2, 0.8, 0.4, and $0.04 \times 2\Delta(0)/\hbar$), respectively. (Used parameters $2\Delta(0)/hc = 25\,\text{cm}^{-1}$, $T_c = 10$ K, and $\sigma_{dc} = 1066\,\Omega^{-1}\,\text{cm}^{-1}$) (after [72]).

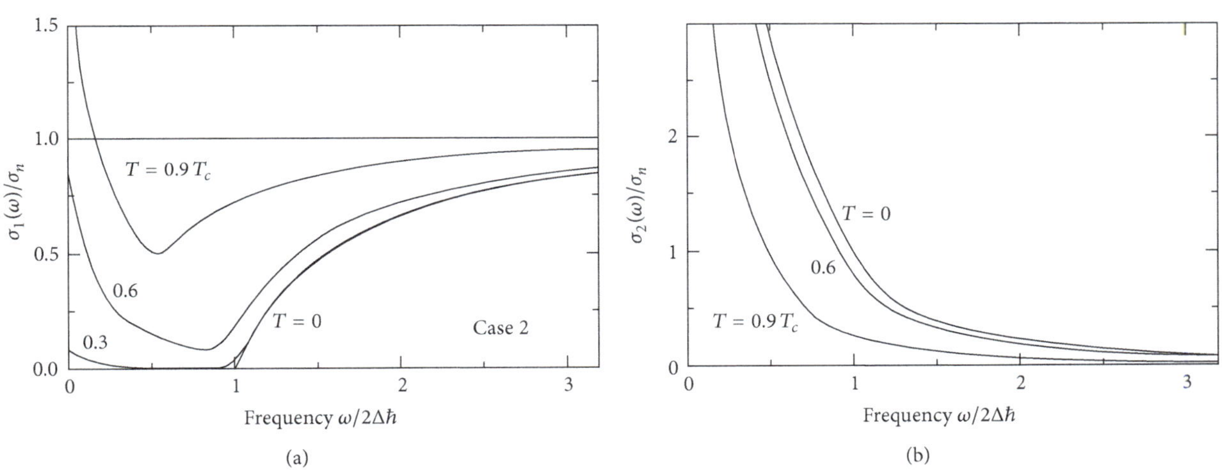

FIGURE 5: Frequency dependence of the conductivity of a superconductor at different temperatures T relative to the transition temperature T_c. (a) The real part $\sigma_1(\omega)/\sigma_n$ exhibits a sharp drop at the superconducting gap 2Δ and vanishes below for $T = 0$. For finite temperatures, the gap gradually moves to lower frequencies; in addition a finite conductivity emerges at low frequencies that increases considerably as $T \to T_c$; σ_1 even exceeds the normal state conductivity in a narrow frequency range right below T_c. (b) The imaginary part of the conductivity $\sigma_2(\omega)/\sigma_n$ diverges for low temperatures as $\sigma_2(\omega) \propto \omega^{-1}$. The kink of $\sigma_1(\omega)$ at $\omega = 2\Delta/\hbar$ also shows up in the imaginary part $\sigma_2(\omega)$ [46].

2.2. Scattering Effects, Length Scales, and Sum Rule. The Equations (1) and (2) were derived by Mattis and Bardeen [27] assuming the infinite scattering time $\tau \to \infty$. In 1967 Nam showed [76, 77], however, that these equations are more general and also valid for $\tau \to 0$. He also extended the application towards the strong coupling regime: $2\Delta/k_B T_c > 3.53$. Nevertheless, in order to describe the electrodynamics of superconductors, different length scales have to be considered as their relative magnitude determines the nature of the superconducting state and also the response to electromagnetic fields. The first length scale is the London penetration depth

$$\lambda_L = \left(\frac{mc^2}{4\pi N e^2}\right)^{1/2} = \frac{c}{\omega_p} \tag{8}$$

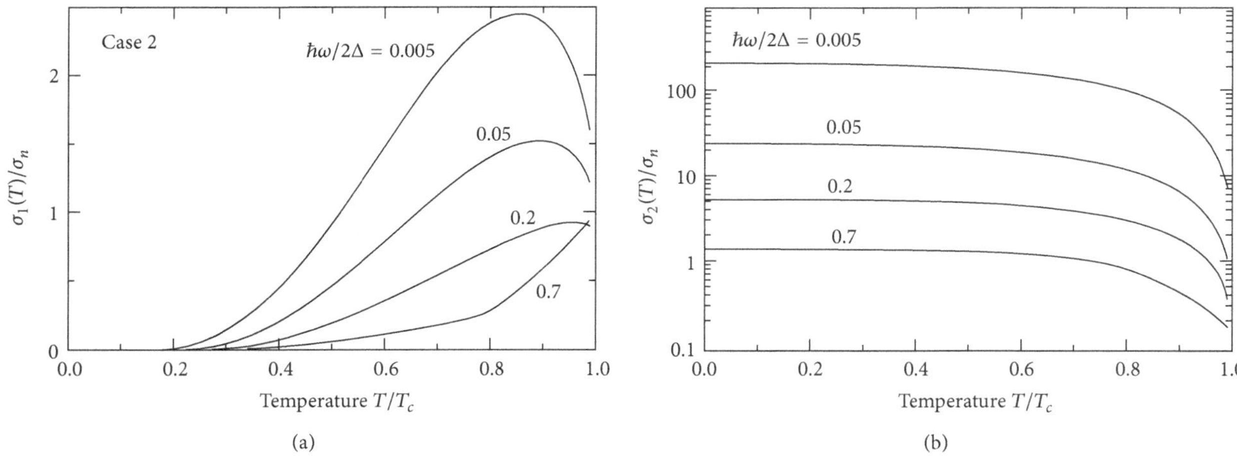

FIGURE 6: Temperature dependence of the conductivity of a superconductor at different frequencies ω with respect to the gap frequency $2\Delta/\hbar$. (a) For low frequencies the conductivity first rises as T decreases below T_c and exhibits a pronounced maximum before it drops to zero for $T \rightarrow 0$. This so-called coherence peak becomes smaller as the frequency increases; eventually $\sigma_1(\omega)/\sigma_n$ drops monotonously to zero. (b) The imaginary part $\sigma_2(\omega)/\sigma_n$ increases continuously with decreasing temperature. In the static limit, it resembles the development of the superconducting gap $\Delta(T)$ [46].

determined through the plasma frequency $\omega_p = (4\pi Ne^2/m)^{1/2}$ by the parameters of the metallic state, that is, the density of charge carriers N, the carrier mass m, and the electronic charge $-e$. The second length scale is the correlation length

$$\xi_0 = \frac{\hbar v_F}{\pi\Delta} \tag{9}$$

describing, crudely speaking, the spatial extension of the Cooper pairs. This is the length scale over which the wave function can be regarded as rigid and cannot respond to a spatially varying electromagnetic field. This parameter is determined through the energy gap Δ, which is linked to critical temperature by the coupling parameter $\alpha = 2\Delta/k_BT_c$. The third length scale is the mean free path of the uncondensed electrons

$$\ell = v_F\tau \tag{10}$$

set by the impurities and lattice imperfections at low temperatures; v_F is the Fermi velocity; τ is the time between two scattering events. For typical metals in the superconducting state, the three length scales are of the same order of magnitude, and, depending on the relative magnitude various types of superconductors are observed, these are referred to as the various limits. The local limit is where $\ell \ll \xi, \lambda$; more commonly it is referred to as the dirty limit defined by $\ell/\xi \rightarrow 0$. In the opposite, so-called clean limit $\ell/\xi \rightarrow \infty$, it is necessary to distinguish the following two cases: the Pippard or anomalous limit, defined by the inequality $\lambda \ll \xi, \ell$ (type I superconductor), and the London limit for which $\xi \ll \lambda, \ell$ (type II superconductor). Figure 7 summarizes the relation between these different length scales and the respective limiting cases.

The relative magnitude of the mean free path with respect to the coherence length has important consequences on the penetration depth, and this can be described using spectral weight arguments. The mean free path is given by $\ell \approx v_F\tau$, while for the coherence length we get $\xi \approx v_F/\Delta$ and consequently in the clean limit $1/\tau < \Delta/\hbar$, while $1/\tau > \Delta/\hbar$ is the so-called dirty limit. In the former case the width of the Drude response in the metallic state, just above T_c, is smaller than the frequency corresponding to the gap $2\Delta/\hbar$, and the opposite is true in the dirty limit.

The spectral weight associated with the excitations is conserved by going from the normal to the broken symmetry states. While we have to integrate the real part of the normal-state conductivity spectrum $\sigma_1^n(\omega)$, in the superconducting phase there are two contributions: one from the condensate of the Cooper pairs $\sigma_1^{\text{coll}}(\omega)$ and one from the single-particle excitations $\sigma_1^{\text{sp}}(\omega)$; thus

$$\int_{-\infty}^{\infty} \left[\sigma_1^{\text{coll}} + \sigma_1^{\text{sp}}\right] d\omega = \int_{-\infty}^{\infty} \sigma_1^n d\omega = \frac{\pi}{2}\frac{Ne^2}{m} \tag{11}$$

assuming that all the normal carriers condense. The arguments for superconductors were advanced by Ferrell et al. [25, 26] who noted that the area A which has been removed from the integral upon going through the superconducting transition

$$\int_{0+}^{\infty} \left[\sigma_1^n - \sigma_1^s\right] d\omega = A \tag{12}$$

is redistributed to give the spectral weight of the collective mode with $\sigma_1(\omega = 0) = A\delta\{\omega\}$. By comparison of the following expression with the Cooper pair response derived in (4):

$$\sigma_1(\omega) = \frac{c^2}{8\lambda_L^2}\delta\{\omega = 0\}, \tag{13}$$

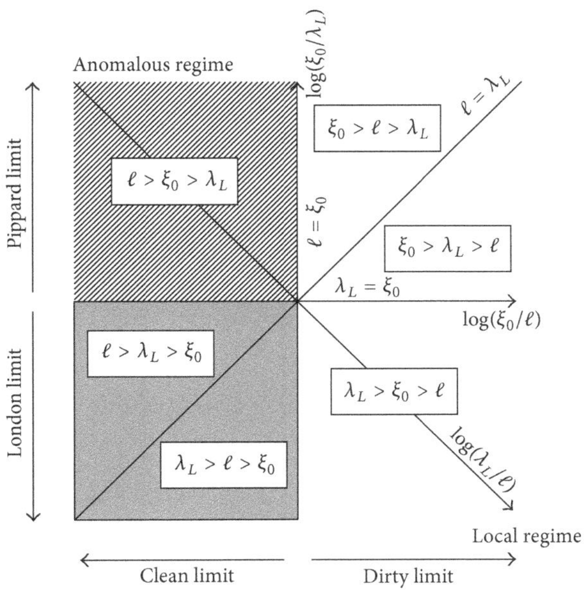

FIGURE 7: Schematic representation of the local, London, and Pippard limits in the parameter space given by the three length scales, the coherence length ξ_0, the London penetration depth λ_L, and the mean free path ℓ. In the local regime ℓ is smaller than the distance over which the electric field changes, $\ell < \xi(0) < \lambda$. When $\ell/\xi(0) \to 0$ the superconductor is in the so-called dirty limit. The opposite situation is in which $\ell/\xi(0) \to \infty$ is the clean limit, in which nonlocal effects are important, and we have to consider Pippard's treatment. This regime can be subdivided if we also consider the third parameter λ. The case $\xi(0) > \lambda$ is the Pippard or anomalous regime, which is the regime of type I superconductors; $\xi(0) < \lambda$ is the London regime, the regime of the type II superconductors [46].

we see that the missing spectral weight A is connected to the penetration depth λ by

$$\lambda = \frac{c}{\sqrt{8A}} \quad \text{or} \quad A = \frac{c^2}{8\lambda}. \tag{14}$$

This relation is expected to hold also at finite temperatures and also for various values of the mean free path and coherence length. In the limit of long relaxation time τ, that is, for $\hbar/\tau \ll \Delta$, all the Drude spectral weight of the normal carriers collapses in the collective mode, giving

$$\sigma_1(\omega = 0) = \frac{\pi N_s e^2}{2m}\delta\{\omega = 0\} \tag{15}$$

and through (4) leading to the London penetration depth λ_L. With increasing $1/\tau$, moving towards the dirty limit, the spectral weight is progressively reduced, resulting in an increase of the penetration depth. In the limit $1/\tau \gg 2\Delta/\hbar$, the missing spectral weight is approximately given by $A \approx \sigma_1(2\Delta/\hbar) = 2Ne^2\tau\Delta/\hbar m$ which can be written as

$$A = \frac{c^2}{2\pi}\frac{\tau\Delta}{\hbar}. \tag{16}$$

Then the relation between the area A and the penetration depth $\lambda(\ell)$—which now is mean free path dependent as $\ell = v_F\tau$—is given by $\lambda^2(\ell) = \lambda_L^2\hbar\pi/(4\tau\Delta)$, which leads to the approximate expression

$$\lambda(\ell) \approx \lambda_L\sqrt{\frac{\xi_0}{\ell}}. \tag{17}$$

3. Experimental Results

3.1. Coherence Peak. In order to determine the temperature dependence of the complex conductivity at frequencies well below the superconducting gap, Klein et al. [37] have performed precise measurements of the complex surface impedance of Nb using cavity perturbation methods [78–80]. A bulk Nb sample of 99.9% purity was used to replace the endplate of a 60 GHz cylindrical copper cavity. An abrupt drop of the bandwidth was observed upon cooling through the 9.3 K transition temperature. In Figure 8(a) the temperature dependence of the surface resistance R_s and surface reactance X_s normalized to the normal state value (defined as the value when the temperature of the sample is slightly above T_c, i.e., ~10 K): $R_n = (0.0538\,\Omega)/Z_0$ (where Z_0 is the free space impedance), corresponding to $\sigma_n(T = 10\,\mathrm{K}) = 0.85 \cdot 10^6\,(\Omega\,\mathrm{cm})^{-1}$. Below T_c, the surface resistance $R_s(T)$ drops rapidly with decreasing temperature, as shown in the logarithmic plot of Figure 8(a). The length scale for the surface reactance is determined by the skin depth in the normal state $\delta = 2270$ Å. $X_s(T)$ saturates at low temperatures, yielding a penetration depth $\lambda(0) = 440$ Å. Using the measured values of R_s and X_s, the temperature dependence of the complex conductivity was evaluated and displayed in Figure 8(b). The coherence peak can be seen as a wide peak below T_c. The solid curves represent the Mattis-Bardeen results [(1) and (2)], while the dashed line corresponds to the Eliashberg theory for strong coupling [37].

Only recently, broadband microwave measurements on Al films succeeded to probe the coherence peak of a superconductor (as shown in Figure 2) both as a function of temperature and frequency. Steinberg et al. [29] employed a broadband microwave Corbino spectrometer [81] where basically a coaxial microwave line is terminated by the sample, in this case, a 550 Å aluminum film on a sapphire substrate. Using a vector network analyzer, the complex reflection coefficient is measured in the spectral range from 45 MHz to 60 GHz, allowing the calculation of the real and imaginary parts of the conductivity after also the calibration measurements have been performed with high precision. The method is most sensitive when the film impedance is close to the waveguide impedance of 50 Ω. For thicker or thinner films different geometries are possible [82]. The specimen is cooled down to 1.1 K and even lower temperatures are available [83]. Figure 9(a) shows data (taken at different frequencies on a film with thickness 500 Å and a room temperature resistivity of $8.7 \cdot 10^{-5}$ Ω cm) of the temperature dependence of the real part of the conductivity. σ_1 is governed by thermally excited charge carriers above and below the gap; their density of states diverges at the gap edge as depicted in Figure 3. Right below T_c, $\Delta(T)$ is small: both the thermal energy and the photon energy $\hbar\omega$ are sufficient to break up Cooper pairs. The imaginary part $\sigma_2(\omega)/\sigma_n$ displayed in the lower frame basically describes the temperature dependence of energy gap, in full agreement with (7).

In accord with theory, the coherence peak is seen as a maximum in $\sigma_1(T)$ at approximately $0.8T_c$ in the low-frequency limit; it becomes smaller with increasing frequency and shifts to higher temperatures in according to (6); this is clearly observed in Figure 9(a). Above 28 GHz [corresponding to $0.2 \cdot \Delta(0)$] the peak is completely suppressed, and for higher frequencies $\sigma_1(T)$ monotonously drops below T_c.

3.2. Energy Gap. In order to explore the energy gap in a superconductor the most straight-forward experiment is a transmission measurement through metallic films in the relevant frequency range around ω_g, and to vary the temperature from above to below T_c, as first has been done by Glover III and Tinkham on lead films in 1956 [10]. In Figure 10(a) the transmission through a NbN film at different temperatures as indicated normalized to the transmission in the metallic state just above $T_c = 13.5$ K [84]. With decreasing temperature a maximum in $\mathrm{Tr}_s/\mathrm{Tr}_n$ is observed that becomes more pronounced and shifts to higher frequencies. As sketched in Figure 1(b) it marks the superconducting energy gap 2Δ; in this case $2\Delta/hc = 35\,\mathrm{cm}^{-1}$.

The effect of the scattering rate on the transmission through superconducting NbN films has been explored experimentally [86] and analyzed on the basis of Leplae's theory [73] that describes strong-coupling materials. By reducing the film thickness from 100 to 20 nm, the critical temperature T_c is lowered from 17 to 14 K, but, more importantly, the energy gap $2\Delta(0)/hc$ is reduced from $52\,\mathrm{cm}^{-1}$ to $39\,\mathrm{cm}^{-1}$. Also the transmission maximum becomes significantly lower. Recently, it was shown [87, 88] that when the superconducting state is suppressed by a magnetic field, this transmission maximum broadens and shifts to lower temperatures. This is in agreement with the fact that the maximum in transmission corresponds to the energy gap. Using the transmission of $13.5\,\mathrm{cm}^{-1}$ gas laser the transmission through a 80 nm NbN film on a Si substrate was measured as a function of temperature for different external fields (Figure 11). With higher frequencies, the transmission peak first becomes more pronounced and shifts to lower temperatures; however, it eventually vanishes at $\omega_g \approx 2\Delta/\hbar$. When a magnetic field is applied parallel to the propagation direction (Faraday geometry), the electric field vector is perpendicular to the external magnetic field. For the Voigt geometry, the external field is chosen parallel to the electric field vector. The different behavior reflects the vortex dynamics in this constraint geometry determined by the thin superconducting film. The high-frequency response of type-II superconductors in the mixed state was extensively studied by Coffey and Clem [89, 90].

The absorption of electromagnetic radiation by a superconductor was directly investigated by measuring the temperature increase due to the deposited energy (bolometric absorption). A sputtered thin film of NbN on a silicon substrate (0.39 mm) was mounted at the bottom of a highly reflecting copper light cone, supported by four strands of 25 μm diameter gold wire of about 1.5 mm long. This gives the bolometer a thermal response time constant of about 30 ms at 10.8 K. The temperature is measured by a small doped silicon thermometer glued to the back of the silicon substrate. The resistance of this thermometer is exponentially decreasing with increasing temperature [85]. In Figure 10(b) the relative power absorption spectra are plotted for a 250 nm NbN superconducting film ($T_c = 13.3$ K) on a 0.39 mm silicon substrate. The curves are taken at 5.3 K and 10.8 K and each normalized to a normal state spectrum taken at $T = 14$ K. With decreasing frequency, the absorption increases and exhibits a maximum around the superconducting energy gap, before it rapidly drops to zero; the behavior is more pronounced for lower temperatures. On the same thin film, reflection measurements have been performed using Fourier-transform spectrometer with a resolution of $0.5\,\mathrm{cm}^{-1}$. Figure 10(c) shows the reflectivity of the NbN film on Si substrate above and below T_c. The reflectivity in the normal state is relatively low (due to the particular film thickness), making the superconductivity behavior very evident. While in the normal state at 15.5 K the reflectivity levels off around $R \approx 0.93$, a pronounced dip is observed in the superconducting state before the reflectivity rapidly rises to 1.0 ± 0.01 for frequencies below $40\,\mathrm{cm}^{-1}$, clearly indicating the superconducting gap frequency. In any case the high-resolution reflection spectra exhibit pronounced interference fringes due to multireflection within the Si substrate. In particular, one can see in the inset that the phase of the fringes at 5.3 K

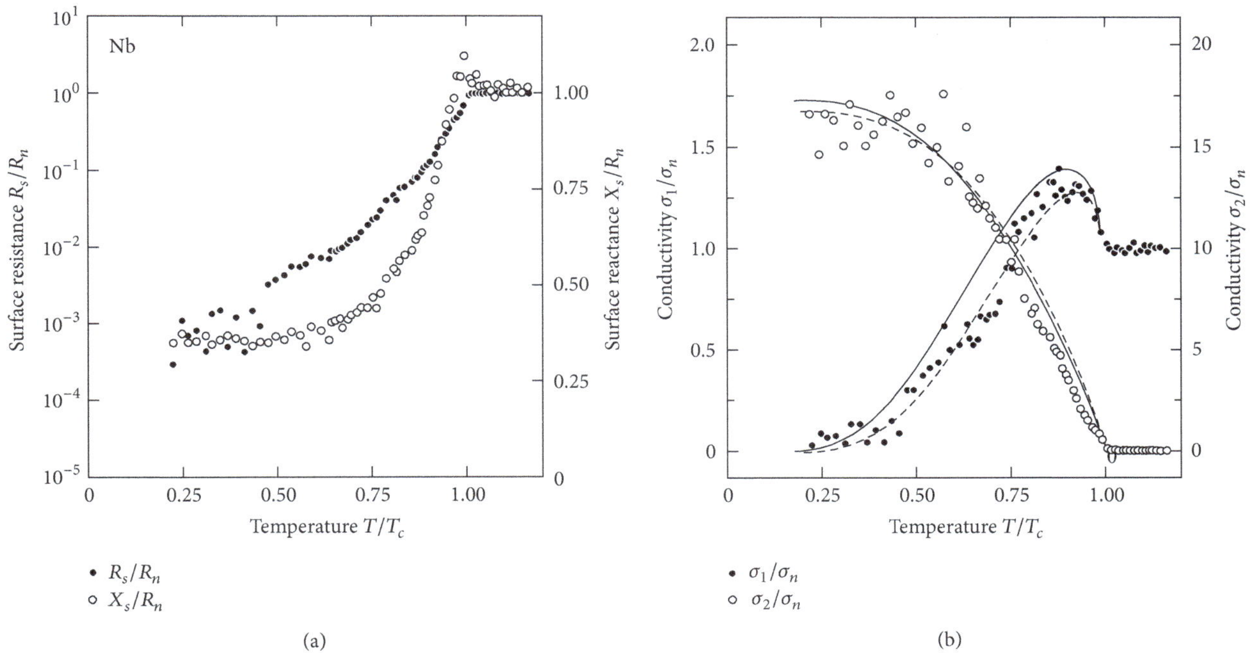

FIGURE 8: (a) Real component $R_s(T)$ (referring to the logarithmic scale on the left axis) and imaginary component $X_s(T)$ (right axis) of the surface impedance of niobium, normalized to the normal state surface resistance R_n measured at 60 GHz as a function of temperature. (b) Temperature dependence of the components of the conductivity $\sigma_1(T)$ and $\sigma_2(T)$ in niobium calculated from the results of the surface impedance shown in (a). Note the enhancement of $\sigma_1(T)$ just below $T_c = 9.2$ K. The full lines are calculated using the Mattis–Bardeen formalism [(1) and (2)] in the local limit, and the dashed lines follow from the Eliashberg theory of a strong coupling superconductor (after [37, 46]).

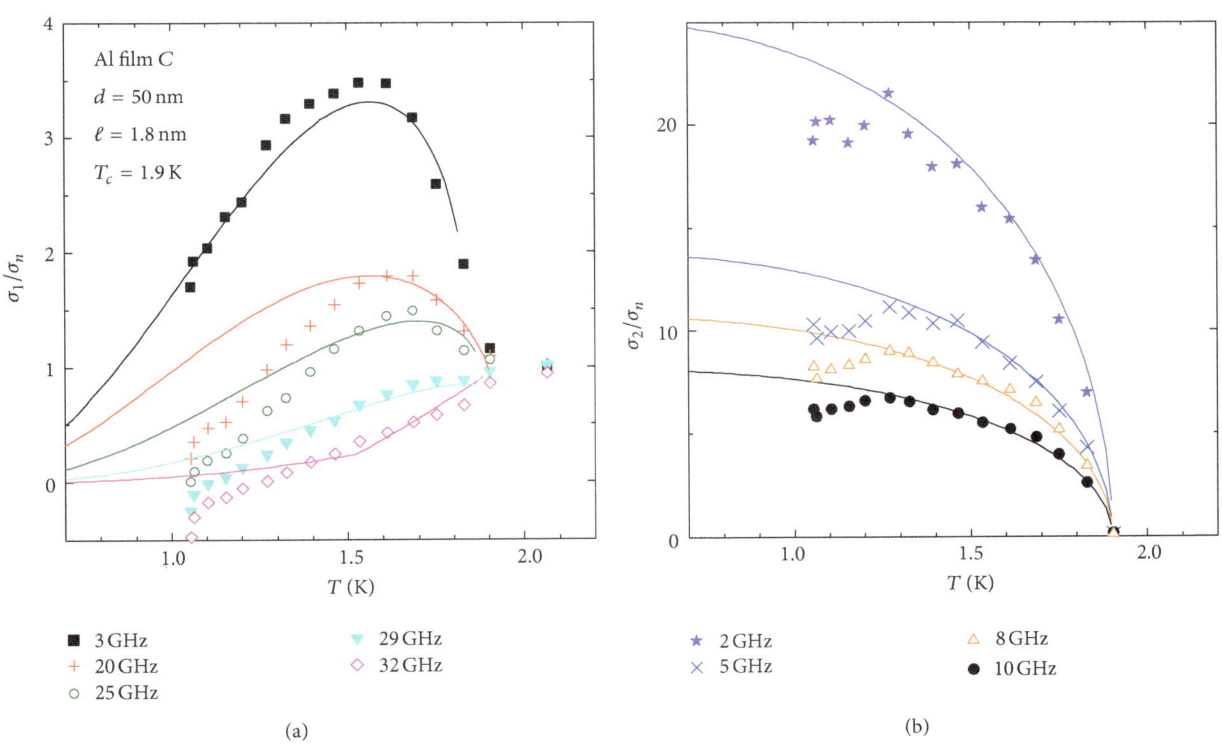

FIGURE 9: Temperature dependence of the (a) real and (b) imaginary parts of the conductivity of aluminum normalized to the normal state conductivity σ_n for selected frequencies. The solid lines are calculated by the BCS theory using $\ell/(\pi\xi) = 0.03$ [28].

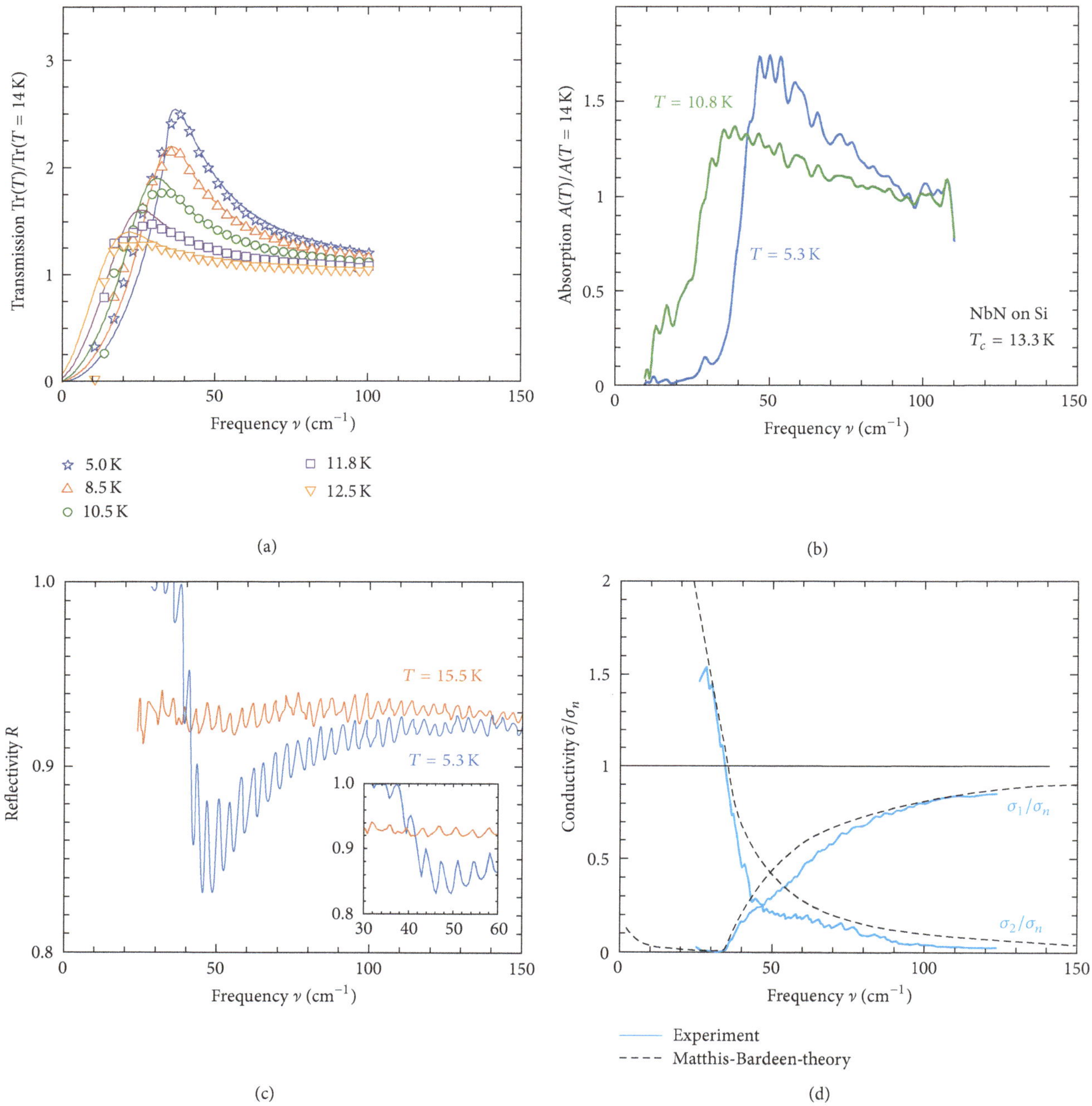

FIGURE 10: (a) Ratio of the infrared transmission through a NbN film in the superconducting state and the normal state for several temperatures; the critical temperature of this film is T_c = 13.5 K (after [84]). (b) The bolometric absorption spectra of a superconducting NbN film at different temperatures below T_c = 13.3 K ratioed with a 14 K spectrum above T_c. The curves have been scaled individually to agree at 110 cm^{-1} (for the purpose of comparison) and overall to give a 5.3 K spectrum, which equals 1.0 at large wave numbers. The resolution is 2 cm^{-1}. (c) Optical reflectivity of a 250 nm NbN film on silicon (d = 0.39 mm) above and below the superconducting transition temperature. The resolution was 0.5 cm^{-1} below 100 cm^{-1} and 2 cm^{-1} above. The inset gives the low-frequency part on a magnified scale in order to demonstrate the shift of the Fabry-Perot resonance pattern: At high frequencies the reflection maxima are in phase, but, for low frequency, the reflectivity in the superconducting state exhibits minima where maxima are observed for $T > T_c$. (d) Real and imaginary parts of the conductivity at T = 5.3 K normalized to the normal state value, as extracted from the reflection data shown above. The dashed lines correspond to the calculations based on Mattis-Bardeen theory, yielding an energy gap $2\Delta/hc$ = 34.5 cm^{-1} (after [85]).

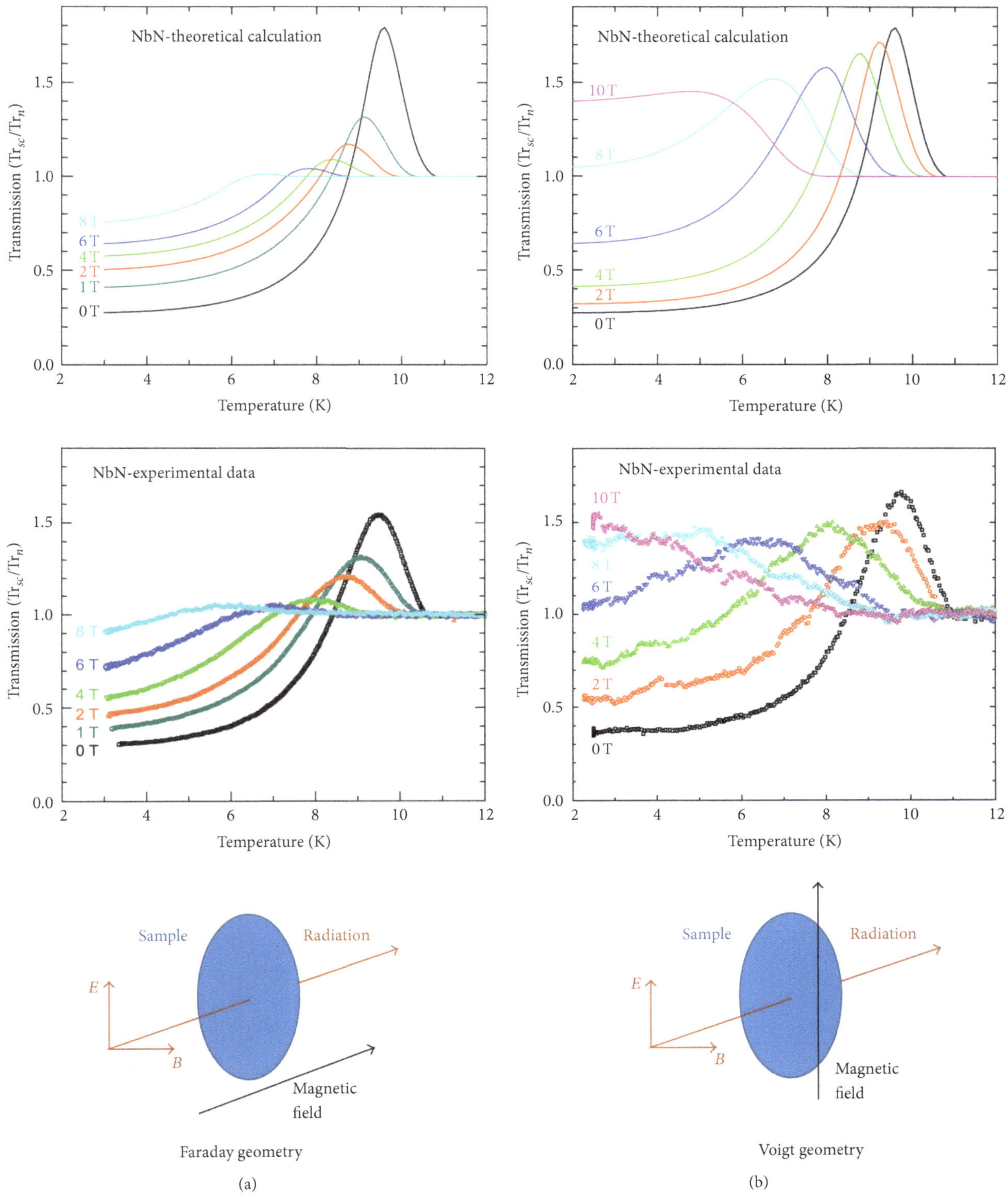

FIGURE 11: Temperature-dependent normal-incidence transmission of 13.5 cm^{-1} radiation through a 80 nm NbN film on a Si substrate of 0.25 mm thickness. The orientation is either in Faraday geometry ($B_0 \perp E$), shown in the left column, or in Voigt geometry ($B_0 \parallel E$) as shown in the right column. The calculations are done using a phenomenological effective medium approach (after [87]).

changes with respect to those at 15.5 K near 46 cm^{-1}. Above 46 cm^{-1} the fringes are in phase, while below they are 180° out of phase. From the lower panel (Figure 10(d)) we see that this is the frequency where the real and imaginary parts of the conductivities are equal at 5.3 K. The metallic reflection turns into a dielectric reflection, governed by the out-of-phase response of the Cooper pairs, leading to a sign change in the reflection coefficient. Using Fresnel's equations in order to describe the complete system of film and substrate, the Fabry-Perot pattern allows the calculation of the real and imaginary parts of the conductivity as displayed in Figure 10(c). No Kramers-Kronig transformation is used here, and thus no extrapolation or other model is required. The results agree very well with the theoretical predictions by Mattis and Bardeen [27], except slightly above ω_g. The superconducting gap is larger than expected from weak-coupling mean-field theory: $2\Delta(0)/k_BT_c = 3.9$; this is in agreement with the strong coupling nature of the Nb compounds.

While the previous experiments have been performed utilizing Fourier-transform spectrometers, Gorshunov et al. employed a THz spectrometer on the basis of tunable coherent monochromatic backward wave oscillators [93–95] to measure 115 Å thin NbN films on sapphire substrate [91, 92]. They demonstrated that for frequencies below 2Δ the error bars are too large to draw conclusions about the superconducting state and in-gap states, for instance. However, if the substrate is covered with a superconducting film on both sides (double-sided film), the Fabry-Perot resonances get much sharper (increase in quality factor) and the conductivity can be evaluated with a much larger precision, allowing even to observe the coherence peak right below the transition temperature. The raw data if the THz transmission are plotted in Figures 12(a) and 12(b) for single and double sided films; the corresponding conductivity spectra are plotted in panel (c) for two different frequencies close to and well below the superconducting gap 2Δ.

3.3. Complex Conductivity. Very similar results have been achieved by Nuss et al. [96] using THz time-domain spectroscopy. The amplitude transmitted through a thin Nb film (T_c = 6.25 K) is plotted in Figure 13 as a function of time. Both the amplitude and phase of the transients change as the film becomes superconducting. In accord with the observations of Glover III and Tinkham [10, 12] and others, the transmission in the superconducting state is higher than in the normal state. Since time-domain spectroscopy actually probes the electric field intensity and the shift in time, the real and imaginary parts of the conductivity can be calculated directly as plotted in Figure 13(b). The agreement with the BCS theory [27] shown by the dashed line is very good.

Alternatively a Mach-Zehnder interferometer allows one to measure both the change in transmission and phase shifts when the electromagnetic radiation passes through a metallic and superconducting film. This method was intensively explored by Gorshunov and collaborators [93–95]. Figure 14 displays the results of a transmission experiment [97] performed on a thin niobium film (thickness 150 Å) on a 0.45 mm thick sapphire substrate using a Mach-Zehnder interferometer. The substrate acts as a Fabry-Perot resonator due to multireflection leading to pronounced interference fringes in the spectra; this effect enhances the interaction of the radiation with the film considerably and thus improves the sensitivity and accuracy. In Figures 14(a) and 14(b) the transmission Tr_F and relative phase shifts ϕ/ν, respectively, through this composite sample are shown as a function of frequency. As the temperature decreases below the superconducting transition T_c = 8.3 K, the transmitted power and phase shifts are modified significantly. Since the properties of the dielectric substrate do not vary in this range of frequency and temperature, the changes observed are due to the electrodynamic properties of the superconductor [97]. When cooling from T = 9 to 6 K, the overall transmission increases for frequencies above 15 cm^{-1} while it drops for lower frequencies. The inset illustrates this behavior nicely. This corresponds to the theoretical predictions (Figure 1(b)) and previous observations. The maximum in the transmission corresponds to the superconducting energy gap 2Δ. Similar changes are observed in the relative phase shift (Figure 14(b)), where a strong decrease is found for $\omega < \omega_g$. The energy gap is evaluated to $2\Delta(0)/hc$ = 24 cm^{-1}. The data can be used to directly calculate the real and imaginary parts of the complex conductivity without using any model or assumption. In Figure 15 the real and imaginary parts of the conductivity of a Nb film are plotted as a function of frequency covering an extremely large range. Since the data are not normalized, the Drude roll-off becomes obvious above 100 cm^{-1}, placing the material into the dirty limit. In all cases it seems to be difficult to actually measure the zero-conductivity well below the gap for very low temperatures. In many cases some residual absorption seems to be present which can either be intrinsic due to some in-gap absorption, or due to inhomogeneities, impurities, surface effects, and so forth.

Thin films of NbN and TaN are perfectly suited for superconducting nanowire single-photon detectors [57, 59] which are built by thin and narrow superconducting stripes typically in a meander-like geometry. The device is driven with a dc bias current slightly below the critical current value. Absorption of a photon locally reduces the critical current below the applied dc bias current; consequently the superconducting state breaks down locally: this generates a normal-conducting spot (hot spot) and increases the resistance. The recovery time is in the picosecond range [63]. Since their performance in the far-infrared strongly depends on the optical properties of the device in this spectral range, the energy gap 2Δ, and the development of $\sigma_2(\omega)$, the information of THz and infrared investigations are highly desired. In particular it is important to study the influence of substrate, film thickness, processing parameters, structuring, and so forth. In Figure 16 the frequency dependence of the real and imaginary parts of the conductivity $\sigma_1(\omega)$ and $\sigma_2(\omega)$ is plotted for selected temperatures [72, 98]. The NbN film is deposited on a 10 mm × 10 mm R-plane-cut birefringent sapphire substrate using reactive dc-magnetron sputtering, which is a well-established and powerful way of film deposition with high demands concerning the performance of single-photon

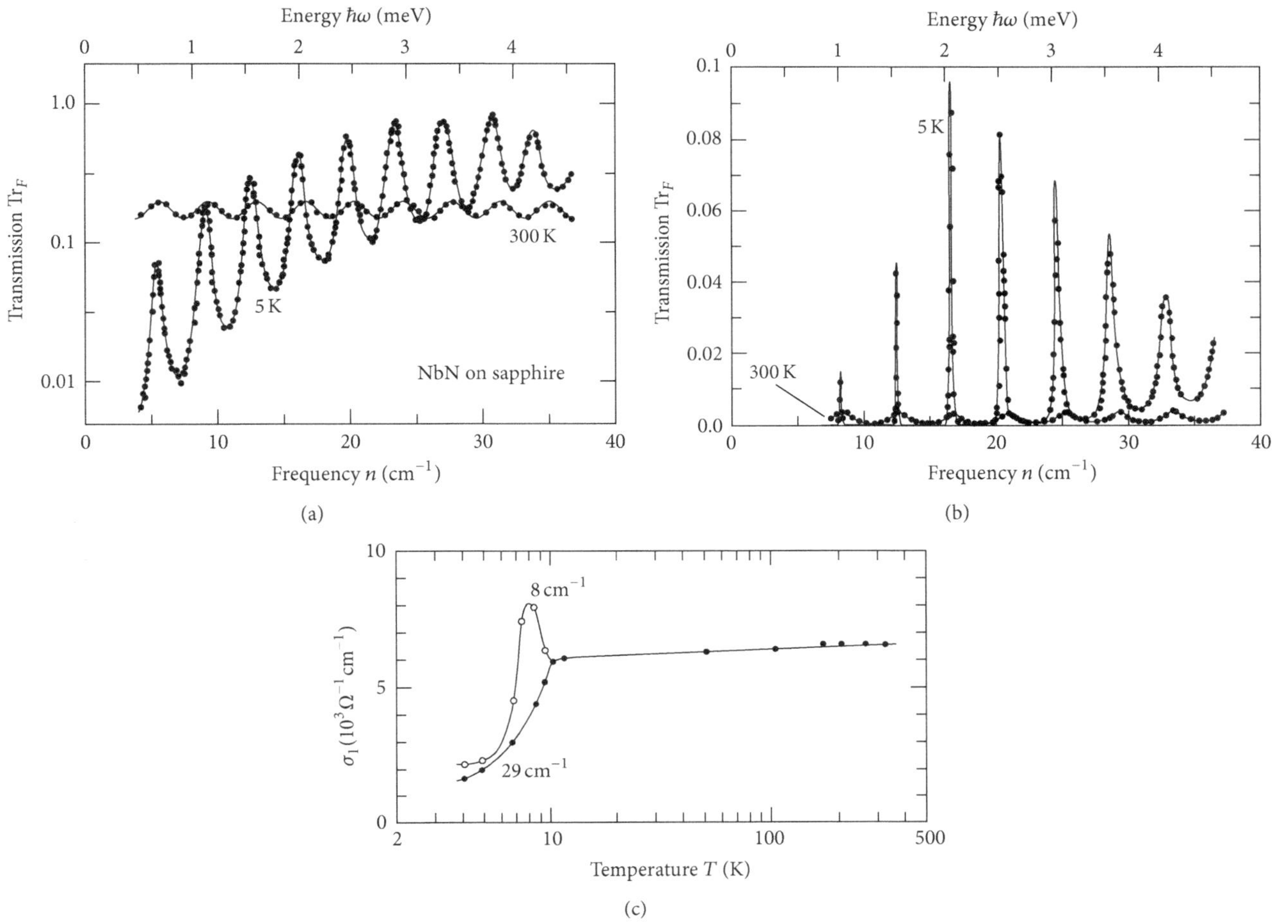

FIGURE 12: (a) Transmission spectra $Tr_F(\nu)$ of an asymmetrical Fabry-Perot resonator, formed by a NbN film ($d_2 = 115$ Å) on a sapphire substrate (0.43 mm), measured at two temperatures above and below critical temperature $T_c \approx 12$ K. (b) Transmission spectra $Tr_F(\nu)$ of a plane parallel sapphire plate (0.41 mm) covered with NbN films on both sides ($d_2 = 250$ Å and $d_4 = 60$ Å), again measured at 300 K and 5 K. The solid lines represent the theoretical fit. (c) For the latter case the temperature dependence of the optical conductivity $\sigma_1(T)$ of NbN film is calculated for two frequencies, 8 cm^{-1} and 29 cm^{-1} (after [91, 92]).

detectors [57, 58]. Typical films have a thickness of $d = 39$ to 47 Å. Since the thickness d is similar to the superconducting coherence length ξ and much smaller than the penetration depth λ, the films are considered ultrathin, that is, quasi-two dimensional, from the superconducting point of view. However, the electron mean free path ℓ is about one order of magnitude smaller than d, which justifies the 3D expressions used in the analysis and places the material in the dirty limit (Figure 7). Depending on the deposition current I_d while sputtering, the critical temperature is lowered in the case of NbN from 12.2 K ($I_d = 100$ mA) to 8.5 K (200 mA) [58, 59, 72, 98, 99]. The degree of disorder can be described by the Ioffe-Regel parameter $k_F\ell$, (with k_F the Fermi wave vector and ℓ the mean free path) ranging from 4.5 to 5.5; that is, in the present case these films are still weakly disordered.

Figure 17(a) displays the temperature dependent energy gap $2\Delta(T)$ of the sample with lowest disorder and highest T_c. Filled and open symbols refer to different main axes of the substrate and coincide within the range of error. Data for each main axis is fitted separately with the BCS expression. For $0.3T_c < T < 0.8T_c$ the measured energy gap is slightly above the BCS curve and for $0.8T_c < T < T_c$ below. Upon decreasing T_c, the energy gap (at 4.2 K) is also reduced from 4.5 to 3.3 meV; compare Figure 17(b). Consequently, $2\Delta(T)$ depends on the amount of disorder and is the greatest for low-disorder films. There are no traces or indications of superconductivity seen in the optical behavior of the normal state among the higher disordered samples as it is observed in tunneling studies [51]. At 4.2 K, $2\Delta(T)$ is reduced as T_c decreases (Figure 17(b)); ratio $2\Delta(0)/k_BT_c$ indicates that upon increasing disorder (decreasing T_c), the ratio $2\Delta(0)/k_BT_c$ which is first reduced from about 4.1, indicating strong-coupling superconductivity, exhibits a minimum and becomes larger again (about 4.5) towards the highest-disorder sample studied.

3.4. Dynamics. The question of how long it takes to break or recombine a Copper pair was considered from early

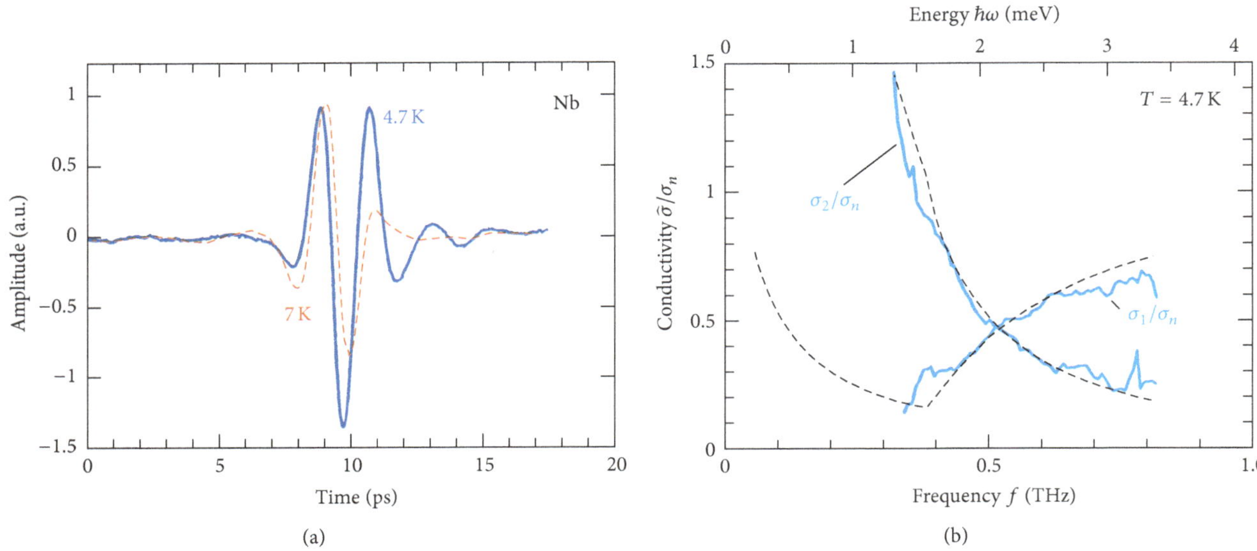

FIGURE 13: (a) THz transients transmitted through a thin niobium film (T_c = 6.25 K) on quartz in the normal state (dashed red line) and the superconducting state (solid blue line). (b) Real and imaginary parts of the normalized complex conductivity, $\sigma_1(\omega)$ and $\sigma_2(\omega)$, of Nb in the superconducting state evaluated from the above data (after [96]). The dashed black line indicates the predictions by the BCS theory using the Mattis-Bardeen equations (1) and (2). The energy gap is given by $2\Delta(0) = 3.8k_BT_c$.

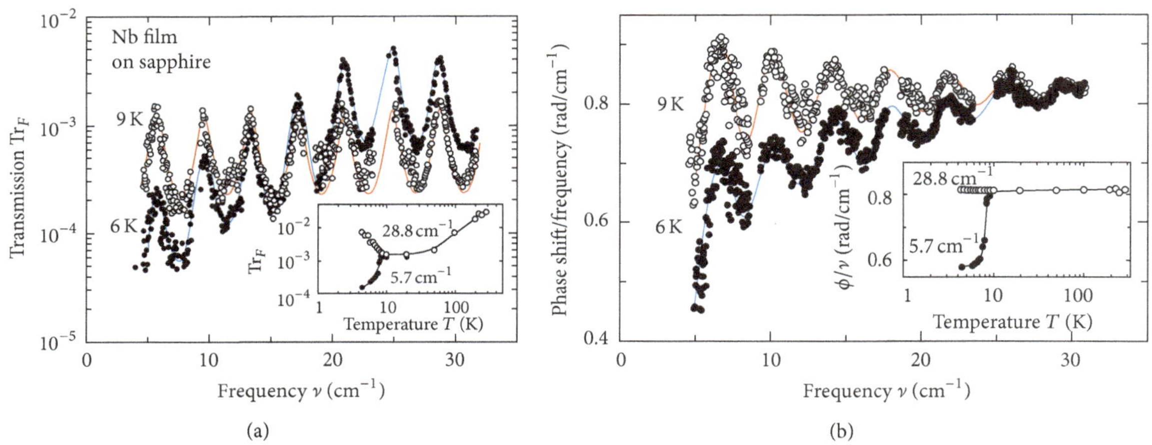

FIGURE 14: Raw data of transmission $\mathrm{Tr}(\omega)$ and relative phase shift $\phi(\nu)/\nu$ spectra of the Nb film on the sapphire substrate. The critical temperature is T_c = 8.3 K. The phase shift is divided by the frequency $\nu = \omega/2\pi c$ for the better perception. The dots represent the measured values; the red lines for 9 K are the fits with the Drude model. The blue lines correspond to the fit using the Mattis-Bardeen equation. In the insets the temperature dependence of Tr and ϕ/ν is displayed for two particular frequencies corresponding to the first and last transmission maxima in measured spectral range (after [97]).

on [100, 101] and already linked to the possibility to use superconductors as quantum detectors in the microwave and THz range. Soon it became clear that not only radiative processes are important (emission of photons), but the emission of phonons dominates [102–104]. From early studies of the nonequilibrium conditions of superconductors [101] it became clear that the recombination of two quasiparticles to a Cooper pair requires the emission of phonons with energy above 2Δ. Rothwarf and Taylor [104] pointed out that reabsorption of high-energy phonons leads to the so-called phonon bottleneck: the recovery of the superconducting state is not governed by the bare recombination of two quasiparticles into the condensate; the return to equilibrium is rather limited by the phonon escape rate. It took some effort before the development of suitable light sources made the experimental investigations possible.

The first time Testardi [105] was able to break Cooper pairs by short laser light pulses (547 nm, 6–40 μs). The experiments are challenging due to the short penetration length of visible light into Pb metal and the heating of the sample. Federici et al. explored the dynamics of the optical response by looking at the photoinduced Cooper-pair breaking in lead

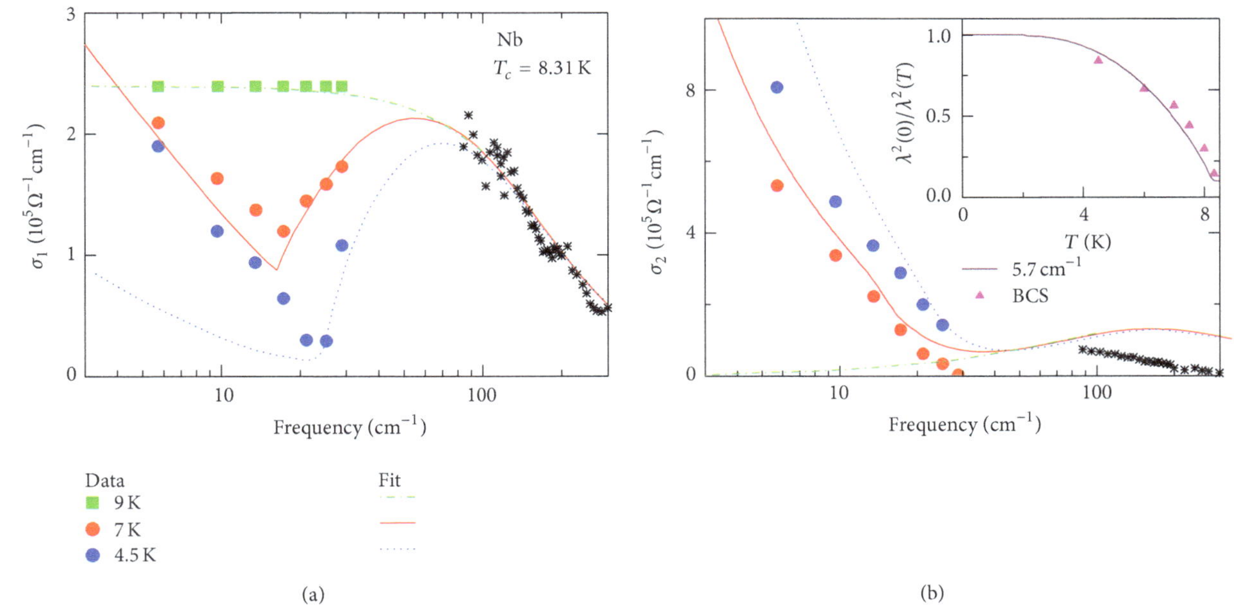

FIGURE 15: Frequency dependence of the real and imaginary parts of the conductivity, $\sigma_1(\omega)$ and $\sigma_2(\omega)$, in niobium at various temperatures T above and below the critical temperature T_c = 8.31 K. The transmission through a 150 Å thick film on sapphire was measured by a Mach-Zehnder interferometer; the stars were obtained by reflection measurements (after [97]). The lines are calculated using the theory of Mattis and Bardeen. The inset shows the temperature dependence of the penetration depth evaluated from the $\sigma_2(T)$ at the lowest frequency.

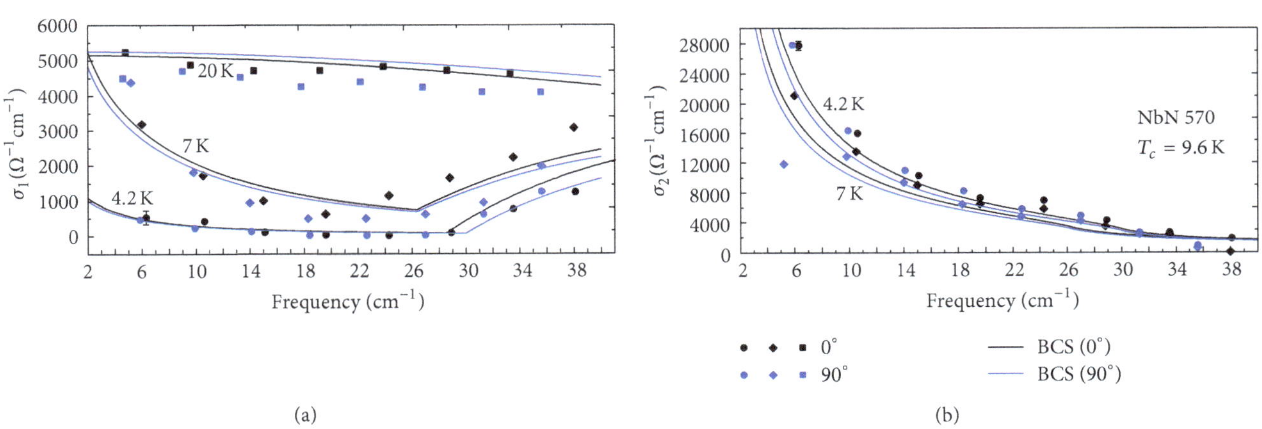

FIGURE 16: (a) Real part and (b) imaginary part of the complex conductivity versus frequency for 4.2 K (dots) and 7 K (diamonds) well below T_c = 9.6 K and 20 K (squares) in the normal state for both main axes of a NbN film. Solid black and blue lines are predictions calculated directly from raw data fits for both axes individually. The kink in σ_1 signaling the energy gap is resolved clearly. The error bar shown is representative for all data (after [98]).

[106]. 100 fs light pulses (630 nm, 1 μJ/cm^2) lead to an abrupt decrease of the transmission maximum within about 1 ps; it takes several nanoseconds before the superconducting state is recovered. This implies that the superconducting gap closes in less than one picosecond after the optical excitation. Kaindl et al. showed [107] that the imaginary part of the optical conductivity in NbN films abruptly changes and then relaxes in an exponential way (Figure 18(c)). The recovery time τ = 580 ps does not depend on the on intensity A_0 of the light pulse.

A more detailed picture is drawn by Demsar and collaborators who succeeded to perform a detailed study on the superconducting state relaxation phenomena in NbN films: both the Cooper-pair breaking and the recovery of the superconducting state depend on the temperature and energy density [108]. In Figure 18(d) the temporal evolution of the superconducting gap is shown after the sample is excited by a femtosecond pulse. The photoinduced change of the gap $\delta\Delta$ (normalized to the equilibrium energy gap at zero temperature) is plotted in Figure 19(a) for various laser

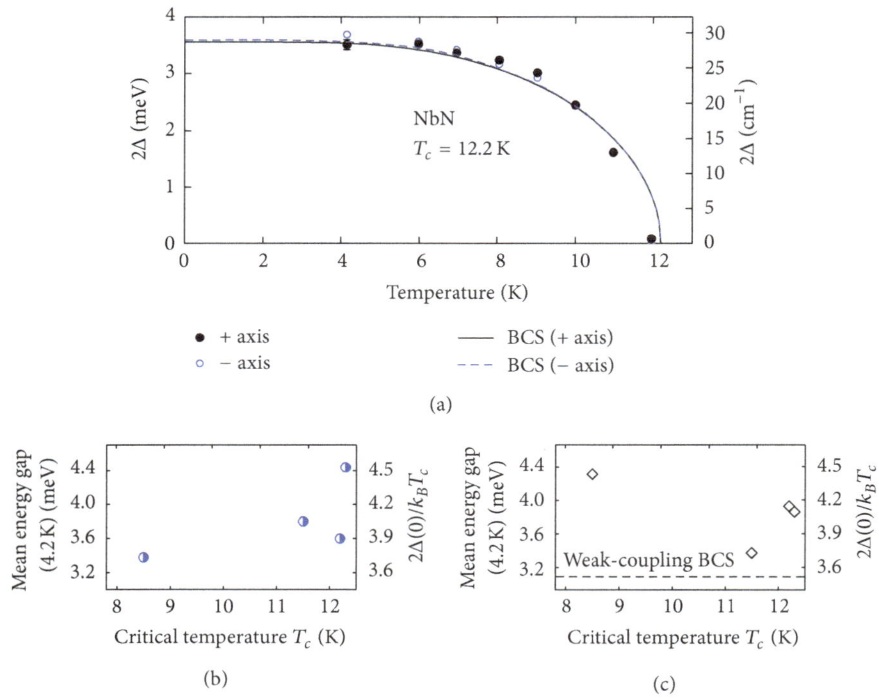

FIGURE 17: (a) Energy gap 2Δ versus temperature of a NbN sample for both main axes of the substrate. Solid lines are the fits according to BCS theory. Results on both main axes yield very similar values of $2\Delta(T)$. The error bar shown is representative for all data. (b) 2Δ at 4.2 K and (c) ratios $2\Delta(0)/k_BT_c$ versus T_c of several NbN films deposited with different parameters, yielding different transition temperatures (after [72]).

intensities. It can be seen that both the Cooper-pair breaking and the recovery of superconductivity depend on intensity A. The long timescale of the Cooper pair breaking implies that photoexcited hot electrons and holes create a high density of high-frequency phonons, which subsequently break Cooper pairs; this process reaches some equilibrium between quasiparticle population and high-frequency phonon population [108]. It is seen from Figure 19(b) that the maximum induced change first increases linearly with excitation intensity (note the logarithmic abscissa) but then saturates when superconductivity is suppressed: $\delta\Delta/\Delta_0 = 1 - \exp\{-A/A_{\text{sat}}\}$, where the temperature dependence of $A_{\text{sat}}(T)$, plotted in the inset, basically follows the superconducting density $\Delta^2(T)$. The superconducting recovery time τ_{rec} depends on temperature T and on intensity A, as shown in Figure 19(c). These findings support the phenomenological model of Rothwarf and Taylor [104] on the phonon bottleneck.

Matsunaga and Shimano [109] investigated the nonlinear dynamics of superconducting state in a NbN film by using intense THz pulses to pump the system which is then probed by time-domain spectroscopy in the same frequency range. After the intense pulse, superconductivity is rapidly switched off with the duration of the THz pulse. In contrast to the previous experiments with visible or infrared pump lasers, the THz pulse resonantly excites the low-energy quasiparticles without the process of phonon emission from hot electrons. The rapid time evolution after the THz pulse is basically independent of intensity implying direct photoinjection of quasiparticles. Figure 20 shows the conductivity spectra obtained at 20 ps after the pump with various intensities. With increasing pump intensity $\sigma_1(\omega)$ below the gap energy significantly increases and exceeds the normal state value (dashed line). At the gap energy of approximately 5 meV the conductivity $\sigma_1(\omega)$ remains always below the metallic behavior. This cannot be explained by the increase of the effective temperature and implies a high density of quasiparticles. For their analysis they apply an effective medium approach of superconducting and normal state phases. The conclusion is a spatially inhomogeneous suppression of the superconducting state on a length scale of 1 μm [110, 111].

4. Outlook

The effect of disorder on superconductivity is an ongoing intriguing topic of research that has puzzled physicists for decades. It was argued by Anderson in a seminal paper [112] in 1959 that s-wave superconductivity is not affected by the presence of mild nonmagnetic disorder, persisting even in polycrystalline or amorphous materials. Experimentally, however, it was found that superconductivity can be destroyed by a sufficiently large degree of disorder [113–118]. Once superconductivity is destroyed, the system undergoes a transition to an insulating state in what has been named the superconductor-insulator transition (SIT).

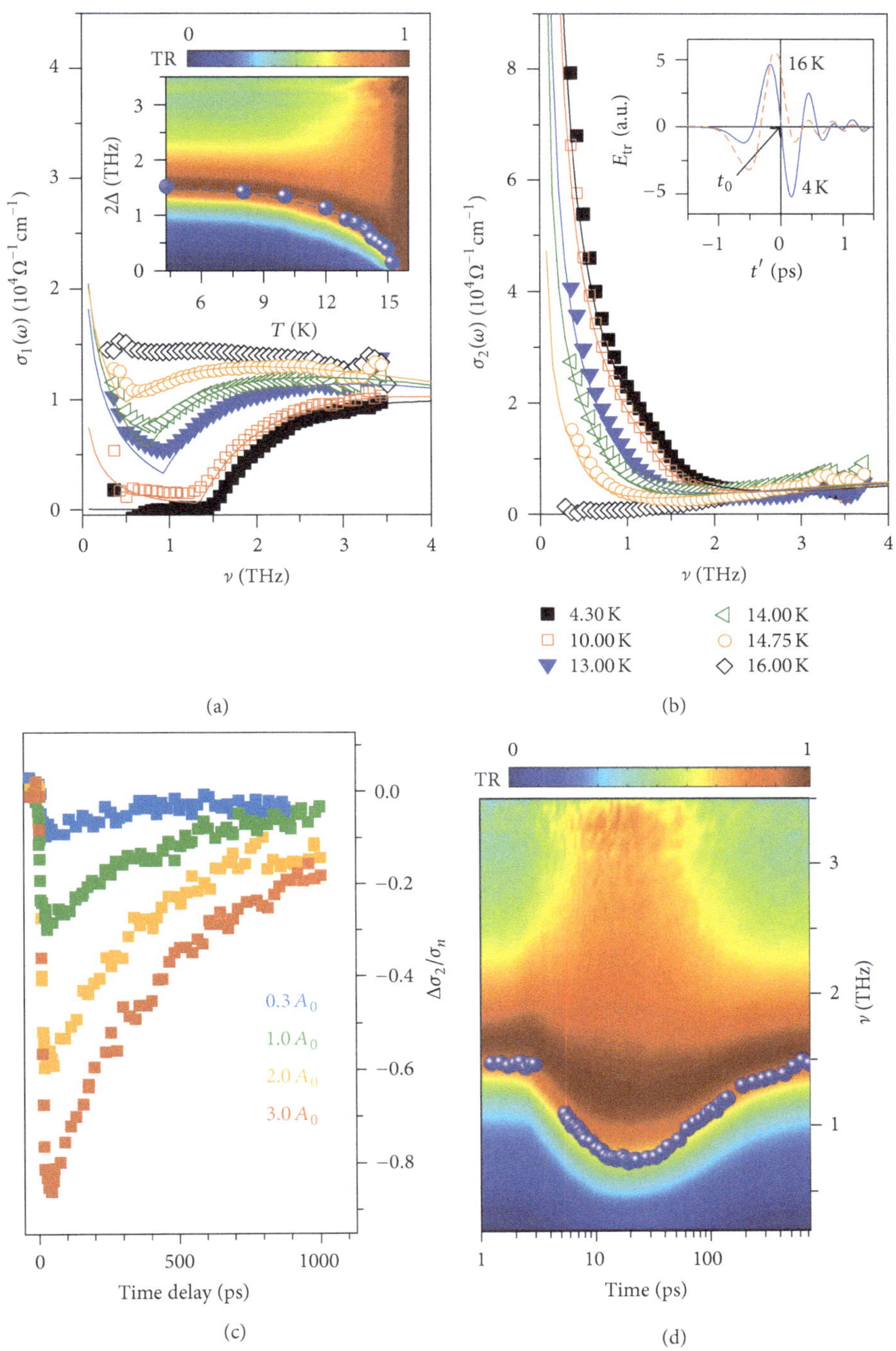

FIGURE 18: Temperature-dependence of the (a) real and (b) imaginary parts of the conductivity spectra obtained on a 15 nm NbN film on MgO substrate (T_c = 15.4 K). The solid lines are fits with the Mattis-Bardeen equations [27, 28]. The inset of panel (a) shows the temperature dependence of the energy gap 2Δ, extracted from fits to $\sigma(\omega)$ (symbols) overlaying the normalized transmissivity ratio, TR. The BCS temperature dependence of Δ is shown by the dashed line. The inset to panel (b) displays the transmitted THz transients through the NbN film below and above T_c; the arrow denotes the time $t' = t_0$ with maximum change in the transmitted electric field (after [108]). (c) The recovery of the imaginary part of the superconducting conductivity for different intensities can be described by a single exponential dependence with τ = 580 ps [107]. (d) Time evolution of normalized transmissivity ratio (TR), recorded at A = 22 mJ/cm^3, together with $\Delta(t)$ extracted from $\sigma_1(\omega, t_d)$ fit with the BCS theory [27, 28] (after [108]).

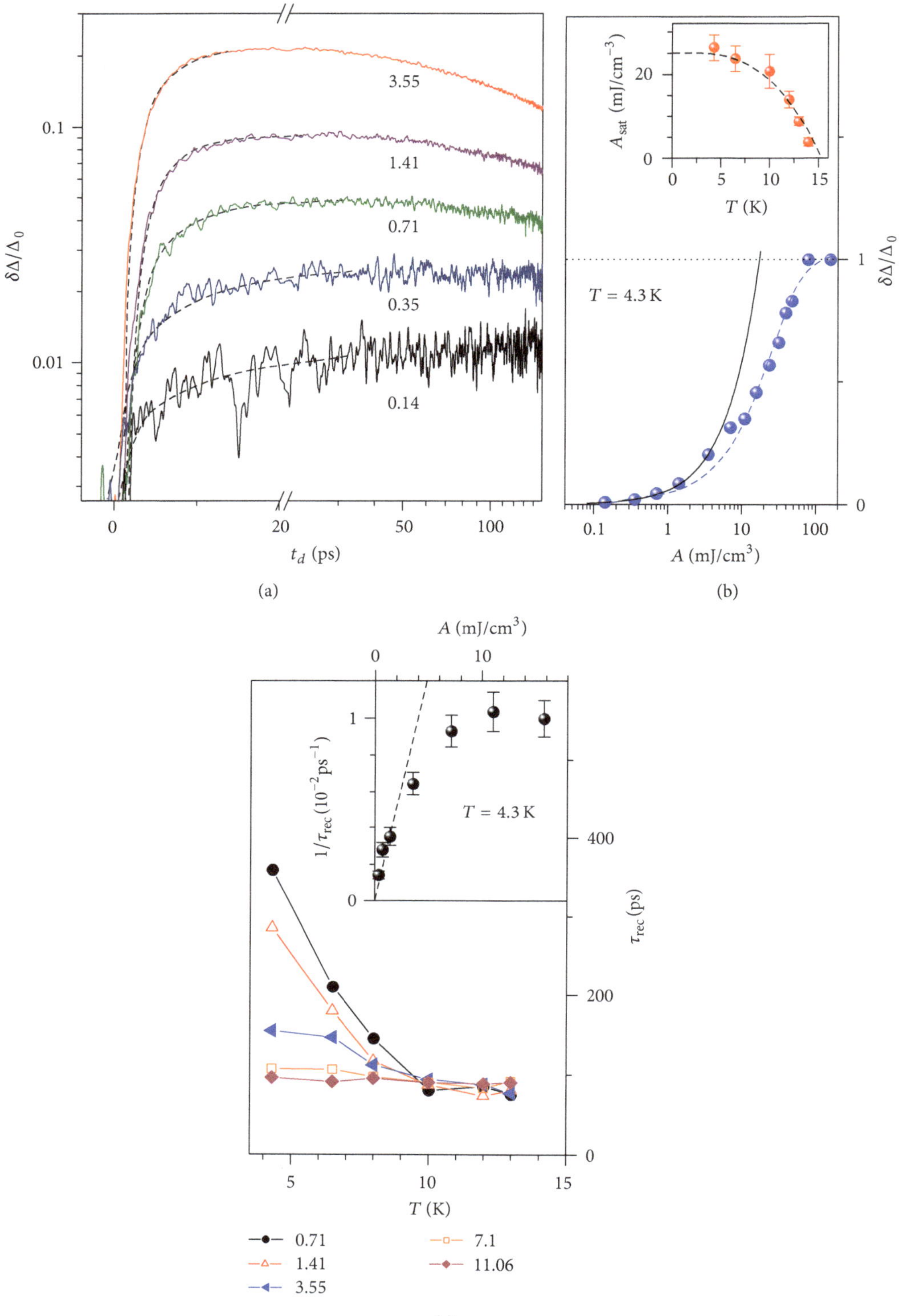

FIGURE 19: (a) The relative change in gap, $\delta\Delta/\Delta_0$ recorded at 4.3 K for various intensities A in mJ/cm^3. The dashed lines are fits to the data with (1). (b) The dependence of $\delta\Delta/\Delta_0$ on intensity A at 4.3 K. The dashed line is a fit to the simple saturation model; the solid line is the linear fit. Inset: the temperature dependence of the saturation energy density, A_{sat}, compared to the T-dependence of Δ^2 (dashed line). Panel (c) shows the T-dependence of the superconducting recovery time, τ_{rec}, for several intensities A (in mJ/cm ^{3}). Inset: $\tau_{\text{rec}}^{-1}(A)$ recorded at 4.3 K. For low intensities A the relaxation rate increases linearly with A (dashed line) (after [108]).

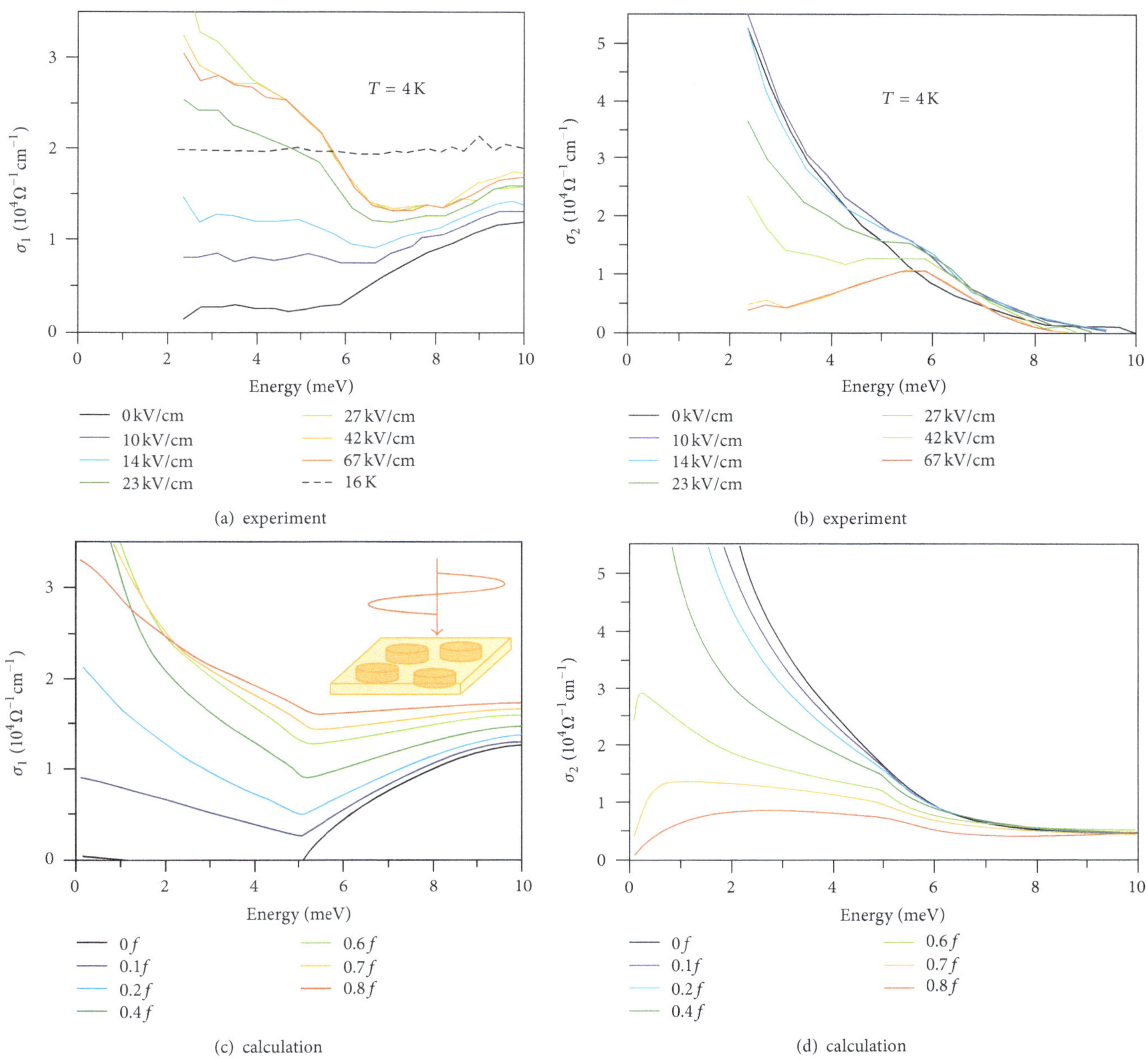

FIGURE 20: The spectra of the (a) real and (b) imaginary parts of the optical conductivity for various THz excitation intensities at 4 K. For reference, $\sigma_1(\omega)$ in the normal metal state at 16 K is plotted as a dotted line. Panel (c) and (d) exhibit the calculated real and imaginary parts of the optical conductivity spectra, respectively, using an effective medium approach. The inset is a schematic image of the inhomogeneously suppressed superconductivity (after [109]).

Interest in this field has been revived recently with the experimental observations of a number of dramatic features near the superconductor-insulator transition in InO, TiN and NbN, such as nonmonotonic temperature dependence of the resistivity, simple activated temperature dependence on the insulating side, large peak in the magnetoresistance, peculiar I-V characteristics, and existence of an energy gap at temperatures above T_c [119–130]. Another reason for renewed interest in this field is that the superconductor-insulator transition may be a very basic realization of a quantum phase transition, that is, a phase transition which occurs at $T = 0$ as a function of some nonthermal parameters (pressure, doping, magnetic field, etc.) and is driven by quantum instead of thermal fluctuations.

The nature of the superconductor-insulator transition and especially that of the insulating phase is still under debate. Recently several indications for the presence of superconductivity in the insulating state have been reported [131–133], and a number of possible explanations have been put forward. One suggestion relates the phenomena to the inhomogeneous nature of the system under study [134, 135]; near the superconductor-insulator transition on both sides of the transition the system is composed of regions of superconducting material embedded in an insulating matrix. In this picture, the transition occurs when the isolated superconducting islands are able to Josephson-couple and allow the many-body wave function to percolate throughout the entire system. There are basic arguments [136, 137] and theoretical considerations [138, 139] that actually anticipate that inherent inhomogeneity has to occur near the superconductor-insulator transition even when the system is structurally uniform and compatible with the underlying disorder (which, near the SIT of a typical superconductor, is fairly strong). Therefore, some inhomogeneity is expected to be present near the superconductor-insulator transition in all systems where the transition temperature does not drop to zero before a substantial disorder has set in (which in 2D is equivalent to the SIT occurring when the sheet resistance $R_\square$ is of the order of the quantum resistance $\hbar/4e^2$).

There are quite a few systems that seem to exhibit transport properties that are consistent with this expectation. An example is the situation in amorphous InO films, one of the systems that showed the peculiar features alluded above. Structural study reveals morphological inhomogeneities on scales of order 100 Å. Transport studies, however, exposed scale dependences up to few microns [136, 137]. Another candidate is films of NbN that exhibit higher T_c and thus are suitable for optical experiments in the THz range.

A second picture invokes the existence of uncorrelated preformed electron pairs which do not constitute a condensate [129, 130, 140, 141] but are characterized by an energy gap that is associated with the pair binding energy. Other models adapt concepts from both pictures [142] and suggest that above T_c and on the insulating site the film is composed of small superconducting islands that are uncorrelated and are too small to sustain bulk superconductivity. Hence, the situation is unclear, and further experimental information is required.

Acknowledgments

The authours thank A. Frydman, B. Gorshunov, E. Heinze, U. Pracht, M. Scheffler, and D. Sherman for the fruitful collaboration and useful discussions. The work was supported by the Deutsche Forschungsgemeinschaft (DFG).

References

[1] F. London, *Superfluids*, vol. 1 of *Macroscopic Theory of Superconductivity*, John Wiley & Sons, New York, NY, USA, 1950.

[2] J. Bardeen, "Theory of the Meissner effect in superconductors," *Physical Review*, vol. 97, no. 6, pp. 1724–1725, 1955.

[3] J. Bardeen, "Theory of Supercondgctivity," in *Handbuch der Physik*, vol. 15, pp. 274–369, Springer, Berlin, Germany, 1956.

[4] H. Welker, "Supraleitung und magnetische Austauschwechselwirkung," *Zeitschrift für Physik*, vol. 114, no. 9-10, pp. 525–551, 1939.

[5] L. N. Cooper, "Bound electron pairs in a degenerate fermi gas," *Physical Review*, vol. 104, no. 4, pp. 1189–1190, 1956.

[6] J. Bardeen, L. N. Cooper, and J. R. Schrieffer, "Theory of superconductivity," *Physical Review*, vol. 108, no. 5, pp. 1175–1204, 1957.

[7] L. C. Hebel and C. P. Slichter, "Nuclear relaxation in superconducting aluminum," *Physical Review*, vol. 107, no. 3, p. 901, 1957.

[8] L. C. Hebel and C. P. Slichter, "Nuclear spin relaxation in normal and superconducting aluminum," *Physical Review*, vol. 113, no. 6, pp. 1504–1519, 1959.

[9] R. W. Morse and H. V. Bohm, "Superconducting energy gap from ultrasonic attenuation measurements," *Physical Review*, vol. 108, no. 4, pp. 1094–1096, 1957.

[10] R. E. Glover III and M. Tinkham, "Transmission of superconducting films at millimeter-microwave and far infrared frequencies," *Physical Review*, vol. 104, no. 3, pp. 844–845, 1956.

[11] M. Tinkham, "Energy gap interpretation of experiments on infrared transmission through superconducting films," *Physical Review*, vol. 104, no. 3, pp. 845–846, 1956.

[12] R. E. Glover III and M. Tinkham, "Conductivity of superconducting films for photon energies between 0.3 and 40kTc," *Physical Review*, vol. 108, no. 2, pp. 243–256, 1957.

[13] M. J. Buckingham, "Very high frequency absorption in superconductors," *Physical Review*, vol. 101, no. 4, pp. 1431–1432, 1956.

[14] G. S. Blevins, W. Gordy, and W. M. Fairbank, "Superconductivity at millimeter wave frequencies," *Physical Review*, vol. 100, no. 4, pp. 1215–1216, 1955.

[15] M. Tinkham and R. E. Glover III, "Superconducting energy gap inferences from thin-film transmission data," *Physical Review*, vol. 110, no. 3, pp. 778–779, 1958.

[16] A. T. Forrester, "Superconducting energy gap inferences from thin-film transmission data," *Physical Review*, vol. 110, no. 3, pp. 776–778, 1958.

[17] M. A. Biondi, A. T. Forrester, M. P. Garfunkel, and C. B. Satterthwaite, "Experimental evidence for an energy gap in superconductors," *Reviews of Modern Physics*, vol. 30, no. 4, pp. 1109–1136, 1958.

[18] M. A. Biondi, M. P. Garfunkel, and A. O. McCoubrey, "Millimeter wave absorption in superconducting aluminum," *Physical Review*, vol. 101, no. 4, pp. 1427–1429, 1956.

[19] M. A. Biondi, M. P. Garfunkel, and A. O. McCoubrey, "Microwave measurements of the energy gap in superconducting aluminum," *Physical Review*, vol. 108, no. 2, pp. 495–497, 1957.

[20] M. A. Biondi, A. T. Forrester, and M. P. Garfunkel, "Millimeter wave studies of superconducting tin," *Physical Review*, vol. 108, no. 2, pp. 497–498, 1957.

[21] M. A. Biondi and M. P. Garfunkel, "Measurement of the temperature variation of the energy gap in superconducting aluminum," *Physical Review Letters*, vol. 2, no. 4, pp. 143–145, 1959.

[22] M. A. Biondi and M. P. Garfunkel, "Millimeter wave absorption in superconducting aluminum. I. Temperature dependence of the energy gap," *Physical Review*, vol. 116, no. 4, pp. 853–861, 1959.

[23] M. A. Biondi and M. P. Garfunkel, "Millimeter wave absorption in superconducting aluminum. II. Calculation of the skin depth," *Physical Review*, vol. 116, no. 4, pp. 862–867, 1959.

[24] M. Tinkham, "Absorption of electromagnetic radiation in superconductors," *Physica*, vol. 24, supplement 1, pp. S35–S41, 1958.

[25] R. A. Ferrell and R. E. Glover III, "Conductivity of superconducting films: a sum rule," *Physical Review*, vol. 109, no. 4, pp. 1398–1399, 1958.

[26] M. Tinkham and R. A. Ferrell, "Determination of the superconducting skin depth from the energy gap and sum rule," *Physical Review Letters*, vol. 2, no. 8, pp. 331–333, 1959.

[27] D. C. Mattis and J. Bardeen, "Theory of the anomalous skin effect in normal and superconducting metals," *Physical Review*, vol. 111, no. 2, pp. 412–417, 1958.

[28] W. Zimmermann, E. H. Brandt, M. Bauer, E. Seider, and L. Genzel, "Optical conductivity of BCS superconductors with arbitrary purity," *Physica C*, vol. 183, no. 1–3, pp. 99–104, 1991.

[29] K. Steinberg, M. Scheffler, and M. Dressel, "Quasiparticle response of superconducting aluminum to electromagnetic radiation," *Physical Review B*, vol. 77, no. 21, Article ID 214517, 2008.

[30] D. M. Ginsberg, P. L. Richards, and M. Tinkham, "Apparent structure on the far infrared energy gap in superconducting lead and mercury," *Physical Review Letters*, vol. 3, no. 7, pp. 337–338, 1959.

[31] D. M. Ginsberg and M. Tinkham, "Far infrared transmission through superconducting films," *Physical Review*, vol. 118, no. 4, pp. 990–1000, 1960.

[32] P. L. Richards and M. Tinkham, "Far-infrared energy gap measurements in bulk superconducting in, Sn, Hg, Ta, V, Pb, and Nb," *Physical Review*, vol. 119, no. 2, pp. 575–590, 1960.

[33] L. H. Palmer and M. Tinkham, "Far-infrared absorption in thin superconducting lead films," *Physical Review*, vol. 165, no. 2, pp. 588–595, 1968.

[34] M. Tinkham, "The electromagnetic properties of superconductors," *Reviews of Modern Physics*, vol. 46, no. 4, pp. 587–596, 1974.

[35] P. Wyder, "Far-infrared and superconductivity," *Infrared Physics*, vol. 16, no. 1-2, pp. 243–251, 1976.

[36] K. Holczer, O. Klein, and G. Grüner, "Observation of the conductivity coherence peak in superconducting Pb," *Solid State Communications*, vol. 78, no. 10, pp. 875–877, 1991.

[37] O. Klein, E. J. Nicol, K. Holczer, and G. Grüner, "Conductivity coherence factors in the conventional superconductors Nb and Pb," *Physical Review B*, vol. 50, no. 9, pp. 6307–6316, 1994.

[38] A. B. Pippard, "Metallic conduction at high frequencies and low temperatures," *Advances in Electronics and Electron Physics*, vol. 6, pp. 1–45, 1954.

[39] A. B. Pippard, "The anomalous skin effect in anisotropic metals," *Proceedings of the Royal Society of London A*, vol. 224, no. 1157, pp. 273–282, 1954.

[40] J. R. Waldram, "Surface impedance of superconductors," *Advances in Physics*, vol. 13, no. 49, pp. 1–88, 1964.

[41] M. Tinkham, *Introduction to Superconductivity*, Mc-Graw Hill, New York, NY, USA, 1st edition, 1975.

[42] M. Tinkham, *Introduction to Superconductivity*, Mc-Graw Hill, New York, NY, USA, 2nd edition, 1996.

[43] M. Tinkham, *Introduction to Superconductivity*, Dover Publication, Mineola, NY, USA, 2nd edition, 2004.

[44] G. Rickayzen, *Theory of Superconductivity*, Interscience Publishers, New York, NY, USA, 1965.

[45] R. D. Parks, Ed., *Superconductivity, Vol. I and II*, Marcel Dekker, New York, NY, USA, 1969.

[46] M. Dressel and G. Grüner, *Electrodynamics of Solids*, Cambridge University Press, Cambridge, UK, 2002.

[47] D. N. Basov, R. D. Averitt, D. van der Marel, M. Dressel, and K. Haule, "Electrodynamics of correlated electron materials," *Reviews of Modern Physics*, vol. 83, no. 2, pp. 471–541, 2011.

[48] D. Sherman, G. Kopnov, D. Shahar, and A. Frydman, "Measurement of a superconducting energy gap in a homogeneously amorphous insulator," *Physical Review Letters*, vol. 108, no. 17, Article ID 177006, 2012.

[49] B. Sacépé, C. Chapelier, T. I. Baturina, V. M. Vinokur, M. R. Baklanov, and M. Sanquer, "Pseudogap in a thin film of a conventional superconductor," *Nature Communications*, vol. 1, no. 9, article 140, 2010.

[50] Y. Mizukami, H. Shishido, T. Shibauchi et al., "Extremely strong-coupling superconductivity in artificial two-dimensional Kondo lattices," *Nature Physics*, vol. 7, no. 11, pp. 849–853, 2011.

[51] S. P. Chockalingam, M. Chand, A. Kamlapure et al., "Tunneling studies in a homogeneously disordered s-wave superconductor: NbN," *Physical Review B*, vol. 79, no. 9, Article ID 094509, 2009.

[52] T. Timusk and B. Statt, "The pseudogap in high-temperature superconductors: an experimental survey," *Reports on Progress in Physics*, vol. 62, no. 1, pp. 61–122, 1999.

[53] N. D. Basov and T. Timusk, "Electrodynamics of high—T_c superconductors," *Reviews of Modern Physics*, vol. 77, no. 2, pp. 721–729, 2005.

[54] U. S. Pracht, M. Scheffler, M. Dressel, D. F. Kalok, C. Strunk, and T. I. Baturina, "Direct observation of the superconducting gap in a thin film of titanium nitride using terahertz spectroscopy," *Physical Review B*, vol. 86, no. 18, Article ID 184503, 2012.

[55] E. F. C. Driessen, P. C. J. J. Coumou, R. R. Tromp, P. J. de Visser, and T. M. Klapwijk, "Strongly disordered tin and nbtin s-wave superconductors probed by microwave electrodynamics," *Physical Review Letters*, vol. 109, no. 10, Article ID 107003, 2012.

[56] M. R. Vissers, M. P. Weides, J. S. Kline, M. Sandberg, and D. P. Pappas, "Identifying capacitive and inductive loss in lumped element superconducting hybrid titanium nitride/aluminum resonators," *Applied Physics Letters*, vol. 101, Article ID 022601, 2012.

[57] D. Henrich, S. Dörner, M. Hofherr et al., "Broadening of hotspot response spectrum of superconducting NbN nanowire singlephoton detector with reduced nitrogen content," *Journal of Applied Physics*, vol. 112, no. 7, Article ID 074511, 2012.

[58] K. Il'in, M. Hofherr, D. Rall et al., "Ultra-thin TaN films for superconducting nanowire single-photon detectors," *Journal of Low Temperature Physics*, vol. 167, pp. 809–814, 2011.

[59] A. Engel, A. Schilling, K. Ilin, and M. Siegel, "Dependence of count rate on magnetic field in superconducting thin-film TaN single-photon detectors," *Physical Review B*, vol. 86, Article ID 140506, 2012.

[60] A. J. Kerman, E. A. Dauler, W. E. Keicher et al., "Kinetic-inductance-limited reset time of superconducting nanowire photon counters," *Applied Physics Letters*, vol. 88, no. 11, Article ID 111116, 2006.

[61] G. Gol'Tsman, O. Okunev, G. Chulkova et al., "Fabrication and properties of an ultrafast NbN hot-electron single-photon detector," *IEEE Transactions on Applied Superconductivity*, vol. 11, no. 1, pp. 574–577, 2001.

[62] J. Zhang, W. Slysz, A. Verevkin et al., "Response time characterization of NbN superconducting single-photon detectors," *IEEE Transactions on Applied Superconductivity*, vol. 13, no. 2, pp. 180–183, 2003.

[63] C. M. Natarajan, M. G. Tanner, and R. H. Hadfields, "Superconducting nanowire single-photon detectors: physics and applications," *Superconductor Science and Technology*, vol. 25, no. 6, Article ID 063001, 2012.

[64] P. B. Miller, "Surface impedance of superconductors," *Physical Review*, vol. 118, no. 4, pp. 928–934, 1960.

[65] A. A. Abrikosov, L. P. Gorkov, and I. M. Khalatnikov, "A Superconductor in a high frequency field," *Soviet Physics: JETP*, vol. 8, no. 1, pp. 182–189, 1959.

[66] A. A. Abrikosov, L. P. Gorkov, and I. M. Khalatnikov, "Analysis of experimental data relating to the surface impedance of superconductors," *Soviet Physics: JETP*, vol. 10, no. 1, pp. 132–134, 1960.

[67] J.-J. Chang and D. J. Scalapino, "Gap enhancement in superconducting thin films due to microwave irradiation," *Journal of Low Temperature Physics*, vol. 29, no. 5-6, pp. 477–485, 1977.

[68] J.-J. Chang and D. J. Scalapino, "Kinetic-equation approach to nonequilibrium superconductivity," *Physical Review B*, vol. 15, no. 5, pp. 2651–2670, 1977.

[69] J.-J. Chang and D. J. Scalapino, "Nonequilibrium superconductivity," *Journal of Low Temperature Physics*, vol. 31, no. 1-2, pp. 1–32, 1978.

[70] D. N. Langenberg and A. L. Larkin, Eds., *Nonequilibrium Superconductivity*, North-Holland, Amserdam, The Netherlands, 1986.

[71] N. Kopnin, *Theory of Nonequilibrium Superconductivity*, Oxford University Press, Oxford, UK, 2009.

[72] U. S. Pracht, E. Heintze, C. Clauss et al., "Electrodynamics of the superconducting state in ultra-thin films at THz frequencies," *IEEE Transactions on Terahertz Science and Technology*, vol. 3, no. 3, pp. 269–280, 2013.

[73] L. Leplae, "Derivation of an expression for the conductivity of superconductors in terms of the normal-state conductivity," *Physical Review B*, vol. 27, no. 3, pp. 1911–1912, 1983.

[74] J.-J. Chang and D. J. Scalapino, "Electromagnetic response of layered superconductors," *Physical Review B*, vol. 40, no. 7, pp. 4299–4305, 1989.

[75] A. J. Berlinsky, C. Kallin, G. Rose, and A.-C. Shi, "Two-fluid interpretation of the conductivity of clean BCS superconductors," *Physical Review B*, vol. 48, no. 6, pp. 4074–4079, 1993.

[76] S. B. Nam, "Theory of electromagnetic properties of superconducting and normal systems. I," *Physical Review*, vol. 156, no. 2, pp. 470–486, 1967.

[77] S. B. Nam, "Theory of electromagnetic properties of strong-coupling and impure superconductors. II," *Physical Review*, vol. 156, no. 2, pp. 487–493, 1967.

[78] O. Klein, S. Donovan, M. Dressel, and G. Grüner, "Microwave cavity perturbation technique: part I: principles," *International Journal of Infrared and Millimeter Waves*, vol. 14, no. 12, pp. 2423–2457, 1993.

[79] S. Donovan, O. Klein, M. Dressel, K. Holczer, and G. Grüner, "Microwave cavity perturbation technique: part II: experimental scheme," *International Journal of Infrared and Millimeter Waves*, vol. 14, no. 12, pp. 2459–2487, 1993.

[80] M. Dressel, O. Klein, S. Donovan, and G. Grüner, "Microwave cavity perturbation technique: part III: applications," *International Journal of Infrared and Millimeter Waves*, vol. 14, no. 12, pp. 2489–2517, 1993.

[81] M. Scheffler and M. Dressel, "Broadband microwave spectroscopy in Corbino geometry for temperatures down to 1.7 K," *Review of Scientific Instruments*, vol. 76, no. 7, Article ID 074702, 2005.

[82] M. Scheffler, S. Kilic, and M. Dressel, "Strip-shaped samples in a microwave Corbino spectrometer," *Review of Scientific Instruments*, vol. 78, no. 8, Article ID 086106, 2007.

[83] K. Steinberg, M. Scheffler, and M. Dressel, "Broadband microwave spectroscopy in Corbino geometry at ^{3}He temperatures," *Review of Scientific Instruments*, vol. 83, no. 2, Article ID 024704, 2012.

[84] R. P. S. M. Lobo, J. D. LaVeigne, D. H. Reitze et al., "Photoinduced time-resolved electrodynamics of superconducting metals and alloys," *Physical Review B*, vol. 72, no. 2, Article ID 024510, 2005.

[85] K. E. Kornelsen, M. Dressel, J. E. Eldridge, M. J. Brett, and K. L. Westra, "Far-infrared optical absorption and reflectivity of a superconducting NbN film," *Physical Review B*, vol. 44, no. 21, pp. 11882–11887, 1991.

[86] D. Karecki, R. E. Peña, and S. Perkowitz, "Far-infrared transmission of superconducting homogeneous NbN films: scattering time effects," *Physical Review B*, vol. 25, no. 3, pp. 1565–1571, 1982.

[87] M. Šindler, R. Tesař, J. Koláček, L. Skrbek, and Z. ŠimŠa, "Far-infrared transmission of a superconducting NbN film," *Physical Review B*, vol. 81, no. 18, Article ID 184529, 2010.

[88] R. Tesař, M. Šindler, K. Ilin, J. Koláček, M. Siegel, and L. Skrbek, "Terahertz thermal spectroscopy of a NbN superconductor," *Physical Review B*, vol. 84, no. 13, Article ID 132506, 2011.

[89] M. W. Coffey and J. R. Clem, "Theory of high-frequency linear response of isotropic type-II superconductors in the mixed state," *Physical Review B*, vol. 46, no. 18, pp. 11757–11764, 1992.

[90] M. W. Coffey and J. R. Clem, "Theory of microwave transmission and reflection in type-II superconductors in the mixed state," *Physical Review B*, vol. 48, no. 1, pp. 342–350, 1993.

[91] B. P. Gorshunov, G. V. Kozlov, A. A. Volkov et al., "Measurement of electrodynamic parameters of superconducting films in the far-infrared and submillimeter frequency ranges," *International Journal of Infrared and Millimeter Waves*, vol. 14, no. 3, pp. 683–702, 1993.

[92] B. P. Gorshunov, I. V. Fedorov, G. V. Kozlov, A. A. Volkov, and A. D. Semenov, "Dynamic conductivity and the coherence peak in the submillimeter spectra of superconducting NbN films," *Solid State Communications*, vol. 87, no. 1, pp. 17–21, 1993.

[93] B. Gorshunov, A. Volkov, I. Spektor et al., "Terahertz BWO-spectrosopy," *International Journal of Infrared and Millimeter Waves*, vol. 26, no. 9, pp. 1217–1240, 2005.

[94] B. P. Gorshunov, A. A. Volkov, A. S. Prokhorov et al., "Terahetz BWO spectroscopy of conductors and superconductors," *Quantum Electronics*, vol. 37, no. 10, pp. 916–923, 2007.

[95] M. Dressel, N. Drichko, B. Gorshunov, and A. Pimenov, "THz spectroscopy of superconductors," *IEEE Journal on Selected Topics in Quantum Electronics*, vol. 14, no. 2, pp. 399–406, 2008.

[96] M. C. Nuss, K. W. Goossen, J. P. Gordon, P. M. Mankiewich, M. L. O'Malley, and M. Bhushan, "Terahertz time-domain measurement of the conductivity and superconducting band gap in niobium," *Journal of Applied Physics*, vol. 70, no. 4, pp. 2238–2241, 1991.

[97] A. V. Pronin, M. Dressel, A. Pimenov, A. Loidl, I. V. Roshchin, and L. H. Greene, "Direct observation of the superconducting energy gap developing in the conductivity spectra of niobium," *Physical Review B*, vol. 57, no. 22, pp. 14416–14421, 1998.

[98] U. Pracht, M. Scheffler, M. Dressel, K. S. Ilin, and M. Siegel,unpublished.

[99] A. Semenov, B. Günther, U. Böttger et al., "Optical and transport properties of ultrathin NbN films and nanostructures," *Physical Review B*, vol. 80, no. 5, Article ID 054510, 2009.

[100] E. Burstein, D. N. Langenberg, and B. N. Taylor, "Superconductors as quantum detectors for microwave and sub-millimeter-wave radiation," *Physical Review Letters*, vol. 6, no. 3, pp. 92–94, 1961.

[101] D. M. Ginsberg, "Upper limit for quasi-particle recombination time in a superconductor," *Physical Review Letters*, vol. 8, no. 5, pp. 204–207, 1962.

[102] J. R. Schrieffer and D. M. Ginsberg, "Calculation of the quasiparticle recombination time in a superconductor," *Physical Review Letters*, vol. 8, no. 5, pp. 207–208, 1962.

[103] A. Rothwarf and M. Cohen, "Rate of capture of electrons injected into superconducting lead," *Physical Review*, vol. 130, no. 4, pp. 1401–1405, 1963.

[104] A. Rothwarf and B. N. Taylor, "Measurement of recombination lifetimes in superconductors," *Physical Review Letters*, vol. 19, no. 1, pp. 27–30, 1967.

[105] L. R. Testardi, "Destruction of superconductivity by laser light," *Physical Review B*, vol. 4, no. 7, pp. 2189–2196, 1971.

[106] J. F. Federici, B. I. Greene, P. N. Saeta, D. R. Dykaar, F. Sharifi, and R. C. Dynes, "Direct picosecond measurement of photoinduced Cooper-pair breaking in lead," *Physical Review B*, vol. 46, no. 17, pp. 11153–11156, 1992.

[107] Experiments on NbN films by R. Kaindl, M. Carnahan, and D. Chemla at the Advanced Light Source of the Lawrance Berkeley National Laboratory. Figure reproduced from N. Gedik: THz spectroscopy of quasiparticle dynamics in complex materials, https://portal.slac.stanford.edu/sites/conf_public/THz_2012_09/presentations/gedik_Frontiers_of_THz_Science.pdf.

[108] M. Beck, M. Klammer, S. Lang et al., "Energy-gap dynamics of superconducting NbN thin films studied by time-resolved terahertz spectroscopy," *Physical Review Letters*, vol. 107, no. 17, Article ID 177007, 2011.

[109] R. Matsunaga and R. Shimano, "Nonequilibrium BCS state dynamics induced by intense terahertz pulses in a superconducting NbN film," *Physical Review Letters*, vol. 109, no. 18, Article ID 187002, 1 page, 2012.

[110] J.-J. Chang and D. J. Scalapino, "New instability in superconductors under external dynamic pair breaking," *Physical Review B*, vol. 10, no. 9, pp. 4047–4049, 1974.

[111] D. J. Scalapino and B. A. Huberman, "Onset of an inhomogeneous state in a nonequilibrium superconducting film," *Physical Review Letters*, vol. 39, no. 21, pp. 1365–1368, 1977.

[112] P. W. Anderson, "Theory of dirty superconductors," *Journal of Physics and Chemistry of Solids*, vol. 11, no. 1-2, pp. 26–30, 1959.

[113] M. Strongin, R. S. Thompson, O. F. Kammerer, and J. E. Crow, "Destruction of superconductivity in disordered near-monolayer films," *Physical Review B*, vol. 1, no. 3, pp. 1078–1091, 1970.

[114] R. C. Dynes, A. E. White, J. M. Graybeal, and J. P. Garno, "Breakdown of eliashberg theory for two-dimensional superconductivity in the presence of disorder," *Physical Review Letters*, vol. 57, no. 17, pp. 2195–2198, 1986.

[115] D. B. Haviland, Y. Liu, and A. M. Goldman, "Onset of superconductivity in the two-dimensional limit," *Physical Review Letters*, vol. 62, no. 18, pp. 2180–2183, 1989.

[116] J. M. Valles Jr., R. C. Dynes, and J. P. Garno, "Electron tunneling determination of the order-parameter amplitude at the superconductor-insulator transition in 2D," *Physical Review Letters*, vol. 69, no. 24, pp. 3567–3570, 1992.

[117] A. Yazdani and A. Kapitulnik, "Superconducting-insulating transition in two-dimensional a-MoGe thin films," *Physical Review Letters*, vol. 74, no. 15, pp. 3037–3040, 1995.

[118] A. M. Goldman and N. Marković, "Superconductor-insulator transitions in the two-dimensional limit," *Physics Today*, vol. 51, no. 11, pp. 39–44, 1998.

[119] M. A. Paalanen, A. F. Hebard, and R. R. Ruel, "Low-temperature insulating phases of uniformly disordered two-dimensional superconductors," *Physical Review Letters*, vol. 69, no. 10, pp. 1604–1607, 1992.

[120] V. F. Gantmakher, M. V. Golubkov, V. T. Dolgopolov, G. E. Tsydynzhapov, and A. A. Shashkin, "Destruction of localized electron pairs above the magnetic-field-driven superconductor-insulator transition in amorphous In-O films," *JETP Letters*, vol. 68, no. 4, pp. 363–369, 1998.

[121] V. Y. Butko and P. W. Adams, "Quantum metallicity in a two-dimensional insulator," *Nature*, vol. 409, no. 6817, pp. 161–164, 2001.

[122] V. F. Gantmakher, S. N. Ermolov, G. E. Tsydynzhapov, A. A. Zhukov, and T. I. Baturina, "Suppression of 2D superconductivity by the magnetic field: quantum corrections vs. the superconductor-insulator transition," *JETP Letters*, vol. 77, no. 8, pp. 424–428, 2003.

[123] N. Hadacek, M. Sanquer, and J.-C. Villégier, "Double reentrant superconductor-insulator transition in thin TiN films," *Physical Review B*, vol. 69, no. 2, Article ID 024505, 2004.

[124] G. Sambandamurthy, L. W. Engel, A. Johansson, and D. Shahar, "Superconductivity-Related Insulating Behavior," *Physical Review Letters*, vol. 92, no. 10, Article ID 107005, 2004.

[125] G. Sambandamurthy, L. W. Engel, A. Johansson, E. Peled, and D. Shahar, "Experimental evidence for a collective insulating state in two-dimensional superconductors," *Physical Review Letters*, vol. 94, no. 1, Article ID 017003, 2005.

[126] M. Steiner and A. Kapitulnik, "Superconductivity in the insulating phase above the field-tuned superconductor-insulator transition in disordered indium oxide films," *Physica C*, vol. 422, no. 1-2, pp. 16–26, 2005.

[127] T. I. Baturina, J. Bentner, C. Strunk, M. R. Baklanov, and A. Satta, "From quantum corrections to magnetic-field-tuned superconductor-insulator quantum phase transition in TiN films," *Physica B*, vol. 359-361, pp. 500–502, 2005.

[128] T. I. Baturina, C. Strunk, M. R. Baklanov, and A. Satta, "Quantum metallicity on the high-field side of the superconductor-insulator transition," *Physical Review Letters*, vol. 98, no. 12, Article ID 127003, 2007.

[129] B. Sacépé, C. Chapelier, T. I. Baturina, V. M. Vinokur, M. R. Baklanov, and M. Sanquer, "Disorder-induced inhomogeneities of the superconducting state close to the superconductor-insulator transition," *Physical Review Letters*, vol. 101, no. 15, Article ID 157006, 2008.

[130] B. Sacépé, T. Dubouchet, C. Chapelier et al., "Localization of preformed Cooper pairs in disordered superconductors," *Nature Physics*, vol. 7, no. 3, pp. 239–244, 2011.

[131] S. M. Hollen, H. Q. Nguyen, E. Rudisaile et al., "Cooper-pair insulator phase in superconducting amorphous Bi films induced by nanometer-scale thickness variations," *Physical Review B*, vol. 84, no. 6, Article ID 064528, 2011.

[132] V. M. Vinokur, T. I. Baturina, M. V. Fistul, A. Y. Mironov, M. R. Baklanov, and C. Strunk, "Superinsulator and quantum synchronization," *Nature*, vol. 452, no. 7187, pp. 613–615, 2008.

[133] S. Poran, E. Shimshoni, and A. Frydman, "Disorder-induced superconducting ratchet effect in nanowires," *Physical Review B*, vol. 84, no. 1, Article ID 014529, 2011.

[134] Y. Dubi, Y. Meir, and Y. Avishai, "Theory of the magnetoresistance of disordered superconducting films," *Physical Review B*, vol. 73, no. 5, Article ID 054509, 2006.

[135] Y. Dubi, Y. Meir, and Y. Avishai, "Nature of the superconductor-insulator transition in disordered superconductors," *Nature*, vol. 449, no. 7164, pp. 876–880, 2007.

[136] D. Kowal and Z. Ovadyahu, "Disorder induced granularity in an amorphous superconductor," *Solid State Communications*, vol. 90, no. 12, pp. 783–786, 1994.

[137] D. Kowal and Z. Ovadyahu, "Scale dependent superconductor-insulator transition," *Physica C*, vol. 468, no. 4, pp. 322–325, 2008.

[138] N. Trivedi, R. T. Scalettar, and M. Randeria, "Superconductor-insulator transition in a disordered electronic system," *Physical Review B*, vol. 54, no. 6, pp. R3756–R3759, 1996.

[139] Y. Imry, M. Strongin, and C. C. Homes, "An inhomogeneous Josephson phase in thin-film and high-T_c superconductors," *Physica C*, vol. 468, no. 4, pp. 288–293, 2008.

[140] M. V. Feigei'man, L. B. Ioffe, V. E. Kravtsov, and E. A. Yuzbashyan, "Eigenfunction fractality and pseudogap state near the superconductor-insulator transition," *Physical Review Letters*, vol. 98, no. 2, Article ID 027001, 2007.

[141] M. V. Feigel'man, L. B. Ioffe, V. E. Kravtsov, and E. Cuevas, "Fractal superconductivity near localization threshold," *Annals of Physics*, vol. 325, no. 7, pp. 1390–1478, 2010.

[142] K. Bouadim, Y. L. Loh, M. Randeria, and N. Trivedi, "Single- and two-particle energy gaps across the disorder-driven superconductor-insulator transition," *Nature Physics*, vol. 7, no. 11, pp. 884–889, 2011.

Structural, Elastic, and Electronic Properties of Antiperovskite Chromium-Based Carbides $ACCr_3$ (A = Al and Ga)

D. F. Shao,[1] W. J. Lu,[1] S. Lin,[1] P. Tong,[1] and Y. P. Sun[1,2]

[1] *Key Laboratory of Materials Physics, Institute of Solid State Physics, Chinese Academy of Sciences, Hefei 230031, China*
[2] *High Magnetic Field Laboratory, Chinese Academy of Sciences, Hefei 230031, China*

Correspondence should be addressed to W. J. Lu; wjlu@issp.ac.cn and Y. P. Sun; ypsun@issp.ac.cn

Academic Editor: Laifeng Li

We theoretically investigated antiperovskite chromium-based carbides $ACCr_3$ through the first-principles calculation based on density functional theory (DFT). The structure optimization shows that the lattice parameter of $ACCr_3$ is basically proportional to the radius of A-site elements. The calculated formation energies show that $AlCCr_3$ and $GaCCr_3$ can be synthesized at ambient pressure and are stable with nonmagnetic ground states. Based on the calculation of elastic constants, some elastic, mechanical, and thermal parameters are derived and discussed. $AlCCr_3$ and $GaCCr_3$ show ductile natures and may have similar thermal properties. From the analysis of the electronic structures, it was found that there are electron and hole bands that cross the Fermi level for $AlCCr_3$ and $GaCCr_3$, indicating multiple-band natures. The Fermi level locates at the vicinity of the density of states (DOSs) peak, which leads to a large DOS at Fermi level dominated by Cr-3d electrons. The band structures of $AlCCr_3$ and $GaCCr_3$ are very similar to those of the superconducting antiperovskite $MgCNi_3$. The similarity may make $AlCCr_3$ and $GaCCr_3$ behave superconductively, which needs to be further investigated in theoretical and experimental studies.

1. Introduction

Recently, antiperovskite compounds AXM_3 (A, main group elements; X, carbon, boron, or nitrogen; M, transition metal) have attracted considerable attention. Due to the high concentration of transition metals in a cell, it can be deduced that interesting properties will be found in the family of compounds. In the antiperovskites family, nickel-based and manganese-based antiperovskites were extensively studied. Abundant physical properties were shown in the two kinds of compounds, such as superconductivity [1–3], giant magnetoresistance (MR) [4, 5], large negative magnetocaloric effect (MCE) [6, 7], giant negative thermal expansion [8, 9], magnetostriction [10], and nearly zero temperature coefficient of resistivity [11, 12]. But there are only a few reports about other 3d-metal-based antiperovskites so far. The difficulty restricting researchers is the exploration of new materials that can be experimentally synthesized. Therefore, theoretical investigations on these potential 3d-metal-based antiperovskites are useful to find the easily prepared stable materials and explore the new physical properties.

In the earlier years, researchers have found that in chromium compounds there are varieties of interesting physical properties. Many of chromium alloys such as Cr-Ru, Cr-Rh, and Cr-Ir alloys show superconductivity [13]. And it was found there is spin density wave antiferromagnetism that coexists with superconductivity in $Cr_{1-x}Re_x$ ($x > 0.17$) [14], $Cr_{1-x}Ru_x$ ($x > 0.17$) [15], and $(Cr_{1-x}Mo_x)_{75}Ru_{75}$ ($x = 0, 0.03, 0.06, \text{and } 0.10$) [16]. For chromium-based antiperovskites, Wiendlocha et al. [17] discussed the possibility of superconductivity in $GaNCr_3$ and $RhNCr_3$, and recent phonon and electron-phonon coupling calculation of $RhNCr_3$ [18] supports Wiendlocha et al.'s prediction. In the present work, we theoretically investigated antiperovskite chromium-based carbides $ACCr_3$ through the first-principles calculation based on density functional theory (DFT). The optimized lattice parameter of $ACCr_3$ is basically proportional to the radius of A-site elements. From the analysis of formation energies, we predict that only $AlCCr_3$ and $GaCCr_3$ can be synthesized at ambient pressure, and they are stable with nonmagnetic ground states. The elastic and electronic properties of the two compounds are specifically discussed.

TABLE 1: The coefficient A_2 in (1) as combinations for second order elastic constant for cubic crystal. η_α denotes the Lagrangian strain tensors in terms of ξ.

Strain type	A_2
$\eta_1 = (\xi, 0, 0, 0, 0, 0)$	C_{11}
$\eta_2 = (\xi, \xi, 0, 0, 0, 0)$	$2C_{11} + 2C_{12}$
$\eta_3 = (\xi, \xi, \xi, 0, 0, 0)$	$C_{11} + 4C_{44}$

$AlCCr_3$ and $GaCCr_3$ show ductile natures and may have similar thermal properties. The electron and hole bands cross the Fermi level, implying the multiple-band nature of $AlCCr_3$ and $GaCCr_3$. The Fermi level locates at the vicinity of the density of states (DOS) peak, which leads to a large DOS at Fermi level dominated by Cr-3d electrons. The bands properties of $AlCCr_3$ and $GaCCr_3$ are very similar to those of superconducting antiperovskite $MgCNi_3$. The similarity may make $AlCCr_3$ and $GaCCr_3$ show superconductivity.

2. Computational Details

The calculations were performed by projected augmented-wave (PAW) [19, 20] method using the ABINIT code [21–23]. The PAW method can lead to very accurate results comparable to other all-electron methods. For the exchange-correlation functional, the generalized gradient approximation (GGA) according to the Perdew-Burke-Ernzerhof [24] parametrization was used. Electronic wavefunctions are expanded with plane waves up to an energy cutoff of 1200 eV. Brillouin zone sampling is performed on the Monkhorst-Pack (MP) mesh [25] of 16 × 16 × 16. The self-consistent calculations were considered to be converged when the total energy of the system was stable within 10^{-6} Ha. Nonmagnetic, ferromagnetic (FM), and antiferromagnetic (AFM) states were tested in the study. For AFM states, we only considered the simplest case: the spins of Cr atom in [111] layers are parallel with each other in the same layer and are antiparallel with spins in the neighboring layers.

The elastic constants are evaluated according to the finite-strain continuum elasticity theory [26, 27]. The strain energy per unit mass can be expressed as a polynomial of the strain parameter ξ [28]:

$$\phi(\xi) = \frac{1}{2}A_2\xi^2 + \frac{1}{6}A_3\xi^3 + O(\xi^4), \quad (1)$$

where the coefficients A_2 and A_3 are combinations of second- and third-order elastic constants of the crystal, respectively. For a cubic structure, there are three independent second-order elastic constants (C_{11}, C_{12}, and C_{44}). Therefore, here we introduced 3 Lagrangian strain tensors labeled as η_α in terms of ξ. η_α and the corresponding coefficients A_2 are listed in Table 1.

Once the single-crystal elastic constants are calculated, the related elastic and mechanic parameters may also be evaluated. Using the Voigt-Reuss-Hill approximation [29], bulk modulus B and shear modulus G of cubic crystal can be defined as

$$B = \frac{C_{11} + 2C_{12}}{3}, \quad (2)$$

$$G = \frac{G_V + G_R}{2}, \quad (3)$$

where the subscripts V and R denote the Voigt, and Reuss average, respectively. In (3),

$$G_V = \frac{C_{11} - C_{12} + 3C_{44}}{5}, \qquad G_R = \frac{5(C_{11} - C_{12})C_{44}}{4C_{44} + 3(C_{11} - C_{12})}. \quad (4)$$

Thus Young's modulus E and Poisson's ratio σ can be obtained using the relations

$$E = \frac{9BG}{3B + G}, \qquad \sigma = \frac{3B - 2G}{2(3B + G)}. \quad (5)$$

Using the calculated bulk modulus B, shear modulus G, and Young's modulus E, the Debye temperature Θ_D can be determined as

$$\Theta_D = \frac{h}{k_B}\left(\frac{3n\rho N_A}{4\pi M}\right)^{1/3} v_m, \quad (6)$$

where h, k_B, and N_A are Planck's, Boltzmann's constants, and Avogadro's number, respectively. ρ is the mass density, M is the molecular weight, and n is the number of atoms in the unit cell. The mean sound velocity is defined as

$$v_m = \left[\frac{1}{3}\left(\frac{2}{v_t^3} + \frac{1}{v_l^3}\right)\right]^{1/3}, \quad (7)$$

where v_l and v_t are the longitudinal and transverse sound velocities, which can be obtained from bulk modulus B and shear modulus G:

$$v_l = \left(\frac{3B + 4G}{3\rho}\right)^{1/2}, \qquad v_t = \left(\frac{G}{\rho}\right)^{1/2}. \quad (8)$$

3. Results and Discussion

3.1. Ground State and Formation Energy. The structures were fully optimized with respect to lattice parameter and atomic positions. The optimized lattice parameters of $ACCr_3$ for different A-site elements are listed in Table 2. The results show that the lattice parameter of $ACCr_3$ is basically proportional to the radius of A-site elements (see Figure 1).

The common technique for producing antiperovskite carbides is the solid-state synthesis from the parent materials.

TABLE 2: Lattice parameter a, magnetic moments per Cr atom M, and the formation energies ΔE of $ACCr_3$.

	a (Å)	ΔE_{NM} (eV/atom)	ΔE_{FM} (eV/atom)	M_{FM} (μ_B/Cr)	ΔE_{AFM} (eV/atom)	M_{AFM} (μ_B/Cr)
$CuCCr_3$	3.869	0.214	0.212	0.396	0.205	1.006
$ZnCCr_3$	3.894	0.125	0.126	0.127	0.115	1.020
$GaCCr_3$	3.891	−0.019	−0.019	0.001	−0.019	0.099
$MoCCr_3$	3.898	0.105	0.105	0.053	0.105	0.240
$AlCCr_3$	3.887	−0.0479	−0.0478	0.034	−0.0478	0.212
$AgCCr_3$	3.936	0.286	0.292	0.544	0.268	1.649
$CdCCr_3$	3.973	0.256	0.257	0.341	0.213	2.102
$SnCCr_3$	3.981	0.082	0.0786	0.400	0.082	0.000
$MgCCr_3$	3.952	0.220	0.219	0.173	0.195	1.824
$InCCr_3$	3.999	0.187	0.189	0.406	0.188	0.006
$YCCr_3$	4.088	0.269	0.268	0.611	0.270	0.000
$LaCCr_3$	4.122	0.496	0.362	1.8101	0.497	0.002
$CaCCr_3$	4.083	0.539	0.541	0.3311	0.539	0.001
$SrCCr_3$	4.155	0.795	0.795	0.402	0.796	0.002
$BaCCr_3$	4.246	1.166	0.998	1.957	1.166	0.002

TABLE 3: Elastic constants and derived quantities for $ACCr_3$ (A = Al and Ga).

	C_{11} (GPa)	C_{12} (GPa)	C_{44} (GPa)	A	C_p (GPa)	B (GPa)	G (GPa)	B/G	E (GPa)	σ	Θ_D (K)
$AlCCr_3$	448.18	297.98	68.31	0.91	229.67	348.04	70.95	4.90	199.30	0.40	532.43
$GaCCr_3$	480.81	314.43	64.63	0.78	249.80	369.89	71.51	4.40	201.53	0.41	497.12

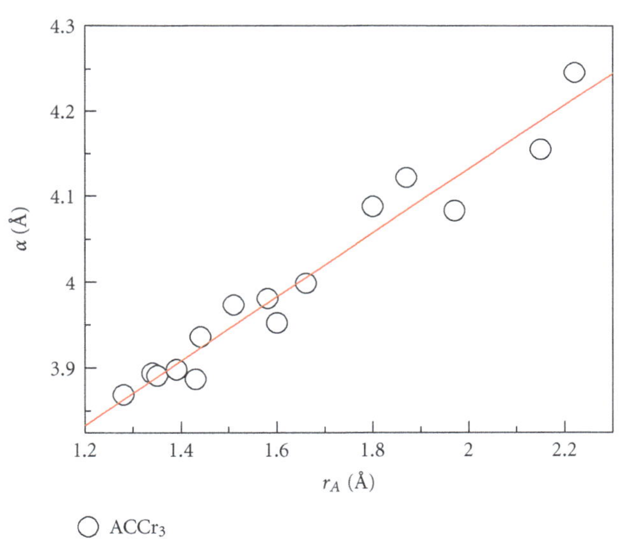

FIGURE 1: Relation between the lattice parameter a and the radius of A-site elements r_A.

Therefore we assumed that the antiperovskite chromium-based carbides $ACCr_3$ could be synthesized in the same routine. We analyze the stability of compounds by calculating the formation energies ΔE, which is defined as the difference between the total energies of initial reagents (elements) and final compounds:

$$\Delta E = E_{tot}^{ACCr_3} - \left(E_{tot}^{A} + E_{tot}^{C} + 3E_{tot}^{Cr}\right). \quad (9)$$

Here, E_{tot}^{A}, E_{tot}^{C}, and E_{tot}^{Cr} are the total energies of the initial reagents.

According to the definition of ΔE, negative values indicate that the formation of compounds from the initial components is energetically favorable and the compounds are stable with respect to a mixture of the initial elements. Conversely, if $\Delta E > 0$, the compounds are unlikely to be synthesized under normal conditions.

The calculated ΔE for the series of A-site elements with different magnetic configurations are listed in Table 2. Except for $AlCCr_3$ and $GaCCr_3$, all other compounds show positive ΔE. For $AlCCr_3$ and $GaCCr_3$, the values of ΔE for different magnetic configurations are very close, and the magnetic moments per Cr atom for the FM and AFM states are very small, which implies the nonmagnetic ground states of $AlCCr_3$ and $GaCCr_3$. From the analysis of formation energies, it seems that $ACCr_3$ can be synthesized only when A = Al and Ga. The elastic and electronic properties of the two compounds will be discussed later.

3.2. Elastic and Mechanical Properties. In order to obtain accurate elastic constants, ξ is varied with a step of 0.0025 in every case for η_α. As an example, the strain energies for $AlCCr_3$ and the fitted polynomials are shown in Figure 2. The calculated elastic constants at ambient pressure and other derived mechanical parameters are listed in Table 3. From the calculated values of the elastic constants, one can find that they satisfy the mechanical stability criteria for a cubic crystal, that is, $C_{11} > C_{12}$, $C_{44} > 0$, and $C_{11} + 2C_{12} > 0$ [30], which indicates that the compounds are mechanically stable.

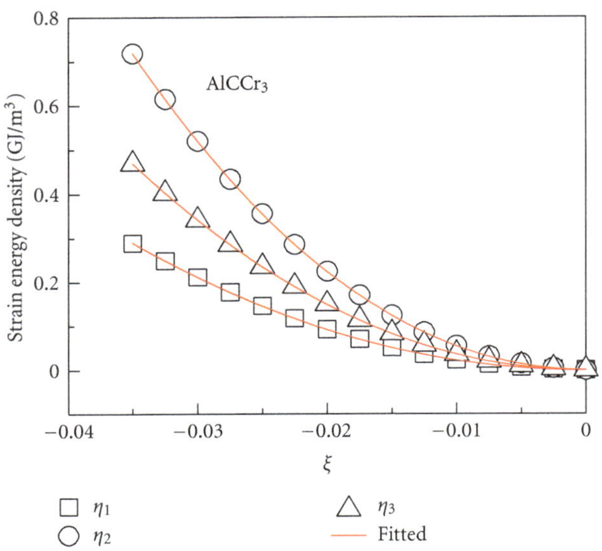

FIGURE 2: The strain-energy relation of $AlCCr_3$.

Cracks in crystals are directly related to the anisotropy of thermal and elastic properties [31]. The elastic anisotropy factor A of cubic crystal can be described as

$$A = \frac{2C_{44}}{C_{11} - C_{12}}. \tag{10}$$

For a complete isotropic system, the anisotropy factor A takes the value of unity, and the deviation from unity measures the degree of elastic anisotropy [32]. The elastic constant C_{44} represents the shear resistance in a [100] direction, which is related to the C–Cr bonding. The values of C_{44} for the two compounds are almost the same, which implies the similar C–Cr bonding nature of the two compounds. $C_{11} - C_{12}$ turns out to be the stiffness associated with a shear in a [110] direction, which can be connected with the bonding between A–Cr bonding. Although A–M bonding is usually very weak in antiperovskite ACM_3, it still can influence the anisotropy of the material. The anisotropic factors A are listed in Table 3. The different A–Cr bondings of the two compounds lead to different anisotropy. For $AlCCr_3$, $A = 0.91$ indicates that $AlCCr_3$ is almost isotropic. However, $GaCCr_3$ seems like more anisotropic ($A = 0.78$).

Young's modulus E defines the ratio between the linear stress and strain. The larger the value of E, the stiffer is the material. In general, as Young's modulus increases, the covalent nature of the compound also increases, which further has an impact on the ductility of the compounds. The values of E for $AlCCr_3$ and $GaCCr_3$ are very close to each other, suggesting the similar covalent nature of the two compounds.

The Cauchy's pressure C_p is defined as the difference between the two particular elastic constants $C_p = C_{12} - C_{44}$. It is considered to serve as an indication of ductility: $C_p > 0$ suggests that the material is expected to be ductile; otherwise, the material is expected to be brittle [33]. As shown in Table 3, the values of C_p for $AlCCr_3$ and $GaCCr_3$ are positive, which is a clear indication for the compounds to be ductile. Another index of ductility is the ratio B/G. According to Pugh [34], the ductile/brittle properties of materials could be related empirically to the ratio B/G. If $B/G > 1.75$, the material would be ductile; otherwise the material behaves in a brittle manner. For $ACCr_3$ (A = Al and Ga), all the calculated B/G ratios are much larger than 1.75, which clearly highlights the ductile nature of $ACCr_3$ (A = Al and Ga). Poisson's ratio generally quantifies the stability of the crystal against shear and takes the value between −1 and 0.5, which are the lower and the upper bounds. The lower bound is where the material does not change its shape, and the upper bound is where the volume remains unchanged. The calculated σ of $ACCr_3$ (A = Al and Ga), listed in Table 3, are very close to the typical σ value of 0.33 for ductile metallic materials [35]. All the parameters clearly show the ductility of $ACCr_3$ (A = Al and Ga).

Using bulk modulus B, shear modulus G, and Young's modulus E, the Debye temperature Θ_D can be calculated. As a matter of fact, a higher Θ_D would imply a higher thermal conductivity associated with the material [36]. For the present calculation, the Θ_D are estimated to be 532.43 K and 497.12 K for $AlCCr_3$ and $GaCCr_3$, respectively. The similar values of Θ_D suggest the similar thermal characteristics of the two compounds.

3.3. Electronic Properties. The calculated electronic band structures along the high symmetry directions in the Brillouin zones together with the total and site-projected l-decomposed DOS at equilibrium lattice parameters for $ACCr_3$ (A = Al and Ga) are shown in Figures 3 and 4. The band structures of the two compounds are very similar. Bands from −13 eV to −10 eV are mainly from the C-2s and Cr-3d states with strong hybridizations characteristic. From −8 eV to 2 eV, there are hybridizations between C-2p and Cr-3d states, and the Cr-3d states contribute dominantly to the bands, which suggests the itinerant nature of Cr-3d electrons. From Figure 4, it can be seen the Fermi level locates at the vicinity of the DOS peak, which leads large DOS at the Fermi level $N(E_F)$ with values of 4.79 states/eV for $AlCCr_3$ and 5.62 states/eV for $GaCCr_3$.

The Fermi surfaces of the two compounds are similar (see Figure 5). There are four bands that cross the Fermi level. Hole pockets surrounding Γ point come from the lower two bands (denoted with red and green colors in Figure 3). Electron pockets surrounding R point come from the upper two bands (denoted with blue and purple colors in Figure 3). The differences of the Fermi surface of the two compounds come from the band denoted with blue color in Figure 3. It crosses the Fermi level at the vicinity of M point for $AlCCr_3$, which makes the electron pockets surround the corners of the Brillouin zones connected to each other at M point. However, for $GaCCr_3$, it crosses the Fermi level in the place between Γ and M points, which forms a cubic cage-like electron pocket surrounding Γ point and connecting with the corner-centered electron pockets. The presence of electron and hole bands crossing the Fermi level indicates the multiple-band natures for $AlCCr_3$ and $GaCCr_3$.

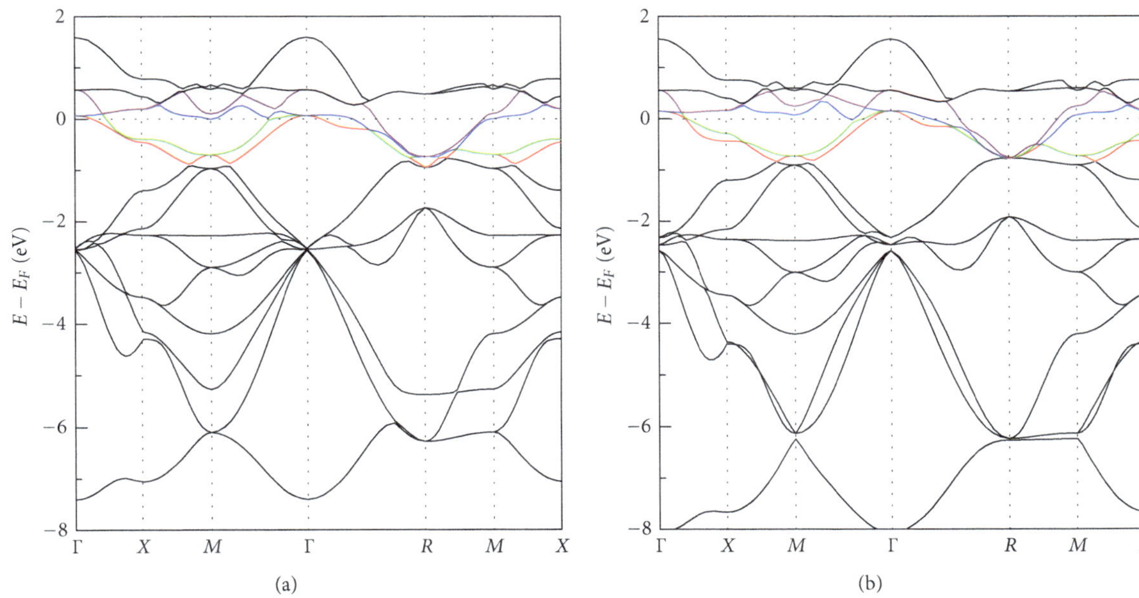

FIGURE 3: The band structures for (a) $AlCCr_3$ and (b) $GaCCr_3$. The colored lines denote the bands crossing the Fermi level.

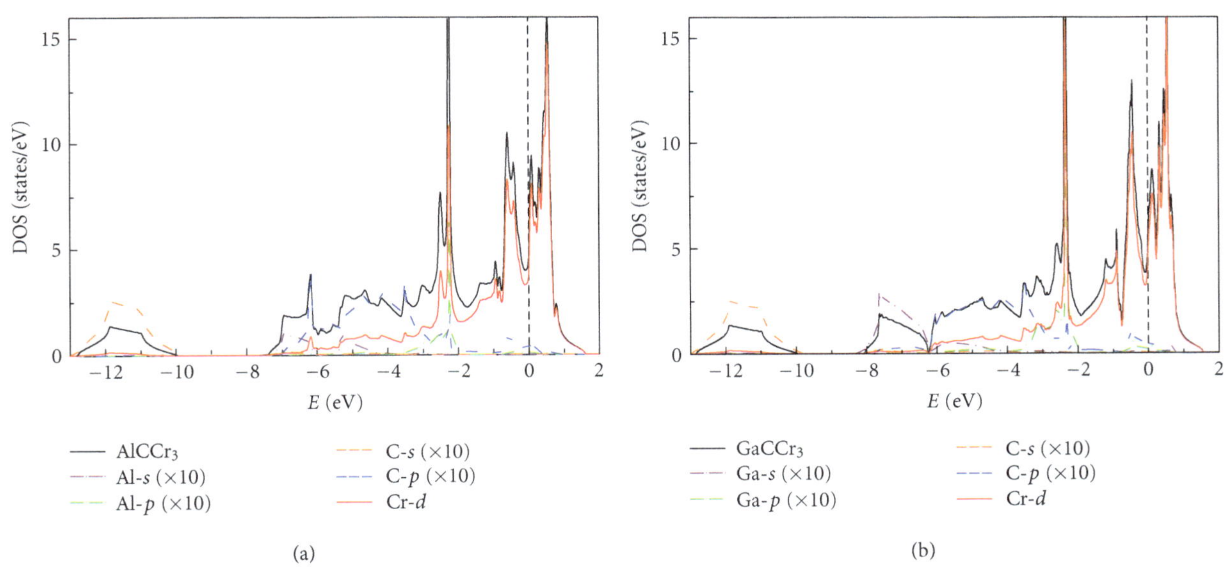

FIGURE 4: The density of states of (a) $AlCCr_3$ and (b) $GaCCr_3$.

In order to understand the bonding nature among the atoms in $AlCCr_3$ and $GaCCr_3$, we plot the contour maps of the charge density of the two compounds in Figure 6. The C–Cr bonds are very strong, which coincides with the strong hybridization between C-2p and Cr-3d electrons shown in DOS figures. The similarity of the bonding nature for the two compounds coincides with the similar Young's modulus E discussed before.

The electronic properties of $AlCCr_3$ and $GaCCr_3$ are very similar to those of the superconducting antiperovskite $MgCNi_3$ [37], for which the Fermi level locates at the vicinity of the DOS peak as well, and there is multiple-band nature present. The similarity possibly makes $AlCCr_3$ and $GaCCr_3$ potential superconductors. According to McMillan's coupling theory [38], the electron-phonon coupling constant λ can be calculated from the following expression $\lambda = \sum_\alpha(\eta_\alpha/M_\alpha\langle\omega_\alpha^2\rangle)$, where M_α is the atomic mass and $\langle\omega_\alpha^2\rangle$ is the averaged phonon frequency and can be approximated as $\langle\omega_\alpha^2\rangle = \Theta_D^2/2$. For superconducting $MgCNi_3$, there is a peak (a van Hove singularity) in DOS just below Fermi level, which leads to a large $N(E_F)$. The $N(E_F)$ is not large enough to lead magnetic instability, but it can lead to a large electron-phonon coupling constant [39]. To the best of our knowledge, except for the superconducting nickel antiperovskites, there are no

FIGURE 5: (a) Brillouin zone of $ACCr_3$. Fermi surfaces of (b) $AlCCr_3$ and (c) $GaCCr_3$.

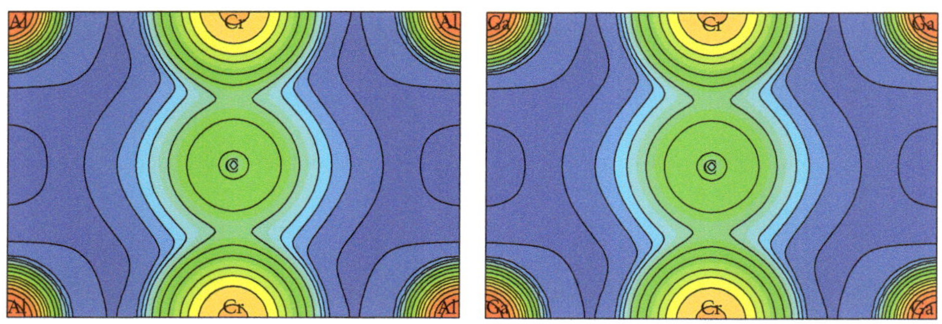

FIGURE 6: The contour map of electron charge density of (a) $AlCCr_3$ and (b) $GaCCr_3$.

other antiperovskites that have such a nature. Wiendlocha et al. [17] pointed out that in chromium systems, chromium atoms have very large values of the McMillan-Hopfield parameters [38], which may lead to a very strong electron-phonon coupling. The recently reported work about $RhNCr_3$ [18] supports Wiendlocha et al.'s prediction. The phonon and electron-phonon coupling calculations show that $RhNCr_3$, which has a large $N(E_F)$, is a strong coupling superconductor with Tc above 16 K. Therefore, we consider that there is possibility for superconductivity appearing in $AlCCr_3$ and $GaCCr_3$. Phonon and electron-phonon coupling calculations will be carried out in the future to confirm the possibility.

4. Conclusion

In conclusion, we theoretically investigated the antiperovskite chromium-based carbides $ACCr_3$ through the first-principles calculation based on density functional theory. The optimized lattice parameter of $ACCr_3$ is basically proportional to the radius of A-site elements. Only $AlCCr_3$ and $GaCCr_3$ have negative formation energies, implying that the two compounds can be synthesized at ambient pressure and are stable with nonmagnetic ground states. $AlCCr_3$ and $GaCCr_3$ show ductile natures, and they may have similar thermal properties. Similar to superconducting antiperovskite $MgCNi_3$, there are electron and hole bands that cross the Fermi level for $AlCCr_3$ and $GaCCr_3$, indicating multiple-band natures. The Fermi level locates at the vicinity of the DOS peak, which leads to a large $N(E_F)$ dominated by Cr-3d electrons. These similarities possibly make $AlCCr_3$ and $GaCCr_3$ show superconductivity, which needs to be further investigated from theoretical and experimental studies.

Acknowledgments

This work was supported by the National Key Basic Research under Contract no. 2011CBA00111 and the National Nature Science Foundation of China under Contract nos. 91222109, 11274311, 51171177, 11174295, U1232139, and 50701042. The calculations were partially performed at the Center for Computational Science, CASHIPS.

References

[1] T. He, Q. Huang, A. P. Ramirez et al., "Superconductivity in the non-oxide perovskite $MgCNi_3$," *Nature*, vol. 411, no. 6833, pp. 54–56, 2001.

[2] M. Uehara, T. Yamazaki, T. Kôri, T. Kashida, Y. Kimishima, and I. Hase, "Superconducting properties of $CdCNi_3$," *Journal of the Physical Society of Japan*, vol. 76, no. 3, Article ID 034714, 2007.

[3] M. Uehara, A. Uehara, K. Kozawa, and Y. Kimishima, "New antiperovskite-type superconductor ZnN_yNi_3," *Journal of the Physical Society of Japan*, vol. 78, no. 3, 2009.

[4] K. Kamishima, T. Goto, H. Nakagawa et al., "Giant magnetoresistance in the intermetallic compound Mn_3GaC," *Physical Review B*, vol. 63, no. 2, Article ID 024426, 2001.

[5] Y. B. Li, W. F. Li, W. J. Feng, Y. Q. Zhang, and Z. D. Zhang, "Magnetic, transport and magnetotransport properties of $Mn_{3+x}Sn_{1-x}C$ and $Mn_3Zn_ySn_{1-y}C$ compounds," *Physical Review B*, vol. 72, no. 2, Article ID 024411, 2005.

[6] T. Tohei, H. Wada, and T. Kanomata, "Negative magnetocaloric effect at the antiferromagnetic to ferromagnetic transition of Mn_3GaC," *Journal of Applied Physics*, vol. 94, no. 3, pp. 1800–1802, 2003.

[7] M. H. Yu, L. H. Lewis, and A. R. Moodenbaugh, "Large magnetic entropy change in the metallic antiperovskite Mn_3GaC," *Journal of Applied Physics*, vol. 93, no. 12, pp. 10128–10130, 2003.

[8] K. Takenaka and H. Takagi, "Giant negative thermal expansion in Ge-doped anti-perovskite manganese nitrides," *Applied Physics Letters*, vol. 87, no. 26, Article ID 261902, pp. 1–3, 2005.

[9] K. Takenaka, K. Asano, M. Misawa, and H. Takagi, "Negative thermal expansion in Ge-free antiperovskite manganese nitrides: tin-doping effect," *Applied Physics Letters*, vol. 92, no. 1, Article ID 011927, 2008.

[10] K. Asano, K. Koyama, and K. Takenaka, "Magnetostriction in Mn_3 CuN," *Applied Physics Letters*, vol. 92, no. 16, Article ID 161909, 2008.

[11] E. O. Chi, W. S. Kim, and N. H. Hur, "Nearly zero temperature coefficient of resistivity in antiperovskite compound $CuNMn_3$," *Solid State Communications*, vol. 120, no. 7-8, pp. 307–310, 2001.

[12] J. C. Lin, B. S. Wang, P. Tong et al., "Tunable temperature coefficient of resistivity in C- and Co-doped $CuNMn_3$," *Scripta Materialia*, vol. 65, no. 5, pp. 452–455, 2011.

[13] B. T. Matthias, T. H. Geballe, V. B. Compton, E. Corenzwit, and G. W. Hull, "Superconductivity of chromium alloys," *Physical Review*, vol. 128, no. 2, pp. 588–590, 1962.

[14] Y. Nishihara, Y. Yamaguchi, T. Kohara, and M. Tokumoto, "Itinerant-electron antiferromagnetism and superconductivity in bcc Cr-Re alloys," *Physical Review B*, vol. 31, no. 9, pp. 5775–5781, 1985.

[15] Y. Nishihara, Y. Yamaguchi, M. Tokumoto, K. Takeda, and K. Fukamichi, "Superconductivity and magnetism of bcc Cr-Ru alloys," *Physical Review B*, vol. 34, no. 5, pp. 3446–3449, 1986.

[16] H. L. Alberts, D. S. McLachlan, T. Germishuyse, and M. Naidoo, "Superconductivity and antiferromagnetism in Cr-Mo-Ru alloys," *Journal of Physics*, vol. 3, no. 12, pp. 1793–1800, 1991.

[17] B. Wiendlocha, J. Tobola, S. Kaprzyk, and D. Fruchart, "Electronic structure, superconductivity and magnetism study of Cr_3GaN and Cr_3RhN," *Journal of Alloys and Compounds*, vol. 442, no. 1-2, pp. 289–291, 2007.

[18] H. M. Tütüncü and G. P. Srivastava, "Phonons and superconductivity in the cubic perovskite Cr_3RhN," *Journal of Applied Physics*, vol. 112, no. 9, Article ID 093914, 2012.

[19] P. E. Blöchl, "Projector augmented-wave method," *Physical Review B*, vol. 50, no. 24, pp. 17953–17979, 1994.

[20] M. Torrent, F. Jollet, F. Bottin, G. Zérah, and X. Gonze, "Implementation of the projector augmented-wave method in the ABINIT code: application to the study of iron under pressure," *Computational Materials Science*, vol. 42, no. 2, pp. 337–351, 2008.

[21] X. Gonze, J. M. Beuken, R. Caracas et al., "First-principles computation of material properties: the ABINIT software project," *Computational Materials Science*, vol. 25, no. 3, pp. 478–492, 2002.

[22] X. Gonze, B. Amadon, P. M. Anglade et al., "ABINIT: first-principles approach to material and nanosystem properties," *Computer Physics Communications*, vol. 180, no. 12, pp. 2582–2615, 2009.

[23] X. Gonze, G. M. Rignanese, M. Verstraete et al., "A brief introduction to the ABINIT software package," *Zeitschrift fur Kristallographie*, vol. 220, no. 5-6, pp. 558–562, 2005.

[24] J. P. Perdew, K. Burke, and M. Ernzerhof, "Generalized gradient approximation made simple," *Physical Review Letters*, vol. 77, no. 18, pp. 3865–3868, 1996.

[25] H. J. Monkhorst and J. D. Pack, "Special points for Brillouin-zone integrations," *Physical Review B*, vol. 13, no. 12, pp. 5188–5192, 1976.

[26] F. Birch, "Finite elastic strain of cubic crystals," *Physical Review*, vol. 71, no. 11, pp. 809–824, 1947.

[27] F. D. Murnaghan, *Finite Deformation of An Elastic Solid*, Dover Publications, New York, NY, USA, 1951.

[28] J. Zhao, J. M. Winey, and Y. M. Gupta, "First-principles calculations of second- and third-order elastic constants for single crystals of arbitrary symmetry," *Physical Review B*, vol. 75, no. 9, Article ID 094105, 2007.

[29] R. Hill, "The elastic behaviour of a crystalline aggregate," *Proceedings of the Physical Society A*, vol. 65, no. 5, pp. 349–354, 1952.

[30] D. C. Wallace, *Thermodynamics of Crystals*, John Wiley & Sons, New York, NY, USA, 1972.

[31] D. J. Green, *An Introduction to the Mechanical Properties of Ceramics*, Cambridge University Press, Cambridge, UK, 1998.

[32] R. E. Newnham, *Properties of Materials; Anisotropy, Symmetry, Structure*, Oxford University Press, New York, NY, USA, 2005.

[33] D. G. Pettifor, "Theoretical predictions of structure and related properties of intermetallics," *Materials Science and Technology*, vol. 8, no. 4, pp. 345–349, 1992.

[34] S. Pugh, "Relations between the elastic moduli and the plastic properties of polycrystalline pure metals," *Philosophical Magazine Series*, vol. 7, no. 45, pp. 823–843, 1954.

[35] J. Haines, J. M. Léger, and G. Bocquillon, "Synthesis and design of superhard materials," *Annual Review of Materials Research*, vol. 31, pp. 1–23, 2001.

[36] V. Kanchana, "Mechanical properties of Ti_3AlX (X = C, N): Ab initio study," *Europhysics Letters*, vol. 87, no. 2, p. 26006, 2009.

[37] S. Mollah, "The physics of the non-oxide perovskite superconductor $MgCNi_3$," *Journal of Physics Condensed Matter*, vol. 16, no. 43, pp. R1237–R1276, 2004.

[38] W. L. McMillan, "Transition temperature of strong-coupled superconductors," *Physical Review*, vol. 167, no. 2, pp. 331–344, 1968.

[39] J. H. Shim, S. K. Kwon, and B. I. Min, "Electronic structures of antiperovskite superconductors $MgXNi_3$ (X = B, C, and N)," *Physical Review B*, vol. 64, no. 18, Article ID 180510, 2001.

21

Heterostructures for Realizing Magnon-Induced Spin Transfer Torque

P. B. Jayathilaka,[1] M. C. Monti,[2] J. T. Markert,[2] and Casey W. Miller[1]

[1] *Department of Physics, University of South Florida, 4202 East Fowler Avenue, Tampa, FL 33620, USA*
[2] *Department of Physics, University of Texas at Austin, 1 University Station C1600, Austin, TX 78712, USA*

Correspondence should be addressed to Casey W. Miller, millercw@usf.edu

Academic Editor: Giancarlo Consolo

This work reports efforts fabricating heterostructures of different materials relevant for the realization of magnon-induced spin transfer torques. We find the growth of high-quality magnetite on MgO substrates to be straightforward, while using transition metal buffer layers of Fe, Cr, Mo, and Nb can alter the structural and magnetic properties of the magnetite. Additionally, we successfully fabricated and characterized Py/Cr/Fe_3O_4 and Fe_3O_4/Cr/Fe_3O_4 spin valve structures. For both, we observe a relatively small giant magnetoresistance and confirm an inverse dependence on spacer layer thickness. Thus, we have shown certain materials combinations that may form the heterostructures that are the building blocks necessary to achieve magnon-induced spin transfer torque devices.

1. Introduction

The prediction [1, 2] and observation of "spin transfer" or "spin-transfer torque" in mechanical point contacts [3], and subsequently in lithographically defined point contacts [4], and nanopillars [5], revealed that spin-polarized currents can manipulate magnetization dynamics. The interest in this area is dominated by the potential for spin-torque-based magnetic random access memory to become a universal memory [6], as well as interesting high-frequency devices based on spin torque oscillators (STOs) [7, 8]. Spin torque research is one of the key areas of spintronics and nanomagnetism today [9], being relevant to both science and technology.

Theoreticians have recently promoted the idea of initiating spin transfer torque by thermal transport from magnons, initially in all-metal systems [11, 12] and recently with insulators [10]. Slonczewski's concept in [10] is based on the ability to convert the spin momentum of magnons in a magnetic insulator into a polarized current in a normal metal, which is then driven by a thermal gradient into a metallic ferromagnet. He showed that the quantum yield of heat-driven in-plane spin-transfer torque can exceed that from electric current in state-of-the-art spin-transfer devices utilizing magnetic tunnel junctions by nearly two orders of magnitude. STOs would benefit greatly from the increased quantum yield deliverable from this magnon-induced spin transfer torque (MISTT) because increasing the usable torque would increase their output power, which is not yet compatible with applications [13–15]. Spin torque-based devices in general would benefit from replacing the high-current densities now necessary for operation, as this causes appreciable heating, as well as vortex nucleation in the free layer via the unavoidable Oersted field. MISTT could address both of these issues, allowing significant improvements in device performance, fabrication requirements, and reliability. The experimental realization of this magnonic spin torque may thus have transformative impact by essentially creating a new class of spin torque devices.

The significant difference between existing spin-transfer devices and Slonczewski's proposal is the origin of the torque on the free layer. Traditionally, the spin-transfer torque comes from direct electrical injection of electrons from a fixed metallic ferromagnet through a normal metal into the free layer [9]. The spin polarization of these electrons mimics that of the fixed ferromagnet, which results in an imposed torque on the free layer magnetization when the

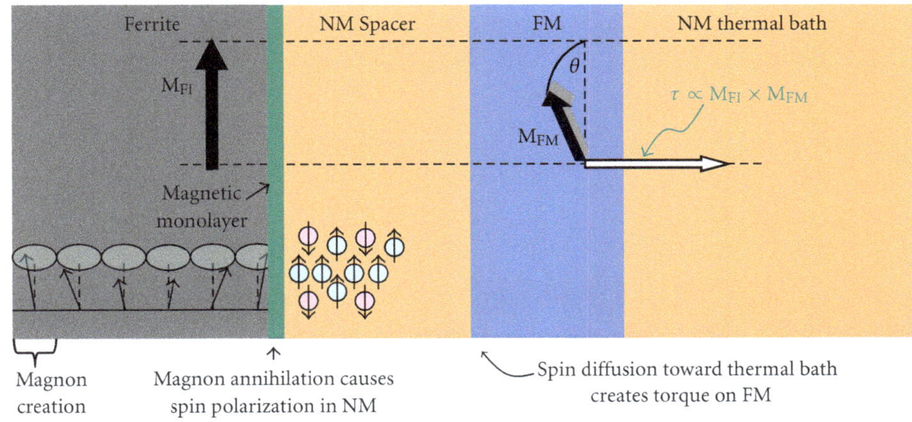

FIGURE 1: Magnon-induced spin transfer torque (MISTT) ala Slonczewski [10]. Magnons are created thermally in the magnetic insulator. Though these magnons are annihilated at the ferrite-normal metal (NM) interface (which contains a monolayer of ferromagnet metal atoms), their spin momentum is transferred to the electrons in the NM spacer. This creates spin accumulation in the NM, which then diffuses toward the NM thermal bath. The spin accumulation current density (more precisely, its time derivative) at the Py interface imposes a torque on the Py magnetization if its magnetization M_{FM} is at some angle relative to the ferrite's magnetization, M_{FI}.

spin polarization axes of the two ferromagnets are not colinear. In the MISTT device concept (shown schematically in Figure 1), spin polarized electrons still impose a torque on the free layer via a normal metal spacer, but there are several critical differences. First, the electrons are driven by a thermal—not electrical—gradient. Second, because the fixed layer in MISTT is a resistive ferri or ferromagnet (such as spinel ferrites), electrons cannot be simply injected, so a more exotic mechanism for polarizing the electrons in the normal metal is required: magnon annihilation. The magnons within the ferrite carry spin momentum, and, being bosons, the magnon population can be increased dramatically through heating. Conversely, these magnons can also be annihilated when they reach the ferrite-normal metal interface. The spin momentum originally carried by the annihilated magnons can be transferred to electrons in the normal metal, causing them to become spin polarized. These spin-polarized electrons then diffuse into the free magnet, leading to a spin-transfer torque.

Given their great potential, we have been exploring the materials combinations that may lead to the successful fabrication of these devices. This article outlines our progress toward the production of structures that may be useful for realizing magnonic spin transfer torque devices.

2. Experimental Section

All the films were grown in an Ar ion sputtering system with a base pressure of 20 nTorr. Ultra-high-purity gases were used for all the deposition. Samples were rotated at a speed of 40 rotations per minute during the deposition to obtain uniform film thickness. To grow different thicknesses of the same material at a single deposition, the rotation of the sample holder was stopped, so that one sample closer to the gun sees more material flux and grows thicker and the samples away from the target get less material flux and grows thinner. This was used to grow "wedges" for thickness-dependent studies. In this study, three different sets of samples were grown. Those are (a) Fe_3O_4 on MgO (100) substrates with transition metal buffer layers (Fe, Cr, Nb, Mo and No buffer), (b) Py/Cr/Fe_3O_4 spin valve with different Cr thicknesses on Si (100) substrates, and (c) Fe_3O_4/Cr/Fe_3O_4 substrates with different Cr thicknesses on MgO (100) substrates. Fe_3O_4 was grown at 300C by rf reactive sputtering from an Fe target in an Ar and O_2 environment. The amount of Ar and O_2 were controlled by the adjusting the flow rates of those gases in to the deposition chamber (20 SCCM Ar and 0.75 SCCM O_2). The total deposition pressure was 10 mTorr and the deposition rate was 0.26 Å/S. Epitaxy was confirmed by X-ray diffraction, and the quality of the Fe_3O_4 was indicated by the quality of the Verwey transition. The Fe_3O_4 films we grew showed the Verwey transition in between 119 K to 125 K, which suggests the stoichiometry was Fe_3O_{4-x}, with x bounded above by 0.001 [16].

3. High-Quality Ferrite Films on MgO Substrates

Since magnetite and MgO have nearly perfect lattice matching (as do most spinel ferrites), it is hard to distinguish them using the principal reflections in XRD because the intensity of the substrate (002) peak drowns out the magnetite's (004) peak. We have used a combination of wide angle XRD to observe the (004) and (008) peaks of the substrate and film [17], along with in-plane XRD (i.e., phi scans) to distinguish the Fe_3O_4 from the substrate peaks. We carried out scans focusing on the (311) and (220) orientations. Figures 2(a) and 2(d) show the results of these scans, which confirm the epitaxial growth of the magnetite on the MgO. Using higher resolution in Figures 2(b), 2(c), 2(e), and 2(f) shows that the magnetite peaks are broadened slightly relative to those of the substrate, which may be expected from finite-size effects. The full width at half maximum (FWHM) for MgO (311),

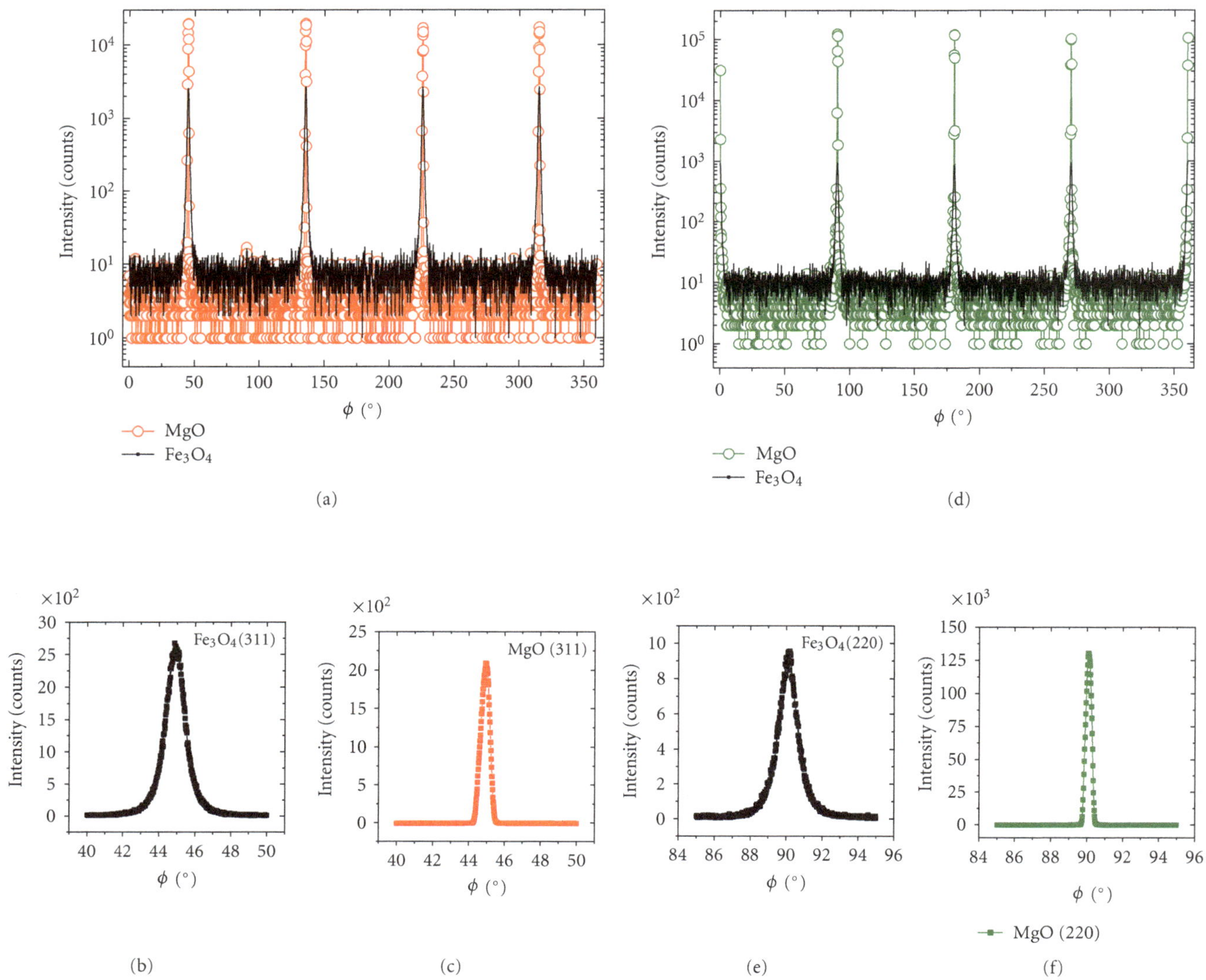

FIGURE 2: XRD Phi scans of Fe_3O_4 on MgO (100) substrates. (a) Complete scan on (220) plane, (b) complete scan on (311) plane, (c) MgO (220) peak, FWHM = 0.38°, (d) Fe_3O_4 (220) Peak, FWHM = 1.36° (e) MgO (311) peak, FWHM = 0.56° (f) Fe_3O_4 (311) peak, FWHM = 1.34°.

Fe_3O_4 (311), MgO (220), and Fe_3O_4 (220) was 0.56°, 1.34°, 0.38°, and 1.36° respectively.

4. Impact of Thin Buffer Layers on Magnetite

In a previous study, we showed that, the magnetite can be grown on 30 Å buffer layers of Fe, Mo, Nb, and Cr [17]. Fe, Mo, and Nb lend themselves to varying degrees of (100) texture, with Nb being the lowest quality; Cr leads to polycrystalline magnetite. The Mo-buffered magnetite has lattice parameters similar to magnetite grown directly on MgO, which is somewhat strained relative to bulk magnetite. The large lattice mismatch between Mo and magnetite appears to lead to the growth of magnetite nanoparticles, not a film per se. The Fe-buffered sample shows the least amount of strain of any sample. This reduced strain appears to reduce the density of antiphase boundaries, as inferred from mobile magnetic domains observed with temperature-dependent MFM. This is in contrast to magnetite grown directly on MgO [18].

Resistivity measurements were carried out to determine the effect of the transition metal buffer layers on the Verwey transition, as shown in Figure 3. In growing, the magnetite on these buffer layers an in situ mask exchange system allowed us to grow the transition metals on separate MgO (100) substrates individually, and then to grow the magnetite simultaneously on all the samples in one deposition run (i.e., preparation took place entirely under vacuum). Since the magnetite is the same on all the samples, any differences in resistivities should be due to the buffer layer. In measuring the resistivity, we attached the samples to a measurement probe and dip the end into a liquid nitrogen bath, allowing the sample temperature to cool from 300 K down to about 100 K in about 30 minutes. Because of the

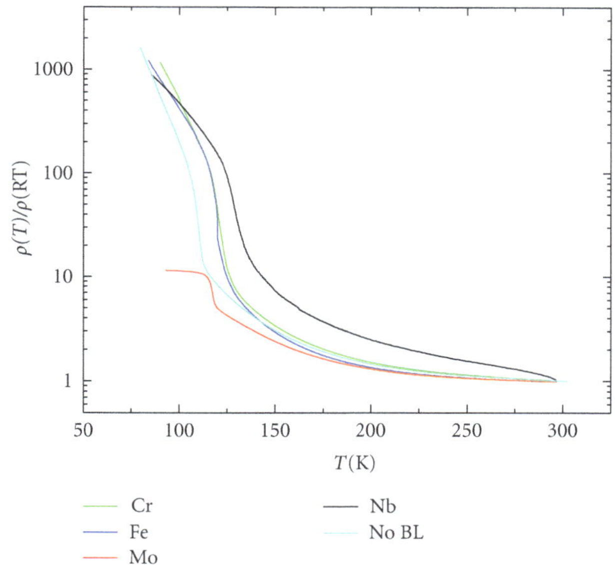

FIGURE 3: Change in resistivity as a function of temperature for magnetite simultaneously grown on different transition metal buffer layers, relative to the room temperature resistivity.

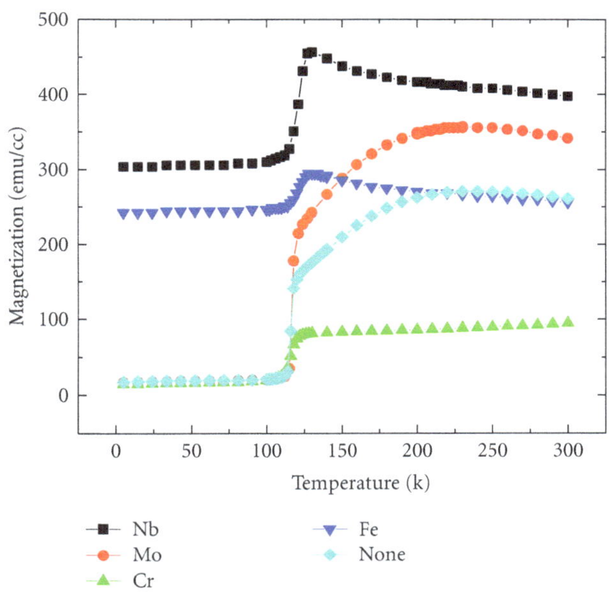

FIGURE 4: Temperature dependent magnetization of the Fe_3O_4 grown on top of different buffer layers. These SQUID magnetometry measurements were performed with a measurement field of 100 Oe applied along the (110) direction of the MgO.

high resistivity, we used 2 point resistivity measurements, making contact to the sample with pressed indium. The resistivity is measured as the sample cools down; the cooling rate is sufficiently slow that no thermal lag is noted with subsequent warming.

The unbuffered magnetite shows the sharpest Verwey transition, with an increase in resistivity of more than an order of magnitude in just a few degrees. Cr, Fe, and Nb buffered samples showed little higher transition temperatures than the unbuffered magnetite, but the transition was not sharp. This shift in the transition temperature may be due to the stress created from these buffer layers on the magnetite layer, or possibly the formation of an interfacial layer between the buffer and the magnetite (e.g., if the buffer became oxidized by reducing the magnetite, then the interfacial magnetite may have its stoichiometry altered). The Mo-buffered sample showed a transition point closer to the unbuffered sample, but the resistance saturates, unlike any other samples we investigated. We hypothesize this being due to the formation of conduction paths at grain boundaries; indeed, force microscopy images show that the Mo-buffered sample's topography is quite granular due to the formation of magnetite nanoparticles rather than a film. Additionally, these grains appear to have magnetic switching events during magnetic force microscopy that are caused by sample-tip interactions. This indicates the formation of grains that are weakly coupled to their neighbors.

The Verwey transition can also be observed through magnetization measurements. Figure 4 shows the magnetization as a function of temperature with a 100 Oe applied field along the (110) direction (along the diagonal of the substrate plane) for each of the samples. We see each sample has a distinct drop in magnetization at the Verwey transition, which is the typical behavior. It is notable that the Fe and Nb-buffered samples' behavior is distinct from measurements made with the field along the (100) direction on the same samples. We previously observed anomalous magnetization peaks in these samples along (100) direction at the Verwey transition [17]. Thus, it seems that the magnetite anisotropy is temperature dependent, leading to a temperature-dependent rotation of the easy axis above and below the Verwey transition; this manifests itself as a peaking magnetization if the moment rotates into the direction sampled by the SQUID. Similar behavior was reported in bulk Magnetite [19]. Measurement with a vector magnetometer should clarify these observations.

Figure 5 shows hysteresis loops of buffered and unbuffered magnetite samples along the (100) direction at temperatures above (150 K) and below (10 K) the Verwey transition. The samples with no buffer, Mo, and Cr buffers all have similar behavior, showing enhanced coercive and saturation fields at low temperatures. There is also a shearing of the low-temperature loops. On the other hand, the Fe- and Nb-buffered samples show an increasing coercivity at low fields, but the loop shape is not as dramatically changed. This could be due to these buffer layers being more prone to oxidation than the other buffer materials used, which may cause the interfacial magnetite to be reduced.

5. Magnetite-Based Spin Valves

The magnetite was introduced into spin valve structures. The Cr spacer layers were grown as a wedge to study the thickness dependence. Current in-plane (CIP) magnetoresistance measurements were carried out at room temperature for $Py/Cr/Fe_3O_4$ and $Fe_3O_4/Cr/Fe_3O_4$ spin valve structures.

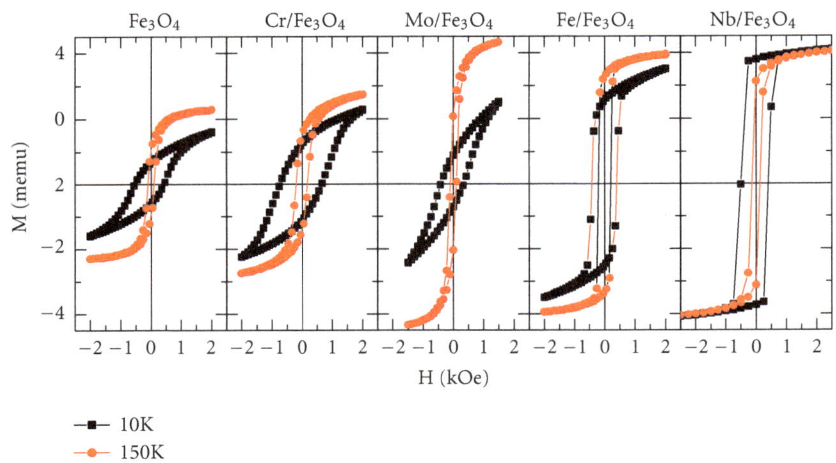

FIGURE 5: Hysteresis loops for Fe_3O_4 grown simultaneously on MgO (100) substrates with and without 30 Å transition metal buffer layers. The SQUID magnetometry measurements were performed with a measurement field of 100 Oe applied along the (100) direction of the MgO.

While CIP measurements are not typically used in spin-torque devices, this method can indicate that spin-dependent scattering is present in the structure. Note that MISTT does not even need current to flow in the device, as the magnons may be generated thermally (e.g., by optical means). We used pseudo-four-point resistance measurements, in which I^+ and V^+ leads are connected together at a single Indium contact (similarly for I^- and V^-) in order to remove effects related to the leads. In Py/Cr/Fe_3O_4 spin valves, the Cr thickness was varied from 41 Å to 107 Å and the Py and Fe_3O_4 thicknesses were kept constant at 300 Å (Figure 6). The GMR was calculated by subtracting the resistance of the saturated parallel state by the resistance of the shoulder in the antiparallel state. Then, the percentage was calculated with respect to the resistance of the saturated parallel state. The magnetoresistance showed a decrease with increasing spacer layer thickness in Py/Cr/Fe_3O_4. A fit of this thickness dependence reveals a decay constant of about 19 Å. This is less than the 45 Å reported for the spin diffusion length of Cr at 4.2 K [20]. Since the thicknesses used for the Cr spacer layer were relatively large, we did not observe any oscillatory coupling behavior as a function of spacer layer thickness. This ensured we were able to switch both ferromagnets separately, and it indicates the existence of a GMR signal, even for large thicknesses of normal metal spacer layers. In addition to the GMR signal there is also an anisotropic magnetoresistance (AMR) peak of Py due to the reversal of magnetization by the applied field. This AMR signal is essentially unavoidable because the metallic Py and Cr have much lower resistivity than the magnetite, so the current flows primarily through these layers. The intensity of this peak and its field range can be reduced by applying a magnetic field during the growth process, and likely by lithographic patterning to take advantage of shape anisotropy [21].

In Fe_3O_4/Cr/Fe_3O_4 spin valves (Figure 7), the Cr thickness was varied from 57 Å to 151 Å. To get different coercive fields from the magnetite, the bottom layer was 500 Å and the top layer was 1000 Å. Here, there are no well-defined parallel and antiparallel states due to the butterfly-like AMR signal intrinsic to the magnetite layers [22]. Thus, we calculated the GMR by taking the shift of the signal relative to the AMR signal. In this case, the magnetoresistance signal also decays exponentially with the spacer layer thickness, but the decay constant is 33 Å. We did not observe any oscillatory coupling due to higher spacer layer thickness.

The GMR values are quite small for both spin valve varieties relative to all-metallic CIP spin valves [23], but fall in the range of GMR reports for magnetite-normal metal-ferromagnetic systems [22, 24]. This is most likely related to a combination of factors, which most likely includes possible oxidation of the Cr at the magnetite interface acting as a barrier to spin-dependent reflections necessary for large GMR, as well as the conductivity mismatch between the magnetite and the metals shunting much of the current in the Py-containing devices. The differences in the two decay constants observed in the two spin valves is probably related to impurities in the spacer (which may be due to residual oxygen from the reactive sputtering to form the magnetite), which are known to impact the spin diffusion length [20], and possibly uncertainty in the spacer layer thickness.

6. Conclusions

We have studied the growth of magnetite on transition metal buffer layers and the formation of magnetite-based spin valves. We found the structural and magnetic properties of magnetite to be strongly dependent on the buffer layer. Despite the differences noted for Cr buffer layers, we were able to grow spin valves that use Cr as a spacer layer and magnetite as the top layer. While these spin valves do not appear to have promise for GMR applications, this work shows we have successfully fabricated heterostructures with

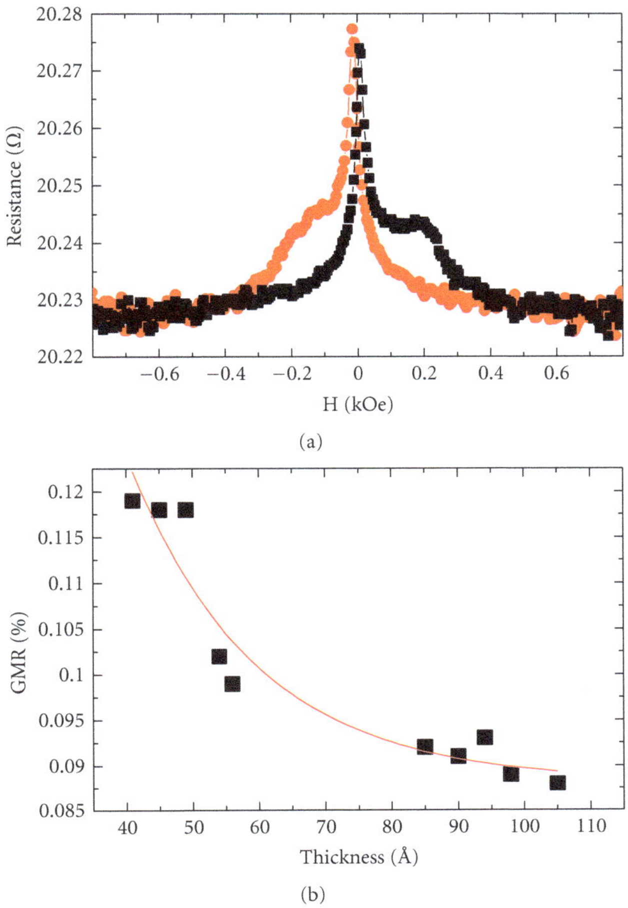

FIGURE 6: (a) Room temperature current in plane magneto resistance measurements on Py(300 Å)/Cr(41 Å)/Fe_3O_4(300 Å) spin valves. The shoulders are the antiparallel state. The AMR peak of the Py is unavoidable because of the conductivity mismatch between the metals and the oxide. (b) GMR as a function of Cr spacer layer thickness.

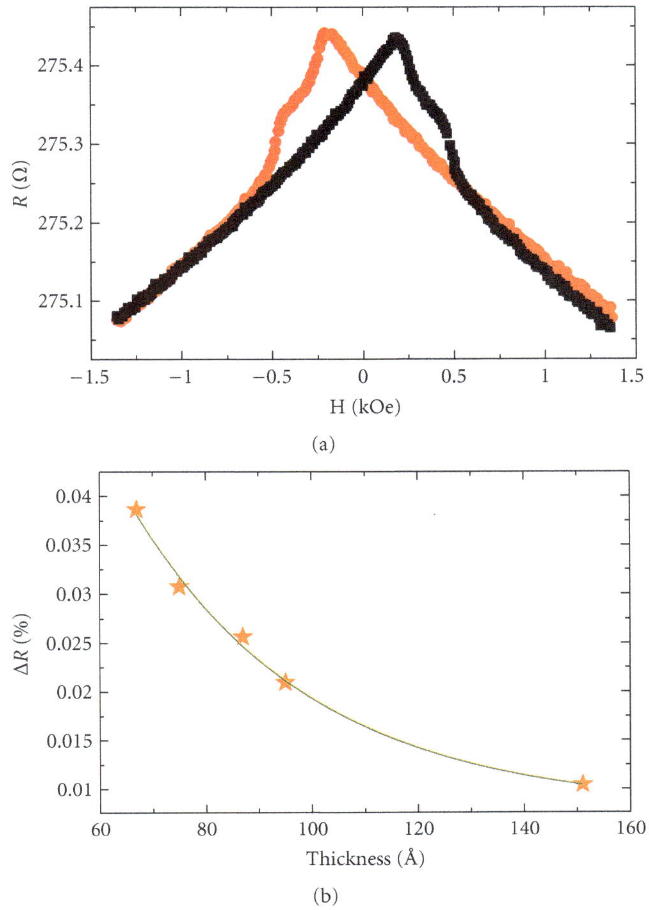

FIGURE 7: (a) Current in plane magnetoresistance measurements on Fe_3O_4(500 Å)/Cr(57 Å)/Fe_3O_4(1000 Å) spin valves. The black (red) data are measurements with increasing (decreasing) fields. (b) Change of GMR as a function of Cr spacer layer thickness; the line is a fit to an exponential decay, with a decay constant of 33 Å.

the potential for growing devices to realize magnon-induced spin transfer torques.

Acknowledgments

The work at USF was supported by the National Science Foundation; the work at UT was separately supported by the National Science Foundation, as well as the Welch Foundation. The authors thank C. Bauer and D. Williams for assistance with some of the measurements.

References

[1] L. Berger, "Emission of spin waves by a magnetic multilayer traversed by a current," *Physical Review B*, vol. 54, no. 13, pp. 9353–9358, 1996.

[2] J. C. Slonczewski, "Current-driven excitation of magnetic multilayers," *Journal of Magnetism and Magnetic Materials*, vol. 159, no. 1-2, pp. L1–L7, 1996.

[3] M. Tsoi, A. G. M. Jansen, J. Bass et al., "Excitation of a magnetic multilayer by an electric current," *Physical Review Letters*, vol. 80, no. 19, pp. 4281–4284, 1998.

[4] E. B. Myers, D. C. Ralph, J. A. Katine, R. N. Louie, and R. A. Buhrman, "Current-induced switching of domains in magnetic multilayer devices," *Science*, vol. 285, no. 5429, pp. 867–870, 1999.

[5] J. A. Katine, F. J. Albert, R. A. Buhrman, E. B. Myers, and D. C. Ralph, "Current-driven magnetization reversal and spin-wave excitations in Co/Cu/Co pillars," *Physical Review Letters*, vol. 84, no. 14, pp. 3149–3152, 2000.

[6] J. Åkerman, "Toward a universal memory," *Science*, vol. 308, no. 5721, pp. 508–510, 2005.

[7] S. I. Klselev, J. C. Sankey, I. N. Krivorotov et al., "Microwave oscillations of a nanomagnet driven by a spin-polarized current," *Nature*, vol. 425, no. 6956, pp. 380–383, 2003.

[8] I. N. Krivorotov, N. C. Emley, J. C. Sankey, S. I. Kiselev, D. C. Ralph, and R. A. Buhrman, "Time-domain measurements of nanomagnet dynamics driven by spin-transfer torques," *Science*, vol. 307, no. 5707, pp. 228–231, 2005.

[9] D. C. Ralph and M. D. Stiles, "Spin transfer torques," *Journal of Magnetism and Magnetic Materials*, vol. 320, no. 7, pp. 1190–1216, 2008.

[10] J. C. Slonczewski, "Initiation of spin-transfer torque by thermal transport from magnons," *Physical Review B*, vol. 82, no. 5, Article ID 054403, 2010.

[11] M. Hatami, G. E. W. Bauer, Q. Zhang, and P. J. Kelly, "Thermal spin-transfer torque in magnetoelectronic devices," *Physical Review Letters*, vol. 99, no. 6, Article ID 066603, 2007.

[12] J. E. Wegrowe, "Spin transfer from the point of view of the ferromagnetic degrees of freedom," *Solid State Communications*, vol. 150, no. 11-12, pp. 519–523, 2010.

[13] A. V. Nazarov, H. M. Olson, H. Cho et al., "Spin transfer stimulated microwave emission in MgO magnetic tunnel junctions," *Applied Physics Letters*, vol. 88, no. 16, Article ID 162504, 2006.

[14] A. M. Deac, A. Fukushima, H. Kubota et al., "Bias-driven high-power microwave emission from MgO-based tunnel magnetoresistance devices," *Nature Physics*, vol. 4, no. 10, pp. 803–809, 2008.

[15] D. Houssameddine, S. H. Florez, J. A. Katine et al., "Spin transfer induced coherent microwave emission with large power from nanoscale MgO tunnel junctions," *Applied Physics Letters*, vol. 93, no. 2, Article ID 022505, 2008.

[16] M. G. Chapline and S. X. Wang, "Observation of the Verwey transition in thin magnetite films," *Journal of Applied Physics*, vol. 97, no. 12, Article ID 123901, 3 pages, 2005.

[17] P. B. Jayathilaka, C. A. Bauer, D. V. Williams, M. C. Monti, J. T. Markert, and C. W. Miller, "Impact of ultrathin transition metal buffer layers on Fe_3O_4 thin films," *Journal of Applied Physics*, vol. 107, no. 9, Article ID 09B101, 2010.

[18] A. K. H. Lee, P. B. Jayathilaka, C. A. Bauer et al., "Magnetic force microscopy of epitaxial magnetite films through the Verwey transition," *Applied Physics Letters*, vol. 97, no. 16, Article ID 162502, 2010.

[19] K. Moloni, B. M. Moskowitz, and E. Dan Dahlberg, "Domain structures in single crystal magnetite below the Verwey transition as observed with a low-temperature magnetic force microscope," *Geophysical Research Letters*, vol. 23, no. 20, pp. 2851–2854, 1996.

[20] J. Bass and W. P. Pratt Jr., "Spin-diffusion lengths in metals and alloys, and spin-flipping at metal/metal interfaces: an experimentalist's critical review," *Journal of Physics Condensed Matter*, vol. 19, no. 18, Article ID 183201, 2007.

[21] P. B. Jayathilaka, C. A. Bauer, D. V. Williams, and C. W. Miller, "Influence of growth field on NiFe, Fe_3O_4, and NiFe/Cr/Fe_3O_4 spin-valves," *IEEE Transactions on Magnetics*, vol. 46, no. 6, pp. 1777–1779, 2010.

[22] S. van Dijken, X. Fain, S. M. Watts, K. Nakajima, and J. M. D. Coey, "Magnetoresistance of Fe_3O_4/Au/Fe_3O_4 and Fe_3O_4/Au/Fe spin-valve structures," *Journal of Magnetism and Magnetic Materials*, vol. 280, no. 2-3, pp. 322–326, 2004.

[23] M. N. Baibich, J. M. Broto, A. Fert et al., "Giant magnetoresistance of (001)Fe/(001)Cr magnetic superlattices," *Physical Review Letters*, vol. 61, no. 21, pp. 2472–2475, 1988.

[24] D. Tripathy, A. O. Adeyeye, and S. Shannigrahi, "Effect of spacer layer thickness on the magnetic and magnetotransport properties of Fe_3O_4/Cu/Ni_{80}/Fe_{20} spin valve structures," *Physical Review B*, vol. 75, no. 1, Article ID 012403, 2007.

Permissions

The contributors of this book come from diverse backgrounds, making this book a truly international effort. This book will bring forth new frontiers with its revolutionizing research information and detailed analysis of the nascent developments around the world.

We would like to thank all the contributing authors for lending their expertise to make the book truly unique. They have played a crucial role in the development of this book. Without their invaluable contributions this book wouldn't have been possible. They have made vital efforts to compile up to date information on the varied aspects of this subject to make this book a valuable addition to the collection of many professionals and students.

This book was conceptualized with the vision of imparting up-to-date information and advanced data in this field. To ensure the same, a matchless editorial board was set up. Every individual on the board went through rigorous rounds of assessment to prove their worth. After which they invested a large part of their time researching and compiling the most relevant data for our readers. Conferences and sessions were held from time to time between the editorial board and the contributing authors to present the data in the most comprehensible form. The editorial team has worked tirelessly to provide valuable and valid information to help people across the globe.

Every chapter published in this book has been scrutinized by our experts. Their significance has been extensively debated. The topics covered herein carry significant findings which will fuel the growth of the discipline. They may even be implemented as practical applications or may be referred to as a beginning point for another development.

The editorial board has been involved in producing this book since its inception. They have spent rigorous hours researching and exploring the diverse topics which have resulted in the successful publishing of this book. They have passed on their knowledge of decades through this book. To expedite this challenging task, the publisher supported the team at every step. A small team of assistant editors was also appointed to further simplify the editing procedure and attain best results for the readers.

Our editorial team has been hand-picked from every corner of the world. Their multi-ethnicity adds dynamic inputs to the discussions which result in innovative outcomes. These outcomes are then further discussed with the researchers and contributors who give their valuable feedback and opinion regarding the same. The feedback is then collaborated with the researches and they are edited in a comprehensive manner to aid the understanding of the subject.

Apart from the editorial board, the designing team has also invested a significant amount of their time in understanding the subject and creating the most relevant covers. They scrutinized every image to scout for the most suitable representation of the subject and create an appropriate cover for the book.

The publishing team has been involved in this book since its early stages. They were actively engaged in every process, be it collecting the data, connecting with the contributors or procuring relevant information. The team has been an ardent support to the editorial, designing and production team. Their endless efforts to recruit the best for this project, has resulted in the accomplishment of this book. They are a veteran in the field of academics and their pool of knowledge is as vast as their experience in printing. Their expertise and guidance has proved useful at every step. Their uncompromising quality standards have made this book an exceptional effort. Their encouragement from time to time has been an inspiration for everyone.

The publisher and the editorial board hope that this book will prove to be a valuable piece of knowledge for researchers, students, practitioners and scholars across the globe.

List of Contributors

Yu Ting Wang, Chih Wei Luo and Takayoshi Kobayashi
Department of Electrophysics, National Chiao Tung University, Hsinchu 30010, Taiwan

Takayoshi Kobayashi
Advanced Ultrafast Laser Research Center, University of Electro-Communications, Chofugaoka 1-5-1, Chofu, Tokyo 182-8585, Japan

Cheng-Chiang Chen, Lung-Chien Chen and Yi-Hsuan Lee
Department of Electro-optical Engineering, National Taipei University of Technology, 1, Section 3, Chung-Hsiao E. Road, Taipei 106, Taiwan

Ya-Fen Wu
Department of Electronic Engineering, Ming Chi University of Technology, Taishan, New Taipei City 243, Taiwan

Marius Schmidt
Physics Department, University ofWisconsin-Milwaukee, WI, USA

Hyein Lim, Sora Ahn, Miryeon Kim, Seungjun Lee and Hyungsoon Shin
Department of Electronics Engineering, Ewha Womans University, 11-1 Daehyun-Dong, Seodaemoon-Gu, Seoul 120-750, Republic of Korea

Mustafa Kurt
Department of Physics, Canakkale Onsekiz Mart University, 17100 Canakkale, Turkey

Hamit Yurtseven
Department of Physics, Middle East Technical University, 06531 Ankara, Turkey

Feng-Ching Lin, Bo-Ru Jian and Wei Lee
Department of Physics, Chung Yuan Christian University, Chung-Li 32023, Taiwan

Po-Chang Wu and Wei Lee
College of Photonics, National Chiao Tung University, Tainan 71150, Taiwan

Wei Lee
Center for Nanotechnology, Chung Yuan Christian University, Chung-Li 32023, Taiwan

Ying Sun
International Center for Materials Nanoarchitectonics (WPI-MANA), National Institute for Materials Science, 1-1 Namiki, Ibaraki, Tsukuba 305-0044, Japan

Yan-Feng Guo, Xia Wang, Jun Li, Clastin I. Sathish and Kazunari Yamaura
Superconducting Properties Unit, National Institute for Materials Science, 1-1 Namiki, Ibaraki, Tsukuba 305-0044, Japan

Yoshihiro Tsujimoto
Materials Processing Unit, National Institute for Materials Science, 1-1 Namiki, Ibaraki, Tsukuba 305- 0044, Japan

Clastin I. Sathish and Kazunari Yamaura
Graduate School of Chemical Sciences and Engineering, Hokkaido University, Hokkaido, Sapporo 060- 0810, Japan

Cong Wang
Department of Physics, Center for Condensed Matter and Materials Physics, Beihang University, Beijing, Haidian 100191, China

R. Todorov, J. Tasseva, V. Lozanova and A. Lalova
Institute of Optical Materials and Technologies "Acad. J. Malinowski," Bulgarian Academy of Sciences, Acad. G. Bonchev Street, Building 109, 1113 Sofia, Bulgaria

Tz. Iliev
Geological Institute, Bulgarian Academy of Sciences, Acad. G. Bonchev Street, Building 24, 1113 Sofia, Bulgaria

A. Paneva
Faculty of Physics, Sofia University "St. Kliment Ohridski," 5 James Boucher Avenue, 1164 Sofia, Bulgaria

Rashed Adnan Islam
Philips Lumileds Lighting Co., 370 W. Trimble Road, San Jose, CA 95131, USA

Shashank Priya
Materials Science and Engineering, Virginia Tech, Blacksburg, VA 24061, USA

Chee-Sung Park and Shashank Priya
Center for Energy Harvesting Materials and Systems (CEHMS), Department of Materials Science and Engineering, Virginia Polytechnic Institute and State University, Blacksburg, VA 24061, USA

Claus Falter and Thomas Bauer
Institut f¨ur Festk¨orpertheorie, Westf¨alische Wilhelms-Universit¨at, Wilhelm-Klemm-Straße 10, 48149 M¨unster, Germany

A. S. Tatarenko and M. I. Bichurin
Institute of Electronic and Information Systems, Novgorod State University, B. S. Peterburgskaya Street 41, Veliky Novgorod 173003, Russia

Shosuke Sasaki
KYOKUGEN (Center for Quantum Science and Technology under Extreme Conditions), Osaka University, 1-3 Machikaneyama, Toyonaka, Osaka 560-8531, Japan

Hamit Yurtseven and Sema S¸en
Department of Physics, Middle East Technical University, 06531 Ankara, Turkey

A. D. Rodrigues, A. J. Chiquito, G. Zanelatto and J. C. Galzerani
Department of Physics, Federal University of S˜ao Carlos, CP 676, 13565-905 S˜ao Carlos, SP, Brazil

A. G.Milekhin, A. I. Nikiforov, V. V. Ulyanov and O. P. Pchelyakov
Institute of Semiconductor Physics, Novosibirsk 630090, Russia

D. R. T. Zahn
Institut f¨ur Physik, Technische Universit¨at Chemnitz, 09107 Chemnitz, Germany

Sławomir Mamica, Maciej Krawczyk, and Jaros ław Wojciech Kłos
Nanomaterials Physics Division, Faculty of Physics, Adam Mickiewicz University, 61-614 Pozna´n, Poland

Vito Puliafito
Department of Matter Physics and Electronic Engineering, University of Messina, V.le F. Stagno D'Alcontres 31, 98166 Messina, Italy

Giancarlo Consolo
Department of Science for Engineering and Architecture, University of Messina, Contrada Di Dio, 98166 Messina, Italy

M. Dressel
Physikalisches Institut, Universit¨at Stuttgart, Pfaffenwaldring 57, 70550 Stuttgart, Germany

D. F. Shao, W. J. Lu, S. Lin, P. Tong and Y. P. Sun
Key Laboratory of Materials Physics, Institute of Solid State Physics, Chinese Academy of Sciences, Hefei 230031, China

Y. P. Sun
High Magnetic Field Laboratory, Chinese Academy of Sciences, Hefei 230031, China

P. B. Jayathilaka and Casey W. Miller
Department of Physics, University of South Florida, 4202 East Fowler Avenue, Tampa, FL 33620, USA

M. C. Monti and J. T. Markert
Department of Physics, University of Texas at Austin, 1 University Station C1600, Austin, TX 78712, USA